Fluid mechanics and
transfer processes

J. D. BISSET

Fluid mechanics and transfer processes

J. M. KAY AND R. M. NEDDERMAN

The right of the
University of Cambridge
to print and sell
all manner of books
was granted by
Henry VIII in 1534.
The University has printed
and published continuously
since 1584.

CAMBRIDGE UNIVERSITY PRESS

CAMBRIDGE

LONDON NEW YORK NEW ROCHELLE

MELBOURNE SYDNEY

Published by the Press Syndicate of the University of Cambridge
The Pitt Building, Trumpington Street, Cambridge CB2 1RP
32 East 57th Street, New York, NY 10022, USA
10 Stamford Road, Oakleigh, Melbourne 3166, Australia

First published 1985

Printed in Great Britain at the University Press, Cambridge

Library of Congress catalogue card number: 85-4211

British Library cataloguing in publication data

Kay, J. M.
Fluid mechanics and transfer processes.
1. Heat—Transmission
I. Title II. Nedderman, R. M.
532'.002466 QC320

ISBN 0 521 30303 6 hard covers
ISBN 0 521 31624 3 paperback

CONTENTS

LIST OF SYMBOLS

a — Radius (m), linear dimension (m), flow area of a stream tube (m^2), a parameter (–), velocity of sound ($m\ s^{-1}$), surface area per unit volume (m^{-1})

A — Area (m^2), a constant (—)

b — A parameter (—), linear dimension (m), width of flow passage (m)

B — A constant (—)

c — Mass concentration ($kg\ m^{-3}$), a coefficient (—)

c_f — Friction factor (—)

c_p — Specific heat at constant pressure ($J\ kg^{-1}\ K^{-1}$)

c_v — Specific heat at constant volume ($J\ kg^{-1}\ K^{-1}$)

C — Discharge coefficient (—), molar concentration ($kmol\ m^{-3}$), a constant (—)

C^* — Saturated molar concentration ($kmol\ m^{-3}$)

C^+ — Dimensionless concentration C/C_τ (—)

C_D — Drag coefficient (—)

C_F — Force coefficient (—)

C_τ — Friction concentration ($kmol\ m^{-3}$)

d — Diameter (m)

D — Diameter (m), drag force (N)

D_e — Equivalent diameter (m)

D_h — Hydraulic mean diameter (m)

$\mathscr{D}$ — Diffusivity ($m^2\ s^{-1}$)

e — Internal energy per unit mass ($J\ kg^{-1}$), exponential number (—)

E — Internal energy (J), effectiveness factor (—)

f — Flow component of velocity ($m\ s^{-1}$), fouling factor ($m^2\ K\ W^{-1}$), a mathematical function

$\mathbf{f}$ — Vector flux of a transferable quantity (various)

F — Force (N), a mathematical function

$\mathbf{F}$ — Force vector (N)

g — Acceleration due to gravity ($m^2\ s^{-1}$)

G — Volumetric gas flow ($m^3\ s^{-1}$)

G^+	Molar gas flow (kmol s^{-1})
h	Heat transfer coefficient (W m^{-2} K^{-1}), height (m), head (m), enthalpy per unit mass (J kg^{-1}), Thiele modulus (—)
h_o	Overall heat transfer coefficient (W m^{-2} K^{-1})
H	Height (m); head (m), enthalpy (J), Henry's law constant (J $kmol^{-1}$)
$H^\dagger$	Henry's law constant (N m^{-2})
i	Square root of minus one (—)
i, j, k	Unit vectors in the x, y, z directions (—)
I	An integral (—)
j_D	j-factor for mass transfer (—)
j_f	Superficial liquid velocity (m s^{-1})
j_f^*	Dimensionless superficial velocity (—)
j_g	Superficial gas velocity (m s^{-1})
j_g^*	Dimensionless superficial gas velocity (—)
j_H	j-factor for heat transfer (—)
k	Thermal conductivity (W m^{-1} K^{-1}), Boltzmann's constant (J K^{-1}), first-order reaction velocity constant (s^{-1}), wave number (m^{-1})
k_2	Second-order reaction velocity constant (m^3 s^{-1} $kmol^{-1}$)
k_g	Gas side mass transfer coefficient (m s^{-1})
k_G	Gas side mass transfer coefficient (kmol N^{-1} s^{-1})
k_L	Liquid side mass transfer coefficient (m s^{-1})
k_y	Gas side mass transfer coefficient (kmol m^{-2} s^{-1})
K	A constant (—), circulation (m^2 s^{-1})
K_g	Overall mass transfer coefficient referred to the gas side (m s^{-1})
K_G	Overall mass transfer coefficient referred to the gas side (kmol N^{-1} s^{-1})
K_L	Overall mass transfer coefficient referred to the liquid side (m s^{-1})
K_y	Overall mass transfer coefficient referred to the gas side (kmol m^{-2} s^{-1})
l	Length (m), mixing length (m)
L	Length (m), volumetric liquid flow rate (m^3 s^{-1}), lift force (N)
L_f	Loss of energy due to friction (J)
$L^\dagger$	Molar liquid flow rate (kmol s^{-1})
m	Mass (kg), mass flow rate (kg s^{-1}), molecular mass (kg), solute flux (kmol m^{-1} s^{-1}), mass interchange rate (kg m^{-2} s^{-1})
M	Mass (kg), momentum (kg m s^{-1}), mass flow rate per unit width (kg m^{-1} s^{-1})
M_r	Relative molecular mass (—)

n	Integer (—), stoichiometric coefficient (—), Richardson and Zaki index (—), mass flux (kg m^{-2} s^{-1})
n_0	Number of molecules per unit volume (m^{-3})
N	Number of tubes (—), molar flux (kmol m^{-2} s^{-1}), rotational speed (rev s^{-1})
N_A	Avogadro's number (kmol^{-1})
O	Order of magnitude (—)
p	Pressure (N m^{-2}), partial pressure (N m^{-2})
p^*	Saturated vapour pressure (N m^{-2})
P	Total pressure (N m^{-2}), power (W)
q	Heat flux (W m^{-2} s^{-1}), turbulent velocity (m s^{-1})
Q	Heat flow (W), volumetric flow (kg s^{-1}), rate of heat supply (W)
Q_m	Molar flow (kmol s^{-1})
Q_s	Heat generation per unit volume (W m^{-3})
r	Radius (m), radial co-ordinate (m)
$\mathbf{r}$	Position vector (m)
R	Radius (m), gas constant (J kmol^{-1} K^{-1}), rate of reaction per unit volume (kmol m^{-3} s^{-1})
R'	Specific gas constant (J kg^{-1} K^{-1})
R_{ij}	Correlation coefficient (—)
$\mathbf{R}$	Vector force or reaction (N)
s	Distance (m), length measured along a stream tube (m), entropy per unit mass (J kg^{-1} K^{-1})
S	Surface area (m^2)
$\mathcal{S}$	Surface area of moving boundary (m^2)
t	Time (s), thickness (m)
T	Temperature (K), torque (N m)
T^+	Dimensionless temperature T/T_τ (–)
T_τ	Friction temperature (K)
u, v, w	Velocity components in the x, y, z directions (m s^{-1})
u', v', w'	Turbulent velocity components (m s^{-1})
$\mathbf{u}$	Velocity vector (m s^{-1})
u_m	Mean velocity in a pipe (m s^{-1})
u^+	Dimensionless velocity u/u_τ (—)
u_τ	Friction velocity (m s^{-1})
U	Velocity (m s^{-1})
v	Velocity (m s^{-1})
V	Volume (m^3), rise velocity (m s^{-1}), spatial volume (m^3), specific volume (m^3 kg^{-1})
$\mathcal{V}$	Material volume (m^3)

w	Velocity (m s^{-1}), velocity of whirl (ms^{-1})
W	Work (J), rate of work per unit mass flow rate (J kg^{-1}), mass flow rate (kg s^{-1})
W_i	Work input per unit mass or rate of work input per unit mass flow rate (J kg^{-1})
W_s	Shear work per unit mass, or rate of shear work per unit mass flow rate (J kg^{-1})
W_τ	Rate of shear work per unit volume of fluid (W m^{-3})
$\bar{W}$	Rate of energy dissipation per unit volume by turbulence (W m^{-3})
x, y, z	Cartesian co-ordinates (m)
y^+	Dimensionless distance y/y_τ (—)
y_τ	Friction distance (m)
Y	Correction factor for multi-pass heat exchangers (—)
Z	Height (m)
α	Partition coefficient (—), thermal diffusivity (m^2 s^{-1}), an angle (—), coefficient (—)
β	Coefficient of thermal expansion (K^{-1}), an angle (—), a coefficient (—)
γ	Ratio of specific heats (—)
δ	Boundary layer thickness (m), film thickness (m), elementary increment (—)
δ_C	Mass transfer boundary layer thickness (m)
δ_G	Gas film thickness (m)
δ_L	Laminar sub-layer thickness (m), liquid film thickness (m)
δ_T	Thermal boundary layer thickness (m)
Δ	Difference operator (—), volumetric rate of strain (s^{-1})
ε	Void fraction (—), rate of energy dissipation per unit mass by turbulence (W kg^{-1}), roughness size (m), small angle (—)
η	Dimensionless film thickness (—), effectiveness factor (—), efficiency (—)
θ	Angular co-ordinate (—), temperature (K), momentum thickness of a boundary layer (m)
θ_C	Concentration thickness (m)
θ_T	Thermal thickness (m)
κ	Transport coefficient (various)
λ	Mean free path (m), latent heat (J kg^{-1}), shape factor (—), depth of flow in a channel (m), wavelength (m), parameter (—)
μ	Absolute viscosity (N m s^{-2})
ν	Kinematic viscosity (m^2 s^{-1})
ν_τ or ν_T	Eddy kinematic viscosity (m^2 s^{-1})

ξ	δ_T/δ (—)
ρ	Density (kg m^{-3})
σ	Mean molecular diameter (m), surface tension (N m^{-1}), standard deviation (—), stress (N m^{-2}), slip factor (—)
τ	Shear stress (N m^{-2}), skin friction stress (N m^{-2})
ϕ	Potential function (m^2 s^{-1}), angular co-ordinate (—) transferable property (various)
φ	Martinelli parameter (—)
Φ	Rate of energy dissipation per unit volume by viscous stresses (W m^{-3})
χ	Martinelli parameter (—)
ψ	Stream function (m^2 s^{-1} or m^3 s^{-1})
Ψ	Potential function describing gravity or body force (Nm)
ω	Angular velocity or radian frequency (rad s^{-1}), vorticity component (s^{-1})
$\boldsymbol{\omega}$	Vorticity vector (s^{-1})
Ω	Angular velocity (rad s^{-1})

DIMENSIONLESS RATIOS

Ec	Eckert number ($U^2/c_p\,\Delta T$)
Fr	Froude number (U^2/gL or $U/\sqrt{(gL)}$)
Gr	Grashof number ($\beta g\,\Delta\theta L^3/\nu^2$)
Ha	Hatta number ($\sqrt{(k\mathscr{D})}/k_L^0$)
Ma	Mach number (U/a)
Nu	Nusselt number (hd/k)
Pe	Peclet number ($\rho Uc_pd/k$ or Ud/α)
Pe′	Modified Peclet number ($UD/\mathscr{D}$)
Pr	Prandtl number ($\mu c_p/k$ or ν/α)
Re	Reynolds number ($\rho Ud/\mu$ or Ud/ν)
Sc	Schmidt number ($\nu/\mathscr{D}$)
St	Stanton number ($h/\rho Uc_p$)
St′	Modified Stanton number (k_L/U or k_g/U)
We	Weber number ($\rho U^2L/\sigma$)

PREFACE

There are many excellent books available which deal with the more specialised aspects of fluid mechanics, heat transfer, and mass transfer, but there is all the more need for an introductory text which can link up the different branches and attempt to present a unifying picture. The present book is intended to meet this requirement and it has been planned as a replacement for an earlier text on *Fluid Mechanics and Heat Transfer* which enjoyed a long run of popularity and which evidently satisfied a demand for a concise introductory work in this field.

The present book is based on lectures, given by both authors at different times, for the Chemical Engineering Tripos course at Cambridge University. It is not written exclusively from a chemical engineering angle, however, and the authors hope that the book will also be useful to students in other branches of engineering at both undergraduate and post-graduate level. We would also like to mention that the text covers the syllabus in Fluid Mechanics and Transfer Processes which is one of the subjects offered in Part I B of the Natural Sciences Tripos at Cambridge.

Some comments on the layout and contents of the chapters may be helpful to the reader. The first two chapters are of an introductory nature and the main themes of the convective transfer of matter, energy, and momentum by the motion of the fluid, and also of transfer by molecular motion, are presented in chapter 2. The basic aspects of fluid dynamics, involving especially the transfer of momentum, are outlined in chapters 3–8 and we have tried to give due weight to the central role of turbulence in fluid motion. Chapter 9 picks up the major theme of the energy equation and energy transfer, and provides a link with chapters 15–18 where both heat and mass transfer are investigated in greater detail. Chapters 10 and 11 provide an introduction to flow in turbo-machinery and to gas dynamics, respectively, and involve the application of both the momentum and energy equations. Chapters 13 and 14 develop the related themes of heat transfer by conduction, and mass transfer by diffusion, from their preliminary presentation in chapter 2. Both heat and mass transfer are then studied in parallel in chapters 15–18. Chapters 19–21 are concerned with certain flow and transfer operations, including two-phase flow, condensation and evaporation, flow in packed beds, and fluidised solids, which are of special interest and importance in chemical and process engineering applications. Finally, chapter 22 introduces the subject of mass transfer with chemical reaction.

Although we have followed a reasonably logical sequence in the presentation of the material, this is not the type of book where the reader is obliged to proceed in a systematic manner from one chapter to the next. By the nature of the subject there are many parallel developments and cross-links between the chapters. The reader could in fact start at almost any point in the book and work backwards and forwards to fill in the picture. We have tried to provide adequate cross-references between the different chapters to make this type of approach feasible.

We have included some short numerical illustrations in the text and we have tried to present the mathematical analysis in such a way that numbers can easily be inserted into the equations at appropriate points. Although design calculations in the real world of engineering have been almost entirely computerised in recent years, it is important that professional engineers should retain the ability to carry out quick-check calculations of a simplified nature. The authors would particularly like to urge their readers to develop this habit and to try putting numbers into the equations as often as possible when studying this book.

The authors have received many helpful comments and suggestions during the preparation of this book from colleagues both in industry and in academic circles. While it would not be possible to mention everyone by name, we would like to record our thanks to Professor J. F. Davidson and to Mr G. O. Jackson for their advice on certain specific matters.

Cambridge

J.M.K.
R.M.N.

1

Fluid properties and fluid motion

1.1 Gases, liquids and solids

Matter can normally exist in one of three possible states, gaseous, liquid or solid, depending on the prevailing temperature and pressure. Under certain conditions a solid may change into a liquid by the process of melting, and a liquid may be converted into a gas by the process of boiling or evaporation. In some cases a solid may change directly to the gaseous state by the process of sublimation. In the reverse direction a gas may be condensed to become a liquid or a solid, and a liquid may be converted to the solid state by being frozen or solidified.

Both gases and liquids have the tendency to flow under the application of force, and they may be classified together as *fluids*. The obvious practical distinction between a fluid and a solid is that a solid normally has a definite shape, whereas a fluid has no shape other than that of its containing vessel. We can state more precisely that the distinguishing feature of a fluid is its *inability to resist shearing forces while remaining in static equilibrium.*

Solid materials may be classified under three main headings: crystalline solids, amorphous solids or glasses, and polymers. A crystalline solid usually has a sharp melting point and will change into the liquid state within a narrow temperature band. An amorphous solid, on the other hand, becomes increasingly soft and plastic when heated and does not generally exhibit a sharp change from the solid to the liquid state. Amorphous solids or glasses may, in some respects, be regarded as super-cooled liquids. While in most engineering situations there is a clear common-sense distinction between a fluid and a solid, there are naturally some borderline cases where a material may exhibit properties appropriate to either of these two states depending on the rate of application of the external forces. An amorphous material such as pitch may behave as a brittle solid under sudden impact but will act like a very viscous liquid when subjected to steady pressure over a long period. The slow-motion fluid behaviour of apparently solid material may also be observed in the realm of geophysics. The earth's mantle is normally described as being solid and is believed to consist of material similar to that of the densest rocks of the earth's crust. However, if the evidence of continental drift is to be explained, one has to consider the possibility of convection currents in the mantle. At the extreme pressures prevailing at a few hundred kilometres depth the mantle rocks may effectively behave as a liquid when viewed on a geological timescale extending over millions of years. The science of *rheology* is concerned with the study of the flow and deformation of matter and may be regarded as one extreme branch of the wider subject of fluid mechanics.

The atomic and molecular structure of matter is well understood and the bulk physical properties of both gases and solids can be satisfactorily explained through an analysis of the thermal energies of atoms and molecules and a study of intermolecular forces. It is more difficult, however, to account for the observed physical properties of liquids on the basis of a molecular model.

We normally regard a fluid as a *continuum* when studying the subject of fluid mechanics and we can generally assume that the physical properties of a very small element of the fluid will be exactly the same as those which apply on a macroscopic scale. It is helpful, however, to keep the ultimate molecular structure in mind, particularly when we consider transfer processes in a fluid, as in chapter 2, and when we investigate the motion of gases at low pressures and densities.

1.2 Physical properties of gases

Gases are compressible and tend to fill any space that is available to them. The volume of a particular mass of any gas is almost infinitely variable and is related to the temperature and pressure by the thermodynamic equation of state. Although we will be primarily concerned with the bulk properties of a gas, the molecular model presented by the kinetic theory of gases is useful in providing an explanation of the main physical properties. An outline of some of the more important results from the kinetic theory of gases is given in appendix 2 for reference purposes.

In its simplest form the kinetic theory assumes that a gas consists of a very large number of 'billiard ball'-type molecules moving independently with velocities of random thermal motion and undergoing collisions with each other. For a gas which is in thermal equilibrium the distribution of the thermal velocities among the large number of molecules is represented by the Maxwellian velocity distribution function, as noted in appendix 2. It can be shown that the average kinetic energy of the molecules depends only on the absolute temperature of the gas and is given by the expression

$$\tfrac{1}{2}m\overline{v^2}=\tfrac{3}{2}kT, \tag{1.2.1}$$

where m is the mass of an individual gas molecule in kg; $\overline{v^2}$ is the mean square velocity of all the molecules; T is the absolute temperature of the gas; k is Boltzmann's constant having a value 1.3806×10^{-23} J K^{-1}. The pressure exerted by a gas can be derived from the kinetic theory by considering the dynamics of molecules which strike a plane surface and bounce off again. In appendix 2 we take the xz plane at $y=0$ as a solid boundary and calculate the rate of change of the y component of the momentum of the molecules hitting the plane. The pressure exerted by the gas molecules on the plane is, by definition, the rate of change of the y momentum of the molecules hitting unit area of the plane and this is found to be

$$p=\frac{mn_0}{3}\overline{v^2}, \tag{1.2.2}$$

where n_0 is the total number of molecules per unit volume. If we now substitute from (1.2.1) for the mean square of the molecular velocity $\overline{v^2}$, we get

$$p=n_0kT. \tag{1.2.3}$$

The density ρ of the gas is related to the number of molecules per unit volume by the expression $\rho=mn_0$ and we can therefore express equation (1.2.3) in the alternative form

$$\frac{p}{\rho}=\frac{k}{m}T, \tag{1.2.4}$$

which is easily recognised as the perfect gas law.

For calculations on a macroscopic scale, it is customary to work in terms of the *relative molecular mass* M_r rather than the actual mass m of a molecule, and it is also convenient in physical chemistry to use the mole or kilomole as a measure of the quantity of a substance. The relative molecular mass of a substance is defined as the ratio of the mass per molecule of the substance to $\frac{1}{12}$ of the mass of an atom of the nuclide carbon 12. The mole is defined as the amount of substance of a system which contains as many elementary entities as there are atoms in 0.012 kg of carbon 12. The number of molecules in a mole is Avogadro's number N_A which has a value of 6.022×10^{23} mol^{-1}.

By definition, the mass, measured in kg, of one kilomole of a substance whose relative molecular mass is M_r is given by

$$M_r = mN_A \times 10^3$$

and hence, substituting for the actual molecular mass m in (1.2.4),

$$\frac{p}{\rho} = \frac{N_A \times 10^3}{M_r} kT = \frac{R}{M_r} T, \tag{1.2.5}$$

where R is defined by $R = N_A k \times 10^3$ and is known as the *universal gas constant.* Using the numerical values already quoted for Avogadro's number N_A and for Boltzmann's constant k, we have $R = 8.314 \times 10^3$ J K^{-1} $kmol^{-1}$. Note also from (1.2.4) and (1.2.5) that $k/m = R/M_r$.

If we introduce the specific volume V, defined as the volume in m^3 of one kg mass of the gas, which is the reciprocal of the density ρ, equation (1.2.5) can be written in the familiar form:

$$pV = \frac{R}{M_r} T = R'T, \tag{1.2.6}$$

where $R' = R/M_r$ is the *specific gas constant* for the substance. The perfect gas law (1.2.6) gives an accurate representation of the relationship between pressure, volume, and temperature for the so-called permanent gases such as hydrogen, oxygen and nitrogen. In general, real gases approximate more closely to perfect gases at higher temperatures and lower pressures.

There are many gases, however, which show significant deviation from the perfect gas law even at normal temperatures and pressures. More complex equations of state can be derived to describe the behaviour of many real gases. The simplest example is the van der Waals equation (1.2.7) which introduces two additional terms to take account of the action of intermolecular forces and also the volume occupied by the molecules themselves.

$$\left(p + \frac{a}{V^2}\right)(V - b) = \frac{R}{M_r} T. \tag{1.2.7}$$

The parameters a and b in this equation are constants for a particular gas. The term involving a/V^2 represents the effect of intermolecular forces, while the parameter b takes account of the volume occupied by the gas molecules themselves.

If we are dealing with substances such as steam and certain hydrocarbons, which may be either in the liquid or vapour phase depending on the prevailing temperature and pressure, the thermodynamic equations of state become more complex still and it is generally more convenient to express the observed physical properties of the substance in digital or tabular form, for example in the form of steam tables.

The viscosity of gases is reviewed in § 1.5 and the other important transfer properties, thermal conductivity and diffusivity, are considered in chapter 2. All three transfer properties can be accounted for with varying degrees of accuracy by means of the kinetic theory of gases. While a full analysis would be beyond the scope of this book, an outline of the more important results is given in appendix 2.

Air is a mixture of gases consisting, in proportion by mass, of 75.52% N_2, 23.15% O_2, 1.28% A, and 0.05% CO_2. For most practical purposes dry air may be treated as a perfect gas having an average relative molecular mass M_r of 28.96. Thus the numerical value of the term R/M_r to be used for dry air in equation (1.2.6) is $(8.314/28.96) \times 10^3 = 287 \text{ J K}^{-1} \text{ kg}^{-1}$.

1.3 Physical properties of liquids

By comparison with gases, liquids are relatively incompressible. A definite mass of a particular liquid will have a definite volume which will vary only slightly with temperature and pressure, and a liquid will exhibit a free surface if a larger space is available than that required to contain its volume.

In terms of the molecular structure of matter, the intermolecular forces in a liquid must be relatively strong and it is this feature which accounts for the observed fact that a liquid occupies a definite volume. To some extent liquids resemble solids in showing ordered molecular patterns over limited regions. In contrast to the solid state, however, the molecules in a liquid have sufficient freedom of movement to enable a liquid to flow and to prevent it from maintaining a definite shape. A liquid qualifies for classification as a *fluid* from the fact that a liquid is unable to sustain a shearing stress while remaining in a state of static equilibrium.

When a solid melts to form a liquid there is usually an expansion of volume which may be of the order of 10% depending on the substance in question. This is consistent with the concept of the molecules having a greater degree of mobility in the liquid state. A notable exception to this general rule, however, is observed in the case of ice and water. The density of water at 0 °C is 1000 kg m^{-3} but the density of ice is 920 kg m^{-3}. The expansion of a liquid with temperature is usually relatively small. For example, the density of water at 15 °C is about 999 kg m^{-3}, and at its boiling point of 100 °C at atmospheric pressure the density is approximately 958 kg m^{-3}.

The physical properties which affect the surface of a liquid are of some interest. It is evident from observation that at a liquid/gas interface the free surface has a tendency to contract to a minimum value. This can be explained on the hypothesis that the liquid/gas interface experiences a *surface tension* as a result of the intermolecular forces acting in the vicinity of the boundary between the two media.

A related phenomenon is the angle of contact which is observed where a liquid/gas interface meets a solid wall or boundary. This angle is the result of a balance between the values of the surface tensions or surface energies of the interfaces which separate the three media.

In chemical engineering applications we frequently encounter liquids in a dispersed form as drops or sprays. Liquid droplets dispersed in a gas, or in another immiscible liquid, are spherical in form and this is consistent with the principle that the surface energy must be a minimum for the given volume of liquid contained within the droplet. In a similar manner, small bubbles of gas dispersed in a liquid are generally of spherical form.

The difference between the pressure inside a droplet or bubble and the pressure of the surrounding fluid is related in a simple manner to the surface tension as indicated in figure 1. If we consider a spherical droplet of radius r and imagine it to be intersected by a plane through its centre, the circumference of the hemispherical surface will be $2\pi r$ and the cohesive force exerted by the surface tension σ will be $2\pi r\sigma$. The pressure inside the drop will exceed the pressure of the surrounding fluid by an amount Δp and the separating force exerted by this pressure difference will be $\pi r^2\ \Delta p$. Hence

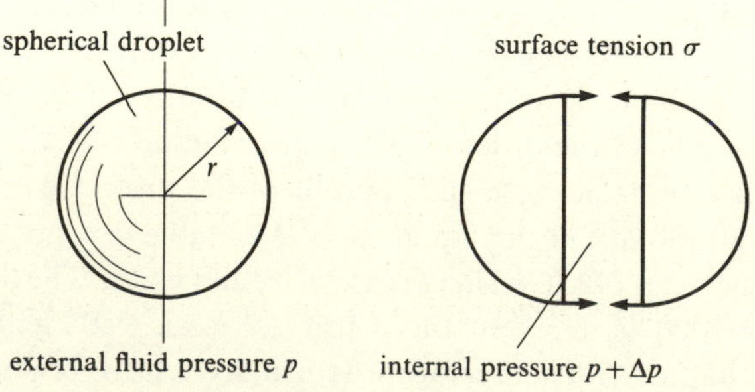

Fig. 1. Surface tension forces on a sphere or droplet.

for equilibrium we must have

$$p=2\frac{\sigma}{r}. \tag{1.3.1}$$

Note that the surface tension σ for a liquid/gas or liquid/liquid interface is expressed as a force per unit length and is measured in N m^{-1} or J m^{-2}. It may equally well be interpreted as the surface energy per unit area of free surface.

The capillary rise which is observed when a vertical tube of small bore is inserted into a liquid is another direct consequence of the surface tension.

1.4 Hydrostatic pressure

As we have already noted, a fluid which is at rest is unable to sustain any shearing stress. It can easily be proved from this basic fact that the direct stress or *pressure* in a static fluid at any particular point must be the same in all directions. In other words, for a fluid in static equilibrium, the internal state of stress at any point is specified completely by the pressure which will generally be a function of spatial position.

For the special case of equilibrium under terrestrial gravity we have the well-established result that the pressure is uniform over any horizontal plane, but that it varies with height or depth in accordance with the relationship:

$$\mathrm{d}p=-\rho g\,\mathrm{d}z, \tag{1.4.1}$$

where ρ is the density of the fluid measured in kg m^{-3}; g is the gravitational acceleration = 9.81 m s^{-2}; and z is the vertical distance measured positive upwards. Equation (1.4.1) is a special case of the general hydrostatic equation relating any applied *body force* **F** per unit mass of fluid to the consequential pressure field, i.e.

$$\rho\mathbf{F}=\operatorname{grad} p. \tag{1.4.2}$$

This simply states that the body force acting on the fluid *per unit volume* ρ**F** is equal to the *pressure gradient*. A surface of constant pressure must be perpendicular at any point to the body force acting at that point. If the body force **F** can be expressed as the gradient of a scalar potential Ψ we can write $\mathbf{F}=-\operatorname{grad}\Psi$, noting that the minus sign arises from the convention that a falling potential gradient corresponds to a positive value for **F**. Equation (1.4.2) then takes the form

$$\operatorname{grad} p=-\rho\operatorname{grad}\Psi. \tag{1.4.3}$$

For the special case of terrestrial gravity acting vertically in the z direction, we have $\Psi = gz$ and it follows that the vector equation (1.4.3) reduces in this case to the following scalar equation for the z direction:

$$\frac{dp}{dz} = -\rho g, \tag{1.4.4}$$

which is identical with (1.4.1).

For a liquid of constant density, equation (1.4.4) may be integrated immediately to give the result:

$$p - p_0 = -\rho g(z - z_0) = \rho g h, \tag{1.4.5}$$

where h is the difference of level, or depth measured from the datum level z_0, as indicated in figure 2.

The SI unit of pressure is the *pascal* or newton per square metre. The pascal is rather small as a unit for practical purposes and an alternative non-SI unit is the *bar* defined by 1 bar $= 10^5$ N m^{-2}. This unit (or its subdivision, the millibar) is frequently employed when the atmospheric pressure is involved since it so happens that 1 bar is nearly equal to the pressure of the 'standard atmosphere' at sea level. In fact, 1 bar $= 0.987$ standard atmospheres.

For the equilibrium under gravity of a gas of variable density, we need further information regarding the temperature distribution in the gas before we can integrate equation (1.4.4). For the special case of an isothermal atmosphere we have $p/\rho = \text{constant} = p_0/\rho_0$ and hence from

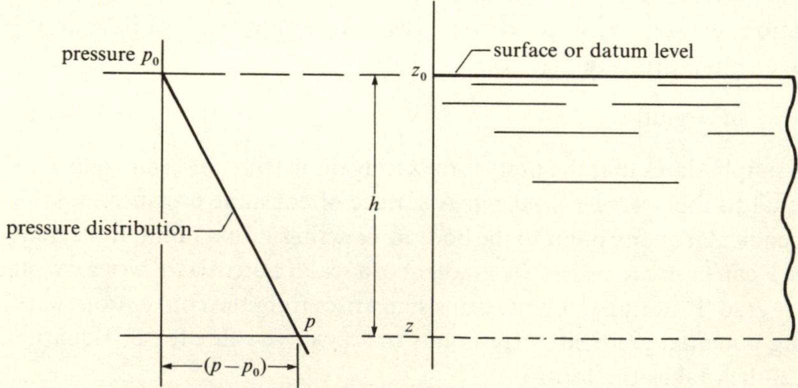

Fig. 2. Variation of hydrostatic pressure with depth.

(1.4.4)

$$\frac{\mathrm{d}p}{p} = -\frac{\rho_0}{p_0} g \, \mathrm{d}z$$

or

$$\log_e \frac{p}{p_0} = -\frac{\rho_0}{p_0} g(z - z_0). \tag{1.4.6}$$

In fact, as we all know from experience, the temperature in the terrestrial atmosphere varies very considerably. The international *standard atmosphere* is defined on the basis of a uniform temperature gradient of $-6.5\,^\circ\mathrm{C}$ per 1000 metres of altitude from sea level to a height of 11 000 m and a constant temperature of $-56.5\,^\circ\mathrm{C}$ for heights above 11 000 m. The standard atmospheric temperature at sea level is taken as $15\,^\circ\mathrm{C}$ and the pressure as $1.0132 \times 10^5\ \mathrm{N\ m^{-2}}$.

1.5 Viscosity

Although there are no shearing forces in a fluid which is in static equilibrium, shearing stresses can be sustained in a fluid which is in motion and the existence of these stresses is associated with the physical property of *viscosity*. Shear stresses in moving fluids are related to transverse velocity gradients, or rate-of-strain components, in much the same way that shear stresses in elastic solids are related to strain components according to the theory of elasticity.

Another important experimental fact associated with the property of viscosity is that, with the exception of a rarefied gas, *a real fluid does not slip at a solid boundary*. When a fluid is flowing parallel to a solid surface the particles of fluid which are immediately adjacent to the boundary are in fact at rest. There must therefore be a region in the flow in which the fluid velocity rises from zero at the solid boundary to the full value of the main stream velocity. In many cases this region is quite thin and is known as a *boundary layer*.

The simplest example of the action of viscosity occurs with shearing flow between two parallel planes. Referring to figure 3, if the lower plane surface is fixed and the upper plane is moving with constant velocity u_0, the velocity distribution for a fluid occupying the space between the two plane surfaces will be a straight line under steady laminar flow conditions, rising from zero at the lower plane to the value u_0 at the upper plane. Newton's law of fluid

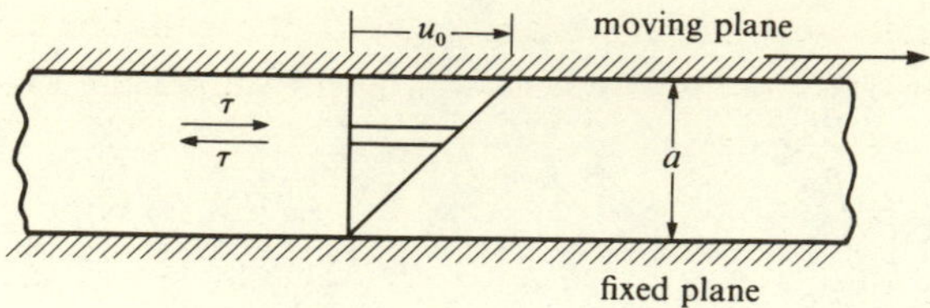

Fig. 3. Uniform shearing motion between two planes.

friction states that the shear stress in this situation is given by

$$\tau = \mu \frac{u_0}{a}, \tag{1.5.1}$$

where μ is the coefficient of viscosity, assumed to be constant. Note that in this case of *uniform shear flow* the shear stress τ is constant.

In more general terms we can consider the two-dimensional case of a fluid stream, having an undisturbed stream velocity u_1, flowing past a flat plate or solid wall. A velocity profile will be established in the boundary-layer region as indicated in figure 4. We will assume that the fluid velocity u is effectively a function only of the co-ordinate y which is the distance measured normal to the solid wall. The value of u increases from zero at $y=0$ to the mainstream velocity u_1 at $y=y_1$. The shear stress at any level y in the flow will then be given by

$$\tau = \mu \frac{\mathrm{d}u}{\mathrm{d}y}. \tag{1.5.2}$$

It will be seen from figure 4 that, whereas the shear stress becomes zero in the outer region of the undisturbed flow for values of $y > y_1$, the stress

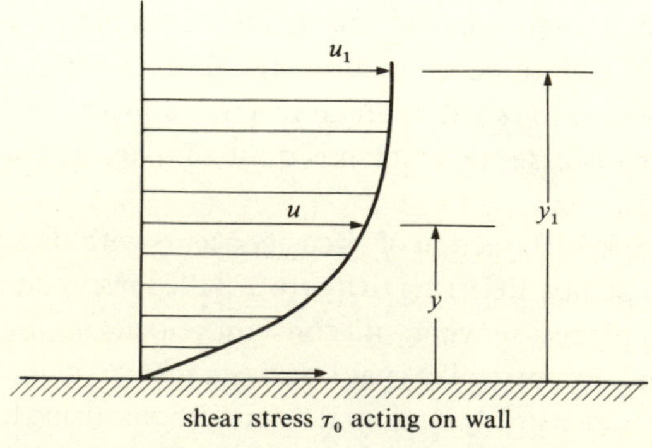

Fig. 4. Velocity profile in a boundary layer.

increases as the velocity gradient steepens towards the wall. The maximum value of the shear stress in this case will occur at the wall where $y=0$ and is given by

$$\tau_0=\mu\left(\frac{\mathrm{d}u}{\mathrm{d}y}\right)_{y=0}. \qquad (1.5.3)$$

Note that this is the *skin-friction force* exerted on the wall per unit area.

The shear stress τ has dimensions of force per unit area and is measured in units of N m^{-2} or kg m^{-1} s^{-2}. It follows from (1.5.1) or (1.5.2) that the coefficient of viscosity is measured in units of N s m^{-2} or kg m^{-1} s^{-1}.

The cgs units of viscosity, the poise and centipoise, are still in common use, however, and values in these units can be converted to SI by the relationships

$$1\ \mathrm{N\ s\ m^{-2}}=10\ \text{poise}=10^3\ \text{centipoise}.$$

It is frequently more convenient, as will become evident in later chapters, to make use of the ratio of the coefficient of viscosity μ to the density of the fluid ρ. This ratio is known as the *kinematic viscosity* $\nu=\mu/\rho$. Since density is measured in units of kg m^{-3}, the kinematic viscosity must have units m^2 s^{-1}.

The viscosity of a gas can be explained in terms of the kinetic theory of gases and this approach is outlined in chapter 2 and appendix 2. It will be appreciated that shear stresses arise in a fluid as a result of the *transport of momentum* by molecular action. The shear stress τ in equation (1.5.1) or (1.5.2) may be interpreted as being equivalent to the rate of transport of the x component of momentum of the fluid across a plane surface, per unit area, defined by $y=$constant.

The basic expression for the coefficient of viscosity of a gas, as derived in chapter 2, takes the form

$$\mu=0.179\,\frac{(kmT)^{\frac{1}{2}}}{d^2}=0.499\,\frac{m\bar{v}}{\sqrt{2}\,\pi d^2}, \qquad (1.5.4)$$

where m is the molecular mass in kg; d is the molecular diameter in m; $\bar{v}$ is the mean value of the magnitude of the molecular velocity; and k is Boltzmann's constant. This result implies that the viscosity of a gas is independent of pressure or density, but that it varies with the square root of the absolute temperature. These conclusions are largely confirmed by direct measurements although it has to be noted that the variation of viscosity with temperature which is actually observed with gases is greater than that predicted by (1.5.4). For example, the viscosity of air at 15 °C is

1.78×10^{-5} N s m^{-2} or kg m^{-1} s^{-1}. Applying the temperature variation given by (1.5.4) we would expect the viscosity of air at 100 °C to be 2.03×10^{-5} N s m^{-2} whereas the actual observed value at 100 °C is about 2.18×10^{-5} N s m^{-2}.

The following table gives the observed values of the viscosity μ, and kinematic viscosity ν at atmospheric pressure, for air over a range of temperature from 0 °C to 100 °C

T	0	20	40	60	80	100 °C
μ	1.71	1.81	1.90	2.00	2.09	2.18×10^{-5} N s m^{-2}
ν	1.32	1.50	1.69	1.88	2.09	2.30×10^{-5} m^2 s^{-1}

The viscosity and kinematic viscosity of some other common gases at 20 °C and 1 atmosphere pressure are tabulated as follows

	μ	ν
H_2	0.872×10^{-5} N s m^{-2}	10.5×10^{-5} m^2 s^{-1}
CH_4	1.09×10^{-5}	1.64×10^{-5}
NH_3	0.98×10^{-5}	1.39×10^{-5}
CO_2	1.48×10^{-5}	0.37×10^{-5}
Argon	2.22×10^{-5}	1.39×10^{-5}

It may be noted that the viscosity of a gas is typically about one hundredth of that of a liquid of comparable molecular mass, while the kinematic viscosity of a gas is about ten times that of a liquid.

By contrast with gases, the coefficient of viscosity of a liquid usually decreases with increasing temperature. The following table gives the viscosity of water over a range of temperature from 0 °C to 100 °C

T	0	20	40	60	80	100 °C
μ	1.79	1.01	0.656	0.469	0.357	0.284×10^{-3} N s m^{-2}
ν	1.79	1.01	0.661	0.477	0.367	0.296×10^{-6} m^2 s^{-1}

Many liquids of low molecular mass have viscosities of the order of 10^{-3} N s m^{-2} but the viscosity tends to increase rapidly with increasing molecular mass. Since most liquids have densities of order 10^3 kg m^{-3}, their kinematic viscosities are typically of order 10^{-6} m^2 s^{-1} or higher. The viscosities of some common liquids at 20 °C are listed in the following table

	μ N s m^{-2}	ν m^2 s^{-1}
Di-ethyl ether	0.233×10^{-3}	0.329×10^{-6}
Benzene	0.652×10^{-3}	0.742×10^{-6}
Water	1.01×10^{-3}	1.01×10^{-6}
Ethanol	1.20×10^{-3}	1.52×10^{-6}
Ethylene glycol	19.9×10^{-3}	17.8×10^{-6}
Sulphuric acid	25.4×10^{-3}	13.9×10^{-6}
Typical lubricating oil	450×10^{-3}	500×10^{-6}
Glycerol	1490×10^{-3}	1180×10^{-6}

The viscous properties of liquids are generally more complex than those of gases. Most liquids of low molecular weight and simple chemical structure behave in a Newtonian manner in that shear stresses are proportional to velocity gradients or rate-of-strain components and Newton's law of fluid friction (1.5.1) is applicable in the case of uniform shear flow. For liquids of large molecular weight and complex molecular structure, however, the relationship between shear stress and shear strain rate is more complicated and such substances are described as *non-Newtonian fluids*. Colloidal liquids, which are dispersions of a particulate phase in a continuous liquid phase, generally exhibit non-Newtonian characteristics.

For many non-Newtonian fluids the shear stress can be related to the transverse velocity gradient or shear-strain rate by an empirical power law of the form

$$\tau = k\left(\frac{\mathrm{d}u}{\mathrm{d}y}\right)^n, \tag{1.5.5}$$

where k and n are constants for the particular fluid considered. Fluids for which n is less than 1.0 are known as pseudoplastic fluids and have apparent viscosities that decrease with increasing shear rates. This behaviour is typical of materials of high molecular mass and is due to the breakdown of molecular entanglements as the shear rate increases. A less common group of materials have values of n greater than 1.0 and are known as dilatant fluids. This type of behaviour is found in concentrated slurries and, as the name suggests, it is often accompanied by an increase of volume in shearing flow.

Another group of substances are known as Bingham plastic materials. These show extreme pseudoplasticity to the extent that they are effectively rigid until the shear stress exceeds a certain critical value or yield stress τ_c.

Their behaviour is described by

$$\frac{du}{dy}=0 \qquad \text{if } \tau<\tau_c$$

$$\mu\frac{du}{dy}=\tau-\tau_c \quad \text{if } \tau>\tau_c.$$

Strictly speaking, Bingham plastic materials are outside our definition of a *fluid*, since they can sustain a limited shear stress while remaining in static equilibrium.

Some materials have viscosities that vary with the duration of flow due to the slow breakdown or build-up of molecular entanglements. Fluids such as paints are known as thixotropic fluids and show a decreasing viscosity with both shear rate and time. Fluids which show an increasing viscosity with time of flow are known as rheopectic fluids.

1.6 Mathematical notation for fluid motion

It is only necessary to look at the movement of water in a river, flowing past the pier of a bridge, for example, to appreciate that fluid motion can be a complex phenomenon. In order to attempt a quantitative study of fluid dynamics, even for a simplified or idealised flow pattern, we require a suitable system of mathematical notation. It is convenient to use the method of analysis which was originally developed for this purpose by the Swiss mathematician Leonhard Euler (1707–83), who may be regarded as the founder of fluid dynamics.

In the Eulerian system we adopt the point of view of an observer who watches the fluid flowing past a fixed frame of reference, and we consider what is happening at the various points in space occupied by the moving fluid at any given instant of time. We define a point in space by means of the position vector $\mathbf{r}$ whose co-ordinates (using a Cartesian framework) are x, y, z. The fluid velocity at the point defined by the vector $\mathbf{r}$, which is observed at time t, is then represented by the velocity vector $\mathbf{u}$ and we can say that

$$\mathbf{u}=f(\mathbf{r}, t). \tag{1.6.1}$$

If we fix attention on one particular point in space and observe the change of velocity which is taking place with respect to time, we will in effect be measuring the *local differential* $\partial\mathbf{u}/\partial t$. It should be noted, however, that this term does not measure the acceleration of an element of fluid at the point considered. The full expression for the acceleration of a fluid element is derived in § 3.1.

If the local differential $\partial\mathbf{u}/\partial t$ is zero at all points in the flow system, we have the special case of *steady flow*, i.e. there is no variation of the flow pattern with time. We can then say simply that

$$\mathbf{u} = f(\mathbf{r}). \tag{1.6.2}$$

It is usually convenient in engineering problems to work in terms of a three-dimensional Cartesian co-ordinate system with unit vectors **i, j, k** in the directions of the x, y, z axes. Referring to appendix 1, the usual convention for representing a vector in terms of its Cartesian components is to say that

$$\mathbf{u} = u_x\mathbf{i} + u_y\mathbf{j} + u_z\mathbf{k}. \tag{1.6.3}$$

The magnitude of the vector **u** can then be represented by the lower case italic letter u. By taking the scalar product of the vector **u** with itself, we get the square of the velocity magnitude, i.e.

$$\mathbf{u}\cdot\mathbf{u} = u^2 = u_x^2 + u_y^2 + u_z^2. \tag{1.6.4}$$

An alternative mathematical convention is to express the magnitude of the vector **u** by means of the modulus sign $|\mathbf{u}|$.

Although the notation of (1.6.3) is the most appropriate for use in connection with vector analysis, there is another well-established convention in which the velocity components in the x, y, z directions are represented by the letters u, v, w. Under this scheme the definition (1.6.3) is replaced by

$$\mathbf{u} = u\mathbf{i} + v\mathbf{j} + w\mathbf{k}. \tag{1.6.5}$$

The advantage of this notation is that we avoid the use of subscripts which helps to clarify the mathematics when working with scalar components rather than with vectors. The disadvantage is that we obviously cannot use the lower case italic letter u for the magnitude of the velocity vector **u** and we must therefore rely on using the modulus sign $|\mathbf{u}|$ or the upper case letter U for this purpose.

Although it is less logical for use in conjunction with vector analysis, the notation of (1.6.5) is widely adopted in textbooks and scientific papers. We have therefore followed this practice in those sections of the present book where we make extensive use of conventional three-dimensional Cartesian analysis, as for example in chapter 5. Elsewhere we occasionally use the notation of (1.6.3) when it seems more appropriate to do so.

It is sometimes convenient to use a different orthogonal co-ordinate system, in particular spherical polar or cylindrical polar co-ordinates. The usual convention is as follows:

spherical polar	co-ordinates	r	θ	ϕ
	velocity components	u_r,	u_θ,	u_ϕ
cylindrical polar	co-ordinates	r	ϕ	z
	velocity components	u_r,	u_ϕ,	u_z

Note, however, that the two rs are not the same. The spherical polar co-ordinate r measures the radial distance from the origin and is related to the x, y, z co-ordinates by $r^2 = x^2 + y^2 + z^2$. It is evidently identical with the magnitude of the position vector $\mathbf{r}$. The cylindrical polar co-ordinate r, on the other hand, measures the distance from the z axis and in this case $r^2 = x^2 + y^2$.

When using tensor notation, the fluid velocity is represented by u_i where i takes the values 1, 2, 3. In tensor analysis the Cartesian co-ordinates are denoted by x_i, or x_1, x_2, x_3 (in place of x, y, z), and the velocity components are u_1, u_2, u_3.

The Eulerian approach leads directly to the concept of a *streamline*. If we view the flow pattern at a particular instant of time t and draw a set of lines which coincide in direction with the velocity vectors at every point, we obtain a set of streamlines. By definition there can be no flow across a streamline. If the flow is unsteady, the streamline picture drawn for a particular instant will give us an instantaneous photograph of the flow pattern. If the flow is steady, the streamline picture remains unchanged with passage of time and the path lines of individual fluid elements will coincide with streamlines.

Referring to figure 5, if the position vector $\mathbf{r}$ traces out a streamline we require that the vector element $\mathrm{d}\mathbf{r}$ is parallel to the velocity vector $\mathbf{u}$ at each point. The mathematical condition is expressed in terms of vector analysis by saying that the vector product of $\mathrm{d}\mathbf{r}$ with $\mathbf{u}$ must be zero, i.e.

$$\mathrm{d}\mathbf{r} \times \mathbf{u} = 0 \tag{1.6.6}$$

and using the notation of (1.6.3) this may be written

$$\begin{vmatrix} \mathbf{i} & \mathbf{j} & \mathbf{k} \\ \mathrm{d}x & \mathrm{d}y & \mathrm{d}z \\ u_x & u_y & u_z \end{vmatrix} = 0.$$

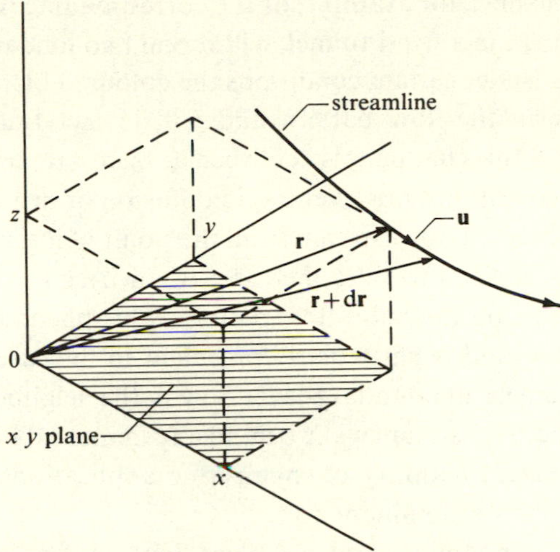

Fig. 5. Vectorial representation of a streamline.

Expressed in co-ordinates, this is equivalent to the condition that

$$\frac{\mathrm{d}x}{\mathrm{d}y}=\frac{u_x}{u_y},\quad \frac{\mathrm{d}y}{\mathrm{d}z}=\frac{u_y}{u_z},\quad \text{and}\quad \frac{\mathrm{d}z}{\mathrm{d}x}=\frac{u_z}{u_x}$$

or

$$\frac{\mathrm{d}x}{u_x}=\frac{\mathrm{d}y}{u_y}=\frac{\mathrm{d}z}{u_z}, \tag{1.6.7}$$

which is the basic equation for a streamline.

The Eulerian method of analysis also leads to the important concept of a *control surface.* This is an imaginary surface, fixed in relation to the Eulerian frame of reference, enclosing a definite volume of the space occupied by the flowing fluid. The control surface may be crossed by the moving fluid over part of its area, or it may be chosen in such a way that in certain regions it coincides with a set of streamlines, or with a solid boundary such as the wall of a tube, across which there can be no flow. A *stream tube* is a particular example of a control surface especially applicable in the case of steady flow, and is used in chapter 3 to derive the basic Euler equation of fluid motion in its simplest 'one-dimensional' form.

1.7 Laminar and turbulent flow

Flow visualisation by means of a thin filament of coloured fluid is a useful technique in experimental fluid dynamics. The application of this technique in a water channel, for example, or the corresponding technique of using a smoke filament in a wind tunnel, will reveal two fundamentally different types of flow. Under certain conditions the coloured filament will remain as a thin line in the flow pattern and will in fact trace out a streamline. Fluid flow of this character is described as being streamlined or *laminar*. Under other conditions, however, rapid diffusion or dispersion of the coloured fluid is observed downstream from the point of insertion and the local fluid motion appears to take the form of an irregular eddying movement superimposed on the main flow pattern. This phenomenon is known as *turbulent flow* and is particularly prevalent in the case of flow through pipes and channels, in boundary-layer flow in the neighbourhood of solid walls, and in the downstream wake behind any immersed body. It is fair to say that in the great majority of engineering applications of fluid dynamics turbulence plays a dominant role.

The distinction between laminar and turbulent flow was first observed by Osborne Reynolds in 1883, who carried out experiments on the flow of fluids through circular pipes. He concluded that an important criterion in determining whether the flow in a pipe would be laminar or turbulent in character was the value of the dimensionless ratio which is now known as the Reynolds number and is defined by

$$Re=\frac{\rho U D}{\mu}, \tag{1.7.1}$$

where D is the diameter of the pipe and U is the mean velocity of the fluid.

The experiments reported by Osborne Reynolds suggested that the flow through a pipe of circular cross-section is laminar if the value of the dimensionless ratio Re is less than about 2000. At higher values of the Reynolds number the flow was generally observed to be turbulent in character. Reynolds also observed, however, that transition to turbulent flow depended on conditions in the pipe entry region and on the degree of care which was exercised in eliminating initial disturbances. Under carefully controlled flow conditions, and with smooth-walled pipes, it was possible for laminar flow to persist at higher values of the Reynolds number.

These conclusions have been confirmed by subsequent investigators and it is generally agreed that there is a critical value for the Reynolds number in the range 2000 to 2100 for flow through smooth pipes. If the Reynolds number for the actual flow is less than this critical value, the flow is extremely stable and remains laminar in character even if disturbances are introduced at the pipe entry. If the Reynolds number of the actual flow is increased above the critical value, however, small disturbances are no longer damped out and the flow is liable to become turbulent. In most actual engineering applications we find that there are extraneous disturbances present in the flow at entry to a pipe, and we must therefore assume that for values of $Re > 2100$, approximately, the flow will usually be turbulent.

By way of numerical illustration, it will be of interest at this point to calculate the critical velocity for (a) water and (b) air at atmospheric pressure and a temperature of 20 °C flowing through a smooth pipe of 5 cm internal diameter as a typical example.

Noting that $Re = \rho UD/\mu = UD/\nu$ and taking values for ν from § 1.5 we have

Water at 20 °C:

$$\nu = 1.01 \times 10^{-6}\ \text{m}^2\ \text{s}^{-1},$$

$$Re = \frac{U \times 0.05}{1.01 \times 10^{-6}} = U \times 4.95 \times 10^4,$$

taking $Re_{crit} = 2000$ we have

$$U_{crit} = 0.04\ \text{m s}^{-1} \quad \text{or} \quad 4\ \text{cm s}^{-1}\ ;$$

Air at 20 °C:

$$\nu = 1.50 \times 10^{-5}\ \text{m}^2\ \text{s}^{-1},$$

$$Re = \frac{U \times 0.05}{1.50 \times 10^{-5}} = U \times 3.33 \times 10^3,$$

taking $Re_{crit} = 2000$ we have

$$U_{crit} = 0.6\ \text{m s}^{-1}\ .$$

It will be appreciated that these values for the critical velocity are very low by normal engineering standards. A typical velocity for water flowing through a 5 cm-diameter tube in a process plant would be of the order of $1\ \text{m s}^{-1}$ and this would correspond to a value of around 5×10^4 for the Reynolds number assuming the same temperature of 20 °C. Turbulent flow is thus the normal situation for water flowing through pipes in any engineering application.

Similarly, for air or gas flow through a tube, a more representative situation would be a velocity of the order of $10\ \text{m s}^{-1}$ and a pressure of the order of 10 atmospheres. Taking these figures as an illustration, and with air at 20 °C, we get a value of 3.33×10^5 for the Reynolds number as a typical figure for air flow in tubes. We are thus entirely in the realm of turbulent flow in this instance.

1.8 Dynamical similarity and the use of dimensionless ratios

We can take a preliminary view at this stage of a concept which plays an important role throughout the subsequent discussion of fluid mechanics. As we have already noted, fluid motion is usually a rather complex phenomenon and despite the resources of mathematical analysis we have to rely heavily on observation and measurement of actual flow patterns. In many practical applications, which are beyond the scope of analytical methods, we need to be able to translate measured results from one flow situation to another. The principle of dynamical similarity and the use of dimensionless ratios provide us with a very useful technique for this purpose.

It is obvious that we can only make a meaningful comparison between two different flow situations if we have complete *geometric similarity* of the physical boundaries of the flow in the two cases. As a practical example we can consider the problem of the steady flow of a fluid with undisturbed stream velocity U past a submerged body such as a sphere of diameter D. We can imagine two different cases, as indicated in figure 6, with different fluids characterised by their physical properties of density ρ and viscosity μ, different stream velocities, and different linear scales measured by the diameters of the two spheres.

Provided the lateral boundaries of the stream are sufficiently far distant from the sphere, so that the width of the stream is effectively infinite, we will satisfy the condition of geometric similarity. We will prove in chapter 5, by analysing the complete equations of motion for a viscous fluid, that the essential condition for the two flow patterns to be similar is that

$$\frac{\rho_A U_A D_A}{\mu_A} = \frac{\rho_B U_B D_B}{\mu_B}. \tag{1.8.1}$$

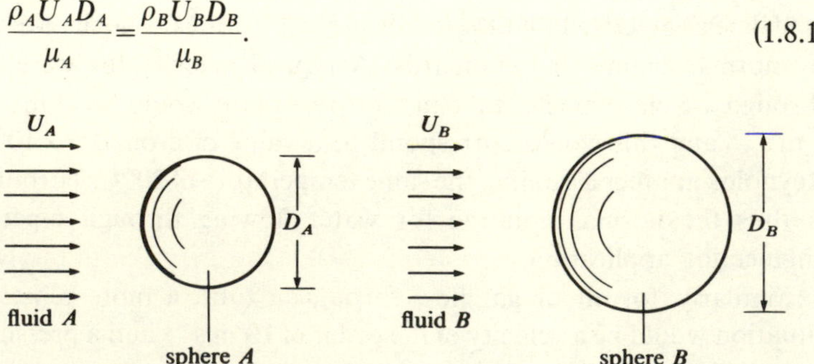

Fig. 6. A sphere in a flowing stream.

This dimensionless ratio will be recognised as the *Reynolds number* defined in this instance in terms of the diameter of the sphere as the representative length.

The physical significance of the Reynolds number is that it provides a measure of the relative magnitude of the *inertial forces*, associated with the acceleration or deceleration of a fluid element, to the *viscous forces* occurring within the flow. It is shown in chapter 3, using Bernoulli's equation, that the maximum local value of the pressure acting on the sphere occurs at the front stagnation point and is equal to $\frac{1}{2}\rho U^2$ measured relative to the static pressure of the undisturbed stream. This pressure, which is directly associated with the deceleration of the fluid stream from its undisturbed stream velocity U to a local value of zero at the front stagnation point, may be regarded as a typical inertial stress. Noting that pressure is force per unit area, we can therefore take $\frac{1}{2}\rho U^2 D^2$ or more simply $\rho U^2 D^2$ as a representative measure of the inertial forces. We have already noted in § 1.5 that shear stresses in a moving fluid are associated with transverse velocity gradients and that the tangential or skin-friction stress exerted at a solid boundary is given by (1.5.3). Since the tangential velocity must be zero at the surface of the sphere, and since the undisturbed stream velocity at some lateral distance from the sphere is U, we can regard $\mu(U/D)$ as a typical viscous stress. We can therefore take μUD as a representative measure of the viscous forces occurring within the flow pattern. We then have

$$Re = \frac{\rho U^2 D^2}{\mu UD} = \frac{\rho UD}{\mu}. \qquad (1.8.2)$$

We have observed that the maximum local pressure acting on the sphere at the frong stagnation point is $\frac{1}{2}\rho U^2$ and we might reasonably expect that, for a given flow pattern, the total force F acting on the sphere would be related by a simple numerical factor to the product of this local maximum pressure multiplied by the projected area $\frac{1}{4}\pi D^2$. In other words, we could define a *force coefficient* by the relationship

$$C_F = \frac{F}{\frac{1}{2}\rho U^2(\pi D^2/4)} \quad \text{or} \quad \frac{F}{\frac{1}{2}\rho U^2 A}. \qquad (1.8.3)$$

Since the form of the flow pattern depends on the value of the Reynolds number, we may conclude that for a given geometry

$$\frac{F}{\frac{1}{2}\rho U^2 A} = f\left(\frac{\rho UD}{\mu}\right), \qquad (1.8.4)$$

i.e. the force coefficient C_F will be a function of the Reynolds number *Re*.

We can arrive at the same conclusion by a different route using the method of dimensional analysis. All physical quantities can be expressed in terms of a limited number of fundamental dimensions. If we are concerned only with mechanical quantities, and if we exclude thermal and electrical phenomena for the time being, we require only three fundamental dimensions, mass, length, and time, for which we will use the symbols $[m]$, $[l]$, $[t]$. All other mechanical quantities such as velocity, acceleration, force, energy, momentum, etc., may be regarded as derived quantities and may be expressed in terms of the three fundamental dimensions.

In the example of flow past a submerged sphere, as illustrated in figure 6, we have five physical quantities to consider. There are four independent variables ρ, μ, D, U, and one dependent variable F. We can assert in the most general terms that the force F must be a function of the four independent variables even if we have no knowledge of the detailed physical mechanism involved. The basic rules of dimensional analysis, which are considered in more detail in chapter 15, state that if we have a functional relationship between n physical quantities, and if these physical quantities can be expressed in terms of p fundamental dimensions, the functional relationship can be reduced to one between $(n-p)$ *dimensionless groups.*

The procedure involved is to select p primary quantities (i.e. the same number as that of the fundamental dimensions) noting that each fundamental dimension must appear at least once among the primary quantities selected. The remaining $(n-p)$ physical quantities are then expressed as dimensionless ratios in terms of the p primary quantities. The final outcome is that we have a functional relationship between $(n-p)$ dimensionless groups.

We can illustrate this procedure by taking the case of flow past a sphere, where we have five physical quantities:

physical quantity	F	ρ	μ	D	U
dimensions	$[mlt^{-2}]$	$[ml^{-3}]$	$[ml^{-1}t^{-1}]$	$[l]$	$[lt^{-1}]$

If we now select ρ, D, U, as the primary quantities we require to express F and μ as dimensionless ratios. For F we can try $F/\rho^a U^b D^c$ and on checking the dimensions we find that $a=1$, $b=2$, and $c=2$, so that the appropriate group or dimensionless ratio for the force F is $F/\rho U^2 D^2$. A similar procedure applied to the fluid viscosity μ gives the result that the appropriate dimensionless ratio is $\mu/\rho UD$. We may equally well take the inverse of the latter ratio, which we then recognise as the Reynolds number

$\rho UD/\mu$. The final conclusion from the principle of dimensional analysis is that

$$\frac{F}{\rho U^2 D^2}=f\left(\frac{\rho UD}{\mu}\right), \tag{1.8.5}$$

which is the same result as (1.8.4).

We have confined ourselves in this preliminary discussion to the case of the steady flow of an incompressible fluid past a submerged object. We can use the same methods, however, to investigate the requirements for dynamical similarity and to determine the appropriate dimensionless ratios in flow problems where there is a free surface or where the fluid is a compressible gas moving with high velocity.

Taking the case of flow with a free surface where gravitational forces and wave-making resistance are involved, we have to introduce the gravitational acceleration g as an additional physical quantity. We will prove in chapter 5 that the form of the flow pattern will depend on the numerical value of the *Froude number Fr* defined by U^2/gD in addition to the value of the Reynolds number. The physical significance of the Froude number is that it provides a measure of the relative magnitude of the *inertial forces* to the *gravitational forces* occurring within the flow. We can take $\rho g D^3$ as a representative gravitational force, and we have already noted that $\rho U^2 D^2$ is a representative measure of the inertial forces. We thus have:

$$Fr=\frac{\rho U^2 D^2}{\rho g D^3}=\frac{U^2}{gD}. \tag{1.8.6}$$

An alternative definition for the Froude number is to take the square root of the ratio given by (1.8.6) and to say that

$$Fr=\frac{U}{\sqrt{(gD)}}. \tag{1.8.7}$$

The latter definition is to be preferred when we are dealing with flow in open channels. As will be shown in chapter 12, the velocity of propagation for long waves in a shallow channel is given by $c=\sqrt{(gh)}$, where h is the depth of the stream flowing in the channel. We thus see that a Froude number defined by (1.8.7) or by $U/\sqrt{(gh)}$ is in the form of a velocity ratio.

When we are dealing with the high-speed flow of a compressible gas, temperature changes become significant and we have to take account of thermal effects. We will see in chapter 11 that the flow pattern under these circumstances is determined primarily by the value of the *Mach number Ma*. The Mach number is defined as the ratio of the undisturbed stream velocity

U to the local value of the velocity of sound waves in the gas. For a perfect gas the velocity of sound a is related to the gas temperature T by $a=\sqrt{(\gamma R'T)}$ where γ is the ratio of the specific heats c_p/c_v. We thus have for the Mach number:

$$Ma=\frac{U}{a}=\frac{U}{\sqrt{(\gamma R'T)}}. \tag{1.8.8}$$

It will be seen that there is a certain analogy between the Froude number as defined by (1.8.7) and the Mach number as defined by (1.8.8).

2

Transfer processes in a fluid

2.1 Transport phenomena

It is a matter of observation that inequalities in the properties of a fluid, such as velocity, temperature, or the concentration of a dissolved substance, tend to get levelled out with the passage of time unless there is some active external agency at work to create and maintain these differences. Processes of equalisation such as these, which involve exchanges between the elements of the fluid, are described under the general heading of *transport phenomena.* We will be especially concerned in this book with the transfer processes involving:

transport of momentum;
transport of energy,
transport of mass.

Although there are important similarities between these three phenomena, it is necessary to recognise that there are also some significant differences. For example, momentum is a vector quantity whereas energy and mass are scalar quantities. The mathematical analysis of momentum transfer is consequently more complex than that of energy or mass transfer. Momentum transfer can only take place in a moving fluid, whereas energy transfer by heat conduction and mass transfer by diffusion can occur both in static and moving fluids.

When a fluid is in motion, the physical properties of the fluid such as temperature or internal energy are transported in the direction of flow as a result of the mass velocity of the fluid. This phenomenon is known as *convection.* If the bulk motion is maintained by means of an externally applied pressure difference, through the agency of a pump or compressor, for example, we use the term *forced convection.* If, on the other hand, the bulk motion is due to density changes and the action of gravity, we use the term *free convection.* The rate of change of a fluid property, following the motion of the fluid, is analysed in §2.2.

We have seen in § 1.5 that the viscosity of a fluid, and the consequential existence of shearing stresses in non-uniform moving fluids, may be interpreted as the transfer by molecular motion of a component of the fluid momentum. There are other physical properties, which may exist in a non-uniform distribution in the fluid, which can be transferred by molecular motion in exactly the same way. If, for example, the temperature of the fluid is not uniform the molecular motion will cause a transfer of internal energy which is interpreted as heat conduction. If the fluid contains a dissolved substance, whose concentration is not uniform, molecular motion will cause a transfer by diffusion of the solute through the fluid. These three molecular transport phenomena may be expressed in their most simple one-dimensional form as experimental laws under the following headings:

Newton's law of fluid friction

$$\text{shear stress} \quad \tau = \mu \frac{\mathrm{d}U}{\mathrm{d}y}. \tag{2.1.1}$$

Fourier's law of heat conduction

$$\text{heat flux} \quad q = -k \frac{\mathrm{d}T}{\mathrm{d}y}. \tag{2.1.2}$$

Fick's law of diffusion

$$\text{mol flux} \quad N = -\mathscr{D} \frac{\mathrm{d}C}{\mathrm{d}y}. \tag{2.1.3}$$

Although these must be regarded primarily as experimental laws based on observation and measurement, they may also be derived on a theoretical basis for the case of a gas using the concepts of the kinetic theory of gases. Further discussion of molecular transport phenomena is pursued in §§ 2.5–2.8.

In the special case of laminar flow, the overall transfer of heat or mass in any flow system may be analysed mathematically in terms of the molecular transport phenomena combined with the bulk motion of the fluid associated with the basic laminar flow pattern. Heat or mass transfer with laminar flow through a long pipe or in a boundary layer, for example, is amenable to analysis on this basis as described in detail in chapter 16.

We have already noted in chapter 1, however, that the motion of a fluid may be either laminar or turbulent in character. In most engineering applications, where we are dealing with flow in the vicinity of solid boundaries and at relatively large values of the Reynolds number, the flow pattern is observed to be turbulent and we will see in chapter 6 that turbulence must be regarded as the normal state of fluid motion rather than the exception. The existence of turbulence introduces an additional transfer mechanism. Since it is impossible to follow out the unsteady eddying motion of the turbulent flow pattern on a microscopic scale, we have to describe the bulk motion of the fluid in terms of the average velocity at any point. Momentum will be transferred across the flow, however, as a result of the small-scale turbulent motion. Similarly heat and mass may be transferred by the same eddying or mixing process. We can describe this transfer mechanism in turbulent flow either as turbulent convection, or by analogy with the molecular transport processes we can introduce the concepts of eddy viscosity, eddy conductivity, and eddy diffusivity. Heat and mass transfer in turbulent flow are investigated in chapter 17.

2.2 Differentiation following the motion of the fluid

In order to investigate the process of convection, we need to derive a basic mathematical expression for the rate of change of a fluid property *following the motion of the fluid.* We can use the Eulerian approach and system of notation, as outlined in § 1.6, for this purpose. We start with the definition of the flow pattern or velocity field represented by (1.6.1), i.e.

$$\mathbf{u}=f(\mathbf{r}, t). \tag{2.2.1}$$

This is a vector equation since the velocity $\mathbf{u}$ is a vector quantity. It may be replaced, however, by three scalar equations for the velocity components u, v, w, using the notation of (1.6.5).

We can employ the same mathematical procedure and framework of reference for the purpose of surveying the distribution in space and time of any other continuous property of the fluid. For simplicity we will consider a scalar property such as the density ρ, the temperature T, or the internal energy e per unit mass. Taking the temperature T as a typical example, we can make a general statement to the effect that the fluid property T will be a function of the position vector $\mathbf{r}$ and time t, i.e.

$$T=f(\mathbf{r}, t), \tag{2.2.2}$$

although the actual form of the function in (2.2.2) will usually be different from that in (2.2.1). Equation (2.2.2) implies that, for any given instant of time t, the scalar quantity T will be a function only of the position vector $\mathbf{r}$ and that surfaces of constant T could theoretically be drawn in the space occupied by the flow field.

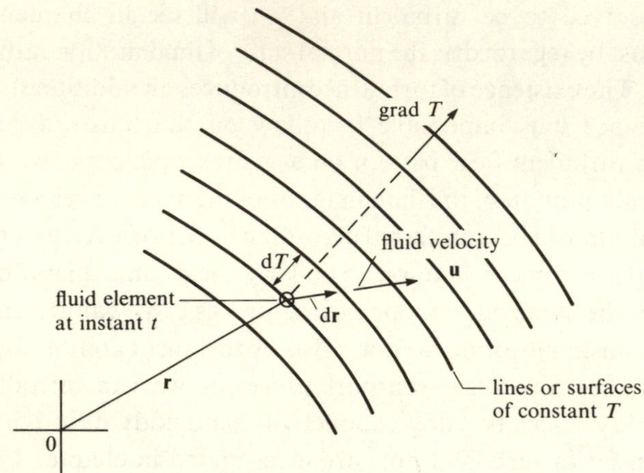

Fig. 7. Contours of constant temperature in a flowing system.

In the simplified two-dimensional picture, or cross-sectional view of a three-dimensional flow pattern, shown in figure 7, lines of constant T are indicated where the plane of the diagram intersects the surfaces of constant T. It will be convenient initially to assume that the spatial distribution of the property is steady, i.e. that it does not vary with time.

We can now identify a particular element of fluid which happens to be at position $\mathbf{r}$ at time t as indicated in figure 7. The fluid element will have a velocity $\mathbf{u}$ as given by (2.2.1). We proceed to follow the motion of this fluid element over an interval of time $\mathrm{d}t$ by moving the position vector $\mathbf{r}$ to $\mathbf{r}+\mathrm{d}\mathbf{r}$, following a path line so as to pinpoint the same element. It is evident that $\mathrm{d}\mathbf{r}$ must be equal to $\mathbf{u}\,\mathrm{d}t$ in this situation. The change in the value of the property T due to this displacement in space is obtained by taking the scalar product of $\mathrm{d}\mathbf{r}$ with grad T, i.e.

$$\mathrm{d}T = \mathrm{d}\mathbf{r}\cdot\nabla T = \mathbf{u}\cdot\nabla T\,\mathrm{d}t.$$

The rate of change of the property T due to displacement, following the motion of the fluid, is therefore given by:

$$\frac{\mathrm{d}T}{\mathrm{d}t} = \mathbf{u}\cdot\nabla T = u\frac{\partial T}{\partial x} + v\frac{\partial T}{\partial y} + w\frac{\partial T}{\partial z} \qquad (2.2.3)$$

using the notation of (1.6.5). The quantity $\mathbf{u}\cdot\nabla T$ is known as the *convective differential.*

In order to obtain the *total differential*, following the motion of the fluid, we have to take account of the possibility that the spatial distribution of the fluid property T may vary with time as implied by (2.2.2). For this purpose we need to add the *local differential* $\partial T/\partial t$ to the convective differential given by (2.2.3). Hence we have:

$$\textit{total differential} \quad \frac{\mathrm{D}T}{\mathrm{D}t}=\frac{\partial T}{\partial t}+\mathbf{u}\cdot\nabla T. \tag{2.2.4}$$

The differential operator $\mathrm{D}/\mathrm{D}t$ employing a capital D is the standard notation for signifying differentiation following the motion of the fluid and will be used in this sense throughout this book. Applying the same procedure to the fluid density, for example, we have the result that

$$\frac{\mathrm{D}\rho}{\mathrm{D}t}=\frac{\partial \rho}{\partial t}+\mathbf{u}\cdot\nabla\rho. \tag{2.2.5}$$

Although the above derivation has been established for a scalar property, the total differential operator is also applicable to a vector quantity such as the fluid velocity $\mathbf{u}$. We will see in § 3.1 that the acceleration of a fluid element is represented by

$$\frac{\mathrm{D}\mathbf{u}}{\mathrm{D}t}=\frac{\partial \mathbf{u}}{\partial t}+\mathbf{u}\cdot\nabla\mathbf{u}. \tag{2.2.6}$$

2.3 Conservation of mass

We can use the concept of a *control surface*, derived from the Eulerian system outlined in § 1.6, to obtain the basic equation governing the conservation of mass in a moving fluid. We can select a surface S of any shape we please, which is *fixed in space* and encloses a volume of space V as indicated in figure 8. In the general case, fluid will be flowing across the surface at different locations. An element of the control surface $\mathrm{d}S$ is shown as a small shaded area in the diagram, with unit normal $\mathbf{n}$, and the fluid velocity at this point on the control surface is indicated by the velocity vector $\mathbf{u}$ which is inclined at angle θ to the unit normal $\mathbf{n}$.

The net rate of outflow of mass across the control surface (i.e. outflow minus inflow) is given by the surface integral:

$$\int_S \rho\mathbf{u}\cdot\mathbf{n}\,\mathrm{d}S \quad \text{or} \quad \int_S \rho u\cos\theta\,\mathrm{d}S.$$

As noted in appendix 1, however, we can use Gauss's theorem to equate the surface integral of the quantity $\rho\mathbf{u}$ with the volume integral of the

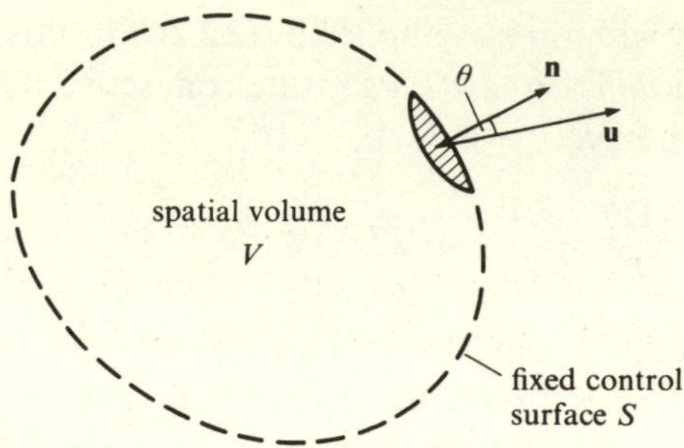

Fig. 8. Fixed control surface S enclosing spacial volume V.

divergence of $\rho\mathbf{u}$, i.e. we can say that:

$$\text{net rate of outflow of mass} = \int_V \text{div}\,(\rho\mathbf{u})\,\mathrm{d}V. \tag{2.3.1}$$

The rate of accumulation of mass within the spatial volume V enclosed by the fixed control surface S is expressed by

$$\frac{\mathrm{d}}{\mathrm{d}t}\int_V \rho\,\mathrm{d}V \quad \text{or} \quad \int_V \frac{\partial \rho}{\partial t}\,\mathrm{d}V.$$

The principle of the conservation of mass requires that the rate of accumulation of mass within the volume plus the net rate of outflow of mass from the surface must be zero, i.e.

$$\int_V \frac{\partial \rho}{\partial t}\,\mathrm{d}V + \int_V \text{div}\,(\rho\mathbf{u})\,\mathrm{d}V = 0. \tag{2.3.2}$$

Since this must be true however small the spatial volume enclosed by the control surface, we can drop the integral sign and say that

$$\frac{\partial \rho}{\partial t} + \text{div}\,(\rho\mathbf{u}) = 0. \tag{2.3.3}$$

This is the basic equation for the conservation of mass in a moving fluid. It is also known as the *equation of continuity* and will be considered further in chapter 3.

We can express (2.3.3) in an alternative form if we make use of the following mathematical identity, which is noted in appendix 1,

$$\text{div}\,(\rho\mathbf{u}) = \mathbf{u}\cdot\nabla\rho + \rho\,\text{div}\,\mathbf{u}. \tag{2.3.4}$$

From (2.3.3) and (2.3.4) we have the result that

$$\frac{\partial \rho}{\partial t} + \mathbf{u} \cdot \nabla \rho + \rho \operatorname{div} \mathbf{u} = 0. \tag{2.3.5}$$

Referring to (2.2.5), it will be seen that the first two terms in (2.3.5) make up the total differential $\mathrm{D}\rho/\mathrm{D}t$ following the motion of the fluid. Hence the mass conservation equation (2.3.3) may be expressed as

$$\frac{\mathrm{D}\rho}{\mathrm{D}t} + \rho \operatorname{div} \mathbf{u} = 0. \tag{2.3.6}$$

The physical meaning of (2.3.6) will become clearer if we write it as

$$\frac{1}{\rho}\frac{\mathrm{D}\rho}{\mathrm{D}t} = -\operatorname{div} \mathbf{u}. \tag{2.3.7}$$

The left-hand side of this equation represents the fractional rate of change of the density of a fluid element following the motion of the fluid. If the fluid element is expanding as it moves, the density will be decreasing and $\mathrm{D}\rho/\mathrm{D}t$ must be negative. The quantity div **u** which appears on the right-hand side of (2.3.7) represents the *rate of dilatation* of a fluid element, or the *volumetric rate of strain*, resulting from the velocity field **u**. If a fluid element is expanding, div **u** must be positive, and similarly if a fluid element is contracting div **u** will be negative. Equation (2.3.7) is thus seen as an almost obvious statement relating the rate of change of density to the volumetric rate of strain in the fluid.

We can clarify the picture even further if we consider the movement of a *material volume* $\mathscr{V}$ which always contains the same material particles or mass of fluid. A material volume $\mathscr{V}$ may be envisaged as being enclosed by a flexible membrane or envelope which moves with the fluid as indicated in figure 9.

Care should be taken to distinguish between a *spatial volume* V enclosed by a fixed control surface, which by definition is constant, and a *material volume* $\mathscr{V}$ which moves with the fluid and always encloses the same mass. The total mass of fluid enclosed by a material volume $\mathscr{V}$ is given by $\int_{\mathscr{V}} \rho \, \mathrm{d}\mathscr{V}$ and this must remain constant. Hence we can say that

$$\frac{\mathrm{d}}{\mathrm{d}t}\int_{\mathscr{V}} \rho \, \mathrm{d}\mathscr{V} = 0. \tag{2.3.8}$$

If we now consider an element of fluid of mass $\rho \, \mathrm{d}\mathscr{V}$ forming part of the material volume, and if we follow its motion noting that the element

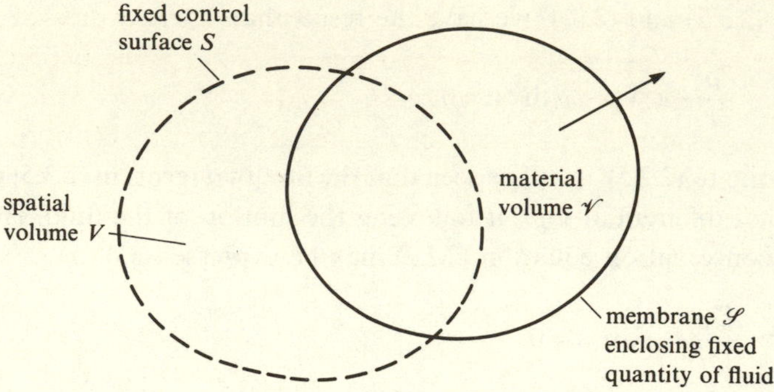

Fig. 9. Distinction between spatial volume V and material volume $\mathscr{V}$.

consists always of the same particles of matter, we have the condition for constancy of mass of the element that

$$\frac{\mathrm{D}}{\mathrm{D}t}(\rho\,\mathrm{d}\mathscr{V})=0$$

or

$$\mathrm{d}\mathscr{V}\frac{\mathrm{D}\rho}{\mathrm{D}t}+\rho\frac{\mathrm{D}}{\mathrm{D}t}(\mathrm{d}\mathscr{V})=0.$$

We can make the substitution $(\mathrm{D}/\mathrm{D}t)(\mathrm{d}\,\mathscr{V})=\operatorname{div}\mathbf{u}\,\mathrm{d}\mathscr{V}$, however, since div **u** is the volumetric rate of strain. The last equation may therefore be written:

$$\frac{\mathrm{D}\rho}{\mathrm{D}t}\mathrm{d}\mathscr{V}+\rho\operatorname{div}\mathbf{u}\,\mathrm{d}\mathscr{V}=0, \tag{2.3.9}$$

which is the same result as (2.3.6). Finally, if we integrate (2.3.9) over the whole material volume, we can say that:

$$\frac{\mathrm{d}}{\mathrm{d}t}\int_{\mathscr{V}}\rho\,\mathrm{d}\mathscr{V}=\int_{\mathscr{V}}\frac{\mathrm{D}\rho}{\mathrm{D}t}\mathrm{d}\mathscr{V}+\int_{\mathscr{V}}\rho\operatorname{div}\mathbf{u}\,\mathrm{d}\mathscr{V}=0. \tag{2.3.10}$$

Note that equation (2.3.10) states the principle of conservation of mass for a material volume of fluid $\mathscr{V}$, whereas (2.3.2) states the same principle as applied to a fixed spatial volume V.

2.4 Conservation and convection of a fluid property

We can extend the argument used in §2.3 to determine the basic conservation law for any specific fluid property, such as the internal energy e or the enthalpy h per unit mass of the fluid. We will use the symbol ϕ to represent the general case of such a property which is expressed *per unit mass* of the substance. Thus the total amount of the property ϕ contained within a material volume $\mathscr{V}$ will be $\int_{\mathscr{V}} \rho\phi \, d\mathscr{V}$.

If we consider the movement of a material volume $\mathscr{V}$ of the fluid, as in figure 9, and if we employ a similar analysis to that used in deriving equation (2.3.10), we can say that the rate of change of the total amount of the property ϕ contained within the material volume will be given by:

$$\frac{d}{dt}\int_{\mathscr{V}} \rho\phi \, d\mathscr{V} = \int_{\mathscr{V}} \frac{D}{Dt}(\rho\phi) \, d\mathscr{V} + \int_{\mathscr{V}} \rho\phi \operatorname{div} \mathbf{u} \, d\mathscr{V}. \tag{2.4.1}$$

Noting that

$$\frac{D}{Dt}(\rho\phi) = \rho\frac{D\phi}{Dt} + \phi\frac{D\rho}{Dt},$$

and making use of the mass conservation equation (2.3.6), we see that equation (2.4.1) reduces to the very simple result:

$$\frac{d}{dt}\int_{\mathscr{V}} \rho\phi \, d\mathscr{V} = \int_{\mathscr{V}} \rho\frac{D\phi}{Dt} \, d\mathscr{V}. \tag{2.4.2}$$

If we apply this, for example, to the internal energy of a material volume of fluid, we can say that:

$$\text{rate of change of internal energy} = \int_{\mathscr{V}} \rho\frac{De}{Dt} \, d\mathscr{V}. \tag{2.4.3}$$

We will use this result in chapter 9 when we apply the first law of thermodynamics to a material volume of fluid in order to derive the complete energy equation.

It will be seen from (2.4.2) or (2.4.3) that the rate of change of the property ϕ *per unit material volume* is given by $\rho(D\phi/Dt)$. The important point to note is that the result is $\rho(D\phi/Dt)$ and *not* $(D/Dt)(\rho\phi)$. Although the two expressions are identical in the special case of an incompressible fluid where the density ρ is constant, the difference between them is significant when we are dealing with a compressible fluid.

The simplest form of the general conservation equation for the property ϕ is obtained by equating the rate of gain of the total amount of the property contained in a material volume, as given by (2.4.2), to the net rate of supply of the same property across the boundary of the volume by molecular transport processes. We will see in §2.9 that transfer by molecular processes across the boundary of an element may be represented mathematically as an equivalent net rate of supply per unit volume. If we express this 'source term' arising from molecular transfer by the symbol Q, noting that Q will in general be a function of position **r** and time t, the basic conservation equation for the property ϕ takes the form:

$$\int_{\mathscr{V}} \rho \frac{\mathrm{D}\phi}{\mathrm{D}t} \mathrm{d}\mathscr{V} = \int_{\mathscr{V}} Q \,\mathrm{d}\mathscr{V} \tag{2.4.4}$$

and, since this must apply however small the material volume $\mathscr{V}$, we can say that

$$\rho \frac{\mathrm{D}\phi}{\mathrm{D}t} = Q. \tag{2.4.5}$$

Under steady-state conditions, when ϕ and Q are functions of **r** only, equation (2.4.5) states simply that the rate of transport by convection must be exactly equal to the net rate of supply through the agency of molecular or other internal transfer processes.

2.5 Transfer by molecular motion in a gas

Having considered the role of convection in the most general terms in §§2.2 and 2.4, it will be appropriate at this stage to take a preliminary look at the mechanism of transfer by molecular motion. It is possible to analyse this basic transfer process in considerable detail when we are dealing with molecular motion in gases. While a full analysis would be quite outside the scope of the present book and would constitute a specialised study in its own right, a restricted and simplified treatment can be helpful in clarifying the relationship between the transport of momentum, energy, and mass by molecular movement. Reference should be made to appendix 2 where we summarise the basic concepts of the *kinetic theory of gases* and outline the derivation of some of the results which are used later in this chapter.

We consider the case of a gas having n_0 molecules per unit volume and a molecular mass m. The density of the gas will therefore be $\rho = mn_0$. We assume that the molecules are moving in a random manner and that the molecular speeds follow the normal Maxwellian distribution as noted in appendix 2 and represented by the function $f(v)$. It should be noted that we use the symbol v, when we are discussing the kinetic theory of gases, to represent the magnitude of the thermal velocity of a molecule irrespective of its direction. Care should be taken not to confuse the molecular speed v, used in §§2.5–2.8 and in appendix 2, with the y component of the fluid velocity for which the same symbol v is used elsewhere in this book.

The number of molecules per unit volume having a thermal velocity whose magnitude lies in the range v to $v+\mathrm{d}v$ is given by $n_0 f(v)\,\mathrm{d}v$. It is shown in appendix 2 that the rate at which molecules within this velocity band are crossing *unit area* of the xz plane, from above to below the plane, is given by:

$$\text{flow rate of molecules in velocity band } \mathrm{d}v = \frac{n_0}{4} v f(v)\,\mathrm{d}v. \qquad (2.5.1)$$

If we now integrate over the whole range of velocities from 0 to ∞ we have for the total flow rate *in one direction*:

$$\text{molecular flow rate per m}^2 = \frac{n_0}{4}\int_0^\infty v f(v)\,\mathrm{d}v = \frac{n_0}{4}\bar{v}, \qquad (2.5.2)$$

where $\bar{v}$ is the mean value of the molecular velocity magnitude and is related to the gas temperature T by the following expression:

$$\bar{v} = \left(\frac{8kT}{\pi m}\right)^{\frac{1}{2}}. \qquad (2.5.3)$$

The simple but important results represented by (2.5.1), (2.5.2), and (2.5.3) provide the key to the investigation of transport phenomena associated with molecular motion in gases.

A further important concept is that of the *mean free path* of a molecule between collisions. This may be defined in different ways and is generally difficult to evaluate. Maxwell's mean free path $\bar{\lambda}$ gives the mean of all paths described in unit time in a gas having only one type of molecule of diameter d, and is expressed by

$$\bar{\lambda} = \frac{1}{\sqrt{2}\,\pi n_0 d^2}. \qquad (2.5.4)$$

It should be noted, however, that the mean free path λ_v for molecules in a particular velocity band v to $v+\mathrm{d}v$ will be a function of the velocity.

It will be obvious that, under steady-state conditions, the flow of molecules from above to below the xz plane, as given by (2.5.2), must be equal and opposite to the flow in the other direction from below to above the plane. In other words, the *net flow* of molecules across the plane must be zero if there is no overall drift of the gas in the y direction. It will be seen, however, from the figure in appendix 2 that molecules crossing the xz plane from above to below will have started their journey from their last collision at some level above the xz plane. Similarly molecules crossing the plane in the opposite direction will have started their journey at some level below the xz plane. If there is a spatial distribution in the y direction of some transferable property, such as internal energy, molecules crossing the xz plane may be assumed to carry a value of the property appropriate to the level at which they experienced their last collision. Under these conditions there will be a net transport of the property across the xz plane as a result of the molecular motion.

2.6 Transport of momentum by molecular motion

We will consider the case of uniform shearing flow of a gas parallel to the xz plane. We assume that the gas has a bulk motion, or fluid stream velocity U, in the x direction as indicated in figure 10, with a constant value for the velocity gradient $\mathrm{d}U/\mathrm{d}y$. This is a similar type of flow to that considered in § 1.5 when we introduced the coefficient of viscosity.

A molecule having velocity v and crossing a small area of the xz plane at the origin will have travelled a certain distance or free path λ from the point of its last collision. Since all directions are equally probable, molecules moving with velocity v, and crossing the xz plane at the origin from below to above, will on the average have travelled from a hemispherical shell of radius $r=\lambda_v$, where λ_v is the mean free path for a molecule moving with velocity v, as indicated in figure 10.

Noting that the ratio of the volume of a hemisphere $\frac{2}{3}\pi r^3$ to its base area πr^2 is $\frac{2}{3}r$, and assuming complete spherical symmetry, we can argue that on average the molecules crossing the xz plane from below to above, at $y=0$,

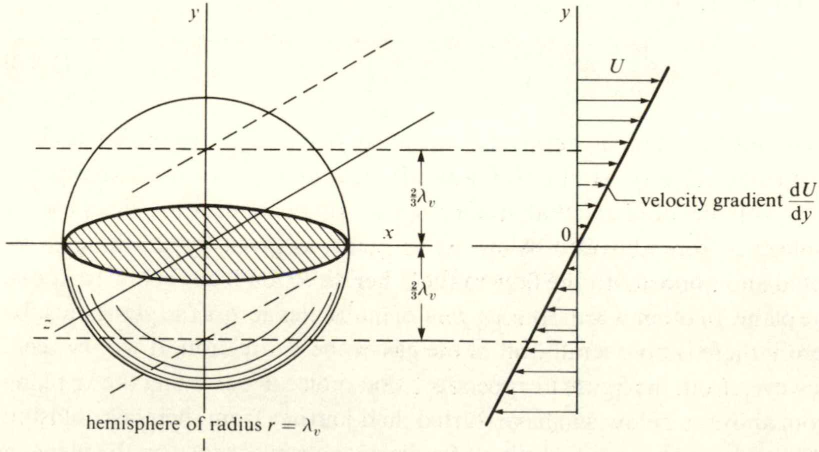

Fig. 10. Mean free path in uniform shearing flow.

will carry a value for the x component of momentum appropriate to the level $y = -\frac{2}{3}\lambda_v$. It follows from (2.5.2) that the rate of transport of the x component of momentum by molecules whose velocity magnitudes are within the band v to $v + \mathrm{d}v$, and which are crossing the xz plane from below to above at $y = 0$, is given by

$$\left(-\frac{2}{3}\lambda_v m \frac{\mathrm{d}U}{\mathrm{d}y}\right)\frac{n_0}{4} vf(v)\,\mathrm{d}v = -\frac{mn_0}{6}\frac{\mathrm{d}U}{\mathrm{d}y}\lambda_v vf(v)\,\mathrm{d}v \qquad (2.6.1)$$

in the direction of y positive. Similarly molecules in the same velocity band crossing the xz plane in the opposite direction, from above to below, will carry a value for the x component of momentum appropriate to the level $y = +\frac{2}{3}\lambda_v$. The rate of transport of the x component of momentum by these molecules moving downwards across the xz plane at $y = 0$ is therefore given by

$$\left(\frac{2}{3}\lambda_v m \frac{\mathrm{d}U}{\mathrm{d}y}\right)\frac{n_0}{4} vf(v)\,\mathrm{d}v = +\frac{mn_0}{6}\frac{\mathrm{d}U}{\mathrm{d}y}\lambda_v vf(v)\,\mathrm{d}v \qquad (2.6.2)$$

in the direction of y negative. The net upward rate of transport of x momentum (in the direction of y positive) is obtained by taking the difference of (2.6.1) minus (2.6.2), i.e.

net rate of transport of momentum per unit area

$$= -\frac{mn_0}{3}\frac{\mathrm{d}U}{\mathrm{d}y}\lambda_v vf(v)\,\mathrm{d}v. \qquad (2.6.3)$$

In order to obtain an expression for the total rate of transfer by all the molecules, moving with different velocities, we have to integrate (2.6.3) over the range of v from 0 to ∞. Hence

$$\text{total net rate of transport} = -\frac{mn_0}{3}\frac{\mathrm{d}U}{\mathrm{d}y}\int_0^\infty \lambda_v vf(v)\,\mathrm{d}v. \qquad (2.6.4)$$

This may be regarded as the basic equation governing the transport of momentum by molecular motion in a gas.

The problem in applying equation (2.6.4) lies in the difficulty of deriving a correct expression for the mean free path between collisions λ_v and in evaluating the integral $\int_0^\infty \lambda_v vf(v)\,\mathrm{d}v$. Among other factors, complications arise through the persistence of velocities after collision. These matters are investigated in detail by Jeans in his book *The Dynamical Theory of Gases* and Chapman & Cowling in *The Mathematical Theory of Non-uniform*

Gases. We will by-pass the problem, however, by simply defining a *transport mean free path* $\bar{\lambda}_t$ such that

$$\bar{\lambda}_t \bar{v} = \int_0^\infty \lambda_v v f(v)\,\mathrm{d}v. \tag{2.6.5}$$

The basic transport equation (2.6.4) can then be written as

rate of transport of momentum per unit area

$$= \text{shear stress } \tau = -\frac{mn_0}{3}\bar{\lambda}_t \bar{v}\frac{\mathrm{d}U}{\mathrm{d}y}. \tag{2.6.6}$$

Although we have taken the case of uniform shearing flow in figure 10 with a constant velocity gradient $\mathrm{d}U/\mathrm{d}y$, the result obtained in (2.6.4) or (2.6.6) is equally applicable to a variable velocity gradient, as for example in a boundary layer, provided the variation is not significant over the length of a molecular free path. This does not represent any significant restriction since for gases at normal densities the free path is very short indeed. For example, with oxygen or nitrogen at atmospheric pressure (1.013×10^5 N m^{-2}) and a temperature of 15 °C the mean free path is of the order of 10^{-7} m.

The more general picture of momentum transfer by molecular action in a boundary layer region is illustrated diagrammatically in figure 11. Note that (2.6.4) and (2.6.6) give the rate of momentum transport from below to above the xz plane in the direction of y positive. The negative sign on the

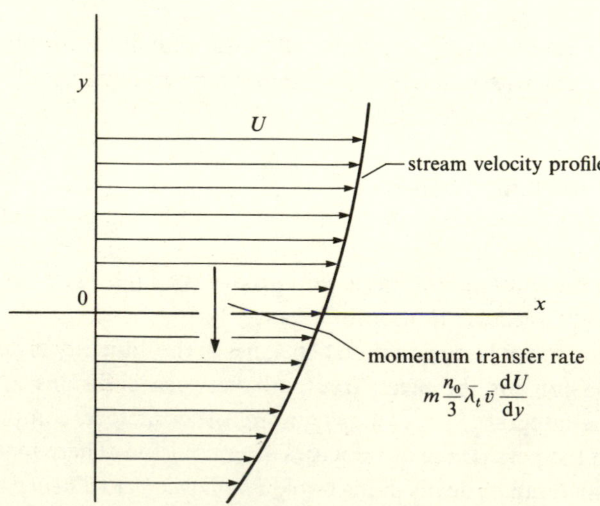

Fig. 11. Momentum transfer by molecular action.

right-hand side occurs because with a positive value for the velocity gradient dU/dy, as indicated in figures 10 and 11, the actual transfer of momentum through molecular motion must be in a downward direction.

From the definition of the coefficient of viscosity μ given in § 1.5 and from equation (2.6.6), we have the result

$$\mu = \tfrac{1}{3} m n_0 \bar{\lambda}_t \bar{v} = \tfrac{1}{3} \rho \bar{\lambda}_t \bar{v}. \tag{2.6.7}$$

The transport mean free path $\bar{\lambda}_t$ introduced in (2.6.5) differs from the ordinary mean free path between collisions $\bar{\lambda}$ but is related to it by a numerical factor which is difficult to calculate. If we take the simplest expression for $\bar{\lambda}$ as given by (2.5.4) and introduce a numerical factor f, we may assume that

$$\bar{\lambda}_t = f\bar{\lambda} = \frac{f}{\sqrt{2}\,\pi n_0 d^2}, \tag{2.6.8}$$

where d is the molecular diameter. The mean value of the molecular velocity magnitude v may be taken from (2.5.3). Hence, substituting in (2.6.7) we get

$$\mu = 0.333 f \frac{m\bar{v}}{\sqrt{2}\,\pi d^2} = 0.1197 f \frac{(kmT)^{\frac{1}{2}}}{d^2}.$$

Various values have been calculated for the correction factor f ranging from 1.382 (Jeans) to 1.497 (Chapman & Cowling). Taking the latter figure the expression for the coefficient of viscosity becomes

$$\mu = 0.499 \frac{m\bar{v}}{\sqrt{2}\,\pi d^2} = 0.179 \frac{(kmT)^{\frac{1}{2}}}{d^2}, \tag{2.6.9}$$

which is the result quoted in (1.5.4).

It may be helpful at this point to put some numbers into the expressions (2.5.3), (2.6.8) and (2.6.9) for the case of hydrogen and oxygen respectively, at atmospheric pressure and a temperature of 15 °C:

	H_2	O_2
molecular mass m	3.345×10^{-27} kg	53.52×10^{-27} kg
molecular diameter d	2.72×10^{-10} m	3.62×10^{-10} m
average molecular speed $\bar{v}$	1740 m s^{-1}	435 m s^{-1}
transport mean free path $\bar{\lambda}_t$	17.8×10^{-8} m	10.1×10^{-8} m
coefficient of viscosity μ	0.887×10^{-5} N s m^{-2}	1.99×10^{-5} N s m^{-2}

2.7 Transport of energy by molecular motion

We can use exactly the same method of analysis outlined in § 2.6 to calculate the rate of transport of energy by molecular motion. We have already noted in (1.2.1) that the average kinetic energy of translation of a molecule of mass m depends only on the thermodynamic temperature of the gas and is given by

$$E=\tfrac{1}{2}m\overline{v^2}=\tfrac{3}{2}kT, \tag{2.7.1}$$

where k is Boltzmann's constant; and T is the thermodynamic or absolute temperature of the gas. Referring to figure 10 in § 2.6, we will have the same picture of molecular motion as before, but we will now assume that there is a temperature gradient $\mathrm{d}T/\mathrm{d}y$ in place of the stream velocity gradient $\mathrm{d}U/\mathrm{d}y$ shown in the diagram.

In place of the momentum mU associated with the fluid stream velocity U, we will now take the average kinetic energy of a monatomic molecule $\tfrac{3}{2}kT$ as the transferable property. We can therefore substitute the quantity $\tfrac{3}{2}k(\mathrm{d}T/\mathrm{d}y)$ in place of $m(\mathrm{d}U/\mathrm{d}y)$ in equation (2.6.4) to give us an expression for the net rate of transport of kinetic energy of translation across the plane $y=0$, i.e.

net rate of transport of energy per unit area

$$=-\frac{1}{2}n_0k\frac{\mathrm{d}T}{\mathrm{d}y}\int_0^\infty \lambda_v v f(v)\,\mathrm{d}v. \tag{2.7.2}$$

This may be regarded as the basic equation governing the transport of kinetic energy by molecular motion in a gas.

It will be convenient to define a transport mean free path exactly as before, using equation (2.6.5), and the basic energy transport equation (2.7.2) can then be written in the more useful form as:

rate of transport of energy per unit area

$$=\text{heat flux } q=-\frac{1}{2}n_0k\bar{\lambda}_t\bar{v}\frac{\mathrm{d}T}{\mathrm{d}y}. \tag{2.7.3}$$

Although we have implied a constant temperature gradient $\mathrm{d}T/\mathrm{d}y$ in connection with figure 10, the result given by (2.7.3) is equally applicable to the case of a variable temperature gradient, provided the variation is not significant over the length of a mean free path. The more general case is shown diagrammatically in figure 12.

Equation (2.7.3) may be compared with the experimental Fourier law of

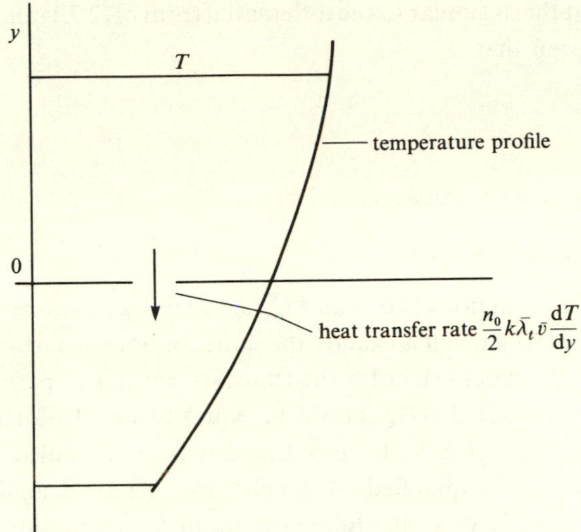

Fig. 12. Heat transfer by molecular action.

heat conduction as quoted in (2.1.2):

$$\text{heat flux } q = -k_h \frac{dT}{dy}. \tag{2.7.4}$$

Note that we use k_h as the symbol for thermal conductivity in this equation (and in the remainder of this chapter) to avoid confusion with the symbol k which is used for Boltzmann's constant. Elsewhere in this book we will follow the normal practice of using the symbol k, without any subscript, to represent thermal conductivity, since we will not require to make further use of Boltzmann's constant in any subsequent chapters.

From (2.7.3) and (2.7.4) it is evident that the thermal conductivity k_h in the Fourier heat conduction law should be related to the molecular properties of the gas by

$$k_h = \tfrac{1}{2} n_0 k \bar{\lambda}_t \bar{v}. \tag{2.7.5}$$

We can also make use of a well-known result from classical thermodynamics which states that the internal energy of a perfect gas is a function of the temperature only, and that $de = c_v \, dT$ where e represents the internal energy *per unit mass of the gas*, and c_v is the specific heat of the gas at constant volume. If, however, we use the symbol E to represent the internal energy *per molecule* as in (2.7.1), the equivalent statement is:

$$dE = mc_v \, dT. \tag{2.7.6}$$

It will be seen that this is similar to the differential form of (2.7.1) and we can draw the conclusion that

$$c_v = \frac{3}{2}\frac{k}{m}. \tag{2.7.7}$$

We can therefore re-write equation (2.7.5) as:

$$k_h = \tfrac{1}{3} m n_0 c_v \bar{\lambda}_t \bar{v} = \tfrac{1}{3} \rho c_v \bar{\lambda}_t \bar{v}. \tag{2.7.8}$$

If we now compare equations (2.6.7) and (2.7.8), and if we assume that the molecular transfer mechanism is exactly the same for both momentum and energy transfer (with equal values for the transport mean free path in each case), we might conclude that k_h should be equal to μc_v. Unfortunately, however, the analysis which we have outlined is over-simplified and the assumptions have to be qualified. A much more detailed analysis by Chapman & Cowling leads to the following result for monatomic gases:

$$k_h = \tfrac{5}{2} \mu c_v. \tag{2.7.9}$$

It will be seen that the exact result (2.7.9) takes the same form as the simple result from (2.6.7) and (2.7.8) but involves a relatively large numerical correction factor. The important point to note, however, is the inter-relationship between viscosity and thermal conductivity.

The specific heat at constant volume c_v and the specific heat at constant pressure c_p are related, in the case of a perfect gas, by a simple numerical ratio γ, i.e. we have $c_p/c_v = \gamma$. We can therefore restate the Chapman & Cowling result (2.7.9) as:

$$k_h = \frac{5}{2}\frac{\mu c_p}{\gamma} \quad \text{or} \quad \frac{\mu c_p}{k_h} = \frac{2}{5}\gamma. \tag{2.7.10}$$

The quantity $\mu c_p/k_h$ is evidently a dimensionless ratio and is known as the *Prandtl number*. It is of special importance in the analysis and correlation of heat transfer by convection in a fluid, and it combines the relevant physical properties of the fluid in a convenient dimensionless group.

For gases with polyatomic molecules the expression for the average kinetic energy of a molecule must be modified to take account of both rotational and vibrational energy. Analysis of this problem leads to the following expression for the Prandtl number of a polyatomic gas in place of (2.7.10):

$$Pr = \frac{\mu c_p}{k_h} = \frac{4\gamma}{9\gamma - 5}. \tag{2.7.11}$$

For air, taking $\gamma = 1.4$, we would therefore expect a numerical value of about 0.74 for the Prandtl number and this figure is in good agreement with other measurements.

2.8 Transport of matter by molecular motion

The simplest example of the diffusion of matter by molecular motion in a gas is the rather restricted case where we have only one size of molecule, as specified by the molecular mass m and the molecular diameter d, but where a proportion of these molecules are 'coloured' or labelled in some way so that they form a special category. We might for example have a proportion of radioactive molecules present in the gas. Provided the radioactivity in question consists of gamma-ray emission (as distinct from alpha or beta particle emission) we can assume that the mechanics of collision will not be significantly affected and we could in principle detect and measure the rate of diffusion of the radioactive molecules.

We will represent the number, per unit volume of gas, of the special category or coloured molecules by n_1 and of the normal molecules by n_2. We will assume a steady-state condition in which the number density n_1 has a spatial distribution which is a function of the y co-ordinate only. If we are to satisfy the condition of constant pressure in the gas we must have

total number of molecules per unit volume

$= n_0 = n_1 + n_2 = \text{constant}.$

It follows that, for constant pressure,

$$\frac{\mathrm{d}n_1}{\mathrm{d}y} = -\frac{\mathrm{d}n_2}{\mathrm{d}y}. \tag{2.8.1}$$

We can use the same method of analysis, which we have already employed for momentum transfer in §2.6 and for energy transfer in §2.7, to calculate the net flux of the special category molecules in the y direction. By analogy with the argument leading up to equation (2.6.6), and referring to figure 10, we have $\mathrm{d}n_1/\mathrm{d}y$ in place of $mn_0(\mathrm{d}U/\mathrm{d}y)$ in equation (2.6.4) and hence the net rate of transport of the special category molecules is given by

$$\text{net flux} = -\frac{1}{3}\frac{\mathrm{d}n_1}{\mathrm{d}y}\int_0^\infty \lambda_v v f(v)\,\mathrm{d}v. \tag{2.8.2}$$

Using the same definition as in (2.6.5) for the transport mean free path $\bar{\lambda}_t$, we can express the basic transport equation for the flux of marked molecules across the xz plane at $y=0$ in the form

$$N_1=-\frac{1}{3}\bar{\lambda}_t\bar{v}\frac{\mathrm{d}n_1}{\mathrm{d}y}. \tag{2.8.3}$$

A similar analysis for the normal category, or type 2, molecules, gives the corresponding flux in the y direction:

$$N_2=-\frac{1}{3}\bar{\lambda}_t\bar{v}\frac{\mathrm{d}n_2}{\mathrm{d}y}. \tag{2.8.4}$$

Noting that the mean free path $\bar{\lambda}_t$ and the mean molecular velocity $\bar{v}$ must be the same for both categories of molecule, in view of their similar size and mass, and making use of the constant pressure condition (2.8.1), we see from (2.8.3) and (2.8.4) that

$$N_1=-N_2 \quad \text{or} \quad N_1+N_2=0, \tag{2.8.5}$$

which is consistent with the steady-state requirement that there is no overall mass transfer.

We can express (2.8.3) in the form

$$N_1=-\mathscr{D}_{12}\frac{\mathrm{d}n_1}{\mathrm{d}y}, \tag{2.8.6}$$

where the diffusion coefficient $\mathscr{D}_{12}=\frac{1}{3}\bar{\lambda}_t\bar{v}$. It is generally more convenient, however, to define the molecular concentration in kmol m^{-3} using the symbol C, and to express the flux in terms of kmol m^{-2} s^{-1}. On this basis (2.8.6) takes the form

$$N_1=-\mathscr{D}_{12}\frac{\mathrm{d}C_1}{\mathrm{d}y}, \tag{2.8.7}$$

which we recognise immediately as *Fick's law of diffusion* (2.1.3). A similar result clearly applies to the diffusion of the normal category molecules, i.e.

$$N_2=-\mathscr{D}_{21}\frac{\mathrm{d}C_2}{\mathrm{d}y}. \tag{2.8.8}$$

Noting that in each case from (2.8.3) and (2.8.4) the diffusion coefficient is given by $\mathscr{D}=\frac{1}{3}\bar{\lambda}_t\bar{v}$ we can conclude from equation (2.6.7) that, on the basis of the simple analysis outlined,

$$\mathscr{D}_{12}=\mathscr{D}_{21}=\frac{\mu}{\rho}=\nu. \tag{2.8.9}$$

As in the case of viscosity and thermal conductivity, however, it is necessary to introduce numerical correction factors into the simple kinetic theory analysis in order to take account of more complex effects such as the persistence of velocities after collision. A corrected result for the diffusion coefficient which applies to the case of two kinds of molecules which are of equal size and mass takes the form

$$\mathscr{D}=1.336\nu. \tag{2.8.10}$$

The ratio $\nu/\mathscr{D}$ is known as the *Schmidt number*. There is an obvious similarity between the role of the Schmidt number in diffusion or mass transfer problems and the corresponding role of the Prandtl number in heat transfer problems.

The analysis becomes more difficult when we consider the important practical case of diffusion in a gas mixture where there are *two different types of molecule* having different masses m_1 and m_2 and different molecular diameters d_1 and d_2. With different values for the mean free paths $\bar{\lambda}_1$ and $\bar{\lambda}_2$, and for the mean molecular velocities $\bar{v}_1$ and $\bar{v}_2$, appropriate to the two different types of molecule, the flow rates given by equations (2.8.3) and (2.8.4) would differ in magnitude and would therefore contradict the steady-state assumption. The requirements of both a steady state and a constant total pressure condition can only now be satisfied if there is a compensating mass motion of the gas mixture with bulk velocity U_0 such that:

$$N_1=n_1U_0-\tfrac{1}{3}\bar{\lambda}_1\bar{v}_1\frac{\mathrm{d}n_1}{\mathrm{d}y}, \tag{2.8.11}$$

$$N_2=n_2U_0-\tfrac{1}{3}\bar{\lambda}_2\bar{v}_2\frac{\mathrm{d}n_2}{\mathrm{d}y}, \tag{2.8.12}$$

and

$$N_1+N_2=0. \tag{2.8.13}$$

This situation is illustrated diagrammatically in figure 13. From (2.8.11), (2.8.12), and (2.8.13) we have

$$(n_1+n_2)U_0-\frac{1}{3}\left(\bar{\lambda}_1\bar{v}_1\frac{\mathrm{d}n_1}{\mathrm{d}y}+\bar{\lambda}_2\bar{v}_2\frac{\mathrm{d}n_2}{\mathrm{d}y}\right)=0$$

and hence the bulk velocity of the gas is given by

$$U_0=\frac{(\bar{\lambda}_1\bar{v}_1-\bar{\lambda}_2\bar{v}_2)}{3(n_1+n_2)}\frac{\mathrm{d}n_1}{\mathrm{d}y}. \tag{2.8.14}$$

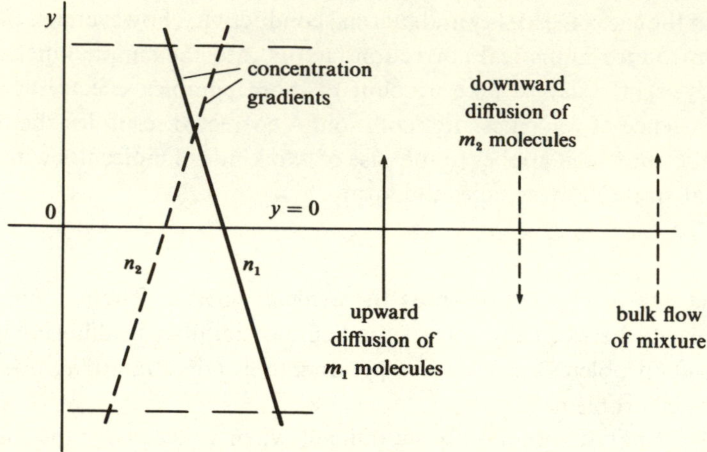

Fig. 13. Mass transfer by molecular action.

We thus have, from (2.8.11)

$$N_1 = -\frac{1}{3}\frac{(n_1\bar{\lambda}_2\bar{v}_2 + n_2\bar{\lambda}_1\bar{v}_1)}{(n_1 + n_2)}\frac{\mathrm{d}n_1}{\mathrm{d}y} \tag{2.8.15}$$

and similarly from (2.8.12)

$$N_2 = -\frac{1}{3}\frac{(n_2\bar{\lambda}_1\bar{v}_1 + n_1\bar{\lambda}_2\bar{v}_2)}{(n_1 + n_2)}\frac{\mathrm{d}n_2}{\mathrm{d}y}. \tag{2.8.16}$$

We therefore conclude that the diffusion coefficient for steady-state diffusion in a two-component gas mixture under constant pressure conditions is given by

$$\mathscr{D}_{12} = \mathscr{D}_{21} = \frac{1}{3}\frac{(n_1\bar{\lambda}_2\bar{v}_2 + n_2\bar{\lambda}_1\bar{v}_1)}{(n_1 + n_2)}. \tag{2.8.17}$$

Note that (2.8.17) reduces to the simple form $\mathscr{D} = \frac{1}{3}\bar{\lambda}\bar{v}$ in the special case of a mixture of molecules of equal size and mass.

It will be appreciated that the main difficulty in using equation (2.8.17) for the evaluation of diffusion coefficients is to determine the appropriate values for the mean free paths $\bar{\lambda}_1$ and $\bar{\lambda}_2$. It may be argued that the rate of diffusion of the m_1 molecules through the m_2 molecules (or vice versa) should depend only on the *collisions between molecules of different types* and should not be affected by collisions between molecules of the same type.

On this hypothesis the appropriate expressions for the mean free paths are

$$\bar{\lambda}_1 = \frac{1}{\pi n_2 \sigma_{12}^2 \sqrt{\left(1+\frac{m_1}{m_2}\right)}} \tag{2.8.18}$$

and

$$\bar{\lambda}_2 = \frac{1}{\pi n_1 \sigma_{12}^2 \sqrt{\left(1+\frac{m_2}{m_1}\right)}}, \tag{2.8.19}$$

where σ_{12} is the mean molecular diameter defined by

$$\sigma_{12} = \tfrac{1}{2}(d_1 + d_2).$$

We can substitute from (2.5.4) for the mean molecular velocities $\bar{v}_1$ and $\bar{v}_2$ and hence from (2.8.17) the diffusion coefficient is given by

$$\mathscr{D} = \frac{1}{3\pi(n_1+n_2)\sigma_{12}^2}\left(\frac{8kT}{\pi}\right)^{\frac{1}{2}} \frac{\sqrt{(m_1+m_2)}}{\sqrt{(m_1 m_2)}}. \tag{2.8.20}$$

This expression may be simplified slightly, noting that $n_0 = n_1 + n_2$ where n_0 is the total number of molecules per unit volume, and using equation (1.2.3) for the total pressure p, i.e.

$$p = n_0 kT.$$

Hence, substituting and re-arranging, we have

$$\mathscr{D} = \frac{1}{3}\left(\frac{2}{\pi}\right)^{\frac{3}{2}} \frac{(kT)^{\frac{3}{2}}}{p\sigma_{12}^2} \sqrt{\left(\frac{1}{m_1}+\frac{1}{m_2}\right)}. \tag{2.8.21}$$

We can also express the actual molecular masses m_1 and m_2 in terms of the relative molecular masses M_{r1} and M_{r2} using the relationship $M_r = mN_A \times 10^3$. Hence taking the numerical values for Boltzmann's constant k and Avogadro's number N_A quoted in chapter 1, we find

$$\mathscr{D} = 2.13 \times 10^{-22} \frac{T^{\frac{3}{2}}}{p\sigma_{12}^2} \sqrt{\left(\frac{1}{M_{r1}}+\frac{1}{M_{r2}}\right)}. \tag{2.8.22}$$

Note that the units of $\mathscr{D}$ are $m^2\ s^{-1}$.

2.9 The basic transport equation for a fluid

We have outlined in §§ 2.6–2.8 the common molecular basis for the transport phenomena described experimentally in terms of Newton's law of

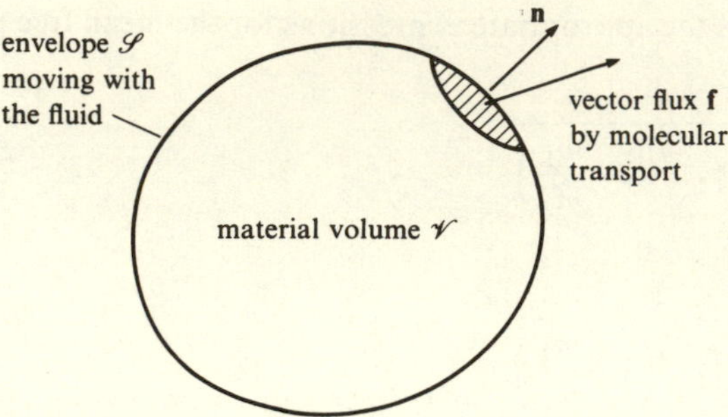

Fig. 14. Moving boundary $\mathscr{S}$ enclosing material volume $\mathscr{V}$.

fluid friction, Fourier's law of heat conduction, and Fick's law of diffusion. So far, however, we have confined ourselves to the simple one-dimensional case with transfer taking place in only one direction. In the general case we have to treat the flux of a transferable property as a vector quantity and the basic molecular transport law may be expressed as

$$\mathbf{f} = -\kappa \operatorname{grad} \phi, \tag{2.9.1}$$

where $\mathbf{f}$ is the vector flux of the transferable property; and κ is the transport coefficient.

If we now consider the three-dimensional case of a material volume of fluid, we can very easily evaluate the rate of supply of the transferable property to the material volume as a result of the molecular transport process. Referring to figure 14, the net rate of transfer *outwards* across the boundary of the volume will be given by the surface integral $\int_{\mathscr{S}} \mathbf{f}\cdot\mathbf{n}\, d\mathscr{S}$. We can use Gauss's divergence theorem as quoted in appendix 1, however, to equate the surface integral of a vector quantity to the volume integral of its divergence. Thus we can say that

$$\int_{\mathscr{S}} \mathbf{f}\cdot\mathbf{n}\, d\mathscr{S} = \int_{\mathscr{V}} \operatorname{div} \mathbf{f}\, d\mathscr{V}. \tag{2.9.2}$$

It follows that the net *inward flow* of ϕ into the material volume is equal to $-\int_{\mathscr{V}} \operatorname{div} \mathbf{f}\, d\mathscr{V}$.

From (2.9.1) and (2.9.2) we see that the net rate of supply into the material volume may be expressed *per unit volume* as

$$-\operatorname{div} \mathbf{f} = \operatorname{div} (\kappa \operatorname{grad} \phi). \tag{2.9.3}$$

Provided the transport coefficient κ is a constant, therefore, we can say that the 'source term' Q or the net rate of supply per unit volume is given by

$$Q=\kappa \operatorname{div} \operatorname{grad} \phi=\kappa \nabla^2\phi. \tag{2.9.4}$$

The basic conservation equation (2.4.5), which expresses the balance between fluid convection and molecular transport, may now be re-stated by substituting for the 'source term' Q from (2.9.4) to give

$$\rho \frac{\mathrm{D}\phi}{\mathrm{D}t}=\kappa \nabla^2\phi. \tag{2.9.5}$$

This may be regarded as the basic transport equation for a moving fluid.

If we now take energy transfer as an example we have the internal energy e as the transferable property ϕ and, in the case of a perfect gas, this is related to the thermodynamic temperature T by $\mathrm{d}e=c_v\,\mathrm{d}T$. It is more convenient to work in terms of the temperature gradient rather than the internal energy gradient and the source term may therefore be expressed as

$$Q=\kappa \nabla^2\phi=k_h\nabla^2 T. \tag{2.9.6}$$

Thus the transport equation (2.9.5) may be written as:

$$\rho c_v \frac{\mathrm{D}T}{\mathrm{D}t}=k_h\nabla^2 T$$

or

$$\frac{\mathrm{D}T}{\mathrm{D}t}=\frac{k_h}{\rho c_v}\nabla^2 T. \tag{2.9.7}$$

This may also be expressed as:

$$\frac{\mathrm{D}T}{\mathrm{D}t}=\alpha \nabla^2 T, \tag{2.9.8}$$

where $\alpha=k_h/\rho c_v$ is known as the thermal diffusivity. Equation (2.9.7) may be compared with the heat conduction equation for a solid or static medium, which is discussed in detail in chapter 13, and which takes the form:

$$\frac{\partial T}{\partial t}=\frac{k_h}{\rho c_v}\nabla^2 T. \tag{2.9.9}$$

In the analysis leading to equation (2.9.7) we have identified the transfer of internal energy with *heat transfer*. This is not strictly correct since heat and internal energy are different physical concepts. The internal energy is a function of the state of a substance whereas heat is not. The two concepts are linked through the first law of thermodynamics but a strict application of this law to a moving fluid must be deferred until chapter 9. In the meantime we can say that equation (2.9.7) is valid provided the mechanical work terms in the first law of thermodynamics associated with the expansion or compression of the fluid element, and also with its distortion, are small compared with the internal energy changes.

If we now turn attention to mass transfer and diffusion, we have for the basic molecular transport law:

$$\mathbf{N} = -\mathscr{D}\,\mathrm{grad}\,C, \tag{2.9.10}$$

where $\mathbf{N}$ is the molar flux, C is the molar concentration, and $\mathscr{D}$ is the diffusion coefficient. The 'source term' Q is therefore given by $\mathscr{D}\nabla^2 C$ and the basic transport equation may then be written as:

$$\frac{\mathrm{D}C}{\mathrm{D}t} = \mathscr{D}\nabla^2 C. \tag{2.9.11}$$

This may be compared with the diffusion equation for a *static medium* which takes the form

$$\frac{\partial C}{\partial t} = \mathscr{D}\nabla^2 C. \tag{2.9.12}$$

The similarity between equations (2.9.8) and (2.9.11), and between equations (2.9.9) and (2.9.12), will be evident. Both the thermal diffusivity α and the diffusion coefficient $\mathscr{D}$ have dimensions $[l^2 t^{-1}]$.

We have necessarily been dealing with generalities in this preliminary chapter, but we have already established the basic principles and equations which govern transfer processes in a fluid. In order to apply this analysis to actual flow situations, and to specific engineering problems, however, we need to have more information about the detailed mechanics of fluid motion and the different types of flow pattern which are found both in the field of the natural sciences and in engineering applications. The next ten chapters of this book are designed to meet this requirement.

3

Fluid dynamics and Euler's equation of motion

3.1 The acceleration of a fluid element

We can apply the method of differentiation following the motion of the fluid to the fluid velocity $\mathbf{u}$ in order to obtain an expression for the acceleration of a fluid element. We start with the basic statement, as in (1.6.1), that the fluid velocity $\mathbf{u}$ is a function of the position in space defined by the position vector $\mathbf{r}$ and also a function of time t, i.e.

$$\mathbf{u}=f(\mathbf{r}, t). \tag{3.1.1}$$

If we use Cartesian co-ordinates, we can write this vector equation in the form of three separate scalar equations for the x, y, and z components of the velocity vector $\mathbf{u}$. Using the notation defined in (1.6.5) for this purpose, we have

$$u=f(x, y, z, t),$$

$$v=f(x, y, z, t),$$

$$w=f(x, y, z, t).$$

We can apply equation (2.2.4) for the total differential following the motion of the fluid to each of the velocity components u, v, w in turn. Thus for the x component of the acceleration we have

$$\frac{\mathrm{D}u}{\mathrm{D}t}=\frac{\partial u}{\partial t}+\mathbf{u}\cdot\nabla u$$

or

$$\frac{\mathrm{D}u}{\mathrm{D}t}=\frac{\partial u}{\partial t}+u\frac{\partial u}{\partial x}+v\frac{\partial u}{\partial y}+w\frac{\partial u}{\partial z}. \tag{3.1.2}$$

Similarly

$$\frac{\mathrm{D}v}{\mathrm{D}t}=\frac{\partial v}{\partial t}+u\frac{\partial v}{\partial x}+v\frac{\partial v}{\partial y}+w\frac{\partial v}{\partial z} \tag{3.1.3}$$

and

$$\frac{\mathrm{D}w}{\mathrm{D}t}=\frac{\partial w}{\partial t}+u\frac{\partial w}{\partial x}+v\frac{\partial w}{\partial y}+w\frac{\partial w}{\partial z}. \tag{3.1.4}$$

It will be noted that the first term on the right-hand side of each of these equations represents the *local differential* or rate of variation of the velocity component with time at a fixed point in space. The remaining group of three terms on the right-hand side of each equation represents the *convective differential* associated with the movement of the fluid element in space.

The three scalar equations (3.1.2) to (3.1.4) may in fact be combined into a single vectorial equation:

$$\frac{\mathbf{Du}}{\mathrm{D}t}=\frac{\partial \mathbf{u}}{\partial t}+\mathbf{u}\cdot\nabla\mathbf{u}, \tag{3.1.5}$$

which is the result anticipated in (2.2.6). It should be noted that the mathematical term $\mathbf{u}\cdot\nabla$ represents the operator

$$\left(u\frac{\partial}{\partial x}+v\frac{\partial}{\partial y}+w\frac{\partial}{\partial z}\right)$$

The velocity $\mathbf{u}$ is the fluid momentum per unit mass, and the quantity $\rho\mathbf{u}$ is the fluid momentum *per unit material volume.* The total momentum of a material volume of fluid $\mathscr{V}$ is given by:

$$\int_{\mathscr{V}} \rho\mathbf{u}\,\mathrm{d}\mathscr{V}.$$

The rate of change of this momentum is simply $(\mathrm{d}/\mathrm{d}t)\int_{\mathscr{V}} \rho\mathbf{u}\,\mathrm{d}\mathscr{V}$. We can deduce from (2.4.2) that

$$\frac{\mathrm{d}}{\mathrm{d}t}\int_{\mathscr{V}} \rho\mathbf{u}\,\mathrm{d}\mathscr{V}=\int_{\mathscr{V}} \rho\frac{\mathbf{Du}}{\mathrm{D}t}\,\mathrm{d}\mathscr{V}. \tag{3.1.6}$$

Thus for a small material volume, i.e. for a *small element of fluid*, we have the result that:

$$\text{rate of change of momentum (per unit volume)}=\rho\frac{\mathbf{Du}}{\mathrm{D}t}. \tag{3.1.7}$$

This simple expression provides the starting point for a study of fluid dynamics.

3.2 Euler's equation of motion

We can obtain the equation of motion for a material element of a fluid by applying Newton's second law, i.e. by equating the rate of change of the momentum of the fluid element as given by (3.1.7) to the net applied force acting on the element.

The net applied force acting on a material element must include the body force, for example gravity, acting directly on the mass of the element, and also the forces acting on the surface of the element which arise from the pressure distribution in the fluid and the action of viscous shearing stresses. A full analysis of the viscous stresses is rather complicated and will be deferred until chapter 5. If we neglect the effect of the viscous stresses, however, we can derive a very simple dynamic equation governing the motion of an *inviscid fluid*.

The components of the applied force acting over the surface of a small element $\mathrm{d}x\,\mathrm{d}y\,\mathrm{d}z$, due to the pressure distribution, can be obtained by reference to figure 15. It will be seen that the x component of the resultant force arising from the pressure distribution, acting on the two opposite faces $\mathrm{d}y\,\mathrm{d}z$ of the element, is given by $-(\partial p/\partial x)\,\mathrm{d}x\,\mathrm{d}y\,\mathrm{d}z$. Hence the x component of the force per unit volume is $-(\partial p/\partial x)$. Similarly the y and z components of the force per unit volume are given by $-(\partial p/\partial y)$ and $-(\partial p/\partial z)$ respectively. We can evidently combine these three components into a single resultant vector force represented by $-\operatorname{grad} p$.

The body force, acting directly on the fluid mass, can be expressed in general terms by the vector **F** per unit mass. Thus the body force acting per unit volume will be $\rho\mathbf{F}$.

Hence, equating the rate of change of momentum of the fluid element to

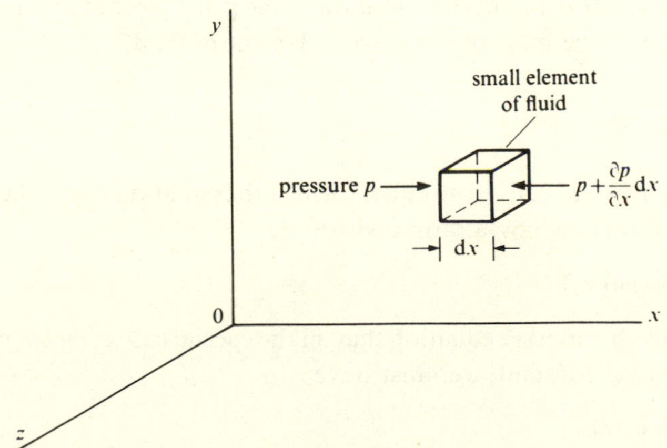

Fig. 15. Force on a small element due to a pressure gradient.

the net applied force acting per unit volume, we have:

$$\rho \frac{\mathrm{D}\mathbf{u}}{\mathrm{D}t} = \rho \mathbf{F} - \operatorname{grad} p, \tag{3.2.1}$$

which is known as *Euler's equation of motion.* For reference purposes we can write out the x component of the equation as follows:

$$\rho \left(\frac{\partial u}{\partial t} + u \frac{\partial u}{\partial x} + v \frac{\partial u}{\partial y} + w \frac{\partial u}{\partial z} \right) = \rho F_x - \frac{\partial p}{\partial x}. \tag{3.2.2}$$

If the acceleration of the fluid is zero at all points, Euler's equation (3.2.1) reduces to the hydrostatic equation (1.4.2):

Hydrostatic equation $\qquad \rho \mathbf{F} - \operatorname{grad} p = 0,$

Euler's equation of motion $\qquad \rho \mathbf{F} - \operatorname{grad} p = \rho \dfrac{\mathrm{D}\mathbf{u}}{\mathrm{D}t}.$

3.3 The equation of continuity

We have seen in § 2.3 that the equation describing the conservation of the total mass of a fluid may be expressed either in the form of (2.3.3) or in the form of (2.3.6). Equation (2.3.3) was derived by applying the principle of the conservation of mass to a fixed volume of space with the conclusion that:

$$\frac{\partial \rho}{\partial t} + \operatorname{div} (\rho \mathbf{u}) = 0. \tag{3.3.1}$$

The alternative form (2.3.6) expressed the same principle, but applies to the conservation of the mass of a material element of fluid, i.e.

$$\frac{\mathrm{D}\rho}{\mathrm{D}t} + \rho \operatorname{div} \mathbf{u} = 0. \tag{3.3.2}$$

It follows mathematically from either form of the equation that if the flow is *steady*, i.e. if there is no variation with time,

$$\operatorname{div} (\rho \mathbf{u}) = 0. \tag{3.3.3}$$

It also follows from either equation that, in the special case of *incompressible flow*, i.e. with ρ constant, we must have

$$\operatorname{div} \mathbf{u} = 0, \tag{3.3.4}$$

which is applicable equally to steady and unsteady flow. We have already seen in §2.3 that the mathematical term div $\mathbf{u}$ represents the rate of dilatation of a fluid element. We may therefore regard equation (3.3.4) as an almost self-evident statement that, for an incompressible fluid, the rate of dilatation, or volumetric rate of strain, must be zero.

Referring to appendix 1, the equation of continuity for an *incompressible fluid* (3.3.4) takes the following form when expressed in Cartesian co-ordinates:

$$\frac{\partial u}{\partial x}+\frac{\partial v}{\partial y}+\frac{\partial w}{\partial z}=0. \tag{3.3.5}$$

The equation of continuity takes a particularly simple form when we are dealing with the special case of steady flow along a stream tube as illustrated in figure 16. Starting at section (1) in the diagram, where the fluid velocity is u_1, we can picture a small closed curve drawn in a plane which is normal to the velocity vector and which encloses a small flow area a_1. The streamlines passing through this curve will form a *stream tube*, and if the curve is drawn small enough the fluid velocity will be effectively uniform over any cross-section of the stream tube. We can picture a *fixed control surface* exactly enclosing this stream tube and extending from section (1) to section (2) in the diagram. Since there can be no flow across a streamline, the law of conservation of mass requires that along the stream tube $\rho a u=$constant, where a is the cross-sectional flow area of the stream tube at any section and u is the corresponding fluid velocity. Hence it is obvious that

$$\rho_1 a_1 u_1=\rho_2 a_2 u_2. \tag{3.3.6}$$

It will be seen that this is a special case of equation (3.3.3). The difference $(\rho_2 a_2 u_2-\rho_1 a_1 u_1)$ represents the net rate of outflow of mass from the control surface and is identical with the surface integral of $\rho\mathbf{u}$ evaluated over the

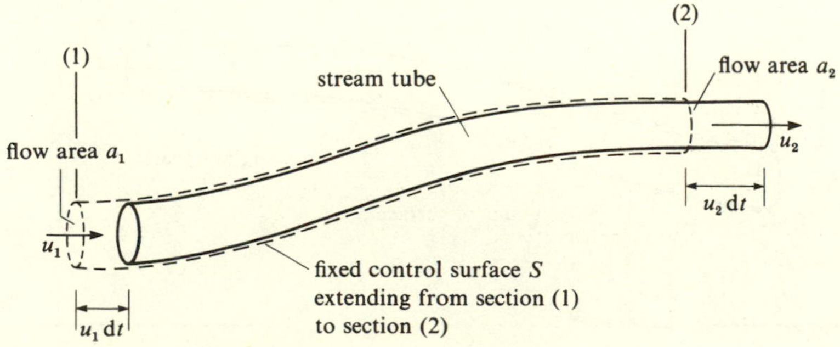

Fig. 16. Steady flow along a stream tube.

control surface. Referring to § 2.3, this surface integral can be equated by Gauss's theorem to the volume integral of div ($\rho\mathbf{u}$). Hence for steady flow, with no net outflow of mass from the control surface, equation (3.3.6) must apply and this is seen to be identical with equation (3.3.3).

3.4 The mechanical energy equation for the steady flow of a frictionless fluid along a stream tube

We can use the same concept of a fixed control surface, which coincides with a stream tube, to apply the principle of the conservation of energy. At this stage, however, we will exclude the effects of heat transfer and changes of internal energy, and we will restrict outselves to *mechanical energy terms*. The full treatment of the energy equation, including thermal effects, will be deferred until chapter 9. We will assume initially that the fluid is incompressible and we will confine ourselves in this section to the special case of a frictionless or inviscid fluid in which the shear stresses are negligible.

Referring to figure 17, the mass-flow rate through the stream tube m may be defined by

$$m=\rho a_1 u_1=\rho a_2 u_2. \qquad (3.4.1)$$

The fluid which is entering the stream tube at section (1) is being pushed by the fluid behind it, and the fluid which is leaving the stream tube at section (2) is pushing the fluid in front of it. The net rate at which work is being done in this manner at sections (1) and (2) *on* the fluid which is contained within the stream tube is given by $(p_1 a_1 u_1 - p_2 a_2 u_2)$ and we can describe

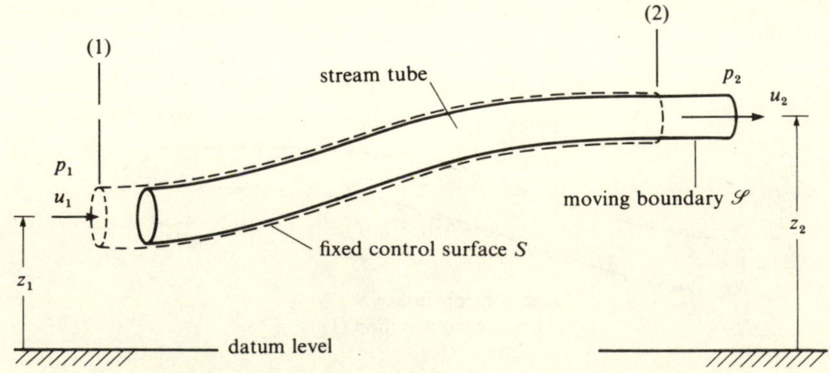

Fig. 17. Steady flow along a stream tube.

this as the *flow work*. Making use of (3.4.1) we can say that:

$$\text{flow work per second} = p_1 a_1 u_1 - p_2 a_2 u_2 = \left(\frac{p_1}{\rho} - \frac{p_2}{\rho}\right) m. \qquad (3.4.2)$$

If we divide equation (3.4.2) by the mass-flow rate m we obtain an expression for the rate at which flow work is being done per unit mass-flow rate. This is identical with the actual net amount of work done by the surrounding fluid at sections (1) and (2) *per unit mass* of fluid passing through the control surface. Thus we can say that

$$\text{flow work per unit mass} = \frac{p_1}{\rho} - \frac{p_2}{\rho}. \qquad (3.4.3)$$

It will be seen that no flow work can be done at the wall of the stream tube since by definition there is no flow across a stream line. It will also be noted that, in the absence of any viscous shear stresses within the fluid, there can be no shear work being done at the wall of the stream tube.

The rate of gain of kinetic energy by the fluid, between entry to the control surface at section (1) and exit at section (2), is

$$m(\tfrac{1}{2}u_2^2 - \tfrac{1}{2}u_1^2).$$

Similarly the rate of gain of potential energy by the fluid due to the change in level from z_1 to z_2 is $mg(z_2 - z_1)$. Hence, equating the rate of gain of kinetic plus potential energy by the fluid between sections (1) and (2) to the rate of flow work, we have

$$m(\tfrac{1}{2}u_2^2 - \tfrac{1}{2}u_1^2) + mg(z_2 - z_1) = p_1 a_1 u_1 - p_2 a_2 u_2$$

or, substituting from (3.4.2) and dividing by m,

$$\left(\frac{u_2^2}{2} - \frac{u_1^2}{2}\right) + g(z_2 - z_1) = \frac{p_1}{\rho} - \frac{p_2}{\rho}. \qquad (3.4.4)$$

This is the simplest form of the *mechanical energy equation* for the steady flow of an incompressible fluid with negligible viscous stresses. It states that the net gain of kinetic plus potential energy by the fluid per unit mass is equal to the flow work per unit mass.

If we re-arrange (3.4.4) by grouping the terms relating to sections (1) and (2), we obtain *Bernoulli's equation*:

$$\frac{p_2}{\rho} + \frac{u_2^2}{2} + gz_2 = \frac{p_1}{\rho} + \frac{u_1^2}{2} + gz_1. \qquad (3.4.5)$$

The application of Bernoulli's equation will be discussed further in §3.5.

We could equally well derive the mechanical energy equation by starting with Euler's equation of motion in its one-dimensional form. Taking equation (3.2.2), with the distance s measured along the axis of the stream tube as the spatial variable in place of the co-ordinate x, and assuming steady-flow conditions, we have

$$\rho u\frac{\mathrm{d}u}{\mathrm{d}s}=\rho F_s-\frac{\mathrm{d}p}{\mathrm{d}s}. \tag{3.4.6}$$

The body force component F_s due to gravity will be given by $F_s=-g(\mathrm{d}z/\mathrm{d}s)$, and Euler's equation of motion for steady flow of a fluid element along the stream tube may therefore be expressed as:

$$u\frac{\mathrm{d}u}{\mathrm{d}s}+g\frac{\mathrm{d}z}{\mathrm{d}s}=-\frac{1}{\rho}\frac{\mathrm{d}p}{\mathrm{d}s}$$

or

$$\frac{\mathrm{d}}{\mathrm{d}s}\left(\frac{u^2}{2}+gz\right)=-\frac{1}{\rho}\frac{\mathrm{d}p}{\mathrm{d}s}. \tag{3.4.7}$$

If the density ρ is constant, this may be integrated along the stream tube from section (1) to section (2) to give

$$\left(\frac{u_2^2}{2}-\frac{u_1^2}{2}\right)+g(z_2-z_1)=\frac{p_1}{\rho}-\frac{p_2}{\rho}, \tag{3.4.8}$$

which is exactly the same result as (3.4.4).

If, however, the density of the fluid is not constant we must express the integration of (3.4.7) in the more general form

$$\int_1^2\frac{\mathrm{d}p}{\rho}+\left(\frac{u_2^2}{2}-\frac{u_1^2}{2}\right)+g(z_2-z_1)=0. \tag{3.4.9}$$

Alternatively we may re-arrange the differential form of the energy equation (3.4.7) by making use of the mathematical identity:

$$\frac{\mathrm{d}}{\mathrm{d}s}\left(\frac{p}{\rho}\right)=\frac{1}{\rho}\frac{\mathrm{d}p}{\mathrm{d}s}-\frac{p}{\rho^2}\frac{\mathrm{d}\rho}{\mathrm{d}s}.$$

Equation (3.4.7) may then be expressed as

$$\frac{\mathrm{d}}{\mathrm{d}s}\left(\frac{u^2}{2}+gz\right)=-\frac{\mathrm{d}}{\mathrm{d}s}\left(\frac{p}{\rho}\right)-\frac{p}{\rho^2}\frac{\mathrm{d}\rho}{\mathrm{d}s}. \tag{3.4.10}$$

Thus for a differential movement ds along the stream tube, the incremental energy change is given by

$$\mathrm{d}\left(\frac{u^2}{2}+gz\right)=-\mathrm{d}\left(\frac{p}{\rho}\right)-\frac{p}{\rho^2}\,\mathrm{d}\rho. \tag{3.4.11}$$

The meaning of the last term on the right-hand side of equation (3.4.10) or (3.4.11) will become clear if we note that the work done *on* a fluid element per unit mass, corresponding to a change $\mathrm{d}V$ in the specific volume, is given by $-p\,\mathrm{d}V$. The specific volume V is related to the density by $V=1/\rho$ and hence we have $\mathrm{d}V=-(1/\rho^2)\,\mathrm{d}\rho$. Thus the work of compression done on the fluid element per unit mass, corresponding to an incremental change $\mathrm{d}\rho$ in the density, is given by

$$\text{compression work per unit mass}=\frac{p}{\rho^2}\,\mathrm{d}\rho. \tag{3.4.12}$$

It will now be apparent that the energy equation (3.4.11) states simply that the gain of kinetic plus potential energy of the fluid element, expressed per unit mass, is equal to the flow work done *on* the element *minus* the work required to compress the element.

If we transfer the flow work term from the right-hand to the left-hand side of equation (3.4.11) we obtain the differential form of Bernoulli's equation for a compressible fluid:

$$\mathrm{d}\left(\frac{p}{\rho}+\frac{u^2}{2}+gz\right)=-\frac{p}{\rho^2}\,\mathrm{d}\rho. \tag{3.4.13}$$

In the special case of a fluid with constant density this result reduces to

$$\mathrm{d}\left(\frac{p}{\rho}+\frac{u^2}{2}+gz\right)=0, \tag{3.4.14}$$

which is simply a re-statement of equation (3.4.5) in differential form.

3.5 Bernoulli's equation and flow measurement

We arrived at Bernoulli's equation in §3.4 by two alternative but equally valid routes involving the principle of the conservation of energy in a steady-flow system and, on the other hand, the direct integration of Euler's equation of motion along a stream tube. It is important to appreciate that the equation is restricted to the idealised case of frictionless flow since we neglected the effect of shear stresses acting at the wall of the stream tube. Despite this limitation, Bernoulli's equation has a wide field of application in flow problems where the effects of viscosity and turbulence on the flow pattern are small. It plays a particularly important role in providing the analytical basis for a number of practical flow-measuring devices.

It will be seen from (3.4.5) or (3.4.14) that for an incompressible fluid Bernoulli's equation may be expressed as

$$\frac{p}{\rho}+\frac{u^2}{2}+gz=\text{constant}. \tag{3.5.1}$$

It must be noted, however, that the equation has only been derived for flow in an individual stream tube of small cross-section and that the value of the constant in (3.5.1) must be assumed to apply only to *one* stream tube or streamline. The special conditions that must be satisfied so that the same constant may apply to all streamlines in the flow pattern are investigated in §4.2.

The term in Bernoulli's equation representing potential energy, associated with terrestrial gravity, is usually only relevant to the flow of liquids where changes of level are involved. If changes of level are not significant equation (3.5.1) reduces to

$$\frac{p}{\rho}+\frac{u^2}{2}=\text{constant}. \tag{3.5.2}$$

We can now apply this equation to the following flow-measuring instruments:

(*a*) *Pitot tube*

When a fluid stream impinges on a flat plate which is held in a position at right angles to the direction of the stream, the point on the central streamline at which the fluid comes to rest as it reaches the plate is known as the *stagnation point*. Similarly when a streamlined body or aerofoil section is located in an airstream, there will be a point on the rounded nose of the section where the streamlines divide on either side and where, locally, the airstream must come to rest relative to the body. The pressure measured at such a point is known as the *stagnation pressure* or *total pressure*.

A Pitot tube is simply an open-ended tube pointing in the direction of the flow and connected to a pressure gauge or manometer, as indicated in figure 18, and is used for the purpose of measuring the stagnation pressure.

Applying Bernoulli's equation in the form of (3.5.2) from section (1) to

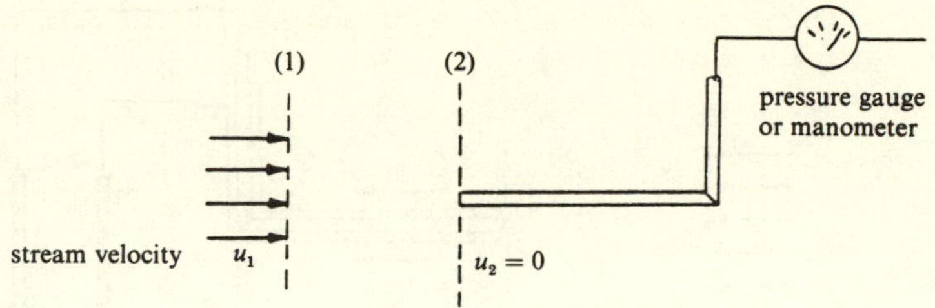

Fig. 18. Pitot tube.

section (2) in figure 18, we have

$$\frac{p_1}{\rho}+\frac{u_1^2}{2}=\frac{p_2}{\rho}+0, \tag{3.5.3}$$

i.e.

$$p_2=p_1+\tfrac{1}{2}\rho u_1^2. \tag{3.5.3}$$

The pressure p_1 in equation (3.5.3) and figure 18 is usually referred to as the 'static pressure', but this term should not be confused with the hydrostatic pressure defined in § 1.4. The term $\frac{1}{2}\rho u_1^2$ in equation (3.5.3) is known as the 'dynamic pressure'. The equation thus states that the total or stagnation pressure measured by the Pitot tube is equal to the sum of the static and dynamic pressures in the free stream.

(*b*) *Pitot-static tube*

The Pitot-static tube consists of a Pitot tube surrounded by a concentric closed outer tube having a circumferential ring of small holes as indicated in figure 19. The inner tube measures the total pressure and the outer tube measures the static pressure. By connecting the two tubes across a manometer or pressure gauge, we can measure the dynamic pressure directly. From (3.5.3) we have

$$p_2-p_1=\tfrac{1}{2}\rho u_1^2$$

and hence

$$u_1=\sqrt{\left[2\frac{(p_2-p_1)}{\rho}\right]}. \tag{3.5.4}$$

Thus, provided we know the density of the fluid, we can calculate the velocity from the measured pressure difference (p_2-p_1).

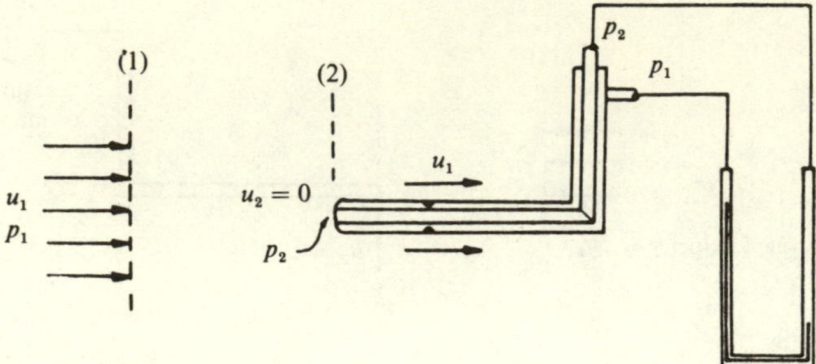

Fig. 19. Pitot-static tube.

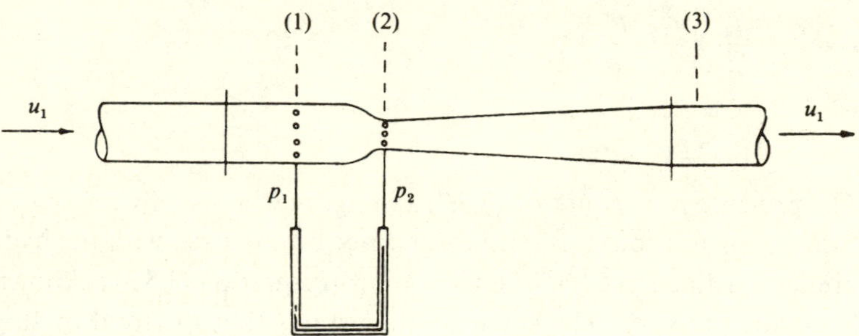

Fig. 20. Venturi tube.

(*c*) *Venturi meter*

A Venturi tube, shown diagrammatically in figure 20, is a device used for measuring the flow rate in a pipe. We make the simplifying approximation that we can deal in terms of an average velocity across any section of the tube. We can then apply the continuity equation (3.3.6) between the entry section (1) and the throat section (2), assuming incompressible flow, in the form

$$a_1 u_1 = a_2 u_2,$$

where a_1 and a_2 are the cross-sectional flow areas at sections (1) and (2) respectively. Bernoulli's equation may be applied to the flow along a streamline in the central region of the tube between sections (1) and (2), i.e.

$$\frac{p_1}{\rho} + \frac{u_1^2}{2} = \frac{p_2}{\rho} + \frac{u_2^2}{2}$$

and hence

$$u_1 = \sqrt{\left[\frac{2(p_1 - p_2)}{\rho(a_1^2/a_2^2 - 1)}\right]}. \tag{3.5.5}$$

From (3.5.5) the theoretical volumetric flow rate Q_t is given by

$$Q_t = a_1 u_1 = a_1 \sqrt{\left[\frac{2(p_1 - p_2)}{\rho(a_1^2/a_2^2 - 1)}\right]}. \tag{3.5.6}$$

In practice, however, the actual flow rate will be slightly less than the theoretical value given by (3.5.6) owing to non-uniformity of the velocity distribution across any section of the tube and also owing to frictional effects on the flow. We therefore introduce an empirical factor or *discharge coefficient C*, and we re-write (3.5.6) in the form:

$$Q = Ca_1 \sqrt{\left[\frac{2(p_1 - p_2)}{\rho(a_1^2/a_2^2 - 1)}\right]}. \tag{3.5.7}$$

The numerical value of the discharge coefficient will depend on the ratio a_1/a_2 and on the value of the Reynolds number at the throat, but for a well-designed Venturi meter it should normally be between 0.96 and 0.98.

The shape of the Venturi tube is important. It will be noted from figure 20 that there is a sharp contraction from section (1) to the throat followed by a gradual expansion between sections (2) and (3). The reason for this is that from (1) to (2) the fluid is being accelerated and the pressure is falling, and there is no difficulty in achieving rapid acceleration of the flow in this region. From (2) to (3), however, the fluid is being decelerated and the pressure is rising. If we try to achieve this pressure recovery too rapidly, the main stream of the fluid is liable to break away from the wall owing to a boundary-layer effect which will be examined in detail in chapter 7. The expanding section of the tube is known as a *diffuser*. The more gradual the expansion in the diffuser the less will be the tendency for flow separation to occur. Ideally, if Bernoulli's equation applies throughout from (1) to (3), there should be complete recovery of pressure and p_3 should therefore approach p_1 in value. In practice, however, there will always be some loss in the Bernoulli energy due to skin friction at the wall of the tube even if there is no separation of flow in the diffuser. Consequently there will be an overall pressure drop and p_3 will be less than p_1.

(*d*) *Orifice plate*

The orifice plate is an alternative to the Venturi meter for

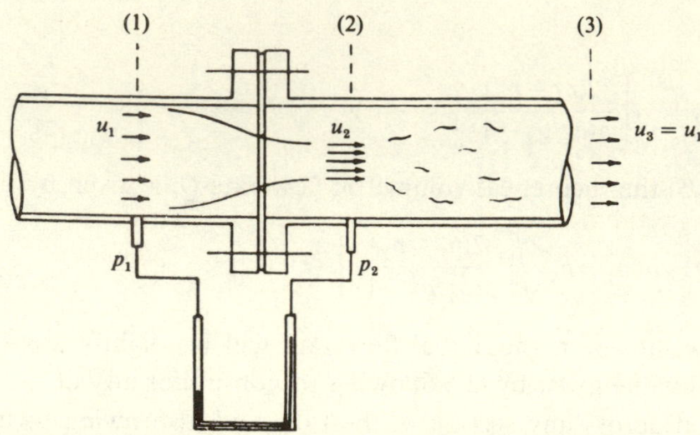

Fig. 21. Orifice plate.

measuring flow rates in a pipeline. As indicated in figure 21, the stream emerges from the sharp-edged orifice in the form of a submerged jet, and there is no attempt to achieve maximum pressure recovery beyond the throat section. Referring to figure 21, let A be the cross-sectional area of the pipe and let a be the area of the orifice. The submerged jet converges to a vena-contracta just downstream from the orifice plate and then breaks up into a disturbed region of eddying and turbulent flow. The ratio of the area of the jet at the vena-contracta to the area a of the orifice is known as the *coefficient of contraction.* For a sharp-edged orifice plate of the type indicated in the diagram, the numerical value of this coefficient is usually within the range 0.62–0.70.

We can apply the continuity and Bernoulli equations between sections (1) and (2) and the analysis is similar to that for the Venturi tube except for the uncertainty regarding the effective value of the flow area at the vena-contracta. The flow rate Q will therefore be given by an expression similar to (3.5.7):

$$Q = CA\sqrt{\left[\frac{2(p_1 - p_2)}{\rho(A^2/a^2 - 1)}\right]}, \tag{3.5.8}$$

but the discharge coefficient C for the orifice plate will include both the effects of non-uniformity of velocity across the pipe and also the effect of the contraction of the jet. The numerical value of C will depend primarily on the area ratio A/a but is usually between 0.60 and 0.62. It will be noted that the orifice plate is arranged with the sharp edge on the upstream side. The purpose of this is to ensure reproducible values for the coefficients of contraction and discharge and to avoid the necessity of calibrating every plate individually.

The orifice plate is convenient from a practical point of view for flow measurement in pipelines. It can easily be introduced at a flanged joint as indicated in figure 21, and the pressure tappings can be located if necessary in the flanges themselves, close on either side of the orifice plate, with little error in the measurement of the effective pressure difference.

3.6 The momentum equation

In § 3.2 we obtained Euler's equation of motion for an inviscid fluid by equating the rate of change of momentum of a material element of the fluid to the net applied force acting on the element. We may therefore describe Euler's equation as the differential form of the *momentum equation* for an inviscid fluid. We could equally well apply Newton's second law of motion to a material volume $\mathscr{V}$ of the fluid and thus obtain the equation of motion in integral form. Making use of (3.1.6) we can say that

$$\int_{\mathscr{V}} \rho \frac{\mathrm{D}\mathbf{u}}{\mathrm{D}t}\,\mathrm{d}\mathscr{V} = \int_{\mathscr{V}} \rho\mathbf{F}\,\mathrm{d}\mathscr{V} - \int_{\mathscr{S}} p\mathbf{n}\,\mathrm{d}\mathscr{S}. \tag{3.6.1}$$

The last term on the right-hand side involving the surface integral of the normal pressure p acting over the boundary of the material volume, may be transformed into a volume integral using the following mathematical relationship:

$$\int_{\mathscr{S}} p\mathbf{n}\,\mathrm{d}\mathscr{S} = \int_{\mathscr{V}} \operatorname{grad} p\,\mathrm{d}\mathscr{V}. \tag{3.6.2}$$

If we make this substitution in equation (3.6.1) we simply obtain Euler's equation of motion (3.2.1) integrated over the material volume $\mathscr{V}$.

If we now consider the motion of a fluid in which viscous stresses are significant we can write down a momentum equation for a material volume $\mathscr{V}$, but it will take the form:

$$\int_{\mathscr{V}} \rho \frac{\mathrm{D}\mathbf{u}}{\mathrm{D}t}\,\mathrm{d}\mathscr{V} = \int_{\mathscr{V}} \rho\mathbf{F}\,\mathrm{d}\mathscr{V} + \int_{\mathscr{S}} \boldsymbol{\sigma}\,\mathrm{d}\mathscr{S}, \tag{3.6.3}$$

where $\boldsymbol{\sigma}$ is the vector stress acting on an element of the boundary $\mathrm{d}\mathscr{S}$. The stress $\boldsymbol{\sigma}$ will have a direct-stress component which is normal to the surface element and also two tangential shearing-stress components. Note that the vector $\boldsymbol{\sigma}$ is drawn outward from the surface element, as in figure 22, and the

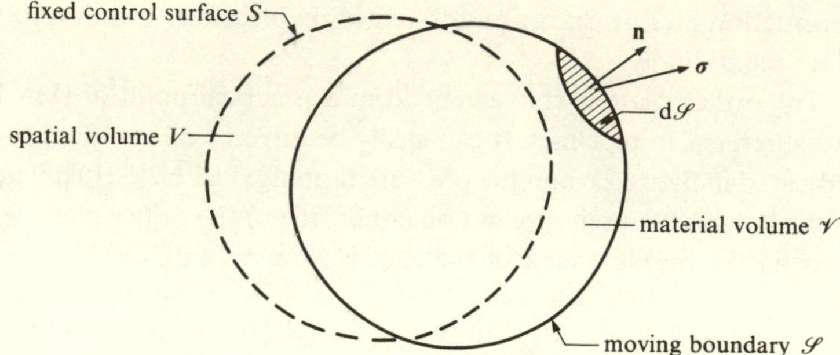

Fig. 22. Fixed control surface S and moving boundary $\mathscr{S}$.

direct-stress component is therefore taken as being positive in tension. The fluid pressure p, however, is always taken as being positive for *compression*, which accounts for the different sign in front of the last term in equations (3.6.1) and (3.6.3). This point will become clear on reference to § 5.2 where we investigate the stresses in a viscous fluid in greater detail.

Further discussion of the equation of motion for a viscous fluid must be deferred until chapter 5. In the meantime, however, we can take an alternative view of the integral form of the basic momentum equation, whether for an inviscid fluid as in (3.6.1) or for a fluid with viscous stresses as in (3.6.3), provided we restrict the situation to *steady-flow conditions*. In figure 22 we consider the movement of the material volume of fluid $\mathscr{V}$ at a certain instant of time t as it passes through a fixed control surface S. We choose the fixed control surface S so that it coincides exactly with the position of the material volume $\mathscr{V}$ at the instant t. It will be self-evident from figure 22 that, *if the flow is steady*, the rate of change at time t of the momentum of the fluid contained within the material volume $\mathscr{V}$ must be equal to the *net rate of outflow of momentum* from the fixed control surface S.

It is also obvious that the surface integral of the stress $\boldsymbol{\sigma}$ acting on the boundary of the material volume $\mathscr{V}$ at the instant t must be identical to the surface integral of the same stress evaluated over the fixed control surface S. Similarly the integral of the body force per unit volume $\rho\mathbf{F}$ evaluated over the material volume $\mathscr{V}$ at time t must be identical to the same integral evaluated over the spatial volume V. We can thus replace equation (3.6.3) with the following statement of the *steady-flow momentum equation* applied to a fixed control surface S:

net rate of outflow of momentum from the control surface

$$= \int_V \rho \mathbf{F}\, \mathrm{d}V + \int_S \boldsymbol{\sigma}\, \mathrm{d}S. \quad (3.6.4)$$

The only body force with which we are normally concerned is that due to gravity and this may be ignored in most practical flow problems if we adopt the convention of measuring the pressure at any point in the fluid relative to the local value of the hydrostatic pressure at that level. In these circumstances the steady-flow momentum equation can be expressed in the very simple form:

$\mathbf{R}$ = net rate of outflow of momentum from the control surface, (3.6.5)

where $\mathbf{R}$ is the resultant force arising from the stresses acting over the entire control surface S, i.e. $\mathbf{R} = \int \boldsymbol{\sigma}\, \mathrm{d}S$.

An exactly similar principle can be established for *angular momentum* in a steady-flow problem, and if we again choose a fixed control surface S we can say that:

$\mathbf{T}$ = net rate of outflow of angular momentum from the control surface, (3.6.6)

where $\mathbf{T}$ is the resultant torque acting on the fluid due to the pressure and shearing-stress components evaluated over the entire control surface. The angular momentum equation is of special importance when we are dealing with the fluid dynamics of turbo-machines.

In the following section, § 3.7, we will apply the steady-flow momentum equation (3.6.5) and its counterpart for angular momentum (3.6.6) to some simple flow problems where it is possible to select a control surface such that the net rate of outflow of momentum can be readily evaluated.

3.7 Applications of the energy and momentum equations

It will be appropriate at this stage to give some simple examples of the application of the continuity, energy, and momentum equations to practical flow problems. The following examples involve flow in pipes or ducts and we immediately encounter a difficulty when we extend the 'one-dimensional' analysis, which is appropriate to flow in a stream tube, to the case of flow in a pipe or duct of variable cross-section. As we have already noted in chapter 1, the relative velocity of a fluid must be zero at a solid boundary and we cannot therefore have a uniform velocity over the entire cross-section of a pipe. Despite this fact there is still a useful role for the 'one-dimensional' approach based on the concept of flow in a stream tube of small cross-sectional area.

A full analysis of the velocity distribution in boundary layers and in pipes will be given in chapters 7 and 8. In the following examples, however, we will simplify the problem by working in terms of a spatial mean velocity averaged over the cross-section of the pipe or duct. On this basis we can apply the continuity equation in the simple form of (3.3.6).

We derived the mechanical energy equation, or Bernoulli equation, for frictionless flow along a stream tube in § 3.4. We can make use of this equation directly in those regions where the effects of viscosity and turbulence are sufficiently small. The extension of the equation to include the full effects of viscous stresses operating within the fluid will be dealt with in chapters 5 and 9. In the meantime we will introduce an additional term, where necessary, into Bernoulli's equation to represent the loss of mechanical energy arising from the internal effects of viscosity and turbulence on the motion of the fluid in a pipe or duct.

(*a*) *Flow in a nozzle*

The flow of water through a nozzle discharging to atmosphere is illustrated in figure 23. The nozzle is connected to the inlet pipe at section (1)

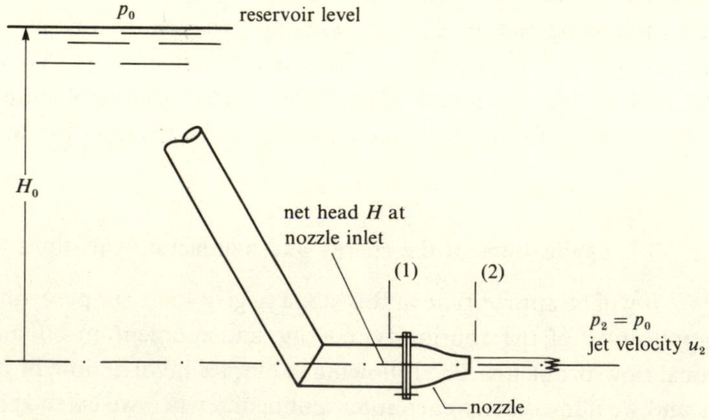

Fig. 23. Flow of water through a Pelton wheel nozzle.

by means of a flanged joint. The total effective head available at the inlet pipe is represented by H. We require to determine the velocity of the jet, the energy flow in the jet, and the separating force acting at the flanged joint. Applying Bernoulli's equation from the reservoir to the free jet at section (2), we can say that

$$\frac{p_0}{\rho}+gH_0-gH_f=\frac{p_0}{\rho}+gH=\frac{p_0}{\rho}+\tfrac{1}{2}u_2^2,$$

where p_0 is the atmospheric pressure and H_f is the loss of head due to friction in the pipeline. Hence we have:

$$u_2=\sqrt{(2gH)}. \tag{3.7.1}$$

If the cross-sectional area of the jet at section (2) is a_2, we have:

$$\text{energy flow in the jet}=\text{mass flow}\times\tfrac{1}{2}u_2^2=\tfrac{1}{2}\rho u_2^3 a_2. \tag{3.7.2}$$

It is usually preferable to work in terms of *gauge pressure*, i.e. the value measured relative to the atmospheric pressure, when dealing with problems in hydraulic engineering. This is equivalent to taking the value of the atmospheric pressure p_0 as zero. The Bernoulli equation for flow through the pipeline and nozzle illustrated in figure 23 may then be expressed as:

$$g(H_0-H_f)=gH=\frac{p_1}{\rho}+\tfrac{1}{2}u_1^2=\tfrac{1}{2}u_2^2.$$

The velocity u_1 in the pipe, where the flow area is a_1, is related to the jet velocity u_2 by the equation of continuity, i.e. $u_1=u_2a_2/a_1$. If we now apply the steady-flow momentum equation to a control surface which exactly encloses the fluid in the nozzle between sections (1) and (2), we have:

resultant force acting over the control surface

$$=p_1a_1-F=\text{net rate of outflow of momentum}=\rho a_1u_1(u_2-u_1),$$

where F is the force exerted on the fluid by the inside surface of the nozzle, and which is equal and opposite to the separating force acting on the nozzle. Hence the separating force is given by:

$$F=p_1a_1-\rho a_1u_1(u_2-u_1). \tag{3.7.3}$$

As a numerical example we can take the nozzle of a Pelton wheel operating under a net head of $H=600$ m. We will take a pipeline internal diameter of 0.8 m at section (1) and a jet diameter of 20 cm. For simplicity we will assume a discharge coefficient of 1.0. The jet velocity from (3.7.1) is then $u_2 = 108.5 \text{ m s}^{-1}$ and the velocity in the pipeline $u_1 = 6.78 \text{ m s}^{-1}$. Taking the density of water as 1000 kg m^{-3}, the mass-flow rate is 3408 kg s^{-1} and hence the energy flow in the jet from (3.7.2) is $20 \times 10^6 \text{ J s}^{-1}$ or 20 MW. The gauge pressure p_1 at section (1) is found to be $5.86 \times 10^6 \text{ N m}^{-2}$ and the separating force F is 2.6×10^6 N.

(*b*) *Sudden enlargement of a pipe*

The problem is illustrated in figure 24. We require to calculate the loss of head due to the sudden enlargement of the pipe from a flow area a at section (1) to a flow area A at section (3). We will assume that the streamlines are parallel at section (2) where the flow is in the form of a submerged jet, and this implies that the pressure p_2 must be uniform across the pipe at this section. From the continuity equation, with constant density, we have

$$au_1 = au_2 = Au_3.$$

Applying Bernoulli's equation from section (1) to section (2), we have

$$\frac{p_1}{\rho} + \frac{u_1^2}{2} = \frac{p_2}{\rho} + \frac{u_2^2}{2}, \quad \text{i.e. } p_1 = p_2.$$

The momentum equation may be applied to a cylindrical control surface extending from section (2) to (3). If we neglect skin friction at the wall of the pipe the resultant force acting over the entire control surface will simply be $(p_2 - p_3)A$, and this can be equated to the net rate of outflow of momentum, i.e.

$$(p_2 - p_3)A = \rho A u_3^2 - \rho a u_2^2.$$

Hence

$$\frac{p_1 - p_3}{\rho} = \frac{p_2 - p_3}{\rho} = u_1^2\left(\frac{a^2}{A^2} - \frac{a}{A}\right). \tag{3.7.4}$$

If we now apply the energy equation between sections (2) and (3) we have to

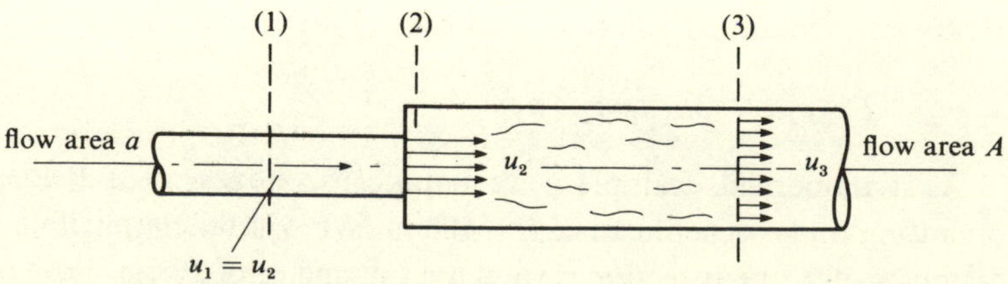

Fig. 24. Sudden enlargement of a pipe.

include a term which represents the loss of mechanical energy in the region where the submerged jet breaks up into irregular eddying motion. For obvious reasons we cannot use the simple form of Bernoulli's equation for frictionless flow in this region. We must therefore express the energy equation as:

$$\frac{p_2}{\rho}+\frac{u_2^2}{2}=\frac{p_3}{\rho}+\frac{u_3^2}{2}+L,$$

where L is the loss of energy per unit mass flow. Hence, substituting for p_3 from (3.7.4), and noting from the continuity equation that $u_2=u_1$ and that $u_3=u_1a/A$, we find that

$$L=\frac{(p_2-p_3)}{\rho}+\left(\frac{u_2^2}{2}-\frac{u_3^2}{2}\right)=\frac{u_1^2}{2}\left(1-\frac{a}{A}\right)^2,$$

i.e.

$$\text{loss of head}=\frac{u_1^2}{2g}\left(1-\frac{a}{A}\right)^2. \tag{3.7.5}$$

It will be seen that in the limiting case where A is large compared with the area a the loss of head will approximate to $u_1^2/2g$, i.e. to the full inlet velocity head.

As a numerical example we can take the case of water flowing in a pipe with velocity $u_1=8.0\ \mathrm{m\ s^{-1}}$. If there is a sudden enlargement to twice the diameter of the entry pipe we have a flow area ratio $a/A=\frac{1}{4}$. The loss of head in these circumstances, calculated from (3.7.5), would be 1.835 m. The pressure difference (p_3-p_1), as given by (3.7.4), is found to be $1.2\times10^4\ \mathrm{N\ m^{-2}}$. Note that p_3 is greater than p_1, but the pressure rise is not as large as it would be if we had a gradual expansion or diffuser section connecting the small-diameter pipe to the larger one. It is easily verified from Bernoulli's equation that, with a gradual expansion and no energy loss, the maximum pressure rise corresponding to a velocity reduction from $8\ \mathrm{m\ s^{-1}}$ to $2\ \mathrm{m\ s^{-1}}$ would be about $3.0\times10^4\ \mathrm{N\ m^{-2}}$.

(c) *Flow in the runner of a water turbine*

Water turbines can be classified as impulse or reaction machines. The characteristic feature of an impulse turbine is that the maximum velocity is developed in passage through the nozzle, with the jet emerging at atmospheric pressure and with a theoretical velocity as given by (3.7.1). Kinetic energy is extracted from the jet as it impinges on the runner, as in the case of the Pelton wheel which is the normal type of impulse turbine. In the case of a reaction turbine, on the other hand, the maximum possible velocity is not developed by the water as it enters the turbine runner and pressure changes occur within the runner itself. There are two main types of reaction turbine commonly used in hydro-electric power plants, the Francis turbine where the flow is mainly radial, and the Kaplan turbine where the flow is in an axial direction. In this example we will consider the flow of water through a radial-flow turbine runner.

In figure 25 velocity diagrams are shown at inlet to the runner at radius r_1, and also at outlet from the runner at radius r_2. The water enters the runner, after passing through fixed or adjustable guide vanes, with *absolute velocity* v_1. It is convenient to resolve this velocity into two components, a tangential component or *velocity of whirl* w_1, and a radial component or

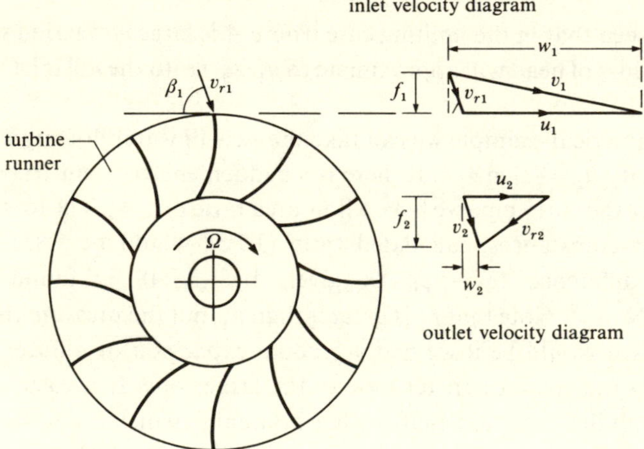

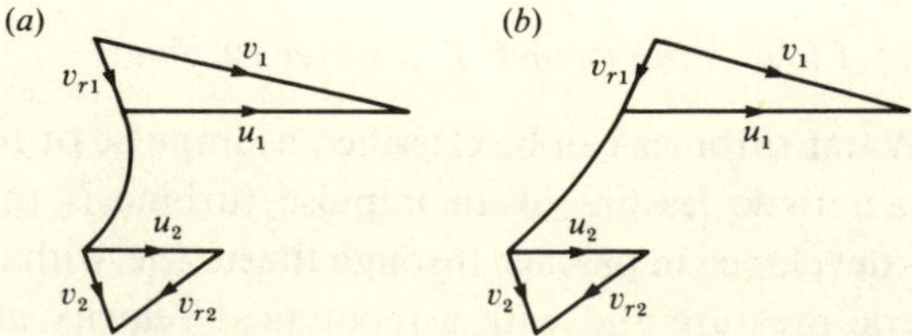

Fig. 25. Velocity triangles for a radial flow turbine runner.

velocity of flow f_1. The peripheral velocity of the runner is denoted by u_1 which is equal to $r_1\Omega$, where Ω is the angular velocity of the runner. The *relative velocity* of the water to the runner at entry is represented by v_{r1}. A similar notation is used for the outlet velocity diagram.

The guide vanes are designed so that the relative velocity v_{r1} is tangential to the blades of the runner at inlet when the turbine is operating at its normal speed. The relative velocity v_{r2} at exit may also be assumed to be approximately tangential to the blades of the runner at the outlet radius r_2. In figure 25 the geometry of the turbine is simplified by representing the flow as being entirely two-dimensional in a radial plane. In reality the flow is essentially three-dimensional in character and although the water enters the runner radially it is usually turned into an axial direction for discharge. The volume flow rate Q at radius r will be given by $2\pi rbf$, where b is the axial width of the flow passage in the runner and f is velocity of flow at radius r. The volume flow must obviously be constant under steady-flow conditions in accordance with the equation of continuity.

The steady-flow angular momentum equation (3.6.6) enables us to calculate the transfer of energy from the water as it flows through the turbine runner. Taking a control surface which exactly encloses the runner, we can say that the torque exerted *by* the water *on* the runner must be equal to the net rate of inflow of angular momentum into the control surface, i.e.

$$T = \rho Q(w_1 r_1 - w_2 r_2). \tag{3.7.6}$$

The power developed by the turbine will then be

$$T\Omega = \rho Q(w_1 u_1 - w_2 u_2) \tag{3.7.7}$$

and the shaft work or energy transferred to the runner per unit mass flow will therefore be

$$W_s = w_1 u_1 - w_2 u_2. \tag{3.7.8}$$

This simple equation expresses the basic fluid dynamics of flow through the rotor of any turbo-machine and it will be discussed further in chapter 10. Since the objective of any turbine design is to extract as much energy as possible from the fluid stream, it will be clear that we usually aim to achieve a minimum value for the absolute velocity v_2 at exit from the runner. The flow component f_2 is determined by the mass-flow requirement and the equation of continuity. The velocity of whirl w_2 at exit, however, can only have an adverse effect on the overall result and we therefore aim at a zero value under normal operating conditions. Hence the equation for the shaft work per unit mass flow reduces still further to $W_s = w_1 u_1$ in these circumstances.

The mechanical energy equation can now be written down for flow through the turbine. If the total effective head available at entry to the turbine is H, and if we neglect frictional losses in the inlet passages, we can apply Bernoulli's equation up to the point of inlet to the runner as follows

$$gH = \frac{p_1}{\rho} + \tfrac{1}{2}u_1^2, \tag{3.7.9}$$

where p_1 is measured as gauge pressure. For flow through the runner we have to take account of the shaft work as given by (3.7.8) and also energy losses due to friction and eddying. Further frictional losses will occur in the outlet passage and draft tube. If the final velocity of the water is u_3 at exit from the draft tube, and if the gauge pressure p_3 is zero (i.e. atmospheric pressure), we can express the energy equation as

$$gH = W_s + L_f + \tfrac{1}{2}u_3^2, \tag{3.7.10}$$

where L_f is the mechanical energy loss per unit mass due to friction and eddying. We can define the *hydraulic efficiency* of the turbine as the ratio W_s/gH.

As a numerical example we can take a Francis turbine operating under a net available head of 180 m. If the rotational speed is to be 300 rpm (i.e. 5 rps), if the blade angle β, at entry to the runner is 90°, and if the hydraulic efficiency is assumed to be 95%, we will calculate the appropriate radius r_1 of the runner. The shaft work per unit mass is given by $0.95 \times gH = 1678\ \mathrm{m^2\,s^{-2}}$. Hence from (3.7.8), noting that $w_1 = u_1$ if $\beta_1 = 90°$, and assuming that $w_2 = 0$, we find that $u_1 = 41.0\ \mathrm{m\,s^{-1}}$ and hence $r_1 = 1.3$ m. If the flow velocity f_1 is 8 $\mathrm{m\,s^{-1}}$, and if the axial width of the flow passage in the runner is 0.3 m at radius r_1, the volume flow rate Q is found to be 19.6 $\mathrm{m^3\,s^{-1}}$. Hence the mass-flow rate is $19.6 \times 10^3\ \mathrm{kg\,s^{-1}}$ and the shaft power is 32.9 MW.

3.8 Vorticity and rotational flow

The vorticity $\boldsymbol{\omega}$ of a fluid is a vector quantity having the same nature as angular velocity and is defined mathematically as the curl of the velocity vector $\mathbf{u}$, i.e.

$$\boldsymbol{\omega} = \operatorname{curl} \mathbf{u} = \nabla \times \mathbf{u}. \tag{3.8.1}$$

Reference to appendix 1 will show that the components of the vorticity, using a Cartesian system and the notation of (1.6.5) for the velocity components, are

$$\omega_x=\left(\frac{\partial w}{\partial y}-\frac{\partial v}{\partial z}\right),\quad \omega_y=\left(\frac{\partial u}{\partial z}-\frac{\partial w}{\partial x}\right),\quad \omega_z=\left(\frac{\partial v}{\partial x}-\frac{\partial u}{\partial y}\right). \tag{3.8.2}$$

The physical significance of vorticity is best understood by imagining a small spherical element of fluid to be suddenly frozen. If the resulting solid sphere has rotation then the fluid has vorticity at the point considered. The numerical magnitude of the vorticity would be equal to *twice* the angular velocity of the solid sphere, as will be seen in the case of the forced vortex considered below.

It is sometimes convenient in practical problems of rotational flow to use cylindrical polar co-ordinates. The components of curl **u** on this system are:

$$\begin{aligned} &r \text{ direction} \quad \omega_r=\left(\frac{1}{r}\frac{\partial u_z}{\partial \phi}-\frac{\partial u_\phi}{\partial z}\right)\\ &\phi \text{ direction} \quad \omega_\phi=\left(\frac{\partial u_r}{\partial z}-\frac{\partial u_z}{\partial r}\right)\\ &z \text{ direction} \quad \omega_z=\left(\frac{u_\phi}{r}+\frac{\partial u_\phi}{\partial r}-\frac{1}{r}\frac{\partial u_r}{\partial \phi}\right)\end{aligned} \tag{3.8.3}$$

A simple example of rotational flow is the case of the *forced vortex* in which the fluid rotates with uniform angular velocity Ω about the z axis as illustrated in figure 26. It will be seen that in this case the fluid rotates like a solid body and the tangential velocity at radius r is simply $u_\phi=r\Omega$. Provided the angular velocity Ω is constant, there will be no tangential component of acceleration in the ϕ direction and Euler's equation of motion (3.2.1) reduces to two simple equilibrium conditions. Since the acceleration of a material element of fluid at radius r must be u_ϕ^2/r, we have the condition for radial equilibrium:

$$\frac{\partial p}{\partial r}=\rho\frac{u_\phi^2}{r}. \tag{3.8.4}$$

For vertical equilibrium we have the ordinary hydrostatic condition as expressed in (1.4.1), i.e.

$$\frac{\partial p}{\partial z}=-\rho g. \tag{3.8.5}$$

Substituting for the tangential velocity u_ϕ in terms of r and Ω, equation

(3.8.4) may be written as

$$\frac{\partial p}{\partial r} = \rho r \Omega^2. \tag{3.8.6}$$

Equations (3.8.5) and (3.8.6) may be integrated directly to give the pressure distribution:

$$\frac{p - p_0}{\rho} = \tfrac{1}{2} r^2 \Omega^2 - g(z - z_0). \tag{3.8.7}$$

Surfaces of constant pressure are thus paraboloids of revolution. If the fluid is a liquid with a free surface, as indicated in figure 26, the free surface must also be a paraboloid.

The Bernoulli energy, as defined by equation (3.4.5) or (3.5.1), can be

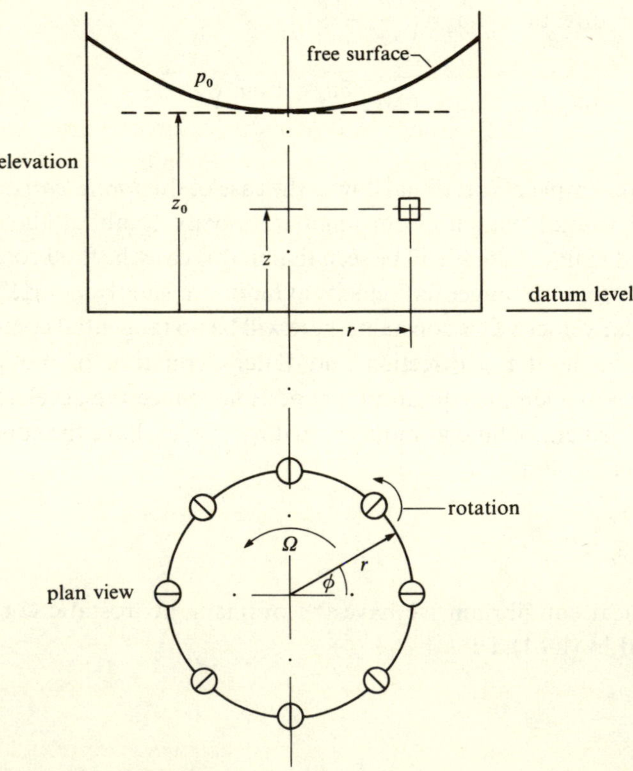

Fig. 26. Forced vortex.

evaluated for a streamline at radius r within the vortex. Substituting for $(p/\rho)+gz$ from (3.8.7) gives the result:

$$\frac{p}{\rho}+\frac{r^2\Omega^2}{2}+gz=\frac{p_0}{\rho}+gz_0+r^2\Omega^2. \tag{3.8.8}$$

It will be seen that the Bernoulli energy is not constant throughout the fluid in a forced vortex but that it increases from streamline to streamline with increasing radius. Flow of this type can only be maintained by the action of some external agency, for example by the rotation of the cylindrical vessel containing the fluid, and could not be established from rest without such external action. Practical engineering examples which approach the conditions of a forced vortex are the centrifuge, the central core of a stirred mixing vessel, and the motion in the impeller of a centrifugal pump under shut-off conditions with the delivery valve closed.

The vorticity in a forced vortex can be evaluated from (3.8.3) noting that u_r and u_z are both zero and that the velocity component u_ϕ is equal to $r\Omega$. Hence the vorticity $\boldsymbol{\omega}$ has only one component, in the z direction, given by

$$\omega_z=\left(\frac{u_\phi}{r}+\frac{\partial u_\phi}{\partial r}\right)=2\Omega. \tag{3.8.9}$$

We thus see that the vorticity is constant in a forced vortex and that its magnitude is twice the angular velocity of the fluid.

The flow pattern in a *free vortex* is entirely different from that of a forced vortex. In a free vortex the Bernoulli energy is uniform for all radii. We can still use equations (3.8.4) and (3.8.5) since these are simply statements of the fundamental laws of dynamics, but we introduce the new condition by differentiating equation (3.5.1) with respect to r at constant z and equating the result to zero, i.e.

$$\frac{1}{\rho}\frac{\partial p}{\partial r}+u_\phi\frac{\partial u_\phi}{\partial r}=0. \tag{3.8.10}$$

Substituting for $\partial p/\partial r$ from (3.8.4), we therefore have

$$\frac{u_\phi}{r}+\frac{\partial u_\phi}{\partial r}=0 \tag{3.8.11}$$

and hence

$$u_\phi r=\text{constant}=c. \tag{3.8.12}$$

For a free vortex, therefore, the *angular momentum* is constant. Figure 27 illustrates the case of a free vortex in a liquid with a free surface. Substituting

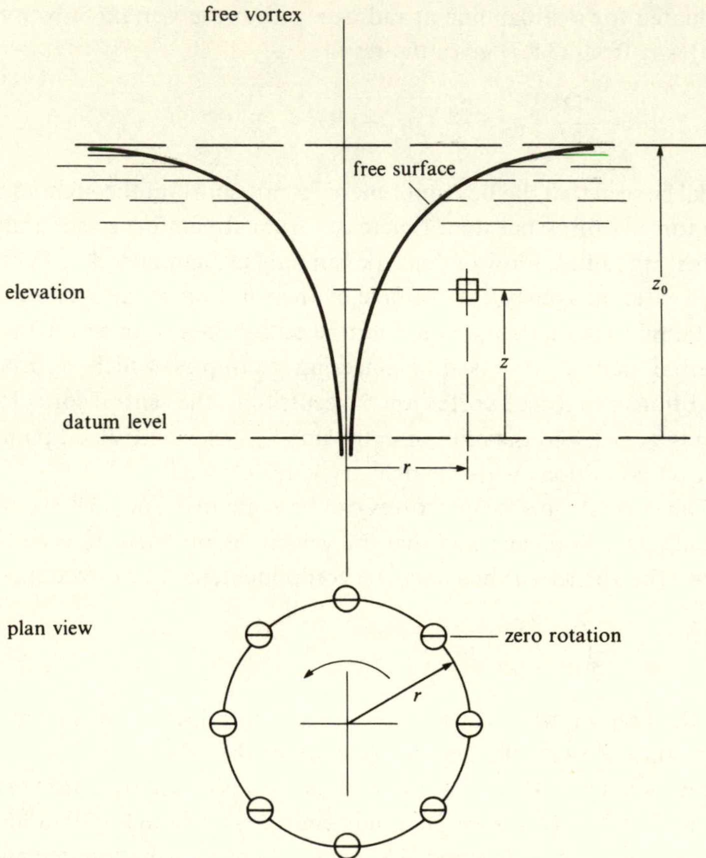

Fig. 27. Free vortex.

from (3.8.12) in (3.8.4), we have for the radial pressure gradient:

$$\frac{\partial p}{\partial r}=\frac{\rho c^2}{r^3}. \tag{3.8.13}$$

The pressure distribution in a free vortex is then obtained by integrating equations (3.8.5) and (3.8.13) with the result that

$$\frac{p-p_0}{\rho}=-\frac{c^2}{2r^2}+g(z_0-z). \tag{3.8.14}$$

The profile of the free surface for a liquid is obtained by putting $p=p_0=$ atmospheric pressure, i.e.

$$z=z_0-\frac{c^2}{2gr^2}. \tag{3.8.15}$$

Equation (3.8.15) implies that the height of the free surface would be $-\infty$ at the axis where $r=0$, and equation (3.8.12) implies that the velocity u_ϕ becomes infinite at $r=0$. Evidently there is a mathematical singularity in this solution at the axis and other physical factors must come into play at small values of r.

We can evaluate the vorticity in a free vortex from (3.8.3) in conjunction with the velocity distribution as given by (3.8.11) or (3.8.12). It will be seen that

$$\omega_z=\left(\frac{u_\phi}{r}+\frac{\partial u_\phi}{\partial r}\right)=0. \tag{3.8.16}$$

At first sight this is a surprising result, but a free vortex is in fact a special case of *irrotational flow* as we will see in chapter 4, and the vorticity of the fluid in a free vortex is zero at all points *except* at the axis $r=0$ where the value becomes infinite. As indicated in figure 27, a fluid element circulating around the centre-line of the vortex will not in fact be rotating about its own axis.

The quantity known as the *circulation* may be defined as the line integral of the tangential velocity component taken once round a closed circuit in the fluid. Thus for a free vortex it follows from (3.8.12) that the circulation K at radius r is given by

$$K=2\pi r u_\phi=2\pi c, \tag{3.8.17}$$

i.e. the circulation is constant for all radii, and we may therefore express the tangential velocity distribution as

$$u_\phi=\frac{K}{2\pi r}. \tag{3.8.18}$$

We will make further use of this result in chapter 4. Practical examples of free vortex motion are the whirlpool and the tornado which occur as natural phenomena.

While the circulation round a free vortex is independent of the radius, the circulation round a forced vortex will vary with radius and its value is seen to be $2\pi r^2\Omega$. Noting that the enclosed area of a circle of radius r is πr^2, and that the vorticity in a forced vortex has a constant value 2Ω, we see that the circulation in this case is equal to the product of the enclosed area and the vorticity. This is a simple example of *Stokes's theorem* which states that the circulation evaluated round a closed curve is equal to the surface integral of the vorticity evaluated over any surface enclosed by the curve.

The concept of a *vortex line* is of importance for an understanding of certain aspects of fluid motion. A vortex line is defined as a line drawn in the fluid so that the tangent at any point has the same direction as the vorticity vector at that point. Thus a vortex line bears the same relationship to the vorticity vector $\boldsymbol{\omega}$ as a streamline does to the velocity vector $\mathbf{u}$. Vortex lines drawn through every point of a small closed curve in the fluid would form a *vortex tube* which is analogous to a stream tube. For a vortex tube or vortex filament of small cross-section, it may be shown that the product of the local magnitude of the vorticity and the cross-sectional area of the tube, known as the strength of the vortex, must have the same value all along the tube. This result is similar in form to the equation of continuity (3.3.6) when applied to an incompressible fluid flowing in a stream tube. It may also be noted that, from the definition of vorticity given by (3.8.1),

$$\operatorname{div} \boldsymbol{\omega} = 0. \tag{3.8.19}$$

This follows from the mathematical fact that div curl $\mathbf{u}$ is identically equal to zero. The similarity between (3.8.19) and the equation of continuity (3.3.4) for an incompressible fluid will be obvious. The physical implication of (3.8.19) is that vorticity cannot be created or destroyed within the interior of a fluid. As we will observe later, vorticity can only be generated by the action of solid boundaries on the fluid through the phenomenon of zero slip.

Kelvin's circulation theorem relates to the persistence of vorticity, once it has been created, in an inviscid fluid. The theorem states that the circulation evaluated round any closed material curve, which moves with the fluid, does not change with the elapse of time. It follows from this that if a flow is initially irrotational, i.e. with zero vorticity at all points, it must remain irrotational. It also follows that, where vorticity exists, vortex lines and vortex tubes move with the fluid and that the strength of a vortex tube remains constant with time. Kelvin's theorem is subject to modification, however, when we are dealing with the motion of a viscous fluid and we will see in §5.6 that vorticity is *diffused* in the fluid by the action of viscosity.

It is important to recognise that vorticity is always present in shear flow in the neighbourhood of solid boundaries. Taking the very simple example of uniform shearing flow, as defined in §1.5 and illustrated in figure 3, it will be seen from (3.8.2) that the vorticity will have a component in the z direction given by

$$\omega_z = -\frac{\partial u}{\partial y} = \frac{\tau}{\mu}. \tag{3.8.20}$$

Although from a macroscopic point of view the flow pattern does not appear to involve rotation, the presence of vorticity can be readily visualised by picturing the fluid as being made up of sheets of small vortices, or layers of roller bearings, lying between the fixed and moving planes.

A similar situation occurs in the case of boundary-layer flow. As will be shown in chapter 7, and as already noted in a preliminary way in § 1.5, the velocity of a fluid moving parallel to a solid wall rises steeply in the boundary-layer region from zero at the wall to the undisturbed mainstream velocity at the outer edge of the boundary layer. The transverse velocity gradient $\partial u/\partial y$ is normally greatest at the wall where $y=0$, and for a Newtonian fluid it is equal to τ_0/μ where τ_0 is the shear stress at the wall. The magnitude of the term $\partial v/\partial x$ by contrast is usually very small. Consequently the vorticity at the wall is given by

$$\omega_z=\left(\frac{\partial v}{\partial x}-\frac{\partial u}{\partial y}\right)\approx -\frac{\partial u}{\partial y}=\frac{\tau_0}{\mu}. \tag{3.8.21}$$

The vorticity decreases in magnitude as one traverses outwards across the boundary layer until it approaches zero at the outer edge where the boundary layer merges into the undisturbed stream. We thus observe again that vorticity is generated by the action of solid boundaries on the flow pattern, through the agency of the zero slip condition, and that boundary layers are also layers of vorticity.

3.9 Flow past blades and aerofoil sections

We will conclude this chapter by applying the basic continuity and momentum equations, together with the concept of circulation, to the important engineering case of flow past a blade element or aerofoil section. It is convenient to start with the two-dimensional picture of flow past a cascade of vanes or blade elements as indicated in figure 28. It will be appreciated that this picture could refer to the flow through a fixed row of blades in an axial-flow turbine or compressor if one imagines a circumferential section through the blading to be laid out flat in the plane of the paper. In figure 28 the absolute velocity of the fluid at entry to the blades is denoted by v_1 and this may be resolved into an axial component v_{x1} and a transverse component v_{y1}. A similar notation is used with subscript 2 to denote the velocity components on the exit side.

In the diagram of figure 28 the direction of the velocity at entry to, and at exit from, the blades is shown in each case as being approximately tangential to the blade profile. It is sometimes assumed in the elementary treatment of flow through turbo-machines that the relative velocity will

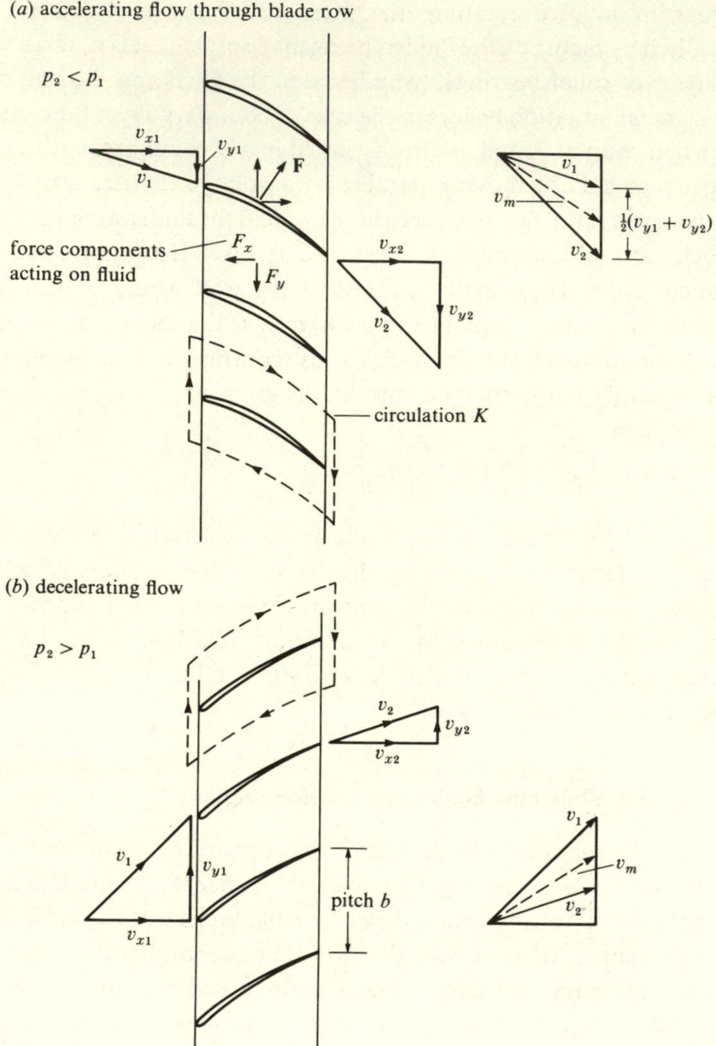

Fig. 28. Flow past a cascade of vanes.

automatically be parallel to the blading in this manner. It will be appreciated, however, that this is an over-simplification. If the blades are very closely pitched, and if the angle of deflection of the fluid stream is small (i.e. if the blades have only a small camber), this assumption may be nearly true, but in general we have to recognise that there is no precise relationship between the blade angles and the relative direction of flow of the fluid. One should also note that, while aerofoil sections usually have a fairly sharp trailing edge, they normally have a rounded nose profile. One purpose of this rounded leading edge is to make the blade section less sensitive to minor variations in the direction of the on-coming fluid stream.

Assuming a constant flow area, i.e. a constant width of passage measured perpendicular to the plane of the diagram, or a constant radial width of passage in the case of an axial-flow machine, it is obvious from the continuity equation for an incompressible fluid that the axial component of velocity must be constant, i.e. $v_{x1}=v_{x2}$. Applying *Bernoulli's equation* to the flow through the blading, assuming negligible skin friction, we have

$$\frac{p_1}{\rho}+\frac{v_1^2}{2}=\frac{p_2}{\rho}+\frac{v_2^2}{2}$$

and therefore

$$p_1-p_2=\tfrac{1}{2}\rho(v_2^2-v_1^2)=\tfrac{1}{2}\rho(v_{y2}^2-v_{y1}^2). \tag{3.9.1}$$

Note that case (*a*) in figure 28 represents accelerating flow with p_2 less than p_1. Case (*b*) represents decelerating flow with p_2 greater than p_1. The following analysis refers to case (*a*).

We can now apply the *momentum equation* to the flow past a single blade by choosing a control surface, indicated by the dotted outline in the diagram, which encloses the spatial volume associated with one blade. It will be seen that this is identical in shape and size with the flow passage *between* a pair of adjacent blades. The fluid pressure distribution acting over the surface of the blade will establish a resulting force on the blade whose components are F_x and F_y in the x and y directions respectively. By the principle of action and reaction, the blade must exert an equal and opposite force on the fluid. For the x direction, therefore, noting that there is no change of axial momentum, we have

$$(p_1-p_2)b-F_x=0 \tag{3.9.2}$$

and for the y direction, equating the net force to the rate of outflow of y momentum, we have

$$F_y=\rho v_{x1}b(v_{y2}-v_{y1}). \tag{3.9.3}$$

From (3.9.1) and (3.9.2), eliminating (p_1-p_2), we can say that

$$F_x=\tfrac{1}{2}\rho(v_{y1}+v_{y2})b(v_{y2}-v_{y1}). \tag{3.9.4}$$

Hence from (3.9.3) and (3.9.4) the two force components may be combined vectorially to give the resultant force F in the form

$$F=\rho v_m b(v_{y2}-v_{y1}). \tag{3.9.5}$$

where v_m is the mean velocity vector, having components v_{x1} in the axial direction, and $\frac{1}{2}(v_{y1}+v_{y2})$ in the y direction, as shown in figure 28. It will be evident from (3.9.3) and (3.9.4) that the force F must be perpendicular to the mean velocity vector v_m.

It will be seen from the diagram that the quantity $b(v_{y2}-v_{y1})$ is the *circulation* evaluated round one of the blades. We have already noted in §3.8 that the circulation is defined as the line integral of the tangential velocity component taken round a closed circuit. If we choose a circuit which coincides with the dotted outline in figure 28, it will be noted that the contributions from the two curved portions cancel out exactly and we are left with the contributions from the two straight portions in the y direction. Hence the circulation K is given by

$$K=b(v_{y2}-v_{y1}). \tag{3.9.6}$$

From (3.9.5) and (3.9.6) we can therefore say that

$$F=\rho v_m K. \tag{3.9.7}$$

This is a special case of a general theorem in aerodynamics, known as the Kutta–Joukowski law, which states that for an aerofoil of infinite span the lift force L per unit span is given by

$$L=\rho UK, \tag{3.9.8}$$

where U is the velocity of the undisturbed airstream and K is the circulation round the aerofoil.

We can now consider the flow past an isolated aerofoil section as indicated in figure 29. An aerofoil may be regarded as a blade profile which is designed to produce circulation and hence a lift force. A typical subsonic aerofoil section usually has a rounded leading edge and a sharp trailing edge and a certain degree of camber as illustrated in the diagram. A curved

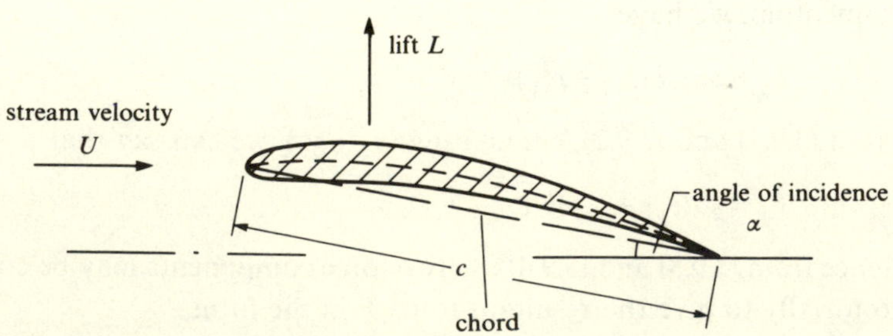

Fig. 29. Flow past an isolated aerofoil.

'profile centre-line' may be drawn midway between the upper and lower surfaces of the profile, and the straight line joining the ends of the profile centre-line is known as the *chord* of the aerofoil. The angle between the chord line and the direction of the undisturbed airstream is the angle of incidence α.

A symmetrical profile with zero camber would obviously have zero lift at zero incidence. A cambered profile, however, will experience a lift force when located in an airstream at zero angle of incidence. In either case if the incidence is increased from zero, the lift force will increase until it reaches a maximum value at the stalling angle. It is convenient to express the lift force L per unit span in terms of a lift coefficient c_L defined by

$$L = c_L \tfrac{1}{2}\rho U^2 c, \tag{3.9.9}$$

where U is the airstream velocity, and c is the chord length. It will be shown in §4.7, when we apply two-dimensional inviscid flow theory to a thin aerofoil section, that the circulation K is related to the angle of incidence α by an expression of the form:

$$K = \pi U c \sin(\alpha + \beta), \tag{3.9.10}$$

where β is a parameter related to the camber of the profile. From (3.9.8), (3.9.9) and (3.9.10), therefore, the lift coefficient is related to the angle of incidence and the parameter β by

$$c_L = 2\pi \sin(\alpha + \beta). \tag{3.9.11}$$

4

Motion of inviscid fluids

4.1 The relevance of inviscid flow theory

In chapter 3 we derived Euler's equation of motion in its general three-dimensional form. We also obtained Bernoulli's equation by integrating Euler's equation along a stream tube under steady flow conditions. We noted in §§ 3.5 and 3.7 that, despite the restriction to idealised frictionless flow conditions, Euler's equation and the related Bernoulli equation can yield useful results when applied with care and discrimination to flow regions where viscous forces are small and where irregular eddying motion and turbulence are effectively absent. In practice this means that the application of inviscid flow theory must be restricted to flow regions which are not close to solid boundaries.

We introduced the important concept of vorticity in § 3.8 and noted that vorticity is generated by the action of solid boundaries on the flow pattern through the phenomenon of zero slip. We will see in chapter 5 that vorticity is diffused from the boundary into the interior of the fluid by the action of viscosity. Although it is theoretically possible for vorticity to exist (without further change) in an inviscid fluid, we can generally identify regions of negligible viscosity with *irrotational flow* conditions, i.e. flow regions where the vorticity $\boldsymbol{\omega}$ is zero.

In many practical cases the overall flow pattern may be divided effectively into two main zones, a *boundary-layer region* where viscous forces are significant and where vorticity is present, and an *external flow region* where viscous forces are negligible and where the flow pattern is nearly irrotational in character. This is especially the case when we are considering flow past thin solid bodies or profiles of streamlined shape such as aerofoil sections. In these situations the theory of inviscid and irrotational flow can give valuable results in predicting and explaining the flow pattern *outside the boundary-layer region.* The nature of the flow inside the boundary layer will be analysed and discussed in some detail in the course of chapters 5, 6, and 7. In this chapter, however, we will confine our attention to the external flow region. In §4.2 we consider the further development of Euler's equation when the flow is irrotational. In §4.3 we review irrotational flow and the concept of a velocity potential for three-dimensional flow patterns. In §§4.4–4.6 we investigate two-dimensional potential flow with special reference to the case of flow past a circular cylinder, and in §4.7 we extend the method to the practical case of flow past a thin aerofoil section or blade element.

4.2 Further development of Euler's equation

Euler's equation for the motion of an inviscid fluid takes the form, as derived in §3.2,

$$\rho \frac{\mathrm{D}\mathbf{u}}{\mathrm{D}t} = \rho \mathbf{F} - \operatorname{grad} p. \tag{4.2.1}$$

If the body force **F** is expressed as the gradient of a scalar potential Ψ, i.e. if $F = -\operatorname{grad} \Psi$, the equation may be written in the form

$$\frac{\mathrm{D}\mathbf{u}}{\mathrm{D}t} = -\nabla\Psi - \frac{1}{\rho}\nabla p. \tag{4.2.2}$$

The total differential $\mathrm{D}\mathbf{u}/\mathrm{D}t$ is defined as in (3.1.5) by

$$\frac{\mathrm{D}\mathbf{u}}{\mathrm{D}t} = \frac{\partial \mathbf{u}}{\partial t} + (\mathbf{u}\cdot\nabla)\mathbf{u}. \tag{4.2.3}$$

We can now make use of the following mathematical identity which is derived in appendix 1:

$$(\mathbf{u}\cdot\nabla)\mathbf{u} = \tfrac{1}{2}\nabla(\mathbf{u}^2) - \mathbf{u} \times \operatorname{curl} \mathbf{u}. \tag{4.2.4}$$

Noting that $\boldsymbol{\omega} = \operatorname{curl} \mathbf{u}$ where $\boldsymbol{\omega}$ is the *vorticity*, we can re-write Euler's equation (4.2.2) using (4.2.3) and (4.2.4) to give the following result:

$$\frac{\partial \mathbf{u}}{\partial t} = \mathbf{u} \times \boldsymbol{\omega} - \tfrac{1}{2}\nabla(\mathbf{u}^2) - \nabla\Psi - \frac{1}{\rho}\nabla p \tag{4.2.5}$$

and hence for *steady flow*, where $\partial \mathbf{u}/\partial t = 0$, we have:

$$\frac{1}{\rho}\nabla p + \nabla\left(\frac{\mathbf{u}^2}{2}\right) + \nabla\Psi = \mathbf{u} \times \boldsymbol{\omega}. \tag{4.2.6}$$

The vector quantity $\mathbf{u} \times \boldsymbol{\omega}$ is always perpendicular to the velocity vector **u**. We can therefore integrate equation (4.2.6) *along a streamline*, following the motion of the fluid under steady-flow conditions, and there will be no contribution from the right-hand side of the equation. Integration along a streamline thus yields the result

$$\int \frac{\mathrm{d}p}{\rho} + \tfrac{1}{2}\mathbf{u}^2 + \Psi = \text{constant}, \tag{4.2.7}$$

where the constant applies only to *the one streamline.*

If the body force $\mathbf{F}$ in equation (4.2.1) is the force due to terrestrial gravity, the scalar potential Ψ is represented by gz, and (4.2.7) then takes the familiar form of Bernoulli's equation for frictionless flow along a streamline as in (3.4.9):

$$\int \frac{\mathrm{d}p}{\rho} + \tfrac{1}{2}\mathbf{u}^2 + gz = \text{constant} \tag{4.2.8}$$

and with constant density ρ we have

$$\frac{p}{\rho} + \tfrac{1}{2}\mathbf{u}^2 + gz = \text{constant}. \tag{4.2.9}$$

For *irrotational flow*, where the vorticity $\boldsymbol{\omega}$ must be zero by definition at all points in the fluid, it will be seen from (4.2.5) that Euler's equation of motion takes the form:

$$\frac{\partial \mathbf{u}}{\partial t} = -\tfrac{1}{2}\,\nabla(\mathbf{u}^2) - \nabla\Psi - \frac{1}{\rho}\,\nabla p \tag{4.2.10}$$

and for an incompressible fluid with ρ constant, this can be written as

$$\frac{\partial \mathbf{u}}{\partial t} = -\nabla\left(\frac{p}{\rho} + \tfrac{1}{2}\mathbf{u}^2 + \Psi\right). \tag{4.2.11}$$

This form of the equation, for unsteady irrotational flow, will be used in § 12.1 for the analysis of surface waves.

If the flow is both irrotational and *steady*, equation (4.2.10) reduces to

$$\frac{1}{\rho}\,\nabla p + \nabla\left(\frac{\mathbf{u}^2}{2}\right) + \nabla\Psi = 0 \tag{4.2.12}$$

and for an incompressible fluid this may be written as

$$\nabla\left(\frac{p}{\rho} + \tfrac{1}{2}\mathbf{u}^2 + \Psi\right) = 0. \tag{4.2.13}$$

Equation (4.2.13) may be integrated directly to give the result for steady incompressible irrotational flow that

$$\frac{p}{\rho}+\tfrac{1}{2}\mathbf{u}^2+\Psi=\text{constant}, \tag{4.2.14}$$

where the constant is now the *same for all streamlines.* If the body force is due solely to gravity, the scalar potential $\Psi=gz$ and (4.2.14) takes the form of Bernoulli's equation

$$\frac{p}{\rho}+\tfrac{1}{2}\mathbf{u}^2+gz=\text{constant}. \tag{4.2.15}$$

4.3 Irrotational flow and velocity potential in three dimensions

A vector field $\mathbf{u}$ is described in mathematical terms as being *conservative* if the vector $\mathbf{u}$ can be expressed as the gradient of a scalar potential ϕ whose value depends only on the position vector $\mathbf{r}$. It is easily proved that curl $\mathbf{u}=0$ if $\mathbf{u}=\text{grad}\,\phi$, and vice versa. When we are dealing with the irrotational flow of a fluid, therefore, where by definition we have curl $\mathbf{u}=0$, the velocity $\mathbf{u}$ can be expressed as the gradient of a *velocity potential* ϕ. It is a matter of convention, however, that we usually introduce a negative sign so that a falling gradient gives a positive value for $\mathbf{u}$ and we therefore define the velocity potential by the statement:

$$\mathbf{u}=-\text{grad}\,\phi. \tag{4.3.1}$$

The equation of continuity for incompressible flow (3.3.4) then takes the form div grad $\phi=0$, which may be written as:

$$\nabla^2\phi=0, \tag{4.3.2}$$

which is known as *Laplace's equation.*

The mathematical analysis of irrotational flow consists in finding solutions to Laplace's equation which will satisfy certain specified boundary conditions. We can apply the condition that the component of the fluid velocity which is normal to the boundary must be zero at all points of a solid surface, but we cannot satisfy the requirement of zero slip. To this extent the solutions are artificial. It should also be kept in mind that all we are doing, when solving Laplace's equation, is to find a velocity distribution which satisfies the equation of continuity and which involves zero vorticity. In order to determine the pressure distribution in the fluid we have to use the Euler or Bernoulli equation.

Laplace's equation is of common occurrence in mathematical physics since it is the basic equation governing the spatial distribution of potential in the theory of electrostatics, magnetism, and Newtonian gravitational fields, where the inverse square law applies to the force of attraction or repulsion between point charges, poles, and masses respectively. It should be noted, however, that whereas in these three instances the potential function has the dimensions of energy and its gradient is a force, the velocity potential as defined by (4.3.1) has dimensions $[l^2t^{-1}]$. The analogy, such as it is, between irrotational fluid flow on the one hand and electrostatics, for example, on the other is best understood by introducing the concept of a *point source* which is the equivalent of a point charge. Since mass cannot normally be created or destroyed, a point source or sink cannot normally exist in the real physical world. The nearest approach in engineering terms to a point source would be a small-diameter pipe delivering fluid under pressure to a small submerged spherical distributor, but this concept presents some obvious practical difficulties. Assuming that such a source could exist, however, with a volume flow rate Q m^3 s^{-1}, and that complete spherical symmetry could be achieved, the radial velocity of the fluid emitted by the source would be given by

$$u_r = \frac{Q}{4\pi r^2}, \tag{4.3.3}$$

where r is the radial distance measured from the point source. This follows from the fact that the surface area of a sphere of radius r is $4\pi r^2$ and it simply represents the inverse square law as it applies to fluid velocity arising from a point source.

Since the velocity, under irrotational flow conditions, is related to the potential gradient by the definition (4.3.1), we can express equation (4.3.3) as

$$\frac{\partial \phi}{\partial r} = -\frac{Q}{4\pi r^2}.$$

If we integrate this expression, taking the velocity potential as zero at infinite radius, we have for a point source:

$$\phi = \frac{Q}{4\pi r}. \tag{4.3.4}$$

An exactly similar result applies in the theory of electrostatics but with the electrostatic charge e replacing the quantity $Q/4\pi$. In line with this analogy the quantity $Q/4\pi$ in (4.3.4) is known as the strength of the source.

If we introduce a negative sign into (4.3.4) we have the potential function for a *point sink*. Sources and sinks may be compared with positive and negative charges in electrostatics and the combination of a source and sink at a small distance δx apart is similar to a *dipole*. The moment of a dipole is defined in electrostatics as the product of the charge e and the distance apart δx. Similarly we can define the magnitude m of a dipole or *doublet* in fluid flow as the product $Q\,\delta x/4\pi$.

Referring to figure 30, and making use of (4.3.4), it will be seen that the potential function for a dipole is given by

$$\phi=\frac{Q}{4\pi}\left(\frac{1}{r_1}-\frac{1}{r_2}\right)=\frac{m}{\delta x}\left(\frac{1}{r_1}-\frac{1}{r_2}\right). \tag{4.3.5}$$

If we now decrease δx and increase Q so that the value of m remains constant, and noting that $r_2=r_1-\delta x\cos\theta$, it is easily verified that the limiting value of the potential ϕ as given by (4.3.5) amounts to

$$\phi=-\frac{m}{r^2}\cos\theta, \tag{4.3.6}$$

which represents the potential function for a doublet of magnitude m located at the origin.

Certain three-dimensional axisymmetrical flow patterns can be constructed graphically by combining sources, sinks, and doublets with uniform streaming motion parallel to the axis. The potential function for a uniform stream from left to right, parallel to the x axis in figure 30, is represented by

$$\phi=-U_0x=-U_0r\cos\theta, \tag{4.3.7}$$

where U_0 is the stream velocity. The combination of a uniform stream and a

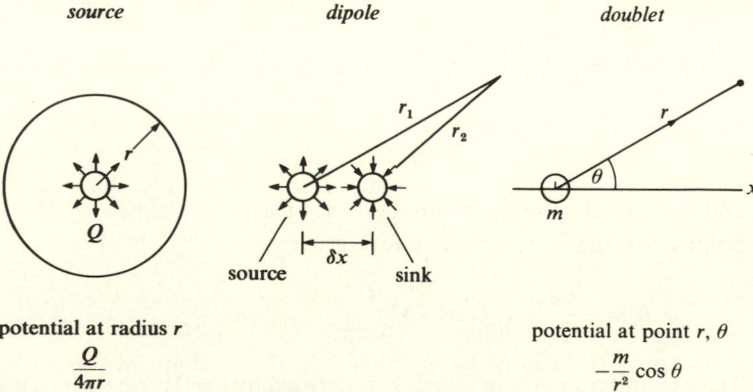

Fig. 30. Potential flow from a source and a dipole.

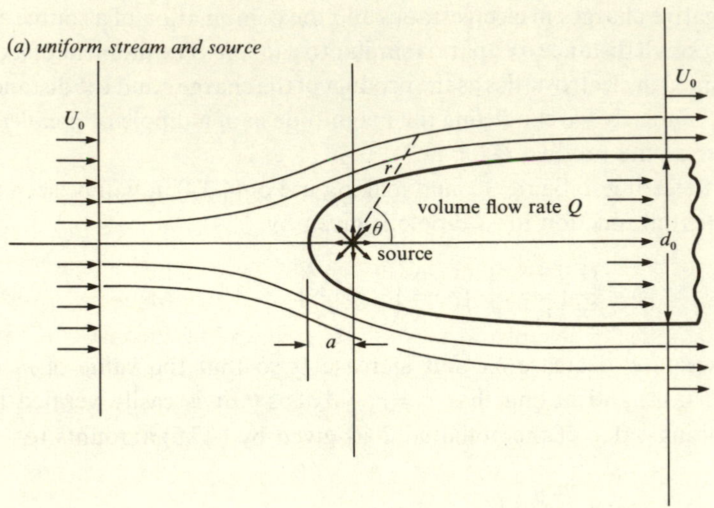

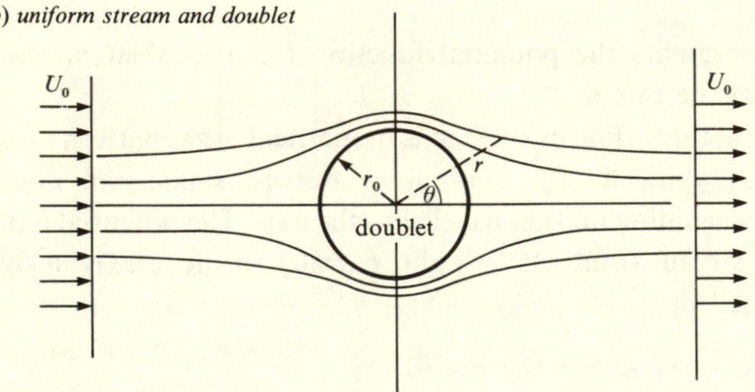

Fig. 31. (*a*) Combination of a uniform stream and a source.
(*b*) Combination of a uniform stream and a doublet.

source at the origin, as shown in figure 31, is therefore represented by

$$\phi = -U_0 r \cos\theta + \frac{Q}{4\pi r} \tag{4.3.8}$$

and the radial velocity component u_r at a point defined by the spherical polar coordinates r, θ, is therefore given by

$$u_r = -\frac{\partial\phi}{\partial r} = U_0 \cos\theta + \frac{Q}{4\pi r^2}. \tag{4.3.9}$$

The axis of symmetry in figure 31 is defined by $\theta = 0$ for points to the right of the origin, and by $\theta = \pi$ for points to the left. It will be seen from (4.3.9) that the radial velocity u_r will be zero at a point on the axis of symmetry specified by $\theta=\pi$ and by $r=a$, where the distance a is given by:

$$a=\left(\frac{Q}{4\pi U_0}\right)^{\frac{1}{2}}. \tag{4.3.10}$$

Referring to figure 31 (*a*), it will be seen that the dividing streamlines passing through this point form an envelope which separates the volume flow associated with the uniform stream from the volume flow originating from the source. Since there can be no flow *across* this surface the envelope could theoretically be replaced by a solid boundary provided we could ignore the real requirement of zero slip at the boundary. The diameter of this envelope d_0 can be calculated by noting that at a sufficient distance downstream, where the streamlines become parallel, the velocity of the fluid inside the envelope must be equal to the mainstream velocity U_0 outside. Hence we have $\frac{1}{4}\pi d_0^2 U_0 = Q$, or

$$d_0=\left(\frac{4Q}{\pi U_0}\right)^{\frac{1}{2}}=4a. \tag{4.3.11}$$

By a similar procedure we can obtain the theoretical flow pattern for a uniform stream flowing past an oval-shaped body of revolution if we combine the velocity potential for a source and a sink, located at a finite distance apart on the axis, with the velocity potential for a uniform stream. In the limiting case, where the distance between the source and the sink becomes infinitesimal, we have the special case of a uniform stream with a doublet of magnitude m at the origin for which the potential function is evidently

$$\phi=-U_0 r\cos\theta-\frac{m}{r^2}\cos\theta. \tag{4.3.12}$$

From (4.3.12), the radial velocity component u_r at any point specified by the co-ordinates r, θ, will be given by

$$u_r=-\frac{\partial\phi}{\partial r}=\left(U_0-\frac{2m}{r^3}\right)\cos\theta. \tag{4.3.13}$$

It will be seen that the radial velocity component will be zero for all values of θ when the spherical polar co-ordinate r has a value such that $r^3=2m/U_0$, or $r=r_0=(2m/U_0)^{\frac{1}{3}}$. We thus conclude that there is zero flow across a spherical surface whose radius is r_0, and that the velocity potential given by (4.3.12) corresponds, for values of r greater than r_0, to the flow of an ideal inviscid fluid stream past a sphere of radius r_0. This flow pattern is illustrated in figure 31 (*b*). If we substitute for m in terms of r_0, equation (4.3.12) may be re-written as

$$\phi = -U_0 r\left(1+\frac{1}{2}\frac{r_0^3}{r^3}\right)\cos\theta. \tag{4.3.14}$$

The streamline pattern predicted by this expression corresponds reasonably well to the observed flow of a real fluid at high values of the Reynolds number over the *front half* of a sphere, i.e. in the region where the boundary layer is thin.

Other axisymmetrical flow patterns may be studied by combining a uniform stream with a distribution of sources and sinks along the axis of symmetry. If the total net strength of the sources and sinks is zero, it can be shown that there will be one set of streamlines, forming a closed surface of revolution, which will enclose the volume flow associated with the sources and sinks. This method, originated by Rankine in 1871, enables a variety of idealised inviscid three-dimensional flow patterns to be analysed.

4.4 Two-dimensional potential flow

Although fluid motion in the real physical world is essentially three-dimensional in character, there are many engineering situations where the flow pattern may reasonably be analysed on the assumption that it is nearly two-dimensional. The obvious example is that of flow past an aerofoil section when the span of the wing or blade is large compared with the chord. It so happens that restriction of the analysis to two dimensions opens up the option of using the mathematical technique of complex numbers, and functions of a complex variable, in solving the two-dimensional form of Laplace's equation.

We can express equation (4.3.2) in cartesian co-ordinates in two dimensions as

$$\frac{\partial^2\phi}{\partial x^2}+\frac{\partial^2\phi}{\partial y^2}=0 \tag{4.4.1}$$

and we recall that this is simply a re-statement of the continuity equation (3.3.4) for incompressible flow which we can write as

$$\frac{\partial u}{\partial x}+\frac{\partial v}{\partial y}=0. \tag{4.4.2}$$

The relationship between the velocity components and the *velocity potential* ϕ is given, as in (4.3.1), by

$$u=-\frac{\partial\phi}{\partial x} \quad \text{and} \quad v=-\frac{\partial\phi}{\partial y}. \tag{4.4.3}$$

We can now introduce a *stream function* ψ such that

$$u=\frac{\partial\psi}{\partial y} \quad \text{and} \quad v=-\frac{\partial\psi}{\partial x}. \tag{4.4.4}$$

The stream function as defined by (4.4.4) automatically satisfies the equation of continuity (4.4.2) whether the flow is irrotational or not. If the flow is irrotational, however, the vorticity must be zero at every point and, in the case of two-dimensional flow, we have from (3.8.2):

$$\omega_z=\frac{\partial v}{\partial x}-\frac{\partial u}{\partial y}=0. \tag{4.4.5}$$

Hence for *irrotational flow*, from (4.4.4) and (4.4.5), we have

$$\frac{\partial^2\psi}{\partial x^2}+\frac{\partial^2\psi}{\partial y^2}=0 \tag{4.4.6}$$

so that the stream function ψ is also governed by Laplace's equation if the flow is irrotational. The stream function ψ is so named because it must have a constant value along a streamline. The equation for a streamline as in (1.6.7), is simply $\mathrm{d}x/u=\mathrm{d}y/v$, or $v\,\mathrm{d}x=u\,\mathrm{d}y$. Hence *along a streamline* we can say that

$$\mathrm{d}\psi=\frac{\partial\psi}{\partial x}\mathrm{d}x+\frac{\partial\psi}{\partial y}\mathrm{d}y=-v\,\mathrm{d}x+u\,\mathrm{d}y=0, \tag{4.4.7}$$

i.e. ψ is constant.

Since the velocity at any point in the flow must be perpendicular to the line of constant potential ϕ passing through that point, it is obvious that streamlines and equipotential lines must form an orthogonal set of curves which intersect each other at right angles.

We will now introduce a change of notation. For the remainder of this chapter we will be concerned exclusively with two-dimensional flow. We will therefore cease to use the letter z as the third cartesian co-ordinate and we will now employ the symbol z to represent the *complex number* defined by

$$z=x+\mathrm{i}y, \tag{4.4.8}$$

where the symbol i denotes the square root of minus one. In accordance with the well-known Argand diagram, the symbol i in (4.4.8) may be interpreted as a unit vector in the y direction. It may equally be interpreted as a mathematical operator which effects a rotation through 90° from the x axis to the y axis.

A further change of notation will be necessary in this chapter when we use a system of two-dimensional polar co-ordinates. Although the normal mathematical convention, as noted in § 1.6, is to employ the symbol ϕ as the angular co-ordinate in a cylindrical polar system, we will be obliged in this chapter to use the symbol θ for this purpose. The reason is simply that we are already using the letter ϕ as the velocity potential.

When working in polar co-ordinates, therefore, we will define the complex number z by

$$z = r(\cos\theta + \mathrm{i}\sin\theta) = r\,\mathrm{e}^{\mathrm{i}\theta}. \tag{4.4.9}$$

It will be obvious that lines defined by r = constant and θ = constant form an orthogonal system of concentric circles and radial lines in the z plane. Similarly lines defined by x = constant and y = constant form an orthogonal or rectangular network in the z plane.

We can now jump a few steps in the mathematical argument and assert that, under two-dimensional irrotational flow conditions, the *complex potential* defined by

$$w = \phi + \mathrm{i}\psi \tag{4.4.10}$$

must be an analytic function of the complex number z defined by (4.4.8) or (4.4.9).

As an illustration of this statement, we can take the case of a two-dimensional line source as indicated in figure 32. The line source located at the origin in the z plane is assumed to emit a volume flow per unit length measured at right angles to the z plane of Q m^2 s^{-1}. The radial velocity at radius r must obviously be $Q/2\pi r$ and hence we can say that

$$u_r = -\frac{\partial\phi}{\partial r} = \frac{Q}{2\pi r}. \tag{4.4.11}$$

This may be integrated directly, noting from the symmetry of the diagram that ϕ is a function of r only, to give the result:

$$\phi_0 - \phi = \frac{Q}{2\pi}\log_e\frac{r}{r_0}$$

and if we put $\phi = 0$ at $r = r_0$ the velocity potential is then given by

$$\phi = -\frac{Q}{2\pi}\log_e\frac{r}{r_0}. \tag{4.4.12}$$

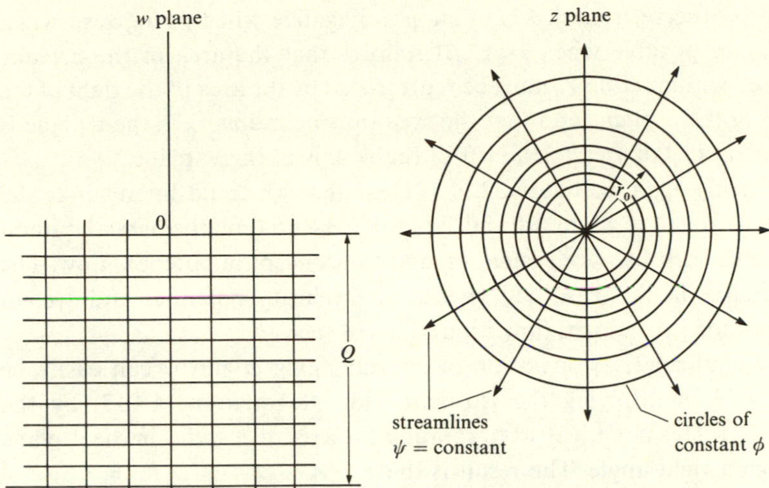

Fig. 32. Conformal transformation of the flow from a line source.

We can extend this result by expressing the complex potential w as the equivalent function of the complex variable z or $r\,e^{i\theta}$, i.e.

$$w=\phi+i\psi=-\frac{Q}{2\pi}\log_e\left(\frac{r\,e^{i\theta}}{r_0}\right). \qquad (4.4.13)$$

This may be re-written as:

$$\phi+i\psi=-\frac{Q}{2\pi}\left(\log_e\frac{r}{r_0}+i\theta\right).$$

If we now equate the real part on each side of this equation we have a re-statement of (4.4.12) for the velocity potential ϕ. If we equate the imaginary parts we have for the stream function ψ:

$$\psi=-\frac{Q}{2\pi}\theta, \qquad (4.4.14)$$

which confirms that streamlines are radial lines emanating from the source as shown in figure 32.

Equation (4.4.13) and figure 32 represent a very simple example of a *conformal transformation* relating a rectangular pattern of lines of constant ϕ and constant ψ in the w plane to another set of orthogonal lines mapping out $\phi=$ constant and $\psi=$ constant in the z plane. It is clear from (4.4.14) that one complete revolution in the z plane (with θ increasing from 0 to 2π) is equivalent to a linear change in the value of ψ from 0 to $-Q$ in the w plane.

We also observe from (4.4.12) that ϕ is negative when $r > r_0$, zero when $r = r_0$, and positive when $r < r_0$. It follows that the area of the z plane enclosed within radius r_0 must be represented by the area to the right of the ψ axis in the w plane, and that the area outside radius r_0 in the z plane is equivalent to the area to the left of the ψ axis in the w plane.

Referring again to figure 32, it is clear that we could interchange the functions ϕ and ψ while still satisfying the essential mathematical criteria associated with Laplace's equation and the concept of potential flow. The interchange in this case will give us the velocity potential and stream function for irrotational flow around a *free vortex.*

The mathematical operation of interchanging ϕ and ψ can easily be effected by multiplying the left-hand side of equation (4.4.13) by the operator i. This is equivalent to rotating the axes of ϕ and ψ in the w plane through a right angle. The result is that:

$$\phi = -\frac{Q}{2\pi}\theta \quad \text{and} \quad \psi = \frac{Q}{2\pi}\log_e \frac{r}{r_0} \tag{4.4.15}$$

and the complex potential w for a free vortex is therefore:

$$w = \phi + \mathrm{i}\psi = \mathrm{i}\frac{Q}{2\pi}\log_e\left(\frac{r\,\mathrm{e}^{\mathrm{i}\theta}}{r_0}\right). \tag{4.4.16}$$

The tangential velocity u_θ at radius r is given by

$$u_\theta = -\frac{1}{r}\frac{\partial\phi}{\partial\theta} = \frac{Q}{2\pi r}$$

and the circulation round the vortex at radius r is

$$K = \int_0^{2\pi} u_\theta r\,\mathrm{d}\theta = Q. \tag{4.4.17}$$

We may therefore re-write the expression (4.4.16) giving the complex potential for a free vortex in terms of the circulation K:

$$w = \mathrm{i}\frac{K}{2\pi}\log_e\left(\frac{r\,\mathrm{e}^{\mathrm{i}\theta}}{r_0}\right) \tag{4.4.18}$$

and the tangential velocity u_θ at radius r may be expressed as

$$u_\theta = \frac{K}{2\pi r}, \tag{4.4.19}$$

which is exactly the same result as we obtained in (3.8.18), noting that we are now using the symbol u_θ in place of u_ϕ in chapter 3. In the above expressions the circulation K is taken as positive when measured in the same direction as the angle θ, i.e. positive in an anti-clockwise sense.

4.5 Potential flow past a cylindrical section

The complex potential for two-dimensional flow past a circular cylindrical section of radius r_0 is represented by

$$w = -U_0\left(z + \frac{r_0^2}{z}\right) \tag{4.5.1}$$

and this may be written alternatively in the form

$$w = -U_0\left(r\,e^{i\theta} + \frac{r_0^2}{r}\,e^{-i\theta}\right).$$

The first term on the right-hand side represents uniform streaming motion with undisturbed velocity U_0 from left to right, as indicated in figure 33, while the second term represents a two-dimensional doublet located at the origin. This may be compared with the equivalent three-dimensional picture of flow past a sphere which was analysed in §4.3.

From (4.5.1), re-grouping the real and imaginary parts on the right-hand side of the equation, we can write

$$\phi + i\psi = -U_0 r\left(1 + \frac{r_0^2}{r^2}\right)\cos\theta - iU_0 r\left(1 - \frac{r_0^2}{r^2}\right)\sin\theta$$

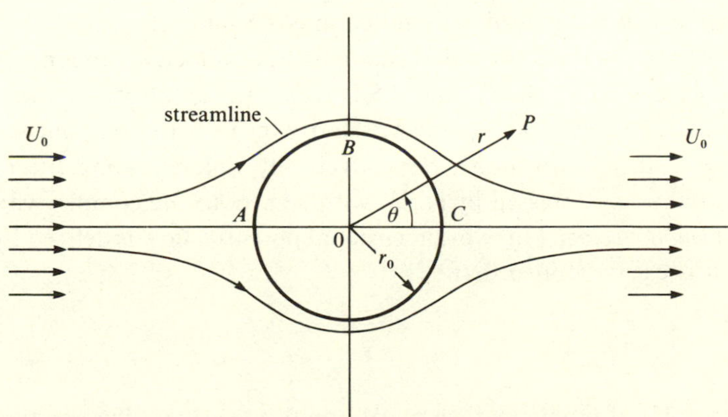

Fig. 33. Potential flow past a cylindrical section.

and hence

$$\phi = -U_0 r\left(1+\frac{r_0^2}{r^2}\right)\cos\theta \tag{4.5.2}$$

and

$$\psi = -U_0 r\left(1-\frac{r_0^2}{r^2}\right)\sin\theta. \tag{4.5.3}$$

From (4.5.3), the streamline $\psi=0$ is evidently defined by

$$\left(1-\frac{r_0^2}{r^2}\right)r\sin\theta=0$$

and this includes the x axis ($\sin\theta=0$ or π) and the circle $r=r_0$. We also observe from (4.5.2) that the radial velocity u_r at any point in the flow is given by

$$u_r = -\frac{\partial\phi}{\partial r} = U_0\left(1-\frac{r_0^2}{r^2}\right)\cos\theta \tag{4.5.4}$$

and this must be zero when $r=r_0$ which confirms the fact that the circle of radius r_0 is a streamline.

The tangential velocity u_θ at any point in the flow is given by

$$u_\theta = -\frac{1}{r}\frac{\partial\phi}{\partial\theta} = -U_0\left(1+\frac{r_0^2}{r^2}\right)\sin\theta \tag{4.5.5}$$

and on the circle $r=r_0$ this will have the value $-2U_0\sin\theta$. If the streamline $r=r_0$ is replaced by a solid boundary, therefore, we must have slipping of the fluid parallel to the surface. While this is theoretically possible in the case of an inviscid fluid, it is inconsistent with the property of viscosity in the case of a real fluid, as already noted in §§ 1.5 and 4.1.

We have obtained the radial and tangential velocity components u_r and u_θ as expressed in (4.5.4) and (4.5.5) from the equation for the velocity potential (4.5.2). We can equally well derive the velocity components u and v in the x and y directions, respectively, by differentiating the complex potential w as expressed by (4.5.1) with respect to the complex variable z. The *complex velocity* in two-dimensional potential flow is defined by $\mathrm{d}w/\mathrm{d}z$ and it is easily verified that

$$\frac{\mathrm{d}w}{\mathrm{d}z} = u - \mathrm{i}\,v. \tag{4.5.6}$$

It is a point of purely mathematical significance that differentiation of the complex potential w with respect to the complex variable z gives the result

$(u-\mathrm{i}\,v)$ and not $(u+\mathrm{i}\,v)$. Thus the complex velocity defined by (4.5.6) is the conjugate (or mirror image in the x axis) of the actual velocity vector $(u+\mathrm{i}\,v)$. The magnitude of the complex velocity, however, is obviously identical to that of the actual velocity vector. We can now apply this technique to equation (4.5.1) with the result that

$$\frac{\mathrm{d}w}{\mathrm{d}z}=u-\mathrm{i}\,v=-U_0\left(1-\frac{r_0^2}{z^2}\right)$$

$$=-U_0\left(1-\frac{r_0^2}{r^2}\,\mathrm{e}^{-2\mathrm{i}\theta}\right)$$

or

$$u-\mathrm{i}\,v=-U_0\left(1-\frac{r_0^2}{r^2}\cos 2\theta\right)-\mathrm{i}\,U_0\frac{r_0^2}{r^2}\sin 2\theta$$

and hence the velocity vector $u+\mathrm{i}\,v$ is represented by

$$u+\mathrm{i}\,v=-U_0\left(1-\frac{r_0^2}{r^2}\cos 2\theta\right)+\mathrm{i}\,U_0\frac{r_0^2}{r^2}\sin 2\theta. \tag{4.5.7}$$

The square of the velocity magnitude is then given by

$$(u^2+v^2)=U_0^2\left(1-\frac{2r_0^2}{r^2}\cos 2\theta+\frac{r_0^4}{r^4}\right). \tag{4.5.8}$$

At the surface of the cylinder, putting $r=r_0$ in (4.5.8), we have

$$(u^2+v^2)=2U_0^2(1-\cos 2\theta)=4U_0^2\sin^2\theta \tag{4.5.9}$$

which is consistent with equation (4.5.5) for the tangential velocity.

We may finally obtain the pressure distribution round the surface of the cylinder by using the result (4.5.9) in conjunction with Bernoulli's equation. If the pressure is p_0 in the undisturbed stream, where the velocity is U_0, the pressure p at any point on the circumference of the circle of radius r_0 will be given by

$$p_0+\tfrac{1}{2}\rho U_0^2=p+\tfrac{1}{2}\rho(u^2+v^2)=p+2\rho U_0^2\sin^2\theta$$

or

$$p-p_0=\tfrac{1}{2}\rho U_0^2(1-4\sin^2\theta). \tag{4.5.10}$$

It is of interest to follow out the variation in velocity and pressure as one moves round the surface of the cylinder from the front stagnation point A in figure 33 to the rear stagnation point C, using equations (4.5.9) and (4.5.10). At point A, where $\theta=\pi$, the velocity is zero and the pressure has a maximum value equal to the stagnation pressure $p_0+\frac{1}{2}\rho U_0^2$. From A to B the velocity increases to reach a maximum value of $2U_0$ at point B where $\theta=\frac{1}{2}\pi$, while the pressure falls until it reaches its minimum value of $p_0-\frac{3}{2}\rho U_0^2$ at point B. A falling pressure gradient in the direction of flow is favourable from the point of view of the boundary layer which remains quite thin over this portion of the surface in a real fluid. Consequently the flow pattern predicted by (4.5.7) corresponds quite closely to the actual observed flow around the front portion of the cylinder outside the boundary layer.

From B to C, however, equations (4.5.9) and (4.5.10) predict a falling velocity and a rising pressure distribution which would theoretically be an exact mirror image of the pattern over the front half. For reasons which will be discussed in chapter 7, an adverse pressure gradient in the direction of flow causes thickening of the boundary layer and leads to instability of flow in the boundary-layer region. This in turn causes separation of flow and leads to the formation of an unsteady eddying wake. The flow pattern around the downstream side of a bluff-shaped object such as a cylindrical section cannot therefore be predicted or analysed by means of potential flow theory.

Despite this severe limitation, the potential function for flow past a cylinder given by (4.5.1) plays an important role as an intermediate mathematical step in developing a two-dimensional theory for thin aerofoil sections. This aspect is briefly outlined in §4.7.

4.6 Flow past a cylindrical section with circulation

We can obtain the complex potential for two-dimensional flow, with undisturbed streaming velocity U_0, past a cylindrical section of radius r_0 with circulation K by combining (4.5.1) and (4.4.18). Hence

$$w=-U_0\left(z+\frac{r_0^2}{z}\right)+\mathrm{i}\frac{K}{2\pi}\log_e\frac{z}{r_0}. \tag{4.6.1}$$

If we separate the real and imaginary parts, we have for the velocity potential ϕ and the stream function ψ

$$\phi=-U_0r\left(1+\frac{r_0^2}{r^2}\right)\cos\theta-\frac{K}{2\pi}\theta \tag{4.6.2}$$

and

$$\psi=-U_0r\left(1-\frac{r_0^2}{r^2}\right)\sin\theta+\frac{K}{2\pi}\log_e\frac{r}{r_0}. \tag{4.6.3}$$

The tangential velocity component u_θ can easily be derived from (4.6.2)

$$u_\theta = -\frac{1}{r}\frac{\partial \phi}{\partial \theta} = -U_0\left(1+\frac{r_0^2}{r^2}\right)\sin\theta + \frac{K}{2\pi r} \tag{4.6.4}$$

and at the surface of the cylinder, on the circle $r=r_0$, this becomes $-2U_0\sin\theta + K/2\pi r_0$. We note that the tangential or circumferential velocity will be zero for values of θ such that

$$\sin\theta = \frac{K}{4\pi U_0 r_0}. \tag{4.6.5}$$

There will normally be two such values of θ which specify the stagnation points as indicated by the angle β in figure 34. Note that the flow pattern in figure 34 is drawn with a clockwise circulation, i.e. a negative value for K.

The complex velocity at any point in the flow is found by differentiating the complex potential given by (4.6.1), i.e.

$$\frac{\mathrm{d}w}{\mathrm{d}z} = u - \mathrm{i}\,v = -U_0\left(1-\frac{r_0^2}{z^2}\right) + \mathrm{i}\,\frac{K}{2\pi z} \tag{4.6.6}$$

or

$$u - \mathrm{i}\,v = -U_0\left(1-\frac{r_0^2}{r^2}\mathrm{e}^{-2\mathrm{i}\theta}\right) + \mathrm{i}\,\frac{K}{2\pi r}\mathrm{e}^{-\mathrm{i}\theta}.$$

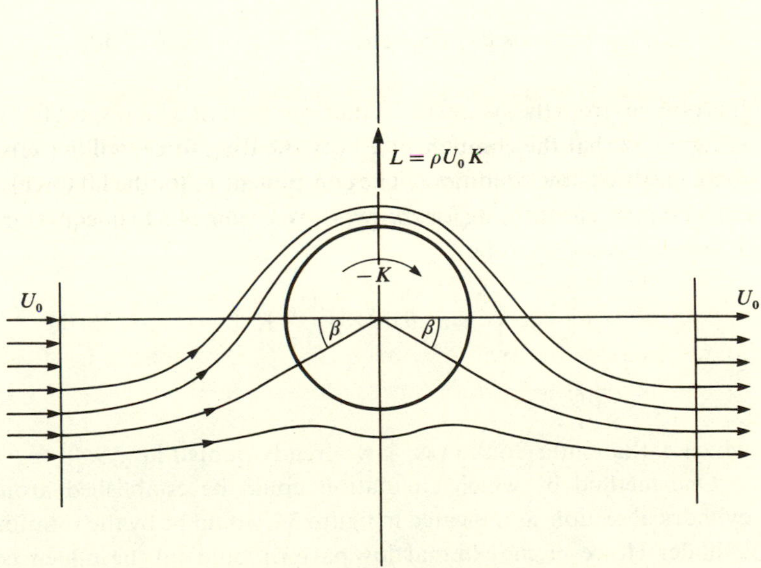

Fig. 34. Potential flow past a cylindrical section with circulation.

The flow pattern is indicated diagrammatically in figure 34. The magnitude of (u^2+v^2) at any point in the flow may be obtained from (4.6.6) and hence, using Bernoulli's equation, we can determine the pressure distribution throughout the flow.

If we confine attention to the pressure distribution round the surface of the cylinder we need to know the value of (u^2+v^2) at points on the circle $r=r_0$. This can be derived, as noted above, from (4.6.6) or more directly from (4.6.4) noting that the radial component of the velocity is zero when $r=r_0$. By either route we have

$$u_\theta^2=\left(-2U_0\sin\theta+\frac{K}{2\pi r_0}\right)^2 \quad \text{at } r=r_0 \tag{4.6.7}$$

and hence from Bernoulli's equation

$$p_0+\tfrac{1}{2}\rho U_0^2=p+\tfrac{1}{2}\rho U_0^2\left(-2\sin\theta+\frac{K}{2\pi r_0 U_0}\right)^2$$

or

$$p-p_0=\tfrac{1}{2}\rho U_0^2\left[1-\left(4\sin^2\theta+\frac{K^2}{4\pi^2 r_0^2 U_0^2}-\frac{2K\sin\theta}{\pi r_0 U_0}\right)\right]. \tag{4.6.8}$$

The resultant force acting on the cylinder can be calculated by integrating the pressure distribution round the surface of the cylinder. The components of the net force are given by

$$F_x=\int_0^{2\pi} p\cos\theta\, r_0\,\mathrm{d}\theta \quad \text{and} \quad F_y=\int_0^{2\pi} p\sin\theta\, r_0\,\mathrm{d}\theta. \tag{4.6.9}$$

It is obvious from the symmetry of the flow pattern with respect to the y axis in figure 34 that the component F_x (or the drag force) will be zero under these idealised flow conditions. The component F_y (or the lift force) is easily calculated by substituting for the pressure p from (4.6.8) in equation (4.6.9). The result is found to be

$$F_y=\rho\frac{U_0}{\pi}K\int_0^{2\pi}\sin^2\theta\,\mathrm{d}\theta=\rho\frac{U_0}{2\pi}K\int_0^{2\pi}(1-\cos 2\theta)\,\mathrm{d}\theta$$

or

$$F_y=\rho U_0 K \tag{4.6.10}$$

which is the Kutta–Joukowski law already quoted in (3.9.8).

One method by which circulation could be established around the cylindrical section, as indicated in figure 34, would be by the rotation of the cylinder. However, the external flow pattern could only be influenced in this way through the action of viscosity in the boundary-layer region. The consequent flow pattern in the neighbourhood of the cylinder for a real fluid would therefore differ from the theoretical velocity distribution represented by (4.6.6) for a hypothetical inviscid fluid. It is impossible to reconcile the condition of zero slip at the surface of a rotating cylinder with the velocity distribution calculated by potential theory. The main importance of the complex potential represented by (4.6.1) and the expression for the lift force (4.6.10) is rather as a mathematical tool which can be used in connection with two-dimensional aerofoil theory.

4.7 Two-dimensional aerofoil theory

We have already seen that the analytic function of z defined by (4.5.1) describes flow past a circular cylinder of radius r_0 as depicted in figure 33. We may regard this equation as representing a conformal transformation from a square-mesh pattern of lines of constant ϕ and constant ψ in the w plane to the set of equipotential lines and streamlines in the z plane associated with flow past a cylindrical contour.

In more general terms we can now consider a similar conformal transformation from the ζ plane (co-ordinates ξ, η) to the z plane (co-ordinates x, y) represented by

$$\zeta = z + \frac{a^2}{z} \qquad (4.7.1)$$

or

$$\xi + \mathrm{i}\eta = r\,\mathrm{e}^{\mathrm{i}\theta} + \frac{a^2}{r}\,\mathrm{e}^{-\mathrm{i}\theta}.$$

Separating the real and imaginary parts gives us

$$\xi = \left(r + \frac{a^2}{r}\right)\cos\theta \qquad (4.7.2)$$

and

$$\eta = \left(r - \frac{a^2}{r}\right)\sin\theta. \qquad (4.7.3)$$

As indicated in figure 35, the circle $r = a$ with centre at the origin in the z plane transforms into a straight line on the ξ axis of the ζ plane extending from $\xi = -2a$ to $\xi = +2a$. We conclude that flow past the cylindrical contour in the z plane is transformed into uniform streaming flow parallel to the flat plate AC of length $4a$ in the ζ plane by means of (4.7.1).

If we now draw a circle of radius a', with centre at $(0, \mathrm{i}\,b)$ in the z plane, passing through the same points $x = -a$ and $x = +a$ indicated by the letters

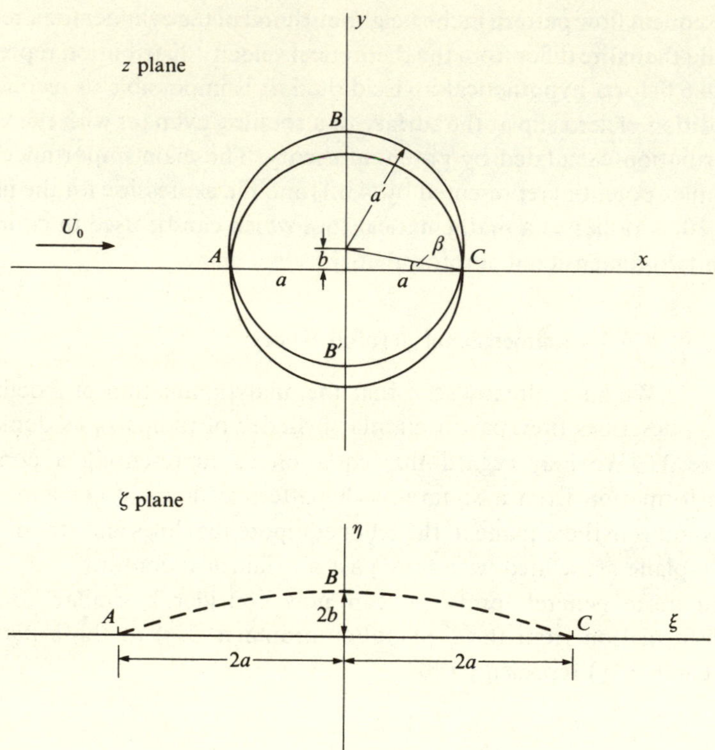

Fig. 35. Conformal transformation of the flow past a cylindrical section.

A and *C* in figure 35, it can be shown analytically from (4.7.2) and (4.7.3) that this circle transforms into a circular arc in the ζ plane extending from $\xi = -2a$ to $\xi = 2a$. This is shown by the dotted curve in figure 35. From the geometry of figure 35 we have

$$a'^2 = a^2 + b^2. \tag{4.7.4}$$

Point *B* in the *z* plane is defined by $x = 0$, $y = (a' + b)$. Hence from (4.7.1) this transforms into a point in the ζ plane given by

$$\zeta = \mathrm{i}(a' + b) + \frac{a^2}{\mathrm{i}(a' + b)}$$

or

$$\zeta = \frac{(a' + b)^2 - a^2}{(a' + b)}\,\mathrm{i}. \tag{4.7.5}$$

Substituting from (4.7.4) in (4.7.5) gives the result that

$$\zeta = 2b\,\mathrm{i}, \tag{4.7.6}$$

which fixes point B in the ζ plane. A similar calculation for point B' in the z plane defined by $x=0$, $y=-(a'-b)$, proves that this point also transforms into the point $\eta = 2b\,\mathrm{i}$, $\xi = 0$, in the ζ plane. Thus both the upper portion ABC and the lower portion $AB'C$ of the circle of radius a' in the z plane transform into the circular arc ABC in the ζ plane.

It will be evident, if figure 35 is compared with figure 34, that for streaming motion parallel to the x axis past the circle of radius a' in the z plane, it will be necessary to add circulation around the cylinder in order to bring the rear stagnation point to coincide with point C in the diagram. The magnitude of this circulation is determined by equation (4.6.5), i.e.

$$K = 4\pi U_0 a' \sin\beta, \tag{4.7.7}$$

where the angle β is defined in figure 35 by $\tan\beta = b/a$. By bringing the rear stagnation point for flow past the cylinder of radius a' in the z plane to point C in this manner, the corresponding flow past the circular arc ABC in the ζ plane will also have a rear stagnation point at the equivalent point C which implies that the flow will be tangential to the curved surface of the circular arc at point C in the ζ plane. In other words, circulation is required in order to achieve tangential flow at the trailing edge of a thin curved plate. The camber of the circular arc is defined by the maximum ordinate $2b$ divided by the chord $4a$, i.e.

$$\text{camber} = \frac{b}{2a} = \tfrac{1}{2}\tan\beta. \tag{4.7.8}$$

We conclude at this point that circulation, and hence lift force, is related to the camber of the curved plate by equations (4.7.7) and (4.7.8).

If we draw a circle of radius a' (slightly larger than a) but with the centre on the x axis at a point defined by $x = -(a'-a)$ in the z plane, the circle will transform by means of equation (4.7.1) into a *symmetrical aerofoil section* in the ζ plane. This transformation is shown in figure 36. We can express the radius a' in terms of a by writing

$$a' = a(1+\varepsilon), \tag{4.7.9}$$

where ε is a small ratio. From the geometry of figure 36 we have the following relationship between a', r, and the angle θ

$$a'^2 = r^2 + (a'-a)^2 - 2r(a'-a)\cos(\pi - \theta). \tag{4.7.10}$$

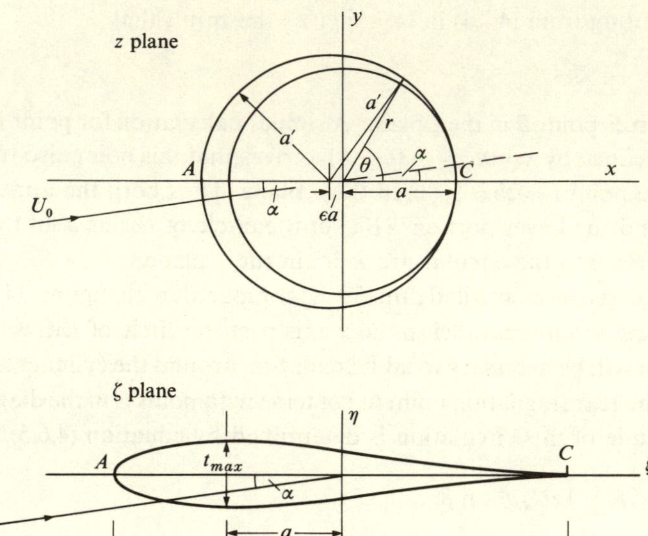

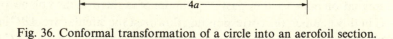

Fig. 36. Conformal transformation of a circle into an aerofoil section.

Substituting for a' from (4.7.9), and neglecting second-order terms involving ε^2, equation (4.7.10) becomes

$$\left(\frac{r}{a}\right)^2+2\frac{r}{a}\varepsilon\cos\theta=1+2\varepsilon$$

and hence, provided ε is small

$$\frac{r}{a}=1+\varepsilon(1-\cos\theta). \tag{4.7.11}$$

If we now substitute for r/a from (4.7.11) in equations (4.7.2) and (4.7.3) we have for the profile of the aerofoil

$$\xi=\left(\frac{r}{a}+\frac{a}{r}\right)a\cos\theta=2a\cos\theta \tag{4.7.12}$$

and

$$\eta=\left(\frac{r}{a}-\frac{a}{r}\right)a\sin\theta=2\varepsilon a(1-\cos\theta)\sin\theta \tag{4.7.13}$$

provided again that ε is small and that we can ignore ε^2. The exact position of the nose of the aerofoil (point A in figure 36) may be found from (4.7.2) by taking $\theta=\pi$ and $r=a(1+2\varepsilon)$, i.e.

$$\xi=-a(1+2\varepsilon)-a(1+2\varepsilon)^{-1}=-2a(1+2\varepsilon^2+\cdots).$$

The trailing edge of the aerofoil (point C in figure 36) is at $\xi=2a$ exactly, thus the chord length is $4a(1+\varepsilon^2)$ to a second order of accuracy.

From (4.7.13) the thickness of the aerofoil at the mid-chord position, where $\theta=\frac{1}{2}\pi$, is found to be $4a\varepsilon$. The maximum thickness, however, will occur when $\cos\theta=-\frac{1}{2}$, i.e. approximately at the quarter-chord position. The value t max is found to be $3\sqrt{3}a\varepsilon$.

If the undisturbed stream is parallel to the axis of the aerofoil, it is obvious from symmetry that the lift force will be zero. If, however, the stream velocity U_0 is inclined at an angle of incidence α to the axis, as indicated in figure 36, it is clear that circulation is required in order to bring the rear stagnation point back to point C in the diagram. The magnitude of the circulation K is determined by equation (4.6.5), i.e. we must have

$$K=4\pi U_0 a' \sin\alpha. \tag{4.7.14}$$

The lift per unit span of the aerofoil is given by (4.6.10). If we define a lift coefficient, as in (3.9.9), by

$$L=c_L\tfrac{1}{2}\rho U_0^2 c, \tag{4.7.15}$$

where $c=$ chord length $\approx 4a$ for the aerofoil section of figure 36, we have from (4.6.10), (4.7.14), and (4.7.15)

$$c_L=2\pi\frac{a'}{a}\sin\alpha=2\pi(1+\varepsilon)\sin\alpha. \tag{4.7.16}$$

Noting that the thickness/chord ratio is $3\sqrt{3}\,\varepsilon/4=1.3\varepsilon$, we can write $\varepsilon=0.77\,t_{\max}/4a$ and hence (4.7.16) may be expressed in the form

$$c_L=2\pi\sin\alpha\left(1+0.77\frac{t_{\max}}{4a}\right).$$

It will now be almost obvious from examination of figures 35 and 36 that we can generate a *cambered aerofoil section* in the ζ plane by means of the transformation equation (4.7.1) if we draw a circle in the z plane of radius a' with centre at $x=(a-a'\cos\beta)$ and $y=a'\sin\beta$, as shown in figure 37. The resulting Joukowski aerofoil profile of figure 37 may be regarded as the symmetrical aerofoil of figure 36 with its profile centre-line bent into the shape of the circular arc of figure 35.

It will also be evident from figure 37 that the circulation necessary to make the rear stagnation point coincide with the trailing edge at point C in

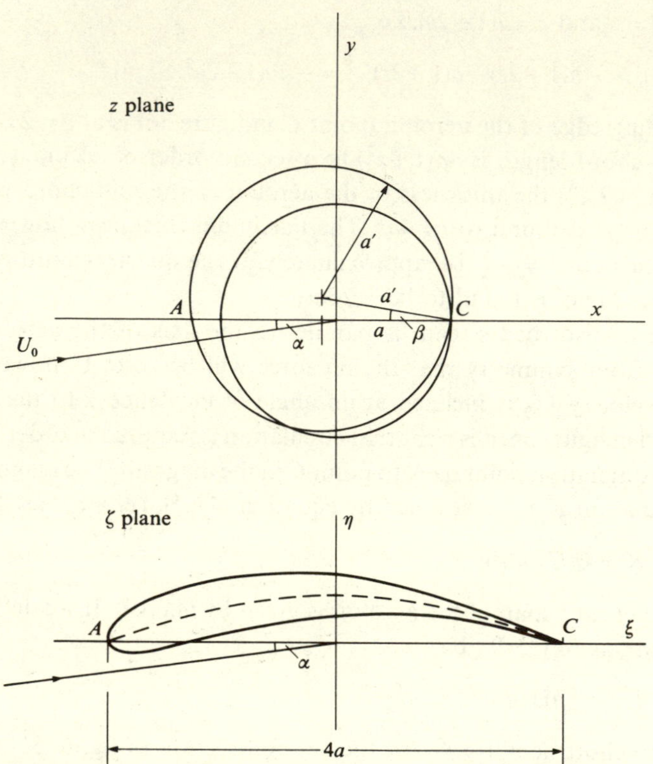

Fig. 37. Conformal transformation of a circle into a cambered aerofoil section.

the diagram is given by

$$K = 4\pi U_0 a' \sin(\alpha + \beta) \tag{4.7.17}$$

and the corresponding lift force L from (4.6.10) will be

$$L = 4\pi\rho U_0^2 a' \sin(\alpha + \beta). \tag{4.7.18}$$

It follows from (4.7.18) that the lift coefficient for a cambered Joukowski-type aerofoil is given by

$$c_L = \frac{L}{\frac{1}{2}\rho U_0^2 \times 4a} = 2\pi \frac{a'}{a} \sin(\alpha + \beta) \tag{4.7.19}$$

and noting from figure 37 that $a(1+\varepsilon)/a' = \cos\beta$ the last equation may be expressed as

$$c_L = \frac{2\pi(1+\varepsilon)\sin(\alpha+\beta)}{\cos\beta}. \tag{4.7.20}$$

In practice for thin aerofoils of small camber ε is small and $\cos\beta \approx 1$. We can therefore usually say that

$$c_L = 2\pi \sin(\alpha+\beta), \tag{4.7.21}$$

which is the result quoted in (3.9.11).

We have used the Joukowski hypothesis, that the rear stagnation point must coincide with the sharp trailing edge, to determine the magnitude of the circulation round the aerofoil. In the absence of circulation, potential flow past a cambered aerofoil, such as that illustrated in figure 37, or past a symmetrical section inclined at an angle of incidence α as shown in figure 36, would require the rear stagnation point to be located on the upper surface of the profile at a short distance from the trailing edge, and the local velocity around the sharp trailing edge would be infinite, which is physically impossible.

It is of interest to consider how the circulation is established round the aerofoil in the first instance. When an aerofoil is accelerated from rest the flow pattern outside the thin boundary-layer region is momentarily potential in character without any circulation. The high transient value of the local velocity round the trailing edge, however, implies that there must be a low pressure zone at the trailing edge and a sharp adverse pressure gradient on the upper surface of the profile between the trailing edge and the temporary rear stagnation point. This leads quickly to separation of the boundary layer and to the creation of a starting vortex close to the trailing edge. As the steady-flow pattern develops, the stagnation point moves back to the trailing edge, thus establishing circulation round the profile, and the starting vortex moves away from the aerofoil to form part of the wake.

Recalling Kelvin's circulation theorem in § 3.8, it will be evident that for an aerofoil of finite span, such as an aeroplane wing, the circulation round the profile must be continued at the extremities as wing-tip vortices in the airstream. These vortices trail back behind the aircraft and ultimately form a closed loop with the starting vortex.

The Joukowski-type aerofoil profiles obtained by the conformal transformation (4.7.1), as illustrated in figures 36 and 37, have the practical disadvantage that the upper and lower surfaces converge at zero angle at the trailing edge. A mathematically more complicated transformation is required to derive an aerofoil profile with a finite angle at the trailing edge. The essential relationship between the lift force (or circulation) and the angle of incidence and camber of the profile, however, may still be expressed by equation (4.7.21) provided the aerofoil profile is thin and of relatively small camber as is usually the case.

It is important to appreciate that potential flow theory is only applicable to two-dimensional flow past an aerofoil section for *small angles of incidence.* At large values of the angle α the flow breaks away from the upper surface at a short distance downstream from the point of maximum velocity and minimum pressure. Under these conditions the aerofoil is said to be *stalled.* A disturbed and eddying wake region is formed behind the aerofoil and the flow pattern no longer conforms to the condition of a thin boundary-layer region with potential flow outside. The stalling angle will depend on the precise geometry of the aerofoil profile but is typically within the range between 10° and 15° for the majority of aerofoil sections.

Although potential flow theory provides a satisfactory account of the lift force, together with its dependence on incidence and camber, it is clearly not possible to draw any conclusions regarding the drag force on the basis of potential flow. The drag of a two-dimensional aerofoil section consists of two components. The *skin-friction drag* is the resultant of the tangential shear stress acting over the surface of the aerofoil and can only be analysed with the help of boundary-layer theory as discussed in chapter 7. The *form drag* is the resultant of the normal pressure distribution integrated over the complete profile. Potential flow theory predicts zero form drag. However, because of the nature of the flow in the boundary-layer region it is not possible in practice to achieve the full theoretical pressure recovery from the minimum pressure point on the profile to the trailing edge or rear stagnation point. Consequently an actual aerofoil section will always have

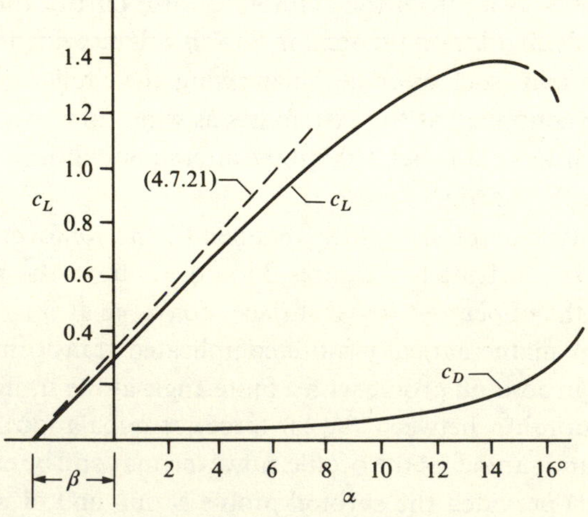

Fig. 38. Lift and drag coefficients for an aerofoil.

measurable form drag and its magnitude may be regarded as a measure of the departure from idealised streamline flow.

It is convenient to define a drag coefficient c_D in the same way as we have defined the lift coefficient, i.e.

$$D = c_D \tfrac{1}{2} \rho U_0^2 c, \tag{4.7.22}$$

where D is the total drag force per unit span of the section, and $c = 4a$ is the chord length as indicated in figure 37. For practical aerofoil profiles the lift and drag are measured in wind-tunnel tests and the results may be recorded graphically in the form of curves of c_L and c_D plotted against the angle of incidence α. A typical instance is shown in figure 38.

5

Equations of motion for a viscous fluid

5.1 The influence of viscosity on the motion of a fluid

We noted in chapter 1 that shear stresses are created in a moving fluid through the physical property of viscosity and that the flow pattern, for a given boundary geometry, is strongly influenced by the magnitude of the kinematic viscosity. In the basic case of the flow of an incompressible fluid, with no free surface, the principal factor which determines the flow pattern is the value of the Reynolds number, defined by $\rho UL/\mu$ or UL/ν where U is a characteristic velocity (e.g. the undisturbed stream velocity) and L is a characteristic length. Thus a small value for the kinematic viscosity ν implies a large value for the Reynolds number and vice versa.

In chapter 4 we considered the motion of an idealised fluid with zero viscosity. In certain aspects, provided we exclude flow near a solid boundary, the theory of irrotational inviscid flow represents the limiting case of the motion of a fluid of very small viscosity. However, all real fluids exhibit the properties both of viscosity (however small) and of zero slip at a solid boundary. Consequently the flow pattern in the neighbourhood of solid boundaries is always influenced by fluid viscosity and, as we shall see in chapters 6 and 7, the nature of the flow in these boundary-layer regions can have a profound influence on the wider flow pattern.

As we have already remarked in § 1.7, two fundamentally different types of fluid motion – laminar and turbulent – are observed in practice. Laminar flow tends to prevail at low Reynolds numbers, while turbulent flow is more probable at high Reynolds numbers. It would be easy to draw the wrong conclusions and to suppose that the basic flow instability which leads to turbulence might be unrelated to viscosity, and that fluid viscosity exercises a purely stabilising effect which tends to maintain laminar flow. While it is true that viscosity acts in such a manner as to damp down small disturbances, we have to recognise the fact that the onset of turbulence is closely associated with the existence of vorticity in a fluid. Vorticity in turn is created in the region of solid boundaries through the phenomenon of zero

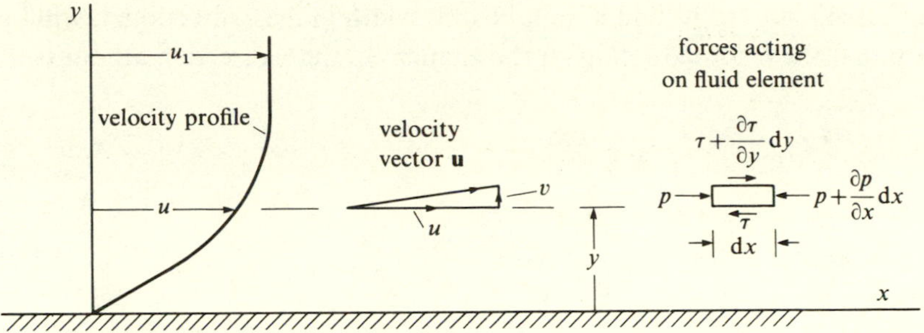

Fig. 39. Flow past a solid boundary.

slip and the action of viscosity. We can thus say that viscosity is largely responsible for the generation of turbulence, as well as for its suppression under certain circumstances.

It is of fundamental importance, therefore, to establish the basic equations of motion for a viscous fluid before proceeding to a discussion of turbulence in chapter 6 and of boundary-layer flow in chapter 7. As a preliminary manoeuvre it may be helpful to see how the Euler equation of motion for an inviscid fluid (3.2.1) could be modified, using a common-sense approach, to take account of the existence of viscous stresses in the fluid. We will start with the case of two-dimensional steady flow parallel to a solid boundary, as indicated in figure 39, and we will confine ourselves to the case of an incompressible fluid. Although we speak of the flow being parallel to the surface, this is strictly true only for the undisturbed stream at some distance from the boundary. As we will see in chapter 7, the thickness of the boundary-layer region increases with distance measured in the direction of the fluid stream. Consequently the local velocity within the boundary layer at any point will in general have two components, the main component u being in the x direction as shown in figure 39 but with a smaller component v in a direction perpendicular to the surface. Thus a local streamline drawn within the boundary-layer region will be inclined at a slight angle to the surface.

We will assume that *velocity gradients* are small in the x direction but relatively large in the y direction. As a first approximation in this particular case, therefore, we can say that the only significant shear stress acting within the fluid is that associated with the transverse velocity gradient $\partial u/\partial y$, i.e.

$$\tau=\mu\frac{\partial u}{\partial y}. \tag{5.1.1}$$

If we now consider a small element of fluid with dimensions $\mathrm{d}x, \mathrm{d}y$, as indicated in figure 39, and if we take unit width in the z direction, it will be seen that the net force acting on the element in the x direction amounts to

$$-\frac{\partial p}{\partial x}\,dx\,dy+\frac{\partial \tau}{\partial y}\,dy\,dx,$$

i.e.

$$\text{net force per unit volume}=-\frac{\partial p}{\partial x}+\frac{\partial \tau}{\partial y}. \tag{5.1.2}$$

Hence substituting from (5.1.1) we have

$$\text{net force per unit volume}=-\frac{\partial p}{\partial x}+\mu\frac{\partial^2 u}{\partial y^2}. \tag{5.1.3}$$

We can now write down the x component of the equation of motion by equating the rate of change of momentum of a material element, per unit volume, to the net force acting on the element, i.e.

$$\rho\frac{\mathrm{D}u}{\mathrm{D}t}=-\frac{\partial p}{\partial x}+\mu\frac{\partial^2 u}{\partial y^2}. \tag{5.1.4}$$

If we divide through by ρ the equation may be written as

$$\frac{\mathrm{D}u}{\mathrm{D}t}=-\frac{1}{\rho}\frac{\partial p}{\partial x}+\nu\frac{\partial^2 u}{\partial y^2} \tag{5.1.5}$$

and for steady flow, with $\partial u/\partial t=0$, it takes the form

$$u\frac{\partial u}{\partial x}+v\frac{\partial u}{\partial y}=-\frac{1}{\rho}\frac{\partial p}{\partial x}+\nu\frac{\partial^2 u}{\partial y^2}. \tag{5.1.6}$$

This is in fact the x component of the equation of motion for a two-dimensional *laminar boundary layer* under steady-flow conditions. Equation (5.1.6) will be found to be identical to (7.2.6) which is derived in a more rigorous manner in chapter 7.

We can easily extend equation (5.1.4) to cover a three-dimensional flow situation where there are transverse velocity gradients in the z direction as well as in the y direction, i.e.

$$\rho\frac{\mathrm{D}u}{\mathrm{D}t}=-\frac{\partial p}{\partial x}+\mu\left(\frac{\partial^2 u}{\partial y^2}+\frac{\partial^2 u}{\partial z^2}\right).$$

In the more general case, however, where the longitudinal velocity gradient $\partial u/\partial x$ may not always be negligible in comparison with the transverse gradients $\partial u/\partial y$ and $\partial u/\partial z$, considerations of mathematical symmetry would suggest that the complete equation of motion for the x direction should take the form

$$\rho\frac{\mathrm{D}u}{\mathrm{D}t}=-\frac{\partial p}{\partial x}+\mu\left(\frac{\partial^2 u}{\partial x^2}+\frac{\partial^2 u}{\partial y^2}+\frac{\partial^2 u}{\partial z^2}\right), \tag{5.1.7}$$

which may be written as

$$\rho\frac{\mathrm{D}u}{\mathrm{D}t}=-\frac{\partial p}{\partial x}+\mu\,\nabla^2 u. \tag{5.1.8}$$

Apart from the appearance of the pressure gradient term on the right-hand side, equation (5.1.8) will be seen to have the same form as the basic transport equation for a moving fluid (2.9.5) which we derived in chapter 2. In this instance the transferable property ϕ takes the form of the x component of momentum per unit mass, which is equal to u. The term $\mu\,\nabla^2 u$ represents the diffusion of momentum through the action of viscosity.

In the general case of three-dimensional motion, similar equations to (5.1.8) must apply to the y component of momentum per unit mass and to the z component. To complete the picture we also need to include the effect of a possible body force $\mathbf{F}$ per unit mass, or $\rho\mathbf{F}$ per unit material volume, as in the case of Euler's equation (3.2.1). Thus we would expect the complete equation of motion for a viscous fluid to take the form

$$\rho\frac{\mathrm{D}\mathbf{u}}{\mathrm{D}t}=\rho\mathbf{F}-\operatorname{grad}p+\mu\,\nabla^2\mathbf{u}. \tag{5.1.9}$$

This is in fact a correct statement of the *Navier–Stokes* equation of motion for an incompressible Newtonian fluid.

If we now compare the hydrostatic equation (1.4.2), Euler's equation of motion (3.2.1), and the Navier–Stokes equation (5.1.9), it will be seen that they form a logical sequence as follows:

	rate of change of momentum		body force		pressure gradient		viscous transport
hydrostatic equation	0	$=$	$\rho\mathbf{F}$	$-$	$\operatorname{grad}p$		
Euler's equation	$\rho\dfrac{\mathrm{D}\mathbf{u}}{\mathrm{D}t}$	$=$	$\rho\mathbf{F}$	$-$	$\operatorname{grad}p$		
Navier–Stokes equation	$\rho\dfrac{\mathrm{D}\mathbf{u}}{\mathrm{D}t}$	$=$	$\rho\mathbf{F}$	$-$	$\operatorname{grad}p$	$+$	$\mu\,\nabla^2\mathbf{u}$

Bearing in mind the conclusions of § 2.9, the Navier–Stokes equation is seen to be an almost obvious extension of Euler's equation of motion. The additional term $\mu \nabla^2 \mathbf{u}$ on the right-hand side simply represents the viscous diffusion of momentum in the fluid through the agency of molecular transport processes.

Although we have arrived at the correct answer in the form of equation (5.1.9) for the viscous motion of an incompressible fluid, the intuitive derivation which we have outlined above does not constitute a mathematical proof. A rigorous derivation of the Navier–Stokes equation in three dimensions involves the analysis of the nine-component stress system, or stress tensor, and the complete investigation of the rate of strain for a fluid element. The full analysis is outlined for reference purposes in §§ 5.2, 5.3, and appendix 3. Although an understanding of the stress system in a fluid is helpful when we come to investigate the energy equation in more detail in § 5.5, and when we discuss turbulence in chapter 6, a complete mastery of the mathematical analysis in these sections is not necessary in order to follow the argument in the later chapters of this book.

5.2 Stresses in a viscous fluid

Stress is defined as force per unit area. Force is a vector quantity and therefore the stress acting on an area of *specified orientation* must also be a vector quantity having three scalar components. For a complete description of the three-dimensional stress system *acting at a point* in a solid or fluid, however, we need to consider three distinct plane surfaces of different orientation, e.g. the planes which are normal respectively to the x, y, and z axes as shown in figure 40. The stress acting on each of these three planes will be a vector quantity having three scalar components. A description of the complete stress system at a point therefore requires the use of nine scalar components which may be set out in tabular form as follows:

	x plane	y plane	z plane
x direction	p_{xx}	τ_{yx}	τ_{zx}
y direction	τ_{xy}	p_{yy}	τ_{zy}
z direction	τ_{xz}	τ_{yz}	p_{zz}

It will be noted that we use a different symbol to distinguish the normal stress components, p_{xx}, etc., from the tangential stress components, τ_{xy}. etc., and the reason for this will become apparent in a moment. The state of stress described in the above table is similar to that which occurs in an elastic solid. Following the usual convention in the theory of elasticity, the

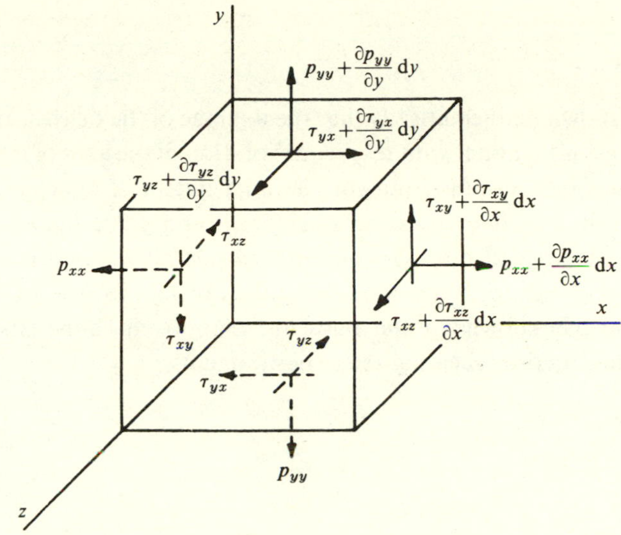

(*Note* that to avoid overcrowding the diagram stresses are omitted from the two faces perpendicular to z-axis.)

Fig. 40. Stresses acting on an element.

normal stress components, p_{xx}, etc., are taken as being positive in sign for tension.

We have already seen that for a fluid at rest the shear stresses are zero, i.e. the stress system is described by a single scalar quantity known as the hydrostatic pressure p. When a fluid ceases to move and relapses into a state of hydrostatic equilibrium, the stress system must accordingly relapse into an isotropic fluid *pressure* which becomes equal to the hydrostatic pressure. In the case of a moving fluid, therefore, it is appropriate to regard the normal stress components p_{xx}, p_{yy}, p_{zz}, as the sum of viscous stress components τ_{xx}, τ_{yy}, τ_{zz}, and an isotropic fluid pressure $-p$, i.e.

$$\left.\begin{aligned} p_{xx}&=\tau_{xx}-p \\ p_{yy}&=\tau_{yy}-p \\ p_{zz}&=\tau_{zz}-p \end{aligned}\right\}. \tag{5.2.1}$$

When the fluid is in motion the fluid pressure p will usually differ from the hydrostatic pressure at the same point in space, but when the fluid ceases to move the viscous stresses vanish and the fluid pressure p becomes equal to the hydrostatic pressure. The fluid pressure in a moving fluid is in fact defined by

$$p = -\tfrac{1}{3}(p_{xx} + p_{yy} + p_{zz}). \tag{5.2.2}$$

It can be shown mathematically that the average of the three normal stress components is invariant with respect to rotation of the axes of reference, so that this definition of fluid pressure is independent of the choice of axes. Note that the negative sign appears in (5.2.1) and (5.2.2) because the fluid pressure p is regarded as being positive for *compression*, whereas the normal stress components, p_{xx}, etc., were defined as being positive for tension.

It can also be shown that the following relationships must exist between the six tangential or shearing-stress components:

$$\left.\begin{aligned} \tau_{xy} &= \tau_{yx} \\ \tau_{yz} &= \tau_{zy} \\ \tau_{zx} &= \tau_{xz} \end{aligned}\right\}. \tag{5.2.3}$$

Thus there are in effect only six independent stress components.

The analysis of stresses in a fluid can be simplified if we use tensor notation. The stress system in a fluid is in fact described by a second-order Cartesian tensor denoted by σ_{ij}. The term σ_{ij} represents the x_j component of the stress vector acting on a plane which is normal to the x_i direction. It will be seen that this notation is in line with the convention adopted in the table of stress components given at the outset, where the first subscript refers to the plane on which the stress is acting while the second subscript specifies the direction of the stress component. Some authors, however, adopt the opposite convention and define the term σ_{ij} as the x_i component of the stress vector acting on a plane which is normal to the x_j direction. It does not actually matter which convention we follow since, in view of (5.2.3), we can say that $\sigma_{ij} = \sigma_{ji}$.

It will be clear from (5.2.1) that the table of stress components may conveniently be divided into two groups as follows:

$$\sigma_{ij} = \underbrace{\begin{bmatrix} -p & 0 & 0 \\ 0 & -p & 0 \\ 0 & 0 & -p \end{bmatrix}}_{\text{fluid pressure}} + \underbrace{\begin{bmatrix} \tau_{xx} & \tau_{yx} & \tau_{zx} \\ \tau_{xy} & \tau_{yy} & \tau_{zy} \\ \tau_{xz} & \tau_{yz} & \tau_{zz} \end{bmatrix}}_{\text{viscous stress components}}. \tag{5.2.4}$$

We can express this more concisely using tensor notation, i.e.

$$\sigma_{ij} = -p\delta_{ij} + \tau_{ij}, \tag{5.2.5}$$

where δ_{ij} is the Kronecker delta having the property that $\delta_{ij}=1$ when $i=j$ and that $\delta_{ij}=0$ when $i \neq j$. The second term on the right-hand side of (5.2.4) or (5.2.5), which represents the viscous stress components, is known as the *deviatoric stress tensor* τ_{ij}.

In order to establish the basic equation of motion we need to know the *resultant force per unit volume* acting on a material element of fluid as a result of the stress system. Referring to figure 40, and making use of (5.2.1), it will be clear that the net force per unit volume due to the fluid pressure distribution (the isotropic part of the stress tensor) is $-\text{grad}\, p$, just as in the case of an inviscid fluid as set out in figure 15 and equation (3.2.1). The resultant force per unit volume due to the viscous stresses (the deviatoric stress tensor) acting over the surface of the element shown in figure 40 may evidently be expressed as a vector with components as follows:

$$x \text{ direction} \quad \mathbf{i}\left(\frac{\partial \tau_{xx}}{\partial x}+\frac{\partial \tau_{yx}}{\partial y}+\frac{\partial \tau_{zx}}{\partial z}\right),$$

$$y \text{ direction} \quad \mathbf{j}\left(\frac{\partial \tau_{xy}}{\partial x}+\frac{\partial \tau_{yy}}{\partial y}+\frac{\partial \tau_{zy}}{\partial z}\right),$$

$$z \text{ direction} \quad \mathbf{k}\left(\frac{\partial \tau_{xz}}{\partial x}+\frac{\partial \tau_{yz}}{\partial y}+\frac{\partial \tau_{zz}}{\partial z}\right).$$

It will now be clear that we can write down the x component of the equation of motion for a viscous fluid as:

$$\rho \frac{\mathrm{D}u}{\mathrm{D}t}=\rho F_x-\frac{\partial p}{\partial x}+\frac{\partial \tau_{xx}}{\partial x}+\frac{\partial \tau_{yx}}{\partial y}+\frac{\partial \tau_{zx}}{\partial z}, \tag{5.2.6}$$

with similar expressions applying for the y and z components.

We can combine the three components of the equation of motion, as represented by (5.2.6), into a single vector equation in the form:

$$\rho \frac{\mathrm{D}\mathbf{u}}{\mathrm{D}t}=\rho \mathbf{F}-\text{grad}\, p+\frac{\partial \boldsymbol{\tau}_x}{\partial x}+\frac{\partial \boldsymbol{\tau}_y}{\partial y}+\frac{\partial \boldsymbol{\tau}_z}{\partial z}, \tag{5.2.7}$$

where the vector quantities $\boldsymbol{\tau}_x$, $\boldsymbol{\tau}_y$, $\boldsymbol{\tau}_z$ represent the viscous stress vectors acting at a point in the fluid on planes which are normal to the x, y, and z directions respectively. It will be seen, for example, that the viscous stress vector $\boldsymbol{\tau}_x$ is defined by

$$\boldsymbol{\tau}_x=\tau_{xx}\mathbf{i}+\tau_{xy}\mathbf{j}+\tau_{xz}\mathbf{k}, \tag{5.2.8}$$

with similar expressions applying for $\boldsymbol{\tau}_y$ and $\boldsymbol{\tau}_z$.

We can express the basic equation of motion (5.2.7) more concisely if we use tensor notation. It then takes the form:

$$\rho \frac{\mathrm{D}u_i}{\mathrm{D}t} = \rho F_i - \frac{\partial p}{\partial x_i} + \frac{\partial \tau_{ij}}{\partial x_i}. \tag{5.2.9}$$

Equation (5.2.7), or its counterpart in tensor notation (5.2.9), represents the most general statement of Newton's second law of motion applied to a material element of fluid. It involves no assumption as to the nature of the viscous stresses or their relationship to the rate-of-strain components.

5.3 Relationship between stress and rate of strain

In order to make use of the basic equation of motion (5.2.7), or (5.2.9), we need to have additional information regarding the nature of the viscous stresses. A comparable situation exists with the analysis of stress in a solid body and the relationship between stress and strain in solids. It is observed experimentally that for some materials, described as elastic solids, the stress is directly proportional to the strain over a limited range of deformation and this is known as Hooke's law. The theory of elasticity is concerned with the analysis of stress and strain in elastic solids within the limited range of deformation over which this linear relationship prevails. We have noted in § 1.5 that in the simple example of uniform parallel shearing motion in a fluid the shearing stress is found experimentally, in the case of ordinary fluids such as air and water, to be proportional to the transverse velocity gradient. It is not unreasonable, therefore, to advance the hypothesis that the stress in a viscous fluid is generally proportional to the *rate of strain*. A fluid where this linear relationship applies between the stress components and the rate-of-strain components is known as a Newtonian fluid.

The relative motion in the neighbourhood of a point in a fluid is analysed briefly in appendix 3 where it is shown that a fluid element may experience expansion (or contraction), shearing, and rotation. The first two of these, i.e. expansion (or contraction) and shearing, may be described as constituting the *rate of strain* of the fluid element and can be represented mathematically by a symmetric second-order Cartesian tensor e_{ij}. It is shown in appendix 3 that

$$e_{ij} = \frac{1}{2}\left(\frac{\partial u_i}{\partial x_j} + \frac{\partial u_j}{\partial x_i}\right). \tag{5.3.1}$$

The components of the rate-of-strain tensor (5.3.1) may be written out in full as follows:

$$e_{xx}=\frac{\partial u}{\partial x},\quad e_{yy}=\frac{\partial v}{\partial y},\quad e_{zz}=\frac{\partial w}{\partial z},$$

$$e_{xy}=e_{yx}=\frac{1}{2}\left(\frac{\partial u}{\partial y}+\frac{\partial v}{\partial x}\right) \tag{5.3.2}$$

$$e_{yz}=e_{zy}=\frac{1}{2}\left(\frac{\partial v}{\partial z}+\frac{\partial w}{\partial y}\right)$$

$$e_{zx}=e_{xz}=\frac{1}{2}\left(\frac{\partial w}{\partial x}+\frac{\partial u}{\partial z}\right)$$

It will be noted that the volumetric rate of strain is given by:

$$\Delta=e_{xx}+e_{yy}+e_{zz}=\frac{\partial u}{\partial x}+\frac{\partial v}{\partial y}+\frac{\partial w}{\partial z}=\text{div}\,\mathbf{u}, \tag{5.3.3}$$

which confirms the comment made in § 2.3 in relation to equation (2.3.7).

Referring again to appendix 3, the linear relationship between the deviatoric stress tensor τ_{ij} and the rate-of-strain tensor e_{ij} can be expressed mathematically as:

$$\tau_{ij}=2\mu(e_{ij}-\tfrac{1}{3}\,\Delta\delta_{ij}), \tag{5.3.4}$$

where δ_{ij} is the Kronecker delta as already noted in § 5.2, and Δ is the volumetric rate of strain or dilatation as given by (5.3.3).

It will be appropriate, for reference purposes, to write out the component parts of the tensor equation (5.3.4) relating stress and rate of strain. For the normal viscous stress components, τ_{xx}, τ_{yy}, τ_{zz}, we have the following results:

$$\tau_{xx}=2\mu\frac{\partial u}{\partial x}-\frac{2}{3}\mu\,\text{div}\,\mathbf{u}$$

$$\tau_{yy}=2\mu\frac{\partial v}{\partial y}-\frac{2}{3}\mu\,\text{div}\,\mathbf{u} \tag{5.3.5}$$

$$\tau_{zz}=2\mu\frac{\partial w}{\partial z}-\frac{2}{3}\mu\,\text{div}\,\mathbf{u}$$

and for the shear-stress components τ_{xy}, etc., we have:

$$\tau_{xy}=\tau_{yx}=\mu\left(\frac{\partial u}{\partial y}+\frac{\partial v}{\partial x}\right)$$

$$\tau_{yz}=\tau_{zy}=\mu\left(\frac{\partial v}{\partial z}+\frac{\partial w}{\partial y}\right) \qquad (5.3.6)$$

$$\tau_{zx}=\tau_{xz}=\mu\left(\frac{\partial w}{\partial x}+\frac{\partial u}{\partial z}\right)$$

It should be noted that equation (5.3.4) is simply mathematical shorthand for equations (5.3.5) and (5.3.6).

5.4 The Navier–Stokes equation

We can now obtain the Navier–Stokes equation of motion for a viscous fluid by substituting in (5.2.6) or (5.2.7) for the stress components as given by (5.3.5) and (5.3.6). The x component of the equation of motion, for example, is obtained from (5.2.6), (5.3.5), and (5.3.6), and after sorting out the individual terms we arrive at the result:

$$\rho\frac{\mathrm{D}u}{\mathrm{D}t}=\rho F_x-\frac{\partial p}{\partial x}+\mu\nabla^2 u+\frac{1}{3}\mu\frac{\partial}{\partial x}(\mathrm{div}\,\mathbf{u}). \qquad (5.4.1)$$

Similar expressions may be written for the y and z components of the equation of motion. Note that we assume μ to be constant.

The three scalar equations, as represented by (5.4.1), may clearly be combined into a single vectorial equation:

$$\rho\frac{\mathrm{D}\mathbf{u}}{\mathrm{D}t}=\rho\mathbf{F}-\mathrm{grad}\,p+\mu\nabla^2\mathbf{u}+\frac{\mu}{3}\,\mathrm{grad}\,\mathrm{div}\,\mathbf{u}. \qquad (5.4.2)$$

This is the Navier–Stokes equation in its complete form for the motion of a viscous Newtonian fluid.

For incompressible flow, with constant density ρ, we have div $\mathbf{u}=0$ from the equation of continuity, and hence in this case the Navier–Stokes equation reduces to

$$\rho\frac{\mathrm{D}\mathbf{u}}{\mathrm{D}t}=\rho\mathbf{F}-\mathrm{grad}\,p+\mu\nabla^2\mathbf{u}, \qquad (5.4.3)$$

which is the form which was anticipated in (5.1.9).

In view of the fact that there is a growing trend towards the use of tensor notation in books and papers dealing with fluid mechanics, it may be helpful for reference purposes to re-state the Navier–Stokes equation for incompressible flow (5.4.3) in the following alternative form:

$$\rho\frac{\partial u_i}{\partial t}+\rho u_j\frac{\partial u_i}{\partial x_j}=\rho F_i-\frac{\partial p}{\partial x_i}+\mu\frac{\partial^2 u_i}{\partial x_j^2}. \tag{5.4.4}$$

The Navier–Stokes equation is valid for both steady and unsteady motion and therefore applies in principle not only to laminar flow conditions but also to turbulent flow, provided we can regard turbulence as an extreme case of unsteady motion. In practice, however, it is not possible to follow the small-scale random movements associated with turbulent motion and the equation cannot therefore be used directly to solve problems in turbulent flow. The consequences of introducing time-average mean velocities together with fluctuating turbulent velocity components into the equation are discussed in chapter 6.

Although exact solutions of the Navier–Stokes equation for laminar flow are possible only in those cases where we have very simple geometric boundary conditions, the equation itself is important for a number of reasons. It enables us to investigate the energy equation in detail, as in § 5.5 and in chapter 9. It provides the basis for a rational discussion of dimensional analysis and dynamical similarity in § 5.8. It helps in the investigation of turbulent flow in chapter 6 and provides the basis for a detailed study of boundary-layer flow in chapter 7.

5.5 The mechanical energy equation for a fluid with viscous stresses

Having obtained the basic equation of motion in its most general form (5.2.7), it will be of interest to derive the corresponding equation for the rate of change of the *kinetic energy of a fluid element* following the motion of the fluid. This can be done by forming the scalar product of each side of equation (5.2.7) with the velocity vector $\mathbf{u}$. Noting that $\mathbf{u}\cdot(\mathrm{D}\mathbf{u}/\mathrm{D}t)=(\mathrm{D}/\mathrm{D}t)\frac{1}{2}\mathbf{u}^2$, we have the result:

$$\rho\frac{\mathrm{D}}{\mathrm{D}t}\left(\frac{1}{2}\mathbf{u}^2\right)=\rho\mathbf{u}\cdot\mathbf{F}-\mathbf{u}\cdot\nabla p+\mathbf{u}\cdot\left(\frac{\partial\boldsymbol{\tau}_x}{\partial x}+\frac{\partial\boldsymbol{\tau}_y}{\partial y}+\frac{\partial\boldsymbol{\tau}_z}{\partial z}\right). \tag{5.5.1}$$

The same result could be obtained by multiplying each of the three component equations, represented by (5.2.6), by the velocity components u, v, and w respectively, and adding the three resulting equations together.

If the body force $\mathbf{F}$ can be expressed as the gradient of a scalar potential in the form $\mathbf{F} = -\text{grad}\,\Psi$, as in the case of the force due to gravity where $\Psi = gz$, and *if the flow is steady*, we can say that

$$\rho\mathbf{u}\cdot\mathbf{F} = -\rho\mathbf{u}\cdot\nabla\Psi = -\rho\frac{\mathrm{D}\Psi}{\mathrm{D}t}. \tag{5.5.2}$$

We can then re-write equation (5.5.1) for steady flow as an equation which expresses the rate of change of the *kinetic plus potential energy* of a material fluid element, i.e.

$$\rho\frac{\mathrm{D}}{\mathrm{D}t}\left(\frac{1}{2}\mathbf{u}^2 + \Psi\right) = -\mathbf{u}\cdot\nabla p + \mathbf{u}\cdot\left(\frac{\partial\boldsymbol{\tau}_x}{\partial x} + \frac{\partial\boldsymbol{\tau}_y}{\partial y} + \frac{\partial\boldsymbol{\tau}_z}{\partial z}\right). \tag{5.5.3}$$

It is obvious that the two terms on the right-hand side of this equation represent the contributions to the rate of change of energy of the fluid arising from the normal pressure distribution and the configuration of the viscous stresses respectively. It is worthwhile investigating these two terms in greater detail, however, in order to establish their exact physical meaning. We will take the term involving the pressure gradient ∇p first of all. We have already introduced the concept of *flow work* in § 3.4. If we consider a material volume of fluid $\mathscr{V}$ contained within a moving boundary surface $\mathscr{S}$, the rate at which flow work is being done at the boundary *on* the fluid within the material volume is given by $-\int_{\mathscr{S}} p\mathbf{u}\cdot\mathbf{n}\,\mathrm{d}\mathscr{S}$. The negative sign arises because the pressure p is taken as positive for compression. It will be seen that this definition of the flow work is consistent with the simple expression (3.4.2) applied to steady flow in a stream tube. Using the divergence theorem we can equate the surface integral of the vector quantity $(p\mathbf{u})$ to the volume integral of its divergence. Thus we can say that:

rate of flow work done on the fluid within volume $\mathscr{V}$

$$= -\int_{\mathscr{S}} p\mathbf{u}\cdot\mathbf{n}\,\mathrm{d}\mathscr{S} = -\int_{\mathscr{V}} \text{div}\,(p\mathbf{u})\,\mathrm{d}\mathscr{V}. \tag{5.5.4}$$

Hence for a small material element of fluid we can say that:

$$\text{rate of flow work per unit volume} = -\text{div}\,(p\mathbf{u}). \tag{5.5.5}$$

We can expand the term div $(p\mathbf{u})$ using the mathematical identity:

$$\text{div}\,(p\mathbf{u}) = \mathbf{u}\cdot\text{grad}\,p + p\,\text{div}\,\mathbf{u}. \tag{5.5.6}$$

Hence we can write:

$$-\mathbf{u}\cdot\nabla p = -\text{div}\,(p\mathbf{u}) + p\,\text{div}\,\mathbf{u}. \tag{5.5.7}$$

We have seen in §§ 2.3 and 5.3 that div **u** represents the volumetric rate of strain or dilatation of a fluid element. It follows that p div **u** represents the rate of work done *by* a fluid element in expanding against the surrounding pressure, and that $-p$ div **u** must therefore be the rate of work done *on* a fluid element by the surrounding fluid. Note that div **u** is negative if the volume of the material element is decreasing, and that $-p$ div **u** will then be positive, which is consistent with the fact that work is being done *on* the fluid element when it is being compressed. It will now be apparent that equation (5.5.7) may be expressed in words by the statement that:

$-\mathbf{u}\cdot\nabla p =$ rate of flow work done on fluid

$\qquad$ − rate of compression work done on fluid.

We can apply a similar method of analysis to the other term on the right-hand side of equation (5.5.3) involving the shear stresses. We start by introducing the concept of *shear work*. This is the rate at which work is being done on a material volume of fluid through the agency of the shear stresses, as given by the deviatoric stress tensor τ_{ij}, acting over the boundary surface. The shear work is evaluated over the boundary of the material volume or element and it is evidently analogous to the flow work associated with the normal pressure or isotropic part of the stress tensor. If we consider a small material element of fluid, we can define the rate of shear work W_τ being done on the element per unit volume, by analogy with (5.5.5), as:

$$W_\tau = \frac{\partial}{\partial x}(\mathbf{u}\cdot\boldsymbol{\tau}_x) + \frac{\partial}{\partial y}(\mathbf{u}\cdot\boldsymbol{\tau}_y) + \frac{\partial}{\partial z}(\mathbf{u}\cdot\boldsymbol{\tau}_z). \tag{5.5.8}$$

We can expand this expression into two parts, noting that

$$\frac{\partial}{\partial x}(\mathbf{u}\cdot\boldsymbol{\tau}_x) = \mathbf{u}\cdot\frac{\partial\boldsymbol{\tau}_x}{\partial x} + \boldsymbol{\tau}_x\cdot\frac{\partial\mathbf{u}}{\partial x}, \quad \text{etc.}$$

Thus we can say that

$$W_\tau = \mathbf{u}\cdot\left(\frac{\partial\boldsymbol{\tau}_x}{\partial x} + \frac{\partial\boldsymbol{\tau}_y}{\partial y} + \frac{\partial\boldsymbol{\tau}_z}{\partial z}\right) + \left(\boldsymbol{\tau}_x\cdot\frac{\partial\mathbf{u}}{\partial x} + \boldsymbol{\tau}_y\cdot\frac{\partial\mathbf{u}}{\partial y} + \boldsymbol{\tau}_z\cdot\frac{\partial\mathbf{u}}{\partial z}\right). \tag{5.5.9}$$

The second group of terms on the right-hand side of (5.5.9) represents the rate at which work is being done by the viscous stresses in *distorting* the fluid element. It will be seen that only velocity gradients, and not the absolute fluid velocity **u**, enter into this group. It is convenient to label this group of terms as the *dissipation function* Φ, since it measures the rate at which mechanical energy is being dissipated through the action of viscosity.

We will see in chapter 9, however, that this irreversible dissipation of mechanical energy re-appears as 'friction heat' or internal energy in the complete energy equation, and does not therefore represent any loss of *total* energy.

We may thus conclude from (5.5.9) that:

$$\mathbf{u}\cdot\left(\frac{\partial \boldsymbol{\tau}_x}{\partial x}+\frac{\partial \boldsymbol{\tau}_y}{\partial y}+\frac{\partial \boldsymbol{\tau}_z}{\partial z}\right)=W_\tau-\Phi, \tag{5.5.10}$$

where W_τ is defined by (5.5.8) and Φ is defined by:

$$\Phi=\boldsymbol{\tau}_x\cdot\frac{\partial \mathbf{u}}{\partial x}+\boldsymbol{\tau}_y\cdot\frac{\partial \mathbf{u}}{\partial y}+\boldsymbol{\tau}_z\cdot\frac{\partial \mathbf{u}}{\partial z}. \tag{5.5.11}$$

It will be seen that (5.5.10) is the counterpart to (5.5.7).

We can now re-state the mechanical energy equation for steady flow (5.5.3) in the more meaningful form:

$$\rho\frac{\mathrm{D}}{\mathrm{D}t}\left(\frac{1}{2}\mathbf{u}^2+\Psi\right)=-\operatorname{div}(p\mathbf{u})+p\operatorname{div}\mathbf{u}+W_\tau-\Phi. \tag{5.5.12}$$

gain of kinetic plus potential energy

= flow work − compression work + shear work − distortion work.

We will make further use of this result in chapter 9 when we derive the complete energy equation taking account of the first law of thermodynamics. It is important to keep in mind that each term in equation (5.5.3) or (5.5.12) is a *rate* of work, or rate of change of energy, *per unit volume* of fluid.

We may note in passing that, for the special case of *constant density*, i.e. when $\operatorname{div}\mathbf{u}=0$, the rate of flow work under steady flow conditions may be written as:

$$-\operatorname{div}(p\mathbf{u})=-\mathbf{u}\cdot\nabla p=-\frac{\mathrm{D}p}{\mathrm{D}t}$$

and we may then transfer the flow work term from the right-hand to the left-hand side of equation (5.5.12) with the result that:

$$\rho\frac{\mathrm{D}}{\mathrm{D}t}\left(\frac{p}{\rho}+\frac{1}{2}\mathbf{u}^2+\Psi\right)=W_\tau-\Phi. \tag{5.5.13}$$

If we now put $\Psi=gz$ to describe the body force due to gravity, it will be clear that equation (5.5.13) is simply an extension of the differential form of Bernoulli's equation (3.4.14). We will see in a moment that, although the rate of shear work W_τ may be either positive or negative, the work of distortion Φ is always positive in value and this causes a steady fall in the Bernoulli energy.

Having obtained the general form of the steady-flow mechanical energy equation in the shape of (5.5.12), it will be appropriate to evaluate the terms associated with the viscous stresses which describe the rate of shear work W_τ, and the rate of distortion work Φ, for the normal case of a Newtonian fluid. Taking the latter function first, we can substitute in (5.5.11) for the vector stress $\boldsymbol{\tau}_x$ from (5.2.8), and for the individual stress components τ_{xx}, τ_{xy}, etc., from (5.3.5) and (5.3.6). Similar substitutions can be made for the other terms $\boldsymbol{\tau}_y$ and $\boldsymbol{\tau}_z$ in equation (5.5.11). After some sorting out of the individual terms, we arrive at the result that:

$$\Phi=\mu\left\{2\left(\frac{\partial u}{\partial x}\right)^2+2\left(\frac{\partial v}{\partial y}\right)^2+2\left(\frac{\partial w}{\partial z}\right)^2+\left(\frac{\partial u}{\partial y}+\frac{\partial v}{\partial x}\right)^2+\left(\frac{\partial v}{\partial z}+\frac{\partial w}{\partial y}\right)^2\right.$$
$$\left.+\left(\frac{\partial w}{\partial x}+\frac{\partial u}{\partial z}\right)^2-\tfrac{2}{3}(\operatorname{div}\mathbf{u})^2\right\}. \quad (5.5.14)$$

We can express (5.5.14) alternatively in terms of the rate-of-strain components as defined by (5.3.2), i.e.

$$\Phi=2\mu\{e_{xx}^2+e_{yy}^2+e_{zz}^2+2e_{xy}^2+2e_{yz}^2+2e_{zx}^2-\tfrac{1}{3}(\operatorname{div}\mathbf{u})^2\}. \quad (5.5.15)$$

or, using tensor notation, we can say that

$$\Phi=2\mu[(e_{ij})^2-\tfrac{1}{3}\Delta^2]. \quad (5.5.16)$$

For an incompressible fluid we have $\operatorname{div}\mathbf{u}=0$ and therefore the last term in (5.5.14), (5.5.15), and (5.5.16) involving Δ^2 or $(\operatorname{div}\mathbf{u})^2$ may be omitted. It will be seen that all the terms in the expression for Φ are then squared terms involving the rate-of-strain components. Thus the numerical value of Φ must always be positive in these circumstances and it will be evident from equation (5.5.12) that the effect of the distortion work is to cause a decrease in the energy of the fluid. The dissipation function for an *incompressible fluid* may be expressed concisely in the form:

$$\Phi=2\mu(e_{ij})^2=\frac{\mu}{2}\left(\frac{\partial u_i}{\partial x_j}+\frac{\partial u_j}{\partial x_i}\right)^2. \quad (5.5.17)$$

We can also obtain an expression for the rate of shear work W_τ by substituting for $\boldsymbol{\tau}_x$, etc., in (5.5.8) from (5.2.8) and making use of the results given by (5.3.5) and (5.3.6) for the deviatoric stress components in the case of a Newtonian fluid. The result for an *incompressible fluid* is that:

$$W_\tau = \mu \operatorname{div}(\mathbf{u}\cdot\nabla\mathbf{u}) + \mu\,\nabla^2(\tfrac{1}{2}\mathbf{u}^2). \tag{5.5.18}$$

The second term on the right-hand side may be interpreted as the diffusion of kinetic energy through the action of viscosity.

We could equally well obtain the mechanical energy equation for an incompressible Newtonian fluid by starting with the Navier–Stokes equation in the form (5.4.3) and taking the scalar product of each term with the velocity vector $\mathbf{u}$. In place of (5.5.1) we then have:

$$\rho\frac{\mathrm{D}}{\mathrm{D}t}\left(\frac{1}{2}\mathbf{u}^2\right) = \rho\mathbf{u}\cdot\mathbf{F} - \mathbf{u}\cdot\nabla p + \mu\mathbf{u}\cdot\nabla^2\mathbf{u}. \tag{5.5.19}$$

It will be seen that the last term on the right-hand side of this equation must be equal to $W_\tau - \Phi$ as already defined and as given by (5.5.17) and (5.5.18). The term may also be expressed in various other mathematical forms. It can easily be proved, for example, that, provided $\operatorname{div}\mathbf{u}=0$:

$$\mu\mathbf{u}\cdot\nabla^2\mathbf{u} = \mu\,\nabla^2(\tfrac{1}{2}\mathbf{u}^2) - \mu\left(\frac{\partial u_i}{\partial x_j}\right)^2, \tag{5.5.20}$$

or alternatively

$$\mu\mathbf{u}\cdot\nabla^2\mathbf{u} = \mu \operatorname{div}(\mathbf{u}\times\boldsymbol{\omega}) - \mu\omega^2, \tag{5.5.21}$$

where $\boldsymbol{\omega}$ is the vorticity vector.

It will be noted that there is a certain similarity between the quantities Φ, $\mu(\partial u_i/\partial x_j)^2$, and $\mu\omega^2$. Each comprises a set of terms which are squares or products of velocity gradients. This point will come up again in § 6.4 when we investigate the mechanical energy equation for turbulent flow.

5.6 The vorticity equation for a viscous fluid

The Navier–Stokes equation for an incompressible viscous fluid takes the form of equation (5.4.3), and dividing through by the density ρ we have:

$$\frac{\mathrm{D}\mathbf{u}}{\mathrm{D}t} = \mathbf{F} - \nabla\left(\frac{p}{\rho}\right) + \nu\,\nabla^2\mathbf{u}. \tag{5.6.1}$$

If the body force $\mathbf{F}$ is expressed as the gradient of a scalar potential Ψ, i.e. if

$F = -\text{grad}\,\Psi$, equation (5.6.1) becomes

$$\frac{D\mathbf{u}}{Dt} = -\nabla\Psi - \nabla\left(\frac{p}{\rho}\right) + \nu\,\nabla^2\mathbf{u}. \tag{5.6.2}$$

We can now make use of the mathematical identity already quoted in (4.2.4) that

$$(\mathbf{u}\cdot\nabla)\mathbf{u} = \tfrac{1}{2}\nabla(\mathbf{u}^2) - \mathbf{u}\times\boldsymbol{\omega}, \tag{5.6.3}$$

where $\boldsymbol{\omega}$ is the vorticity vector defined by $\boldsymbol{\omega} = \text{curl}\,\mathbf{u}$. Substituting from (5.6.3) for the convective part of the total differential term $D\mathbf{u}/Dt$ in (5.6.2), we have the result that

$$\frac{\partial\mathbf{u}}{\partial t} = \mathbf{u}\times\boldsymbol{\omega} - \nabla\left(\frac{p}{\rho} + \frac{1}{2}\mathbf{u}^2 + \Psi\right) + \nu\,\nabla^2\mathbf{u}. \tag{5.6.4}$$

It will be noted that, up to this point, we have carried out exactly the same mathematical manoeuvre with the Navier–Stokes equation that we made with Euler's equation in § 4.2. Equation (5.6.4) is the viscous flow equivalent of equation (4.2.5) for inviscid flow.

If we now take the curl of both sides of equation (5.6.4) we have the vorticity equation for a viscous fluid:

$$\frac{\partial\boldsymbol{\omega}}{\partial t} = \nabla\times(\mathbf{u}\times\boldsymbol{\omega}) + \nu\,\nabla^2\boldsymbol{\omega}. \tag{5.6.5}$$

It will be noted that the gradient terms on the right-hand side of equation (5.6.4) have disappeared since curl grad ϕ is always zero. We can further clarify the vorticity equation (5.6.5) by making use of the following mathematical identity, as quoted in appendix 1,

$$\text{curl}\,(\mathbf{a}\times\mathbf{b}) = (\mathbf{b}\cdot\nabla)\mathbf{a} - (\mathbf{a}\cdot\nabla)\mathbf{b} + \mathbf{a}\,\text{div}\,\mathbf{b} - \mathbf{b}\,\text{div}\,\mathbf{a}.$$

Hence we have the result that

$$\text{curl}\,(\mathbf{u}\times\boldsymbol{\omega}) = (\boldsymbol{\omega}\cdot\nabla)\mathbf{u} - (\mathbf{u}\cdot\nabla)\boldsymbol{\omega}, \tag{5.6.6}$$

since div $\mathbf{u} = 0$ for an incompressible fluid, and div $\boldsymbol{\omega} = 0$ as already noted in (3.8.19).

Substituting from (5.6.6) in equation (5.6.5) gives the result:

$$\frac{\partial\boldsymbol{\omega}}{\partial t} = (\boldsymbol{\omega}\cdot\nabla)\mathbf{u} - (\mathbf{u}\cdot\nabla)\boldsymbol{\omega} + \nu\,\nabla^2\boldsymbol{\omega}, \tag{5.6.7}$$

which may equally well be written in the form:

$$\frac{\mathrm{D}\boldsymbol{\omega}}{\mathrm{D}t}=(\boldsymbol{\omega}\cdot\nabla)\mathbf{u}+\nu\,\nabla^2\boldsymbol{\omega}. \tag{5.6.8}$$

Equation (5.6.8) expresses the fact that the rate of change of the vorticity $\mathrm{D}\boldsymbol{\omega}/\mathrm{D}t$, following the motion of the fluid, is equal to the sum of the two terms $(\boldsymbol{\omega}\cdot\nabla)\mathbf{u}$ and $\nu\,\nabla^2\boldsymbol{\omega}$ which appear on the right-hand side. The first term represents interaction between vorticity components and velocity gradients, while the second term represents the molecular or viscous diffusion of vorticity within the fluid.

In the special case of *two-dimensional flow*, the vorticity vector $\boldsymbol{\omega}$ will always be normal to the plane of the flow and consequently the interaction term $(\boldsymbol{\omega}\cdot\nabla)\mathbf{u}$ will be zero. Thus for a two-dimensional flow system, where for example there is no velocity component w in the z direction at any point in the fluid, equation (5.6.8) reduces to

$$\frac{\mathrm{D}\boldsymbol{\omega}}{\mathrm{D}t}=\nu\,\nabla^2\boldsymbol{\omega}. \tag{5.6.9}$$

This will be recognised immediately as the standard form of the *diffusion equation* for a moving fluid. Since (5.6.9) applies only to two-dimensional flow, we have only one component of vorticity and the vector $\boldsymbol{\omega}$ therefore reduces to a single scalar component. If, for example, the flow pattern is described in terms of the co-ordinates x and y, so that the fluid velocity at any point is normal to the z axis, we will only have the vorticity component ω_z to consider and equation (5.6.9) then takes the form:

$$\frac{\mathrm{D}\omega_z}{\mathrm{D}t}=\nu\,\nabla^2\omega_z, \tag{5.6.10}$$

which is the equation for the diffusion of vorticity in a two-dimensional boundary layer.

Reverting to the general three-dimensional case represented by equation (5.6.8), it will be of interest to investigate more closely the physical meaning of the interaction term $(\boldsymbol{\omega}\cdot\nabla)\mathbf{u}$ which plays an important role in the development of turbulence. Referring to figure 41, we can consider the movement of a short element of a vortex line PQ in time $\mathrm{d}t$.

It will be recalled from § 3.8 that a vortex line coincides at every point with the local axis of rotation of the fluid, and that in the case of inviscid flow vortex lines move with the fluid. Points P and Q in figure 41 are located on a vortex line at a distance $\mathrm{d}l$ apart. The line element may be represented

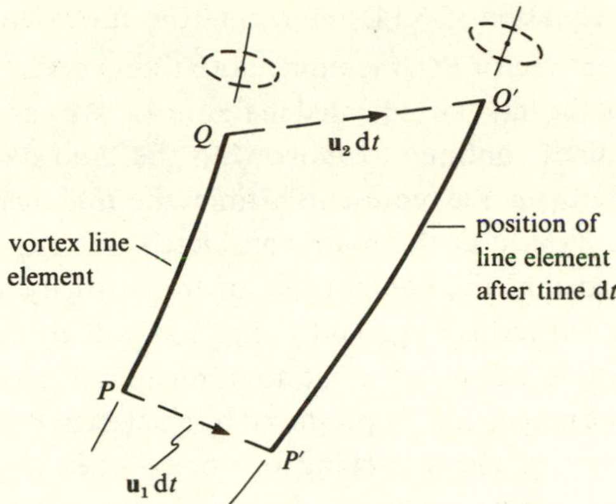

Fig. 41. Movement of a short element of a vortex line.

by the vector $\mathrm{d}\mathbf{l}$. This has magnitude $\mathrm{d}l$ and a direction coinciding with that of the vorticity vector $\boldsymbol{\omega}$ whose magnitude is ω. Thus

$$\mathrm{d}\mathbf{l}=\frac{\mathrm{d}l}{\omega}\boldsymbol{\omega}. \tag{5.6.11}$$

We will assume that an element of fluid located at point P on the vortex line has a velocity $\mathbf{u}_1$ at the instant considered, and that an element of fluid at point Q has velocity $\mathbf{u}_2$. After a short interval of time $\mathrm{d}t$ the fluid element initially at P will have moved to a point P' in the diagram, where $\overline{PP'}=\mathbf{u}_1\,\mathrm{d}t$. Similarly an element initially at Q will have moved to point Q' in the same interval of time, where $\overline{QQ'}=\mathbf{u}_2\,\mathrm{d}t$. By the usual rules of vector analysis we can say that

$$\mathbf{u}_2-\mathbf{u}_1=(\mathrm{d}\mathbf{l}\cdot\nabla)\mathbf{u}. \tag{5.6.12}$$

Thus the rate of change of the vortex line element vector $\mathrm{d}\mathbf{l}$, following the motion of the fluid, must be given by

$$\frac{\mathrm{D}}{\mathrm{D}t}(\mathrm{d}\mathbf{l})=\mathbf{u}_2-\mathbf{u}_1=(\mathrm{d}\mathbf{l}\cdot\nabla)\mathbf{u}. \tag{5.6.13}$$

Substituting for $\mathrm{d}\mathbf{l}$ on the right-hand side of (5.6.13) from (5.6.11) gives

$$\frac{\mathrm{D}}{\mathrm{D}t}(\mathrm{d}\mathbf{l})=\frac{\mathrm{d}l}{\omega}(\boldsymbol{\omega}\cdot\nabla)\mathbf{u}$$

or

$$\frac{1}{\mathrm{d}l}\frac{\mathrm{D}}{\mathrm{D}t}(\mathrm{d}\mathbf{l})=\frac{1}{\omega}(\boldsymbol{\omega}\cdot\nabla)\mathbf{u}. \tag{5.6.14}$$

The left-hand side of equation (5.6.14) represents the fractional rate of change of the line element vector $\overline{PQ}$, including both tilting and stretching, following the motion of the fluid as indicated in figure 41. We can thus say that, provided vortex lines continue to move with the fluid despite the effects of viscosity, the term $(\boldsymbol{\omega}\cdot\nabla)\mathbf{u}$ represents ω times the *fractional rate of change of a vortex line element* at the point considered.

It will be appreciated that the components of the vorticity vector $\boldsymbol{\omega}$ consist of velocity gradient terms as given by (3.8.2). It will also be noted from § 3.8 that a vortex line induces an associated solenoidal velocity field which is analogous to the magnetic field produced by an electric current in a conductor. Thus the tilting and stretching of vortex lines in a three-dimensional flow system, as represented by the term $\boldsymbol{\omega}\cdot\nabla\mathbf{u}$ in the vorticity equation, will result in a continuously changing pattern of velocity gradients and velocity components in the fluid. As we shall see in chapter 6, the vorticity interaction term represents the main process by which infinitesimal instabilities or small extraneous disturbances occurring in a laminar flow pattern are transformed into three-dimensional turbulence.

It is also of interest to consider the energy balance in rotational flow. A small spherical element of fluid has an angular velocity equal to $\frac{1}{2}\boldsymbol{\omega}$ as noted in § 3.8. Thus a small element of fluid would have kinetic energy of rotation equal to $\frac{1}{8}I\boldsymbol{\omega}^2$ if it behaved as a solid sphere, where I is the moment of inertia for a sphere rotating about an axis through its centre. Although it would be misleading to press the analogy with solid body rotation too far, it is fair to say that the rotational kinetic energy per unit volume of fluid must be related to the quantity $\rho\boldsymbol{\omega}^2$.

The concept of a small *vortex tube*, formed from a group of vortex lines drawn through a closed curve, was introduced in § 3.8 and it was noted that the strength of the vortex tube, defined as the product of the vorticity and the cross-sectional area, remains constant in accordance with Kelvin's theorem for an inviscid fluid. Stretching or extension of the vortex lines forming the wall of the tube will cause a reduction in the cross-sectional area and will thus intensify the vorticity $\boldsymbol{\omega}$ and will increase the energy of rotation contained within the vortex tube. Although the dynamics of the problem becomes slightly more complicated in the case of a viscous fluid owing to the presence of the viscous diffusion term $\nu\,\nabla^2\boldsymbol{\omega}$ in equation (5.6.5), it is still true to say that the stretching of vortex lines causes an intensification of the vorticity and results in an increase of the rotational energy. Conversely, contraction of vortex lines leads to a decrease in the rotational energy.

In § 5.5 we formed an equation for the kinetic energy $\frac{1}{2}\mathbf{u}^2$ per unit mass of fluid by taking the scalar product of the velocity vector **u** with the equation of motion. We can form a similar equation for the quantity $\frac{1}{2}\boldsymbol{\omega}^2$, which is related to the energy of rotation, by taking the scalar product of the vorticity vector $\boldsymbol{\omega}$ with the vorticity equation (5.6.8). The result is simply:

$$\boldsymbol{\omega}\cdot\frac{\mathrm{D}\boldsymbol{\omega}}{\mathrm{D}t}=\frac{\mathrm{D}}{\mathrm{D}t}\left(\frac{1}{2}\boldsymbol{\omega}^2\right)=\boldsymbol{\omega}\cdot(\boldsymbol{\omega}\cdot\nabla)\mathbf{u}+\nu\boldsymbol{\omega}\cdot\nabla^2\boldsymbol{\omega}. \tag{5.6.15}$$

The last term on the right-hand side may be expanded, using exactly the same mathematical procedure as in (5.5.20) for the corresponding term $\mathbf{u}\cdot\nabla^2\mathbf{u}$, so that we can say

$$\nu\boldsymbol{\omega}\cdot\nabla^2\boldsymbol{\omega}=\nu\,\nabla^2(\tfrac{1}{2}\boldsymbol{\omega}^2)-\nu\left(\frac{\partial\omega_i}{\partial x_j}\right)^2. \tag{5.6.16}$$

We may thus re-write equation (5.6.15) for the rate of change of the quantity $\frac{1}{2}\boldsymbol{\omega}^2$, following the motion of the fluid, as:

$$\frac{\mathrm{D}}{\mathrm{D}t}\left(\frac{1}{2}\boldsymbol{\omega}^2\right)=\boldsymbol{\omega}\cdot(\boldsymbol{\omega}\cdot\nabla)\mathbf{u}-\nu\left(\frac{\partial\omega_i}{\partial x_j}\right)^2+\nu\,\nabla^2(\tfrac{1}{2}\boldsymbol{\omega}^2). \tag{5.6.17}$$

The first term on the right-hand side represents the rate of change of $\frac{1}{2}\boldsymbol{\omega}^2$ caused by the distortion of vortex lines. It will be positive if the net effect on the interaction term $\boldsymbol{\omega}\cdot\nabla\mathbf{u}$ in equation (5.6.8) is the stretching or extension of vortex lines. The second term on the right-hand side of (5.6.17) represents the viscous dissipation of rotational energy and is analogous to the corresponding term which appeared in the energy equation in § 5.5. It will be seen that this term always causes a reduction in the rotational energy since $(\partial\omega_i/\partial x_j)^2$ is always positive. The last term on the right-hand side of (5.6.17) represents the viscous diffusion of the quantity $\frac{1}{2}\boldsymbol{\omega}^2$.

The essential fact to emerge from equation (5.6.17) is that vortex energy increases in those regions where the rate of extension of the fluid in the direction of the vorticity vector $\boldsymbol{\omega}$ is positive, i.e. in those regions where vortex lines are being stretched, but that the effect of viscosity is to decrease or diffuse the energy. Under steady-state conditions there must be a balance, taken over a material volume of fluid, between the generation of the quantity $\frac{1}{2}\boldsymbol{\omega}^2$ by vortex line extension and the reduction or diffusion of the same quantity through the action of viscosity.

5.7 Dynamical similarity

The Navier–Stokes equation enables us to investigate more precisely the conditions under which two or more instances of flow with

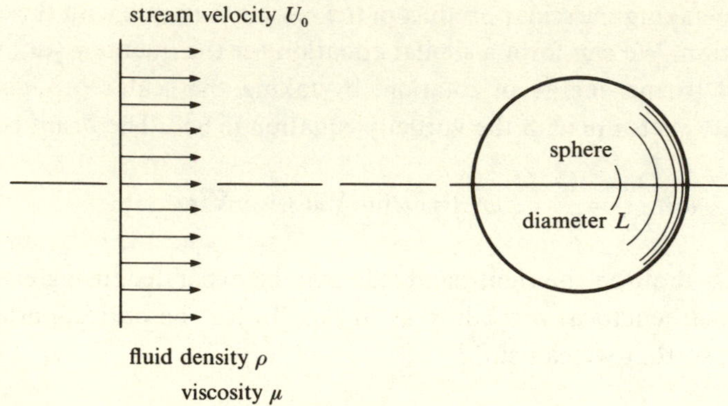

Fig. 42. Solid sphere in a flowing stream.

geometrically similar boundaries, but with different linear scales and different stream velocities and fluid properties, may have similar flow patterns. The method is applicable in principle to any flow system, but it will help to clarify the argument if we picture a specific example such as the three-dimensional case of a fluid with undisturbed stream velocity U_0 flowing past a solid sphere of diameter L as indicated in figure 42.

We will confine ourselves to the incompressible flow of a Newtonian fluid and we can therefore use the Navier–Stokes equation in the form (5.4.3). If we take the x component of this equation, and divide through by ρ, we have:

$$\frac{\partial u}{\partial t}+u\frac{\partial u}{\partial x}+v\frac{\partial u}{\partial y}+w\frac{\partial u}{\partial z}=F_x-\frac{1}{\rho}\frac{\partial p}{\partial x}+\nu\left(\frac{\partial^2 u}{\partial x^2}+\frac{\partial^2 u}{\partial y^2}+\frac{\partial^2 u}{\partial z^2}\right). \quad (5.7.1)$$

The first step is to re-state this equation in dimensionless form and for this purpose we define the diameter of the sphere L as a reference length and the undisturbed stream velocity U_0 as a reference velocity. We can then introduce dimensionless ratios for the velocity components, the space co-ordinates, time, and the fluid pressure. These are defined as follows:

$$\begin{aligned} &u'=\frac{u}{U_0} \quad v'=\frac{v}{U_0} \quad w'=\frac{w}{U_0} \\ &x'=\frac{x}{L} \quad y'=\frac{y}{L} \quad z'=\frac{z}{L}\,. \\ &t'=\frac{U_0 t}{L} \quad p'=\frac{p}{\rho U_0^2} \end{aligned} \quad (5.7.2)$$

We will assume that the body force **F** in equation (5.4.3) arises only as a result of terrestrial gravity. If the flow is entirely submerged, with the fluid stream totally enclosed by solid boundaries, we can omit the effect of gravity and measure the pressure at any point relative to the value of the hydrostatic pressure at that point, i.e. we can work in terms of gauge pressure. If, on the other hand, there is a free surface in the flow system (for example with flow past a partially submerged sphere, or past a sphere which is held at a level just below the surface of the fluid) we have to retain the term **F** in the equation of motion and the components of the gravitational body force will be:

$$F_x=0, \quad F_y=0, \quad F_z=-g. \tag{5.7.3}$$

If we now substitute in (5.7.1) from (5.7.2) we get:

$$\frac{U_0^2}{L}\frac{\partial u'}{\partial t'}+\frac{U_0^2}{L}\left(u'\frac{\partial u'}{\partial x'}+v'\frac{\partial u'}{\partial y'}+w'\frac{\partial u'}{\partial z'}\right)$$
$$=-\frac{1}{\rho}\rho\frac{U_0^2}{L}\frac{\partial p'}{\partial x'}+\frac{\nu U_0}{L^2}\left(\frac{\partial^2 u'}{\partial x'^2}+\frac{\partial^2 u'}{\partial y'^2}+\frac{\partial^2 u'}{\partial z'^2}\right).$$

Dividing through by U_0^2/L, this may be written as

$$\frac{\mathrm{D}u'}{\mathrm{D}t'}=-\frac{\partial p'}{\partial x'}+\frac{\nu}{U_0L}\left(\frac{\partial^2 u'}{\partial x'^2}+\frac{\partial^2 u'}{\partial y'^2}+\frac{\partial^2 u'}{\partial z'^2}\right). \tag{5.7.4}$$

The only term in this dimensionless equation which involves the *magnitude* of the linear scale L, the stream velocity U_0, and the fluid properties μ and ρ, is the dimensionless group on the right-hand side which we immediately recognise as the reciprocal of the Reynolds number, i.e.

$$\frac{\nu}{U_0L}=\frac{\mu}{\rho U_0L}=\frac{1}{Re}. \tag{5.7.5}$$

For given boundary conditions (determined by the geometry of the problem and the requirement of zero relative velocity at any solid surface), the solution of equation (5.7.4), and consequently also the full detail of the flow pattern, will depend only on the numerical value of the *Reynolds number*, *Re*.

The individual stress components, such as τ_{xx}, τ_{xy}, etc., which appear in the stress tensor may also be expressed as dimensionless stress coefficients using the substitutions of (5.7.2). For example

$$\frac{\tau_{xy}}{\rho U_0^2}=\frac{1}{Re}\left(\frac{\partial u'}{\partial y'}+\frac{\partial v'}{\partial x'}\right). \tag{5.7.6}$$

We may thus conclude that, in the case of steady submerged flow, the velocity ratios u/U_0, etc., the local pressure coefficient $p/\rho U_0^2$, and the individual viscous stress-component coefficients are functions of x/L, y/L, z/L, and Re only. In other words, at *corresponding points* in geometrically similar flow systems, the velocity ratios and the stress-component coefficients are functions of the Reynolds number only.

In the case of the flow of a liquid with a *free surface*, we have to take into account the action of gravity and introduce the additional term $F_z = -g$ into the z component of the equation of motion (assuming that the z axis is vertical) which thus takes the form:

$$\frac{\mathrm{D}w}{\mathrm{D}t} = -g - \frac{1}{\rho}\frac{\partial p}{\partial z} + \nu\left(\frac{\partial^2 w}{\partial x^2} + \frac{\partial^2 w}{\partial y^2} + \frac{\partial^2 w}{\partial z^2}\right). \tag{5.7.7}$$

If we substitute as before from (5.7.2) and divide through by U_0^2/L we get:

$$\frac{\mathrm{D}w'}{\mathrm{D}t'} = -\frac{gL}{U_0^2} - \frac{\partial p'}{\partial x'} + \frac{\nu}{U_0 L}\left(\frac{\partial^2 w'}{\partial x'^2} + \frac{\partial^2 w'}{\partial y'^2} + \frac{\partial^2 w'}{\partial z'^2}\right). \tag{5.7.8}$$

which is the dimensionless form of the equation of motion for the vertical direction. For given boundary conditions the solution of this equation will depend on the numerical values of the two dimensionless ratios gL/U_0^2 and ν/U_0L. The latter ratio is the reciprocal of the *Reynolds number* as in equation (5.7.4), while the former ratio is the reciprocal of the dimensionless ratio U_0^2/gL which is known as the Froude number Fr.

The physical meaning of the Reynolds number will become clear from equation (5.7.4). The value of the Reynolds number provides a measure of the relative magnitude of the inertial or acceleration terms on the left-hand side of the equation compared with the viscous stress terms on the right-hand side. Thus the Reynolds number is essentially a force ratio. Similarly the Froude number provides a measure of the relative magnitude of the inertial forces to the gravity force in equation (5.7.8).

It will be noted that we have adopted the definition U_0^2/gL for the Froude number in this chapter since this is the form in which it emerges in (5.7.8) as a force ratio on an equivalent basis to the Reynolds number. This is in line with the definition given in (1.8.6), but the reader is reminded that there is an alternative convention, which we will use in chapter 12, in which the Froude number is defined as a velocity ratio $U_0/\sqrt{(gL)}$.

5.8 Laminar flow between parallel walls and in a circular pipe

A simple, although rather idealised, example of laminar motion is the case of the two-dimensional flow of a viscous fluid between parallel

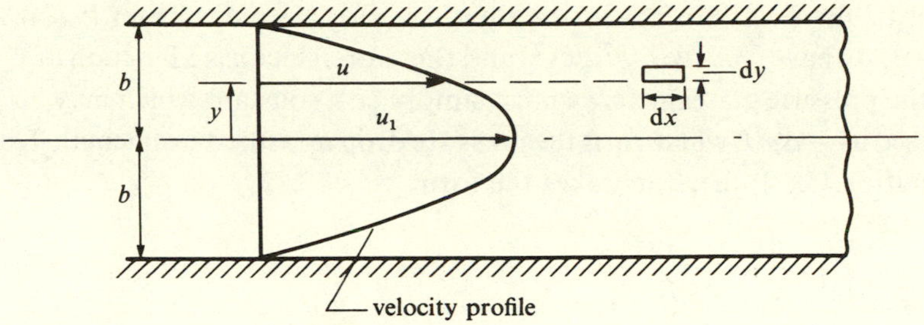

Fig. 43. Laminar flow between parallel walls.

walls as indicated in figure 43. The equations of motion for an incompressible viscous fluid take the following form, from (5.4.3), in the case of steady two-dimensional flow:

$$u\frac{\partial u}{\partial x}+v\frac{\partial u}{\partial y}=-\frac{1}{\rho}\frac{\partial p}{\partial x}+\nu\left(\frac{\partial^2 u}{\partial x^2}+\frac{\partial^2 u}{\partial y^2}\right) \tag{5.8.1}$$

and

$$u\frac{\partial v}{\partial x}+v\frac{\partial v}{\partial y}=-\frac{1}{\rho}\frac{\partial p}{\partial y}+\nu\left(\frac{\partial^2 v}{\partial x^2}+\frac{\partial^2 v}{\partial y^2}\right). \tag{5.8.2}$$

We also have the equation of continuity for incompressible flow:

$$\frac{\partial u}{\partial x}+\frac{\partial v}{\partial y}=0. \tag{5.8.3}$$

The boundary conditions are that $u=0$ and $v=0$ when $y=\pm b$ for all values of x.

The flow pattern in the entry region of a two-dimensional duct is discussed in §8.1. In the meantime we will confine attention to the special case of *fully developed flow* in a long duct at a considerable distance from the duct entry. In these circumstances we can say that the longitudinal velocity gradient $\partial u/\partial x=0$, that the transverse velocity component $v=0$, and that the velocity component u is a function of y only.

Under these conditions the equations of motion reduce to

$$-\frac{1}{\rho}\frac{\partial p}{\partial x}+\nu\frac{\partial^2 u}{\partial y^2}=0 \tag{5.8.4}$$

and

$$-\frac{1}{\rho}\frac{\partial p}{\partial y}=0. \tag{5.8.5}$$

It follows from (5.8.5) that the pressure p must be constant across any section of the duct, normal to the flow, i.e. $p=f(x)$ only. From (5.8.4), however, we have $\partial p/\partial x=\mu(\partial^2 u/\partial y^2)$ and therefore, since u is a function of y only, the pressure gradient $\partial p/\partial x$ must simply be a constant which may be expressed as $-\Delta p/L$ where Δp is the pressure drop measured over length L.

Equation (5.8.4) therefore takes the form

$$\mu\frac{\mathrm{d}^2 u}{\mathrm{d}y^2}=\frac{\mathrm{d}p}{\mathrm{d}x}=-\frac{\Delta p}{L}. \tag{5.8.6}$$

We could arrive at the same result more directly, without invoking the Navier–Stokes equations (5.8.1) and (5.8.2), by considering the equilibrium of a small rectangular element of fluid $\mathrm{d}x\ \mathrm{d}y$ as indicated in figure 44. Under fully developed steady-flow conditions, the fluid element has zero acceleration and the net force due to the pressure gradient must be exactly balanced by the net force due to the viscous stresses. We thus have the requirement that

$$\frac{\partial\tau}{\partial y}\,\mathrm{d}y\,\mathrm{d}x=-\frac{\partial p}{\partial x}\,\mathrm{d}x\,\mathrm{d}y.$$

Substituting $\tau=-\mu(\partial u/\partial y)$ (noting that $\partial u/\partial y$ is negative in figure 44) we have the equilibrium condition

$$\mu\frac{\partial^2 u}{\partial y^2}=\frac{\partial p}{\partial x}$$

which is identical with (5.8.4).

Equation (5.8.6) can be integrated with respect to y, noting that $\mathrm{d}u/\mathrm{d}y=0$ at the centre of the duct where $y=0$, to give:

$$\frac{\mathrm{d}u}{\mathrm{d}y}=-\frac{\Delta p}{\mu L}y$$

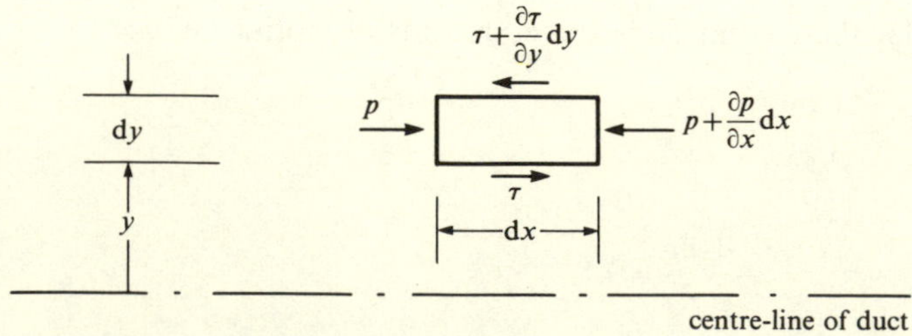

Fig. 44. Force balance on an element of fluid.

and hence

$$u = -\frac{\Delta p}{2\mu L}y^2 + B.$$

The velocity u must be zero, however, at $y = \pm b$ and hence the constant of integration is given by $B = (\Delta p/2\mu L)b^2$. Thus the equation for the velocity profile is

$$u = \frac{\Delta p}{2\mu L}(b^2 - y^2). \tag{5.8.7}$$

This may be expressed in the alternative form as a *velocity defect* relationship:

$$u_1 - u = \frac{\Delta p}{2\mu L}y^2 \tag{5.8.8}$$

where $u_1 = (\Delta p/2\mu L)b^2$ is the maximum velocity at the centre-line.

The total volumetric flow rate per unit width is given by

$$Q = \int_{-b}^{+b} u\,\mathrm{d}y = \frac{2}{3}\frac{\Delta p}{\mu L}b^3 \tag{5.8.9}$$

and hence the mean velocity u_m, averaged over the cross-section of the duct, is given by

$$u_m = \frac{Q}{2b} = \tfrac{2}{3}u_1. \tag{5.8.10}$$

We can obtain a similar solution to the corresponding problem of the fully developed steady flow of a viscous fluid in a long pipe of circular cross-section. As in the case of the two-dimensional duct, the pressure p must be constant over any cross-section of the pipe, and the longitudinal pressure gradient may be expressed as $\Delta p/L$ where Δp is the pressure drop measured over length L. Referring to figure 45, we can consider the equilibrium under fully developed steady-flow conditions of a cylindrical element of fluid of radius r and length L. The total viscous force acting over the surface of the cylindrical element will be

$$2\pi rL\tau = -2\pi rL\mu\frac{\mathrm{d}u}{\mathrm{d}r}.$$

Thus for equilibrium between the force due to the pressure gradient and the

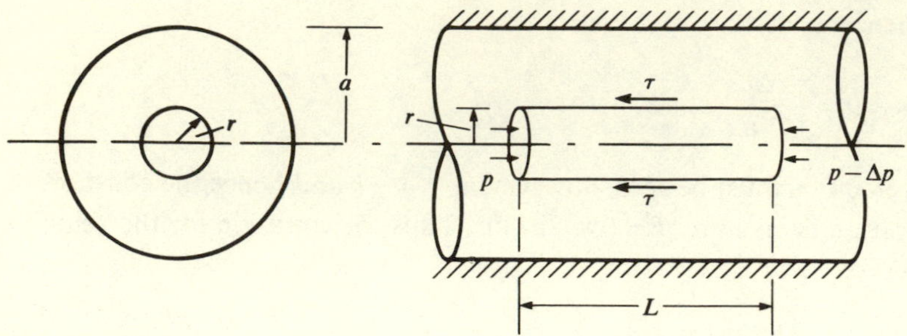

Fig. 45. Laminar flow in a pipe of circular cross-section.

viscous force we must have

$$-2\pi r L\mu \frac{du}{dr} = \pi r^2 \, \Delta p$$

or

$$\frac{du}{dr} = -\frac{\Delta p}{2\mu L} r. \qquad (5.8.11)$$

Integrating between the limits $r = a$ (where $u = 0$) to r, we have

$$\int_0^u du = -\int_a^r \frac{\Delta p}{2\mu L} r \, dr$$

and hence the velocity profile is given by

$$u = \frac{\Delta p}{4\mu L}(a^2 - r^2), \qquad (5.8.12)$$

which may be compared with the corresponding result (5.8.7) for the two-dimensional duct.

From (5.8.12) the maximum velocity u_1 at the centre-line of the pipe is given by $u_1 = (\Delta p/4\mu L)a^2$. Hence the velocity profile, expressed as a velocity-defect relationship, takes the form

$$u_1 - u = \frac{\Delta p}{4\mu L} r^2. \qquad (5.8.13)$$

Alternatively, the profile may be expressed as a velocity ratio, i.e.

$$\frac{u}{u_1} = 1 - \frac{r^2}{a^2}. \qquad (5.8.14)$$

The total volumetric flow rate is given by

$$Q = \int_0^a 2\pi r u \, \mathrm{d}r = \frac{\Delta p}{8\mu L} \pi a^4 \tag{5.8.15}$$

and hence the mean velocity u_m, averaged over the cross-section of the pipe, is given by

$$u_m = \frac{Q}{\pi a^2} = \frac{\Delta p}{8\mu L} a^2 = \tfrac{1}{2} u_1. \tag{5.8.16}$$

It follows from (5.8.16) that

$$\Delta p = \frac{8\mu u_m L}{a^2} = \frac{32\mu u_m L}{D^2}, \tag{5.8.17}$$

where D is the internal diameter of the pipe. Equation (5.8.17) is known as the *Hagen–Poiseuille law* relating to laminar viscous flow in a long tube.

5.9 Flow past a sphere

A more complicated example of the motion of a viscous fluid is the three-dimensional case of flow past a sphere as indicated in figures 42 and 46. The Navier–Stokes equation of motion for the steady flow of an incompressible fluid with zero body force, from (5.4.3), takes the form:

$$\rho \mathbf{u} \cdot \nabla \mathbf{u} = -\operatorname{grad} p + \mu \nabla^2 \mathbf{u}. \tag{5.9.1}$$

This equation cannot be solved, however, using normal analytical methods

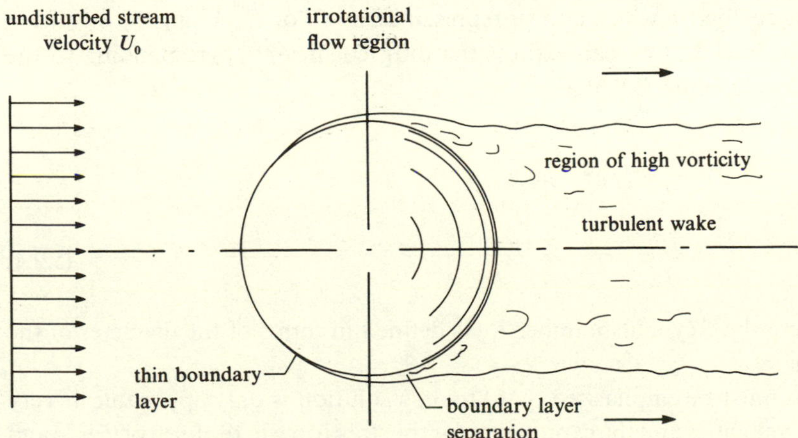

Fig. 46. Flow past a sphere.

in the case of flow past a sphere with the boundary conditions that the velocity vector **u** should be zero at all points on the surface of the sphere.

If we consider the restricted case of flow at very low velocities or *small values of the Reynolds number*, however, we can ignore the inertia terms which comprise the convective differential on the left-hand side of (5.9.1) and thus reduce the equation of motion to the simpler form

$$\mu \nabla^2 \mathbf{u} = \text{grad } p. \tag{5.9.2}$$

It will be seen that this equation is identical in form with equations (5.8.4) and (5.8.6) for fully developed viscous flow in a long two-dimensional duct. It simply states that, with negligible inertial forces, the net force due to the pressure gradient must be balanced at each point in the fluid by the net force due to the viscous stresses.

Stokes's solution for viscous flow past a sphere at small values of the Reynolds number may be obtained by expressing equation (5.9.2), together with the continuity equation, in spherical polar coordinates and fitting the boundary conditions that the velocity components must be zero at the surface of the sphere where $r = a$. Although the details of the mathematical analysis are rather involved, the final result for the net force or drag acting on the sphere is very simple and takes the form:

$$F = 6\pi\mu a U_0, \tag{5.9.3}$$

where U_0 is the undisturbed stream velocity, and a is the radius of the sphere.

As we have already noted in § 1.8, it is convenient to express the force or drag acting on a body in the form of a dimensionless force coefficient or drag coefficient which may be represented by C_F or C_D. Using the definition given in (1.8.3) we can express the drag coefficient corresponding to the Stokes formula (5.9.3) as

$$C_D = \frac{F}{\frac{1}{2}\rho U_0^2 \pi a^2} = \frac{12\mu}{\rho u_1 a}$$

or

$$C_D = \frac{24}{Re}, \tag{5.9.4}$$

where the Reynolds number Re is defined in terms of the diameter of the sphere.

It must be emphasised that Stokes's solution is only applicable at very low velocities and the expressions for the drag force in the form of (5.9.3) and (5.9.4) are only valid for values of the Reynolds number less than about 2.

At larger values of the Reynolds number the inertial terms in equation (5.9.1) exercise an increasing influence on the flow dynamics. Figure 46 presents a simplified picture of the nature of the flow pattern which is observed at relatively large values of the Reynolds number. We can divide the flow into two main zones (*a*) a region of high vorticity comprising a thin boundary layer extending over the front half of the sphere and a turbulent wake on the downstream side of the sphere, and (*b*) an external irrotational flow region outside the boundary layer and wake. The irrotational flow pattern over the front half of the sphere, outside the boundary layer, corresponds quite closely to the velocity distribution which is calculated from equation (4.3.14). Over the front half of the sphere the fluid experiences a falling pressure from the front stagnation point to the point of minimum pressure at the 90° position. Provided the Reynolds number is relatively large, the boundary layer will be thin and the favourable pressure gradient has a stabilising effect on the flow which tends to remain laminar. Once the minimum pressure point is passed, however, the fluid in the boundary-layer region experiences an adverse pressure gradient. For reasons which will be discussed in chapter 7, an adverse pressure gradient eventually causes separation of the boundary layer from the surface and leads to the formation of an irregular eddying or turbulent wake.

Laminar separation generally occurs at only a very short distance behind the minimum pressure point. At sufficiently large values of the Reynolds number, however, transition to turbulent flow will take place in the boundary layer before laminar separation can occur. In these circumstances the turbulent boundary layer will adhere to the surface of the sphere for a slightly greater distance beyond the minimum pressure point before turbulent separation occurs. The characteristic features of flow in both laminar and turbulent boundary layers are discussed in chapter 7.

It is possible to divide the total drag acting on the sphere into two components, skin-friction drag and form drag. The *skin-friction drag* is given by the integral of the tangential shearing stress evaluated over the entire surface of the sphere. The *form drag* is represented by the integral of the normal pressure evaluated over the whole surface. Separation of the flow, with the formation of an eddying or turbulent wake as indicated in figure 46, causes failure to achieve the full pressure recovery over the downstream half of the sphere which would be theoretically possible with an inviscid fluid. It is this effect which is primarily responsible for the existence of form drag.

The object of streamlining is the reduction of form drag. This can be achieved by avoiding the use of bluff-shaped bodies such as spheres and by designing shapes with long tapering extensions on the downstream side to avoid the occurrence of large local adverse pressure gradients. By this means the separation of the boundary layer can be prevented or delayed so that the wake region is greatly reduced in size.

6

Turbulent flow

6.1 The occurrence of turbulence in fluid motion

We have already noted in § 1.7 that turbulence is a characteristic feature of fluid motion in many commonly encountered flow conditions, in pipes and channels and in boundary-layer regions in the vicinity of solid walls. We have also observed that turbulence is more probable at high values of the Reynolds number, i.e. in fluids whose kinematic viscosity ν is small and where flow velocities are relatively large. As we have seen in § 5.7, the Reynolds number measures the relative magnitude of inertia forces compared with viscous forces in the fluid, and it is not surprising that at low values of the Reynolds number small disturbances or instabilities are more readily damped down by the viscous stresses. Conversely at high values of the Reynolds number small disturbances are more likely to be amplified and thus develop into full-scale turbulent motion.

With some flow geometries, for example flow through a long pipe or tube of circular cross-section, a fairly well-defined critical value of the Reynolds number is observed, above which the flow is normally turbulent and below which it is normally laminar. In other cases, such as in boundary-layer flow adjacent to a flat plate, we do not generally find that there is a single critical value marking the division between the two different types of motion. Transition from laminar to turbulent flow in boundary layers is discussed further in chapter 7 and we will simply remark at this stage that transition is a complex phenomenon and that it is influenced, among other factors, by the presence of small eddies or disturbances in the supposedly 'undisturbed' stream at the outside of the boundary layer or at entry to the duct.

It is significant that the regions in which turbulent motion is typically observed, e.g. flow in pipes, ducts, and boundary layers, and also in the downstream wake in the case of flow past solid bodies, are all regions of high vorticity. It is natural therefore to look for a link between vorticity and turbulence. We noted in § 5.6 that the term $\boldsymbol{\omega}\cdot\nabla\mathbf{u}$ which appears in the vorticity equation (5.6.8) represents the interaction between vorticity and velocity gradients in a three-dimensional flow pattern, and that it describes the stretching and tilting of vortex lines. It will be recalled from § 3.8 that the vorticity is itself related to the velocity field by the expression $\boldsymbol{\omega} = \text{curl}\,\mathbf{u}$ and that the components of the vorticity are made up from velocity gradient terms. The net result of the interaction term in the vorticity equation is that the appearance of a small extraneous disturbance in a three-dimensional flow pattern, where vorticity is present, must initiate a process of redistribution of local velocity gradients and hence of velocity components. It is probable that this process is the main cause of the development of turbulence in flow regions where vorticity is present.

A slightly different approach to the problem of transition from laminar to turbulent flow can be followed by considering the question of the inherent stability or instability of a laminar flow pattern. The mathematical theory of hydrodynamic stability is based on the concept of introducing an infinitesimal sinusoidal disturbance into the laminar flow solution and analysing the subsequent development of the flow pattern as governed by the Navier–Stokes equation. If the infinitesimal disturbance is amplified the laminar flow pattern can be regarded as being unstable, although this does not necessarily mean that turbulent motion must follow as a result. The mathematical difficulties of this approach are considerable and it is usually necessary to linearise the equations of motion, i.e. to neglect terms involving products of very small quantities. Most of the solutions which have been obtained by this method refer to stability with respect to two-dimensional disturbances. The results derived from the theory, for example Schlichting's calculations on the stability of the laminar boundary layer, generally give the upper limit for the possible maintenance of laminar flow. If extreme care is taken to eliminate extraneous disturbances in the flow, as in wind-tunnel measurements on boundary-layer flow where special precautions are taken to reduce the background level of turbulence, the theoretical calculations are largely supported by experimental evidence. In the great majority of normal engineering flow situations, however, we encounter finite disturbances in the main fluid stream and the critical values for the Reynolds number, above which turbulent flow is usually observed, are significantly lower than those predicted by stability calculations.

Because we naturally start with an investigation of laminar viscous flow, as in chapter 5 of this book, since it is amenable to mathematical analysis, and because we observe a limited region of laminar flow in the boundary layer around the leading edge of a thin profile, we tend to speak of transition *to* turbulence with the implication that turbulent flow represents an exceptional situation. In reality the reverse of this view is nearer to the truth.

We have to recognise the fact that, in flow regions where vorticity is present, *turbulence is the normal state of fluid motion.* Laminar viscous flow is an abnormal state which only prevails at low values of the Reynolds number when the viscous forces are sufficiently large to damp out any small disturbances in the flow pattern.

6.2 The nature and analysis of turbulence

The distinctive feature of turbulent flow is the apparently random nature of the small-scale motion of the fluid. The simplest example of turbulent motion in a fluid is represented by the case of a nearly uniform stream, for instance in a wind tunnel, where small disturbances or vortices are introduced artificially by locating a grid of wires or small-diameter rods arranged on a regular square-mesh pattern in a plane at right angles to the direction of the stream. The eddies which are formed in the fluid passing through the grid are dispersed in the downstream zone with the result that a small-scale random fluctuating velocity pattern is superimposed on the original undisturbed mainstream velocity. Turbulent motion which is distributed uniformly in the volume of the fluid in this manner is described as *homogeneous turbulence.*

When analysing turbulent motion it is convenient to express the instantaneous velocity components of the fluid as the sum of a macroscopic or mean velocity and a fluctuating turbulent velocity, i.e.

$$\left.\begin{aligned} u&=\bar{u}+u' \\ v&=\bar{v}+v' \\ w&=\bar{w}+w' \end{aligned}\right\} \qquad (6.2.1)$$

where $\bar{u}$, $\bar{v}$, $\bar{w}$, are the mean velocity components (averaged over an interval of time at a particular point in space defined by the co-ordinates x, y, z) and u', v', w', are the random fluctuating turbulent components. The reader should naturally not confuse the symbols u', v', w', used for the turbulent velocity components in this chapter with the same symbols which were used in a completely different sense in § 5.7 to represent dimensionless velocity ratios.

An alternative convention, which is used by many authors, is to represent the mean velocity components by the capital letters U, V, W, and the fluctuating turbulent velocity components by the lower case letters u, v, w. This has the advantage of simplicity but may lead to confusion since the letters u, v, w, are normally used to represent the absolute velocity components. On balance we prefer to adopt the notation as given in (6.2.1).

Care must be taken in defining the mean velocity components. The evaluation must be carried out over an interval of time Δt which is sufficiently large in comparison with the typical period of fluctuation of the turbulent component, and a common-sense approach would suggest that the mean value $\bar{u}$, for example, could be defined by

$$\bar{u} = \frac{1}{\Delta t} \int_{t-\frac{1}{2}\Delta t}^{t+\frac{1}{2}\Delta t} u \, dt.$$

A more precise mathematical definition, however, would be

$$\bar{u} = \lim_{\Delta t \to \infty} \frac{1}{\Delta t} \int_{t-\frac{1}{2}\Delta t}^{t+\frac{1}{2}\Delta t} u \, dt. \tag{6.2.2}$$

If the value of $\bar{u}$ is independent of the instant t (i.e. the mid-point of the interval Δt) we can describe the mean flow as being steady or statistically stationary, as indicated in figure 47 (*a*), and no problem arises in obtaining the mean value.

If, however, the macroscopic motion of the fluid is varying with time, as for example in figure 47 (*b*), a different method should be adopted for

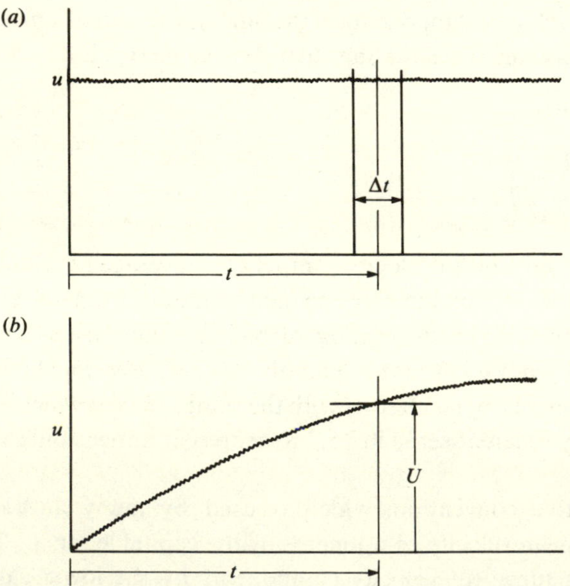

Fig. 47. (*a*) Fluctuating velocities in steady mean flow. (*b*) Fluctuating velocities in varying mean flow.

obtaining a mean value and the correct mathematical procedure in this situation is to use an *ensemble average.* As an example we could consider the case of a wind tunnel being started up from zero stream velocity at time $t=0$ and with the stream velocity increasing with time to reach a value U at time t as indicated in figure 47 (*b*). To obtain an ensemble average for the velocity at time t we would strictly have to conduct a series of experiments, starting the wind tunnel from rest on each occasion, and measuring the instantaneous velocity at the corresponding time t. The ensemble average would be the mean of all these measurements. Provided the macroscopic velocity is changing only slowly with time, however, we can obtain a valid mean value more easily by using the definition given in (6.2.2) and by choosing an interval of time Δt which is small in relation to the timescale of change for the macroscopic velocity while still being large in relation to the typical period of the turbulent fluctuations.

We will see shortly that the mean values of the squares of the fluctuating velocity components, $\overline{u'^2}, \overline{v'^2}, \overline{w'^2}$, play an important part in the analysis of turbulent motion. It is convenient also to define a turbulent velocity q such that

$$q^2 = u'^2 + v'^2 + w'^2 \tag{6.2.3}$$

and the mean value of q^2 is then given by

$$\overline{q^2} = \overline{u'^2} + \overline{v'^2} + \overline{w'^2}. \tag{6.2.4}$$

The rms value $(\overline{q^2})^{\frac{1}{2}}$ provides a measure of the *intensity* of the turbulence. The kinetic energy of the turbulent motion, per unit volume of fluid, is given by $\frac{1}{2}\rho\overline{q^2}$.

If we consider the case of turbulence in a wind-tunnel stream, generated by means of a grid positioned at right angle to the airstream, the newly formed eddies or vortices immediately downstream from the grid will have certain directional properties. However, as these eddies disperse, and as the vorticity components become re-distributed, the velocity pattern of the resulting turbulent motion will gradually lose any preferential directional features. There is a general tendency in this type of flow situation for the mean-square values of the turbulent velocity components to become equal to each other. Turbulence is described as being *isotropic* when $\overline{u'^2} = \overline{v'^2} = \overline{w'^2}$. It is particularly important to appreciate that *turbulence is always three-dimensional* in character whether or not it is fully isotropic.

It will be shown in § 6.3 that the effect of turbulence on the macroscopic or mean motion of the fluid is to transport momentum by means of the velocity fluctuations, and that this is equivalent to the introduction of additional stress components into the equations of motion which are known as the *Reynolds stresses.* The normal stress components take the form $-\rho\overline{u'^2}$, $-\rho\overline{v'^2}$, $-\rho\overline{w'^2}$, while the tangential or shear-stress components are $-\rho\overline{u'v'}$, $-\rho\overline{v'w'}$, $-\rho\overline{w'u'}$. The shear-stress terms depend in each case on the magnitudes of the two turbulent velocity components and on the correlation between them. The *coefficient of correlation* between two turbulent velocity components u'_i *and* u'_j, measured at the same point in space, may be defined by:

$$R_{ij}=\frac{\overline{u'_i u'_j}}{(\overline{u_i'^2}\ \overline{u_j'^2})^{\frac{1}{2}}} \tag{6.2.5}$$

but it should be noted that we do not apply the summation rule for tensor notation in this instance.

As an example of (6.2.5) we can take $u'_i=u'$, and $u'_j=v'$, so that the coefficient of correlation between the turbulent velocity components u' and v' is represented by

$$R_{12}=\frac{\overline{u'v'}}{(\overline{u'^2}\ \overline{v'^2})^{\frac{1}{2}}}. \tag{6.2.6}$$

With zero correlation between u and v we would have $\overline{u'v'}=0$ and hence $R_{12}=0$. With complete correlation we would have $R_{12}=1$.

Because of the random nature of the turbulent velocities the analysis of turbulence involves the use of statistical methods. The statistical theory of turbulence was first presented in a series of papers by G. I. Taylor in 1935 and has subsequently been developed and extended by G. K. Batchelor and others. It is inevitable that the mathematical methods involved in a statistical approach are of a rather difficult and specialised character, and it would be impossible to present a summary in this chapter. All we can do, within the space available, is to mention some of the more important concepts which have emerged from the statistical theory and which have a bearing on the main area of engineering interest relating to turbulent shear flow in pipes, channels, and boundary layers.

The correlation between turbulent velocity components measured at different points in space, for example at points specified by the position vector $\mathbf{x}$ and the vector $\mathbf{x}+\mathbf{r}$, is of fundamental importance in the statistical theory. A velocity correlation function or correlation tensor may be defined, for two points separated by the space vector **r**, by:

$$\overline{u_i'(\mathbf{x})u_j'(\mathbf{x}+\mathbf{r})}.$$

Although the symbol $R_{ij}(\mathbf{r})$ is often used for this function, we prefer to use this symbol for the correlation coefficient defined by

$$R_{ij}(\mathbf{r})=\frac{\overline{u_i'(\mathbf{x})u_j'(\mathbf{x}+\mathbf{r})}}{(\overline{u_i'^2(\mathbf{x})}\;\overline{u_j'^2(\mathbf{x}+\mathbf{r})})^{\frac{1}{2}}}. \tag{6.2.7}$$

If, for example, we take $u_i'=u'$ and $u_j'=v'$, and if we put $\mathbf{r}=0$, we have the correlation coefficient between the two components u' and v' measured at the same point in space as given by (6.2.6).

If we now take $u_i'=u_j'=u'$, and if we take for the space vector **r** a separation in space by a distance y measured along the y axis, we obtain the correlation coefficient between the same velocity component u' measured at different points along the y axis, i.e.

$$R_{11}(y)=\frac{\overline{u'(0)u'(y)}}{(\overline{u'^2(0)}\;\overline{u'^2(y)})^{\frac{1}{2}}}. \tag{6.2.8}$$

The correlation coefficient, as given by (6.2.8), can be evaluated for a particular turbulent flow from direct measurements of the turbulent velocity component u' obtained at different points on the y axis (for example traversing a turbulent stream in a wind tunnel) using a hot-wire anemometer. One would expect that the correlation coefficient would decrease from the value 1.0 as the distance of separation y between the two measurement points is increased from zero. This is confirmed by experimental results as indicated in figure 48 where $R_{11}(y)$ is plotted against the distance y.

The distribution of the coefficient $R_{11}(y)$ with distance y must be related to the 'size' of the eddies which constitute the turbulent motion. It is difficult to define precisely what is meant by the *diameter* of an eddy, but, however we may attempt to define it, a high degree of correlation must exist between the velocity components at two points which are close together when compared with this notional diameter. Equally the correlation is likely to be small if the distance of separation y is many times the eddy diameter. If the experimental measurements are plotted out as in figure 48, and if $R_{11}(y)$ falls to zero at $y=Y$, we could define a length L_y which provides a measure of the size of eddy or the *scale of turbulence*, i.e.

$$L_y=\int_0^\infty R_{11}(y)\,\mathrm{d}y\approx\int_0^Y R_{11}(y)\,\mathrm{d}y. \tag{6.2.9}$$

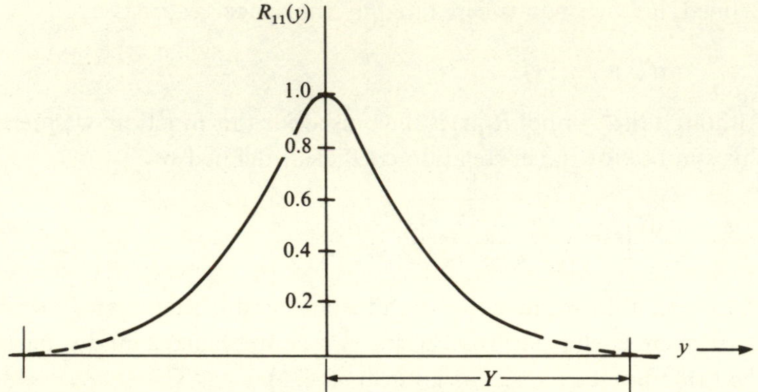

Fig. 48. Variation of the correlation coefficient with distance.

In his first paper on the statistical theory of turbulence, G. I. Taylor established the following expression for the rate of dissipation of energy, per unit volume, in *isotropic turbulence*:

$$\bar{W}=7.5\mu\overline{\left(\frac{\partial u'}{\partial y}\right)^2}. \tag{6.2.10}$$

He also proved that the mean value $\overline{(\partial u'/\partial y)^2}$ is related to the curvature of the $R_{11}(y)$ curve at $y=0$, and that

$$\overline{\left(\frac{\partial u'}{\partial y}\right)^2}=2\,\overline{u'^2}\,\underset{y\to 0}{\text{limit}}\left(\frac{1-R_{11}(y)}{y^2}\right). \tag{6.2.11}$$

Hence, by defining a length λ such that

$$\frac{1}{\lambda^2}=\underset{y\to 0}{\text{limit}}\left(\frac{1-R_{11}(y)}{y^2}\right), \tag{6.2.12}$$

we have the result that

$$\bar{W}=15\mu\left(\frac{\overline{u'^2}}{\lambda^2}\right). \tag{6.2.13}$$

The length λ may be regarded as a measure of the diameters of the smallest eddies which are responsible for the dissipation of energy.

In the case of non-homogeneous turbulent flow, such as we encounter in boundary-layer regions, there is usually a wide range of eddy size. The main process of momentum transfer, which controls the overall dynamics of the *mean flow* pattern, appears to be governed by relatively large eddies whose associated fluctuating velocity products u'^2, $u'v'$, $u'w'$, etc., which appear in the Reynolds stress terms, are correspondingly large. Within the turbulent motion itself, the larger eddies generate smaller eddies, and these in turn generate what is usually described as small-scale or *micro-turbulence* which becomes increasingly isotropic in character. It appears that there is a continuous process of grinding down of large eddies into smaller ones. At the risk of over-simplification in describing an intrinsically complex process, it is helpful to think in terms of an *energy cascade*. Work is done by the mean flow against the Reynolds stresses, thus transferring kinetic energy to the larger eddies. Energy is then transferred from large eddies to smaller eddies by a complex dynamical process. The smaller eddies tend towards nearly isotropic small-scale turbulence and finally transfer kinetic energy into the internal energy of the fluid by the molecular transfer process which we describe in terms of viscosity.

Later developments of the statistical theory of turbulence have involved Fourier analysis of the velocity field and the use of a *spectrum tensor* which is a Fourier transform of the velocity–correlation tensor. A related concept, which is of special importance, is the energy spectrum function $E(k)$ which represents the contribution to the kinetic energy of the turbulence (expressed per unit mass of fluid) per unit width of the wave-number band at level k. As in other applications involving the terminology of wave motion, the wave-number k is inversely proportional to the wavelength and we may therefore regard the quantity $1/k$ as giving a measure of the 'diameter of an eddy'. The total kinetic energy of the turbulence, expressed per unit mass of fluid, is given by

$$\tfrac{1}{2}\overline{q^2} = \tfrac{1}{2}(\overline{u'^2} + \overline{v'^2} + \overline{w'^2}) = \int_0^\infty E(k)\,\mathrm{d}k. \tag{6.2.14}$$

We can distinguish between the wave-number range of the larger eddies which contain the greater part of the kinetic energy, and the wave-number range of the smaller eddies which are responsible for the process of energy dissipation through the action of viscosity.

According to the theory of local similarity, or *universal equilibrium*, advanced by Kolmogorov in 1941, the smallest eddies are statistically steady and isotropic in character and their Fourier coefficients are independent of those of the larger energy-containing eddies. Provided the Reynolds number of the flow is sufficiently large, we can say that the small-scale (high wave-number) energy-dissipating eddies will exhibit a common universal structure which applies to all types of turbulent flow. The rate of energy flow per unit mass, which we will represent by the symbol ε or $\bar{W}/\rho$, is determined by the motion of the larger eddies and their interaction with the mean flow. The small-scale eddies must then adjust themselves to achieve the required rate of energy dissipation ε per unit mass, typically through the term $7.5\nu\overline{(\partial u'/\partial y)^2}$ derived by G. I. Taylor for isotropic turbulence as in (6.2.10), or more generally through the expression $\nu\overline{(\partial u_i/\partial x_j)^2}$ as noted in (6.4.9).

We can thus say (to quote G. K. Batchelor) that the motion associated with the equilibrium range of wave numbers is uniquely determined statistically by the two quantities ε and ν. The effect of any variation in the values of these two parameters is simply to influence the typical length and velocity scales of the small-scale turbulence. Applying the normal method of dimensional analysis, we find that the Kolmogorov scale of length l_ε and velocity scale v_ε are represented by:

$$l_\varepsilon = \left(\frac{\nu^3}{\varepsilon}\right)^{\frac{1}{4}} \quad \text{and} \quad v_\varepsilon = (\nu\varepsilon)^{\frac{1}{4}}. \tag{6.2.15}$$

It will be seen from (6.2.15) that $(l_\varepsilon v_\varepsilon)/\nu = 1$, i.e. the Reynolds number of the energy-dissipating eddies, when expressed in terms of the Kolmogorov scales is exactly unity. It also follows from (6.2.15) that the energy-dissipation rate per unit mass of fluid is given by

$$\varepsilon = \nu\left(\frac{v_\varepsilon^2}{l_\varepsilon^2}\right). \tag{6.2.16}$$

The relationship between the Taylor length λ, as defined by (6.2.12), and the Kolmogorov length l_ε is of some interest. The Taylor expression for the energy-dissipation rate (6.2.13), when expressed per unit mass of fluid, takes the form:

$$\varepsilon = \frac{\bar{W}}{\rho} = 15\nu\overline{\left(\frac{u'^2}{\lambda^2}\right)}, \tag{6.2.17}$$

and it should be noted that this is strictly applicable only to fully isotropic turbulence. The similarity between (6.2.16) and (6.2.17) will be obvious. It follows that, for isotropic turbulence,

$$\frac{v_\varepsilon^2}{l_\varepsilon^2} = 15\overline{\left(\frac{u'^2}{\lambda^2}\right)}. \tag{6.2.18}$$

If we now define a Reynolds number for the small-scale isotropic turbulence using the Taylor length λ and the rms velocity $(\overline{u'^2})^{\frac{1}{2}}$ we can write

$$Re_T = \frac{(\overline{u'^2})^{\frac{1}{2}}\lambda}{\nu} \tag{6.2.19}$$

and hence

$$Re_T^2 = \frac{\overline{u'^2}\,\lambda^2}{\nu^2} = 15\,\frac{(\overline{u'^2})^2}{\nu\varepsilon} = 15\,\frac{(\overline{u'^2})^2}{v_\varepsilon^4}. \tag{6.2.20}$$

It follows that:

$$\frac{v_\varepsilon^2}{\overline{u'^2}} = 3.873\, Re_T^{-1} \tag{6.2.21}$$

and from (6.2.18),

$$\frac{l_\varepsilon}{\lambda} = 0.508\, Re_T^{-\frac{1}{2}}. \tag{6.2.22}$$

6.3 The equations of motion and the Reynolds stresses

There is no reason to suppose that the basic equations of fluid motion derived in chapter 5 should not apply equally to turbulent as to non-turbulent flow provided we could follow all the random fluctuations in the flow pattern. In this sense turbulent motion could be regarded as an extreme case of unsteady fluid flow. In practice, however, an observer is unable to see all the detail of the turbulent motion and most measurements of the fluid velocity, for example, will in reality be measurements of time-averaged mean velocities. It will be helpful, therefore, to insert the expressions for the instantaneous velocity components as defined by (6.2.1) in the equation of continuity and also in the basic equation of motion (5.2.6) or (5.2.7). For simplicity we will confine attention to the case of an incompressible fluid and we will omit the terms involving body forces such as gravity.

The continuity equation for the unsteady flow of an incompressible fluid is given by (3.3.4) or (3.3.5), i.e.

$$\frac{\partial u}{\partial x} + \frac{\partial v}{\partial y} + \frac{\partial w}{\partial z} = 0. \tag{6.3.1}$$

If we now substitute for the instantaneous velocity components from (6.2.1) we can write the equation as:

$$\left(\frac{\partial \bar{u}}{\partial x}+\frac{\partial \bar{v}}{\partial y}+\frac{\partial \bar{w}}{\partial z}\right)+\left(\frac{\partial u'}{\partial x}+\frac{\partial v'}{\partial y}+\frac{\partial w'}{\partial z}\right)=0. \tag{6.3.2}$$

If we now take the time average (or ensemble average) of (6.3.1), we have the result that

$$\frac{\partial \bar{u}}{\partial x}+\frac{\partial \bar{v}}{\partial y}+\frac{\partial \bar{w}}{\partial z}=0, \tag{6.3.3}$$

since it is easily verified mathematically that $\overline{(\partial u/\partial x)}=(\partial \bar{u}/\partial x)$. If we now subtract (6.3.3) from (6.3.2) we obtain:

$$\frac{\partial u'}{\partial x}+\frac{\partial y'}{\partial y}+\frac{\partial w'}{\partial z}=0. \tag{6.3.4}$$

Thus the normal form of the equation of continuity (3.3.4) applies separately to both the mean flow (6.3.3) and to the fluctuating turbulent motion (6.3.4).

The most general form of the equation of motion for a fluid with viscous stresses is expressed by (5.2.7). The x component of this equation (5.2.6) may be written as follows, noting from (5.2.1) that $p_{xx}=\tau_{xx}-p$:

$$\rho\frac{\partial u}{\partial t}+\rho u\frac{\partial u}{\partial x}+\rho v\frac{\partial u}{\partial y}+\rho w\frac{\partial u}{\partial z}=\frac{\partial p_{xx}}{\partial x}+\frac{\partial \tau_{yx}}{\partial y}+\frac{\partial \tau_{zx}}{\partial z}. \tag{6.3.5}$$

We must now carry out a simple mathematical manoeuvre so that we can avoid having to deal with terms such as $\rho v(\partial u/\partial y)$ which involve the product of a velocity component and a velocity gradient. If we take the equation of continuity (6.3.1) and multiply it by ρu we get the equally valid equation:

$$\rho u\left(\frac{\partial u}{\partial x}+\frac{\partial v}{\partial y}+\frac{\partial w}{\partial z}\right)=0. \tag{6.3.6}$$

We can now add (6.3.6) to equation (6.3.5) to give the result:

$$\rho\frac{\partial u}{\partial t}+\rho\frac{\partial (u^2)}{\partial x}+\frac{\partial (uv)}{\partial y}+\frac{\partial (uw)}{\partial z}=\frac{\partial p_{xx}}{\partial x}+\frac{\partial \tau_{yx}}{\partial y}+\frac{\partial \tau_{zx}}{\partial z}. \tag{6.3.7}$$

Each term in the equation is now a simple partial differential and we can regroup the terms so that the equation takes the form:

$$\rho\frac{\partial u}{\partial t}=\frac{\partial}{\partial x}(p_{xx}-\rho u^2)+\frac{\partial}{\partial y}(\tau_{yx}-\rho uv)+\frac{\partial}{\partial z}(\tau_{zx}-\rho uw). \tag{6.3.8}$$

Substituting for the instantaneous velocity components from (6.2.1) gives:

$$\rho \frac{\partial}{\partial t}(\bar{u}+u') = \frac{\partial}{\partial x}(p_{xx} - \rho\bar{u}^2 - \rho u'^2 - 2\rho\bar{u}v')$$
$$+\frac{\partial}{\partial y}(\tau_{yx} - \rho\bar{u}\bar{v} - \rho u'v' - \rho\bar{u}v' - \rho u'\bar{v})$$
$$+\frac{\partial}{\partial z}(\tau_{zx} - \rho\bar{u}\bar{w} - \rho u'w' - \rho\bar{u}w' - \rho u'\bar{w}). \qquad (6.3.9)$$

If we take the time or ensemble average of this equation we get:

$$\rho \frac{\partial \bar{u}}{\partial t} = \frac{\partial}{\partial x}(\bar{p}_{xx} - \rho\bar{u}^2 - \rho\,\overline{u'^2}) + \frac{\partial}{\partial y}(\bar{\tau}_{yx} - \rho\bar{u}\bar{v} - \rho\,\overline{u'v'})$$
$$+\frac{\partial}{\partial z}(\bar{\tau}_{zx} - \rho\bar{u}\bar{w} - \rho\,\overline{u'w'}). \qquad (6.3.10)$$

If this equation is compared with (6.3.8) it will be seen that the basic form of the equation of motion is still applicable to turbulent flow, with mean velocity components in place of instantaneous velocity components, provided the stress terms are modified as follows:

$(\overline{p_{xx}} - \rho\,\overline{u'^2})$ in place of p_{xx},

$(\overline{\tau_{yx}} - \rho\,\overline{u'v'})$ in place of τ_{yx},

$(\overline{\tau_{zx}} - \rho\,\overline{u'w'})$ in place of τ_{zx}.

The new quantities $\rho\,\overline{u'^2}$, $\rho\,\overline{u'v'}$, $\rho\,\overline{u'w'}$ are the *Reynolds stresses* and they express mathematically the physical fact that momentum is transferred by the turbulent motion of the fluid.

The above analysis can obviously be repeated for the other two components of the equation of motion. The complete stress tensor for turbulent flow may then be expressed as:

$$\begin{bmatrix} (p_{xx} - \rho\,\overline{u'^2}) & (\tau_{yx} - \rho\,\overline{u'v'}) & (\tau_{zx} - \rho\,\overline{u'w'}) \\ (\tau_{xy} - \rho\,\overline{u'v'}) & (p_{yy} - \rho\,\overline{v'^2}) & (\tau_{zy} - \rho\,\overline{v'w'}) \\ (\tau_{xz} - \rho\,\overline{u'w'}) & (\tau_{yz} - \rho\,\overline{v'w'}) & (p_{zz} - \rho\,\overline{w'^2}) \end{bmatrix}.$$

6.4 The mechanical energy equation for turbulent flow

We obtained the mechanical energy equation for non-turbulent viscous flow in §5.5 by forming the scalar product of the basic equation of motion with the velocity vector **u**. This is equivalent mathematically to multiplying each of the three component equations in the x, y, and z directions by the appropriate velocity component u, v, and w respectively and then adding the three scalar equations. We can follow a similar procedure to obtain an equation for the energy of turbulent motion. The most convenient starting point will be the Navier–Stokes form for the x component of the basic equation of motion, i.e.

$$\frac{\partial u}{\partial t}+u\frac{\partial u}{\partial x}+v\frac{\partial u}{\partial y}+w\frac{\partial u}{\partial z}=-\frac{1}{\rho}\frac{\partial p}{\partial x}+\nu\,\nabla^2 u. \tag{6.4.1}$$

For simplicity we have omitted the term involving body force and we confine ourselves to the case of incompressible flow, i.e. the density ρ is constant. It will be noted that equation (6.4.1) follows directly from (6.3.5) after substituting the Newtonian expressions for the viscous stresses and dividing through by ρ. We now make the substitution $u=\bar{u}+u'$, etc., for the instantaneous velocity components in (6.4.1) and multiply the whole equation by the turbulent component u'. Hence

$$\begin{aligned}
&u'\frac{\partial}{\partial t}(\bar{u}+u')+u'(\bar{u}+u')\frac{\partial}{\partial x}(\bar{u}+u')\\
&\quad+u'(\bar{v}+v')\frac{\partial}{\partial y}(\bar{u}+u')+u'(w+w')\frac{\partial}{\partial z}(\bar{u}+u')\\
&=-\frac{1}{\rho}u'\frac{\partial}{\partial x}(\bar{p}+p')+\nu u'\,\nabla^2(\bar{u}+u').
\end{aligned} \tag{6.4.2}$$

Note that p' represents the fluctuating pressure increment associated with the turbulent motion. Multiplying out the terms in (6.4.2) and taking a time or ensemble average gives the following result:

$$\begin{aligned}
&\frac{\partial}{\partial t}(\tfrac{1}{2}\overline{u'^2})+\bar{u}\frac{\partial}{\partial x}(\tfrac{1}{2}\overline{u'^2})+\overline{u'^2}\frac{\partial\bar{u}}{\partial x}+\overline{u'^2\frac{\partial u'}{\partial x}}\\
&\quad+\bar{v}\frac{\partial}{\partial y}(\tfrac{1}{2}\overline{u'^2})+\overline{u'v'}\frac{\partial\bar{u}}{\partial y}+\overline{u'v'\frac{\partial u'}{\partial y}}\\
&\quad+\bar{w}\frac{\partial}{\partial z}(\tfrac{1}{2}\overline{u'^2})+\overline{u'w'}\frac{\partial\bar{u}}{\partial z}+\overline{u'w'\frac{\partial u'}{\partial z}}\\
&=-\frac{1}{\rho}\overline{u'\frac{\partial p'}{\partial x}}+\nu\,\overline{u'\,\nabla^2 u'}.
\end{aligned} \tag{6.4.3}$$

The last column or group of terms on the left-hand side may be simplified by adding $(\frac{1}{2}\overline{u'^2})$ times the equation of continuity in the form (6.3.4), which amounts to zero, giving the result that

$$\overline{u'^2\frac{\partial u'}{\partial x}}+\overline{u'v'\frac{\partial u'}{\partial y}}+\overline{u'w'\frac{\partial u'}{\partial z}}=\frac{1}{2}\left[\frac{\partial}{\partial x}(\overline{u'u'^2})+\frac{\partial}{\partial y}(\overline{v'u'^2})+\frac{\partial}{\partial z}(\overline{w'u'^2})\right].$$

We can also make a substitution similar to that used in (5.5.7) for the pressure gradient term on the right-hand side of (6.4.3), i.e.

$$-\frac{1}{\rho}\overline{u'\frac{\partial p'}{\partial x}}=-\frac{\partial}{\partial x}\left(\overline{\frac{p'}{\rho}u'}\right)+\overline{\frac{p'}{\rho}\frac{\partial u'}{\partial x}}.$$

We can now write equations similar to (6.4.3) relating to the y and z components of the equation of motion, and if these are added to (6.4.3) we will get a single equation for the kinetic energy of the turbulent motion. Making use of the substitutions noted above together with the equation of continuity, and using the expression (6.2.4) for the mean square turbulent velocity $\overline{q^2}$, we arrive at the following result after adding the three equations and sorting out the terms:

$$\begin{aligned}&\frac{\partial}{\partial t}(\tfrac{1}{2}\overline{q^2})+\bar{\mathbf{u}}\cdot\nabla(\tfrac{1}{2}\overline{q^2})\\&\quad=-\overline{u_i'u_j'}\frac{\partial\bar{u}_i}{\partial x_j}-\frac{\partial}{\partial x_j}\left(\overline{\frac{p'}{\rho}u_j'}+\tfrac{1}{2}\overline{q^2u_j'}\right)\\&\qquad+\nu[\overline{u'\,\nabla^2u'}+\overline{v'\,\nabla^2v'}+\overline{w'\,\nabla^2w'}].\end{aligned}\tag{6.4.4}$$

It is important to understand the physical meaning of each of the terms in equation (6.4.4). The left-hand side obviously represents the rate of change of the kinetic energy of the turbulence *following the mean motion* of the fluid and could be written as $(\mathrm{D}/\mathrm{D}t)(\frac{1}{2}\overline{q^2})$. The first term on the right-hand side represents the rate at which work is being done by the mean flow against the Reynolds stresses. We have written this term in tensor notation to save space, but reference to the component equation (6.4.3) will show that it comprises nine terms of the type $\overline{u'v'}(\partial\bar{u}/\partial y)$. The second term on the right-hand side represents turbulent 'flow work' against the fluctuating pressure p' plus the transport of turbulent kinetic energy by the fluctuating velocity components. The final group of terms on the right-hand side represents work being done by the turbulent velocity components against the viscous stresses associated with the turbulent motion and may be compared with the equivalent term $\mu\mathbf{u}\cdot\nabla^2\mathbf{u}$ which appeared in equation (5.5.19). We can rewrite this group, using tensor notation, as $\nu[\overline{u_i'(\partial^2 u_i'/\partial x_j^2)}]$. It will in fact be appropriate to restate the complete equation (6.4.4) in tensor notation for reference:

$$\frac{\partial}{\partial t}(\overline{\tfrac{1}{2}q^2})+\bar{u}_j\frac{\partial}{\partial x_j}(\overline{\tfrac{1}{2}q^2})$$

$$=-\overline{u_i'u_j'}\frac{\partial \bar{u}_i}{\partial x_j}-\frac{\partial}{\partial x_j}\left(\overline{\frac{p'}{\rho}u_j'}+\overline{\tfrac{1}{2}q^2u_j'}\right)+\nu\,\overline{u_i'\frac{\partial^2 u_i'}{\partial x_j^2}}. \quad (6.4.5)$$

An important application of this equation relates to flow in a boundary-layer region. For two-dimensional mean flow in a thin turbulent boundary layer, equation (6.4.5) reduces to

$$\frac{\mathrm{D}}{\mathrm{D}t}(\overline{\tfrac{1}{2}q^2})=-\overline{u'v'}\frac{\partial \bar{u}}{\partial y}-\frac{\partial}{\partial y}\left(\overline{\frac{p'}{\rho}v'}+\overline{\tfrac{1}{2}q^2v'}\right)+\nu\,\overline{u_i'\frac{\partial^2 u_i'}{\partial x_j^2}}. \quad (6.4.6)$$

In certain circumstances the middle term on the right-hand side of (6.4.5) or (6.4.6) may be relatively unimportant and the equation may then be simplified by omitting this term. In the simplified form of (6.4.5) or (6.4.6) the equation appears as an energy balance stating that the rate of gain of turbulent kinetic energy by the fluid, per unit mass, is equal to the rate of supply of energy from the work of the mean motion against the Reynolds stresses *minus* the rate of dissipation through the work of the turbulent velocity components against the local viscous stresses. It should be noted that because $\overline{u_i'u_j'}(\partial\bar{u}_i/\partial x_j)$ is normally negative, the term $-\overline{u_i'u_j'}(\partial\bar{u}_i/\partial x_j)$ will have a positive numerical value and will normally represent a supply of energy *from* the mean motion *to* the turbulent motion. The term $\nu\,\overline{u_i'(\partial^2u_i'/\partial x_j^2)}$, on the other hand, has a negative numerical value and therefore represents a loss of energy through viscous dissipation.

This last term $\nu\,\overline{u_i'(\partial^2u_i'/\partial x_j^2)}$ calls for further examination. It can be rearranged mathematically in various ways in a similar manner to the corresponding term for non-turbulent viscous flow in (5.5.20) and (5.5.21). For example we can say that

$$\nu\,\overline{u_i'\frac{\partial^2 u_i'}{\partial x_j^2}}=\nu\frac{\partial^2}{\partial x_j^2}(\tfrac{1}{2}\overline{u_i'^2})-\nu\overline{\left(\frac{\partial u_i'}{\partial x_j}\right)^2}, \quad (6.4.7)$$

which is directly comparable with (5.5.20). It can be proved, however, that the exact expression for the rate of dissipation of turbulent energy through viscous action, is:

$$\varepsilon=\tfrac{1}{2}\nu\overline{\left(\frac{\partial u_i'}{\partial x_j}+\frac{\partial u_j'}{\partial x_i}\right)^2}. \tag{6.4.8}$$

The energy dissipation rate ε per unit mass is thus exactly equivalent to the energy dissipation rate Φ (expressed per unit volume) as given by equation (5.5.17) for non-turbulent viscous flow.

In practice the term $\nu(\partial^2/\partial x_j^2)(\tfrac{1}{2}\overline{u_i'^2})$, which appears in (6.4.7), can usually be neglected. It may also be shown that there is little difference between the expression for ε as given by (6.4.8) and the term $\nu\overline{(\partial u_i'/\partial x_j)^2}$ which appears in (6.4.7). Consequently we can say that, for most practical purposes, the turbulent energy dissipation rate per unit mass of fluid is given by:

$$\varepsilon=-\nu\,\overline{u_i'\frac{\partial^2 u_i'}{\partial x_j^2}}=\tfrac{1}{2}\nu\overline{\left(\frac{\partial u_i'}{\partial x_j}+\frac{\partial u_j'}{\partial x_i}\right)^2}=\nu\overline{\left(\frac{\partial u_i'}{\partial x_j}\right)^2}. \tag{6.4.9}$$

Under *isotropic conditions* the following relationships apply between the mean squares of the velocity gradients:

$$\overline{\left(\frac{\partial u'}{\partial x}\right)^2}=\overline{\left(\frac{\partial v'}{\partial y}\right)^2}=\overline{\left(\frac{\partial w'}{\partial z}\right)^2}$$

and

$$\overline{\left(\frac{\partial u'}{\partial y}\right)^2}=\overline{\left(\frac{\partial u'}{\partial z}\right)^2}=\overline{\left(\frac{\partial v'}{\partial x}\right)^2}=\overline{\left(\frac{\partial v'}{\partial z}\right)^2}=\overline{\left(\frac{\partial w'}{\partial x}\right)^2}=\overline{\left(\frac{\partial w'}{\partial y}\right)^2}.$$

By examining the relationships between the other terms which appear in the expression (6.4.8) involving products of velocity gradients, G. I. Taylor proved that the rate of energy dissipation in isotropic turbulence is given by

$$\varepsilon=15\nu\overline{\left(\frac{\partial u'}{\partial x}\right)^2}=7.5\nu\overline{\left(\frac{\partial u'}{\partial y}\right)^2} \tag{6.4.10}$$

or, expressed per unit volume, with $\bar{W}=\rho\varepsilon$, we have

$$\bar{W}=15\mu\overline{\left(\frac{\partial u'}{\partial x}\right)^2}=7.5\mu\overline{\left(\frac{\partial u'}{\partial y}\right)^2},$$

which is the result quoted in (6.2.10). Noting from this analysis that $\overline{(\partial u'/\partial x)^2}=\tfrac{1}{2}\overline{(\partial u'/\partial y)^2}$, we obtain exactly the same result as in (6.4.10) if we take $\varepsilon=\nu\overline{(\partial u_i'/\partial x_j)^2}$.

6.5 Turbulent shear flow

The transfer processes that occur with the turbulent flow of a fluid parallel to a solid wall are of special relevance in chemical, mechanical, and aeronautical engineering. It is necessary to distinguish between *boundary-layer flow* or *external flow* where there is only one wall to be considered, and where the fluid is supposed to extend to an effectively infinite distance in a direction at right angles to the wall, and the alternative case of *flow in a pipe or channel* where the fluid stream is completely enclosed by the walls of the channel and which may be described as *internal flow*. There are some common features to both types of flow, however, and in this section it will be appropriate to establish a useful concept, based mainly on considerations of dimensional analysis, which is known as the *law of the wall*. The specific application to boundary-layer flow will be discussed in more detail in chapter 7, while turbulent flow in pipes and channels will be investigated in chapter 8.

We will start by assuming a rather simple model for the flow system as indicated in figure 49 consisting of three main layers. In the main turbulent flow layer viscous stresses are insignificant in comparison with the Reynolds stresses. Very close to the wall, however, the turbulent motion must be damped down by the action of viscosity and the shear stress must ultimately be transmitted to the wall in the same manner as in a laminar boundary layer, i.e. by the product of the coefficient of viscosity and the gradient of the mean velocity. We can describe this very thin layer as the

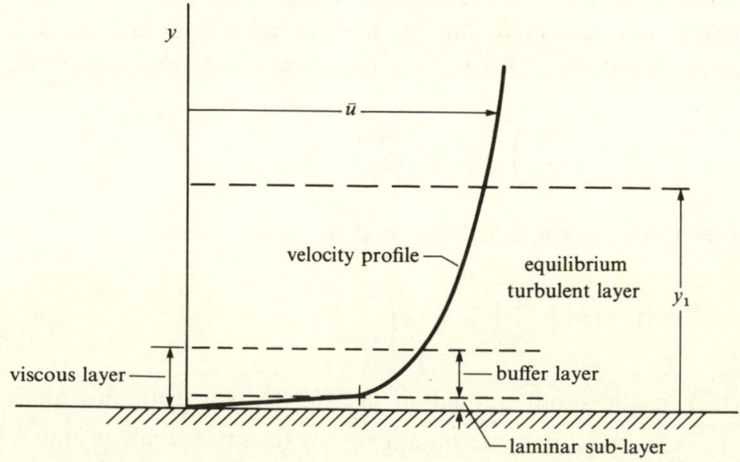

Fig. 49. Mean velocity profile in turbulent boundary layer flow.

laminar sub-layer. It is reasonable to assume that there must be an intermediate zone or *buffer layer* where both viscous shear stresses and Reynolds stresses are of the same order of magnitude.

The diagram shows a typical turbulent mean velocity profile such as might be observed in a boundary-layer region. We assume that the mean velocity $\bar{u}$ is a function of y only. We can start by specifying the shear stress τ in each of the three layers indicated on the diagram:

laminar sub-layer $\quad \tau = \mu \dfrac{\partial \bar{u}}{\partial y}, \qquad (6.5.1)$

buffer layer $\quad \tau = \mu \dfrac{\partial \bar{u}}{\partial y} - \rho\, \overline{u'v'}, \qquad (6.5.2)$

equilibrium turbulent layer $\quad \tau = -\rho\, \overline{u'v'}. \qquad (6.5.3)$

We now make the simplifying assumption that the shear stress τ remains constant through the three layers provided we confine our study of the turbulent zone to a relatively thin 'equilibrium' layer, i.e. we restrict our discussion to $y < y_1$ where y_1 is an arbitrary distance which is only a small fraction of the total width of the turbulent stream. We can then say that

$$\tau \approx \tau_0 = \mu \left(\frac{\partial \bar{u}}{\partial y}\right)_0, \qquad (6.5.4)$$

where τ_0 is the shear stress at the wall.

It will be evident from (6.5.3) that, if $\tau = \tau_0$, we can express the mean value of the velocity product $-u'v'$ in the Reynolds stress term as

$$-\overline{u'v'} = \frac{\tau_0}{\rho} = u_\tau^2 \qquad (6.5.5)$$

where u_τ is a notional 'friction velocity' which is defined by

$$u_\tau = \sqrt{\left(\frac{\tau_0}{\rho}\right)}. \qquad (6.5.6)$$

We can then define dimensionless ratios for the mean velocity of the fluid $\bar{u}$ and for the distance y as follows:

$$\bar{u}^+ = \frac{\bar{u}}{u_\tau} = \frac{\bar{u}}{\sqrt{(\tau_0/\rho)}}, \qquad (6.5.7)$$

$$y^+ = \frac{u_\tau y}{\nu} = \frac{y\sqrt{(\tau_0/\rho)}}{\nu}. \qquad (6.5.8)$$

In the laminar sub-layer, if τ is constant, we have from (6.5.1):

$$\frac{\partial \bar{u}}{\partial y}=\frac{1}{\nu}\frac{\tau_0}{\rho}=\frac{1}{\nu}u_\tau^2$$

and hence, integrating from $y=0$ to level y within the laminar sub-layer, we have

$$\frac{\bar{u}}{u_\tau}=\frac{u_\tau y}{\nu} \tag{6.5.9}$$

or

$$\bar{u}^+=y^+. \tag{6.5.10}$$

We can now turn our attention to the equilibrium turbulent layer which lies outside the laminar and buffer layers. The energy balance in this region will be given by equation (6.4.6). If the mean flow is steady and if $\bar{u}$ does not vary with x, the left-hand side of (6.4.6) must be zero. If we neglect the middle term on the right-hand side of the equation, and making use of (6.4.9), we can express the energy equation in the very simple form:

$$-\overline{u'v'}\frac{\partial \bar{u}}{\partial y}=\varepsilon, \tag{6.5.11}$$

which states that production of turbulent energy by the working of the Reynolds stresses must equal the dissipation of energy by the small-scale turbulence acting through the agency of viscosity.

It is important to keep in mind that we are confining attention to an equilibrium turbulent layer of limited thickness whose outer distance from the wall y_1 is chosen in a rather arbitrary way but which is certainly small compared with the total width of the fluid stream. If we now examine the three physical quantities $-\overline{u'v'}$, $\partial\bar{u}/\partial y$, and ε which appear in the energy equation (6.5.11) it can be argued that these should depend only on the five variables τ_0, ρ, ν, y, and y_1. Taking the energy dissipation rate ε, for example, we can say that

$$\varepsilon=f(\tau_0, \rho, \nu, y, y_1).$$

This relationship between six physical quantities can be reduced, through the normal method of dimensional analysis, to a functional relationship between three dimensionless ratios, i.e. we can say that

$$\frac{\varepsilon y}{(\tau_0/\rho)^{\frac{3}{2}}}=f\left(\frac{y(\tau_0/\rho)^{\frac{1}{2}}}{\nu}, \frac{y}{y_1}\right). \tag{6.5.12}$$

We have already noted in the concluding remarks of § 6.2, however, that the energy dissipation rate ε is largely independent of the magnitude of the viscosity ν. We may also argue that the dimensionless ratio for the energy dissipation rate should be independent of the arbitrary ratio y/y_1 within the range of the relatively thin equilibrium layer which we are considering. It follows that the function in (6.5.12) must be a constant and we may therefore say that

$$\frac{\varepsilon y}{u_\tau^3}=\text{constant}=\frac{1}{k}. \tag{6.5.13}$$

Making use of the energy equation (6.5.11), and substituting from (6.5.6) and (6.5.13), gives

$$u_\tau^2\frac{\partial \bar{u}}{\partial y}=\varepsilon=\frac{u_\tau^3}{ky}$$

or

$$\frac{\partial \bar{u}}{\partial y}=\frac{u_\tau}{ky}. \tag{6.5.14}$$

Noting from (6.5.7) that the velocity ratio u^+ is defined by $\bar{u}/u_\tau$, and using the definition of y^+ from (6.5.8), we see that (6.5.14) may equally well be expressed in the form

$$\frac{\partial \bar{u}^+}{\partial y^+}=\frac{1}{ky^+}. \tag{6.5.15}$$

This can be integrated to give the final result for the mean velocity distribution in the equilibrium turbulent layer

$$\bar{u}^+=\frac{1}{k}\log_e y^+ +\text{constant}, \tag{6.5.16}$$

which is known as the logarithmic form of the *law of the wall.*

The real justification for assigning the status of a law to this result lies in the fact that it is well supported by experimental observation of the mean velocity distribution in turbulent shear layers, at least over a limited range of the ratio y^+ extending from the outer edge of the viscous layer into the main turbulent stream. The experimentally determined value for the constant k, usually known as Karman's constant, is about 0.4, and for the other constant a value of about 5.5 is usually quoted. Thus the practical version of (6.5.16) may be stated as

$$\bar{u}^+=2.5\log_e y^+ +5.5. \tag{6.5.17}$$

It may be noted that the outer edge of the viscous layer (comprising both the laminar sub-layers and the buffer zone) corresponds to a value of about 30 for y^+.

6.6 Velocity distribution for turbulent flow in a pipe or channel

The analysis of turbulent shear flow in the vicinity of a wall as outlined in §6.5, was based on the simplifying assumption that the shear stress could be regarded as being constant in the equilibrium turbulent layer as depicted in figure 49. While this assumption may be approximately correct for a zone of limited thickness in the case of external boundary-layer flow, it cannot be valid for fully developed internal flow in a long pipe where the shear stress must vary in a linear manner with radius. However, the method of dimensional analysis which was used to derive equation (6.5.16) may also be employed to derive a more general expression for the velocity distribution in the neighbourhood of a wall without invoking the condition of constant shear stress. Turbulent flow in pipes will be considered in more detail in chapter 8 but it will be useful to take a preliminary view of the problem at this stage.

In comparison with the parabolic profile for laminar viscous flow in a pipe, a typical turbulent mean velocity profile is relatively flat over the main central area of the cross-section of the pipe but displays a sharp decline or bend in the profile quite close to the pipe wall and a steep gradient at the boundary as indicated in figure 50.

We can make an arbitrary division of the velocity profile into a turbulent shear layer or wall layer, where the velocity distribution is strongly

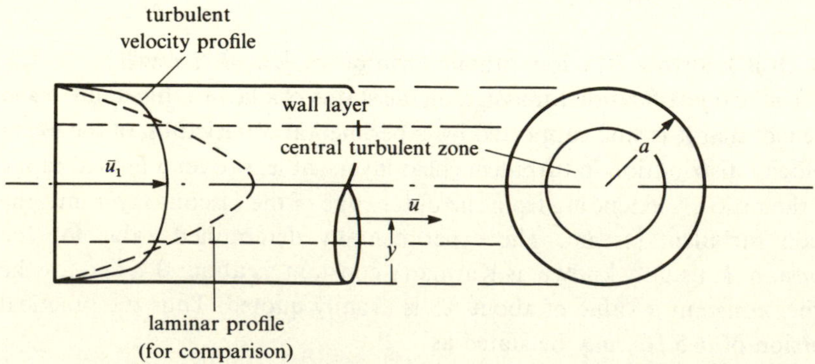

Fig. 50. Laminar and turbulent velocity profiles in a pipe.

influenced by the existence of the pipe wall, and a central turbulent zone where the velocity profile is relatively flat as indicated in figure 50.

In the wall layer, which is broadly equivalent to the equilibrium turbulent layer of figure 49, we can argue that the velocity $\bar{u}$ at distance y from the wall must be a function of the four independent variables, τ_0, ρ, ν, and y, but that it will not be influenced significantly by the pipe diameter. We therefore assume that

$$\bar{u}=f(\tau_0, \rho, \nu, y)$$

and using the normal method of dimensional analysis this relationship between five physical quantities can be reduced to a functional relationship between two dimensionless ratios, i.e.

$$\frac{\bar{u}}{\sqrt{(\tau_0/\rho)}}=f\left(\frac{y\sqrt{(\tau_0/\rho)}}{\nu}\right)$$

or, writing $u_\tau=\sqrt{(\tau_0/\rho)}$ as in (6.5.6),

$$\frac{\bar{u}}{u_\tau}=f\left(\frac{u_\tau y}{\nu}\right) \tag{6.6.1}$$

and using the definitions of (6.5.7) and (6.5.8) this may be written as

$$\bar{u}^+=f(y^+). \tag{6.6.2}$$

This expression is quite general and may be applied to the wall region for turbulent flow either in a pipe or a boundary layer. We have not implied that the shear stress τ at distance y from the wall need be constant. The quantity τ_0 is simply the shear stress at the wall where $y=0$. It will be noted that the logarithmic form (6.5.16) of the law of the wall is a special case of the more general relationship (6.6.2).

If we now turn attention to the central region of the flow we can invoke the principle that, for fully developed steady flow in a long pipe at a sufficient distance from the pipe entry, the mean velocity profile must be the same from section to section and may therefore be expressed in the form:

$$\frac{\bar{u}}{\bar{u}_1}=f\left(\frac{y}{a}\right)=f\left(\frac{a-r}{a}\right), \tag{6.6.3}$$

where $\bar{u}_1$ is the maximum flow velocity at the centre-line; a is the radius of the pipe; and $y=(a-r)$ is distance measured from the wall of the pipe. It may also be noted that the principle of profile continuity represented by (6.6.3) could equally well be expressed as a dimensionless relationship for the *velocity defect* $\bar{u}_1 - \bar{u}$, i.e.

$$\frac{\bar{u}_1 - \bar{u}}{\bar{u}_1} = F\left(\frac{r}{a}\right). \tag{6.6.4}$$

The *mean-flow velocity* $\bar{u}_m$ is defined as the total volume flow rate divided by the flow area πa^2. If the form of the velocity profile is known, the value of $\bar{u}_m$ can be obtained from (6.6.3) by integration as follows:

$$\bar{u}_m = \frac{1}{\pi a^2} \int_0^a 2\pi r \bar{u}\, \mathrm{d}r = \frac{2}{a^2} \int_0^a r\bar{u}\, \mathrm{d}r$$

or

$$\frac{\bar{u}_m}{\bar{u}_1} = \frac{2}{a^2} \int_0^a r f\left(\frac{a-r}{a}\right) \mathrm{d}r = A, \tag{6.6.5}$$

where the ratio A is a constant determined by the form of the profile.

One possible mathematical form which might be employed to give an approximate representation of the velocity profile is the simple power law:

$$\frac{\bar{u}}{\bar{u}_1} = \left(\frac{y}{a}\right)^{\frac{1}{n}} = \left(\frac{a-r}{a}\right)^{\frac{1}{n}}. \tag{6.6.6}$$

It must be noted, however, that this mathematical form is open to the objection that it cannot satisfy the boundary conditions for the velocity profile either at $y=0$ or at the centre-line where $y=a$. It will be seen from (6.6.6) that the velocity gradient is given by

$$\frac{\mathrm{d}\bar{u}}{\mathrm{d}y} = \frac{\bar{u}_1}{ny}\left(\frac{y}{a}\right)^{\frac{1}{n}}. \tag{6.6.7}$$

As $y \to 0$ the gradient tends to an infinite value if $n > 1$. This is not significant, however, since the turbulent velocity profile can only be expected to apply outside the viscous sub-layer. A more serious criticism is the fact that the velocity gradient at the centre-line is finite according to (6.6.7) whereas in reality it should clearly be zero. Despite this difficulty the power law form can be used to give a reasonably good representation of the observed velocity distribution, as we will see subsequently, over the greater part of the cross-section of the pipe.

It is usual to define the skin-friction coefficient c_f for flow in pipes in terms of the mean flow velocity $\bar{u}_m$, so that:

$$\tau_0 = c_f \tfrac{1}{2} \rho \bar{u}_m^2. \tag{6.6.8}$$

It follows from the definition of the 'friction velocity' u_τ given by (6.5.6) that

$$\frac{\bar{u}_m}{u_\tau} = \sqrt{\frac{2}{c_f}}. \tag{6.6.9}$$

The earliest systematic studies of the friction loss for turbulent flow through smooth pipes were made by Blasius, who arrived at the following experimental relationship between the skin-friction coefficient c_f and the Reynolds number Re defined by $\bar{u}_m D/\nu$

$$c_f = 0.079\ Re^{-\frac{1}{4}}. \tag{6.6.10}$$

If we use a Reynolds number defined in terms of the pipe radius a, however, noting that $D = 2a$, the expression (6.6.10) for c_f becomes

$$c_f = 0.0664\left(\frac{\bar{u}_m a}{\nu}\right)^{-\frac{1}{4}}. \tag{6.6.11}$$

Substituting in equation (6.6.9) from (6.6.11) gives

$$\frac{\bar{u}_m}{u_\tau} = 5.487\left(\frac{\bar{u}_m a}{\nu}\right)^{\frac{1}{8}} = 5.487\left(\frac{\bar{u}_m}{u_\tau}\right)^{\frac{1}{8}}\left(\frac{u_\tau a}{\nu}\right)^{\frac{1}{8}},$$

i.e.

$$\left(\frac{\bar{u}_m}{u_\tau}\right)^{\frac{7}{8}} = 5.487\left(\frac{u_\tau a}{\nu}\right)^{\frac{1}{8}}$$

or

$$\frac{\bar{u}_m}{u_\tau} = 6.993\left(\frac{u_\tau a}{\nu}\right)^{\frac{1}{7}}. \tag{6.6.12}$$

We can express the velocity ratio $\bar{u}/u_\tau$ in the form

$$\frac{\bar{u}}{u_\tau} = \frac{\bar{u}}{\bar{u}_1} \times \frac{\bar{u}_1}{\bar{u}_m} \times \frac{\bar{u}_m}{u_\tau}. \tag{6.6.13}$$

Noting from (6.6.5) that $\bar{u}_1/\bar{u}_m = 1/A$, and substituting for $\bar{u}/\bar{u}_1$ from (6.6.6) and for $\bar{u}_m/u_\tau$ from (6.6.12), gives the result:

$$\frac{\bar{u}}{u_\tau} = \left(\frac{y}{a}\right)^{\frac{1}{n}} \times \frac{6.993}{A}\left(\frac{u_\tau a}{\nu}\right)^{\frac{1}{7}}$$

or

$$\frac{\bar{u}}{u_\tau} = \left(\frac{y}{a}\right)^{\frac{1}{n}-\frac{1}{7}} \times \frac{6.993}{A}\left(\frac{u_\tau y}{\nu}\right)^{\frac{1}{7}}. \tag{6.6.14}$$

If this form of the equation for the velocity profile is to be compatible with the requirements of (6.6.2) for the velocity distribution in the wall layer, i.e. if

$\bar{u}/u_\tau$ in this region is to be independent of the pipe radius a, we must have $n=7$. From (6.6.5) and (6.6.6) we can evaluate the ratio A with the result that $A=2[n/(n+1)-n/(2n+1)]=0.817$ if $n=7$. Hence equation (6.6.14) becomes

$$\frac{\bar{u}}{u_\tau}=8.56\left(\frac{u_\tau y}{\nu}\right)^{\frac{1}{7}} \tag{6.6.15}$$

or, if the constant is rounded up to the value 8.6, we have

$$\bar{u}^+=8.6\,y^{+\frac{1}{7}}, \tag{6.6.16}$$

where $\bar{u}^+$ and y^+ are defined as in (6.5.7) and (6.5.8).

The power law velocity profile represented by (6.6.16) must be regarded as a semi-empirical expression based on the Blasius formula (6.6.10) for the skin-friction coefficient for turbulent flow in smooth pipes. It satisfies the two essential requirements, however, of profile continuity as stated by (6.6.3) and of dimensional analysis in the wall layer as stated by equation (6.6.2). Noting that $\bar{u}=\bar{u}_1$ at $y=a$, the velocity profile (6.6.15) or (6.6.16) may equally well be expressed as

$$\frac{\bar{u}}{\bar{u}_1}=\left(\frac{y}{a}\right)^{\frac{1}{7}}. \tag{6.6.17}$$

It is found that the one-seventh power law profile gives good agreement with experimental observations of the velocity distribution for turbulent flow in smooth pipes at values of the Reynolds number up to about 10^5, with the reservation that (6.6.17) implies a finite velocity gradient at $y=a$, whereas the actual velocity profile must in reality be flat at the centre-line with $\mathrm{d}\bar{u}/\mathrm{d}y=0$ at $y=a$.

Despite the different mathematical appearance, the logarithmic form of the law of the wall (6.5.17) and the one-seventh power law profile (6.6.16) give quite close numerical agreement over a fairly wide range of values of y^+, as will be seen from the table:

y^+	30	50	100	200	500	1000	2000
$2.5\log_e y^+ + 5.5$	14.00	15.28	17.01	18.75	21.03	22.77	24.50
$8.6\,y^{+\frac{1}{7}}$	13.98	15.04	16.60	18.33	20.90	23.07	25.47
ratio	0.999	0.984	0.976	0.978	0.994	1.013	1.040

Turbulent flow in pipes will be discussed in greater detail in chapter 8 but it may be helpful in the meantime to present a numerical example using the analysis outlined in this section. We will take the case of water at 20 °C flowing through a 5 cm diameter pipe at a mean flow velocity of 2 m s^{-1}.

We start by evaluating the Reynolds number. From § 1.5 we can take a value for the kinematic viscosity of $\nu = 1.01 \times 10^{-6}\ \mathrm{m^2\ s^{-1}}$. Hence $Re = (2 \times 0.05/1.01) \times 10^6 = 0.99 \times 10^5$ which is just about the top end of the range in which we can use the Blasius formula (6.6.10). Evaluating the skin-friction coefficient from (6.6.10) we have $c_f = 0.004\,44$. Hence $\tau_0/\rho = c_f \times \frac{1}{2}u_m^2 = 0.004\,44 \times \frac{1}{2} \times 4 = 0.008\,88$, and the 'friction velocity' $u_\tau = \sqrt{(\tau_0/\rho)} = 0.0942\ \mathrm{m\ s^{-1}}$.

As noted in § 6.5, the outer edge of the viscous sub-layer is defined approximately by a value of $y^+ = 30$. The corresponding value of $y = (30 \times 1.01/0.0942) \times 10^{-6} = 3.21 \times 10^{-4}$ m or 0.321 mm. There is no precise definition for the distance y_1 to the outer edge of the *wall layer* but it would be reasonable to assume a figure of about one-fifth of the pipe radius as an indication of the wall layer thickness. At this level the fluid velocity according to (6.6.17) is nearly 80% of the maximum velocity at the centre line where $y = a$. Taking $y = 0.5\ \mathrm{cm} = 0.005$ m, we have $y^+ = (0.005 \times 0.0942)/(1.01 \times 10^{-6}) = 466$. At the centre-line of the pipe, where $y = 0.025$ m, we have $y^+ = 2330$.

The range of eddy size to be found in the wall layer can be related directly to the linear dimensions of the layer itself. Noting that the wall layer extends from about $y = 0.3$ mm to about $y = 5$ mm, we can assume that the sizes of the large eddies, which interact most strongly with the mean flow, will also fall within this range. By contrast the typical size of the small eddies, which are responsible for the dissipation of the kinetic energy of the turbulence through the action of the viscous stresses, is represented by the Kolmogorov length scale l_ε defined in (6.2.15).

The energy dissipation rate per unit mass, ε, can be calculated from (6.5.13) taking a value of 0.4 for the constant k. At the outer edge of the viscous sub-layer, or inner edge of the turbulent wall layer, we have $y^+ = 30$, $y = 3.21 \times 10^{-4}$ m, and $\varepsilon = u_\tau^3/ky = 6.5\ \mathrm{m^2\ s^{-3}}$.

The following table summarises the situation at three different levels in the turbulent wall layer: at $y^+ = 30$, at an intermediate level where $y^+ = 100$, and at the nominal outer edge of the wall layer where $y = 5$ mm and $y^+ = 466$.

y^+		30	100	466
y	m	3.21×10^{-4}	1.07×10^{-3}	5.0×10^{-3}
ε	$\mathrm{m^2\ s^{-3}}$	6.5	1.95	0.418
l_ε	m	2.00×10^{-5}	2.70×10^{-5}	3.96×10^{-5}
ratio	y/l_ε	16	40	126

6.7 Eddy viscosity and mixing length models for turbulent motion

The concept of eddy viscosity or eddy diffusivity has been widely used in the past for the correlation of experimental measurements of turbulent motion. It can still serve a limited but useful role despite the fact that the concept is derived from a physically erroneous analogy between the motion of fluid elements in turbulent flow and the motion of molecules as depicted by the kinetic theory of gases. It is postulated in this model that discrete particles or lumps of fluid are moving in a random manner, retaining their own identity and carrying with them certain transferable properties such as momentum and energy, with each movement taking place over an average distance or 'mixing length' which is analogous to the mean free path between molecular collisions. The analogy is fundamentally unsound because fluid elements do not behave like molecules or solid particles. Fluid elements interact continuously with the surrounding fluid, and it also has to be noted that the size of turbulent eddies is typically *not small* compared with the geometry of the mean flow.

Despite these objections, it is appropriate to give a brief outline of the main features of mixing length models. In the following discussion we will assume a two-dimensional mean flow pattern as in the case of the analysis of turbulent shear flow in §6.5.

The basic equation for the transport of momentum by molecular motion in a gas is given by (2.6.4) or (2.6.6) and may be expressed in the form

$$\tau = \tfrac{1}{3}\rho \bar{\lambda}_t \bar{v} \frac{\mathrm{d}\bar{u}}{\mathrm{d}y}, \tag{6.7.1}$$

where $\bar{\lambda}_t$ is the transport mean free path, and $\bar{v}$ is the mean value of the magnitude of the *molecular* velocity. Note carefully that equation (6.7.1) is expressed in the terminology of §§2.5 and 2.6 and that $\bar{v}$ in (6.7.1) does *not* represent a fluid velocity component.

The basic hypothesis of the mixing length model of turbulence is that we can form a similar *gradient transport equation* to express the rate of transport of momentum by turbulent motion, for example

$$\tau = \tfrac{1}{3}\rho l_t \tilde{q} \frac{\mathrm{d}\bar{u}}{\mathrm{d}y}, \tag{6.7.2}$$

where $\tilde{q}$ is the root mean square value of the turbulent velocity q as defined by (6.2.3), and l_t is a length analogous to the transport mean free path in (6.7.1). Alternatively we can define a *mixing length* l' in such a manner that we can express the gradient transport equation in the following more convenient form:

$$\tau = \rho l' \tilde{v}' \frac{\mathrm{d}\bar{u}}{\mathrm{d}y}, \tag{6.7.3}$$

where $\tilde{v}'$ is the root mean square value of the turbulent velocity component v'. Care should again be taken to note the difference in terminology between equations (6.7.1) and (6.7.3).

We could equally well express the gradient transport equation (6.7.3) in the form:

$$\frac{\tau}{\rho} = \nu_T \frac{\mathrm{d}\bar{u}}{\mathrm{d}y}, \tag{6.7.4}$$

where ν_T may be described as the *kinematic eddy viscosity*. While it will be seen from (6.7.3) and (6.7.4) that $\nu_T = l'\tilde{v}'$, which is directly analogous to equation (2.6.7), it is preferable simply to regard equation (6.7.4) as defining the kinematic eddy viscosity. Making use of the definition of the Reynolds stress $\tau = -\rho\,\overline{u'v'}$ as in (6.5.3), therefore, we may re-define the quantity ν_T by the statement that:

$$\nu_T = \frac{\tau}{\rho} \frac{1}{\mathrm{d}\bar{u}/\mathrm{d}y} = -\frac{\overline{u'v'}}{\mathrm{d}\bar{u}/\mathrm{d}y}. \tag{6.7.5}$$

It must be emphasised, however, that unlike the molecular viscosity the eddy viscosity is *not* a constant.

In the special case of a wall layer with constant shear stress, such as we considered in § 6.5, we have $\tau = \tau_0$ and (6.7.5) may then be written in the form

$$\frac{\mathrm{d}\bar{u}}{\mathrm{d}y} = \frac{u_\tau^2}{\nu_T}. \tag{6.7.6}$$

Comparing equations (6.5.14) and (6.7.6), it will be seen that the logarithmic form of the law of the wall (6.5.16) will follow if

$$\nu_T = k u_\tau y. \tag{6.7.7}$$

A more specific form of the mixing length model was postulated by Prandtl involving the additional assumption of complete correlation between the fluctuating velocity components u' and v', so that

$$-\overline{u'v'} = \overline{u'^2} = \overline{v'^2}. \tag{6.7.8}$$

Referring to figure 51, if we picture a small lump of fluid moving from level

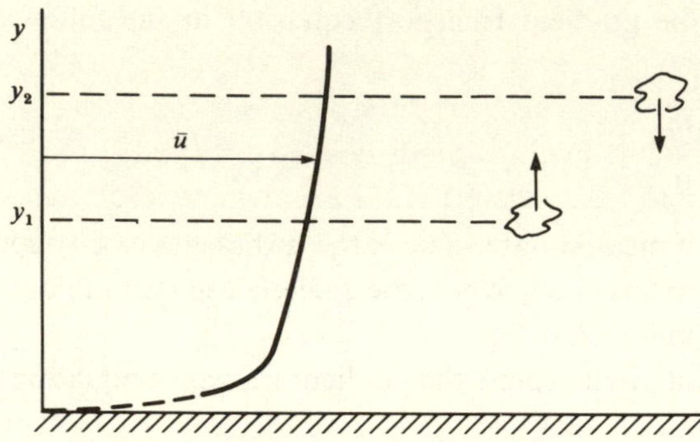

Fig. 51. Prandtl's mixing length model.

y_1 to level y_2 in a turbulent shear layer and if momentum is conserved in the process, the lump will carry with it the mean x momentum appropriate to level y_1. The excess of x momentum at level y_2 due to this movement may then be expressed as

$$\rho u' = -(y_2 - y_1)\frac{\mathrm{d}}{\mathrm{d}y}(\rho\bar{u}). \tag{6.7.9}$$

If this expression is multiplied by the instantaneous fluctuating velocity component v', and if a mean value is then taken over a short interval of time, we arrive at the Reynolds stress or rate of transfer of x momentum across unit area by the turbulent motion, i.e.

$$\tau = -\rho\,\overline{u'v'} = \overline{v'(y_2 - y_1)}\frac{\mathrm{d}}{\mathrm{d}y}(\rho\bar{u}), \tag{6.7.10}$$

which is similar in form to (6.7.3). From (6.7.9) we have

$$u' = -(y_2 - y_1)\frac{\mathrm{d}\bar{u}}{\mathrm{d}y}$$

and therefore, using (6.7.8),

$$-\overline{u'v'} = \overline{u'^2} = \overline{(y_2 - y_1)^2}\left(\frac{\mathrm{d}\bar{u}}{\mathrm{d}y}\right)^2,$$

so that equation (6.7.10) for the shear stress takes the form:

$$\tau = -\rho\,\overline{u'v'} = \rho\,\overline{(y_2 - y_1)^2}\left(\frac{\mathrm{d}\bar{u}}{\mathrm{d}y}\right)^2 \tag{6.7.11}$$

or $$\tau=\rho l^2\left(\frac{d\bar{u}}{dy}\right)^2, \tag{6.7.12}$$

where l is the Prandtl mixing length defined by $l^2=\overline{(y_2-y_1)^2}$. Strictly speaking, in order to take account of the usual convention that the shear stress τ is positive when the gradient $d\bar{u}/dy$ is positive, and vice versa, we should express the last result as

$$\tau=\rho l^2\left|\frac{d\bar{u}}{dy}\right|\left(\frac{d\bar{u}}{dy}\right) \tag{6.7.13}$$

and this is the form which is usually quoted for the shear stress according to the *Prandtl momentum transfer theory*. It should be noted that the Prandtl mixing length l is not the same as the more general form of the mixing length in (6.7.3).

For flow near a wall in a turbulent boundary layer, Prandtl assumed that the mixing length l is related to the distance y by

$$l=ky. \tag{6.7.14}$$

Hence in the special case of a *constant-stress shear layer*, from equation (6.7.12), we can say that

$$\sqrt{\frac{\tau_0}{\rho}}=l\left(\frac{d\bar{u}}{dy}\right)=ky\left(\frac{d\bar{u}}{dy}\right) \tag{6.7.15}$$

or
$$\frac{d\bar{u}}{dy}=\frac{u_\tau}{ky}, \tag{6.7.16}$$

which is identical to equation (6.5.14) and leads directly to the logarithmic form of the law of the wall (6.5.16). It will be seen in fact that (6.7.7) and (6.7.14) are simply alternative ways of expressing the same hypothesis. In a zone of constant shear stress we can say that

$$v_T=l'\bar{v}'=lu_\tau=kyu_\tau, \tag{6.7.17}$$

which clarifies the relationship between the kinematic eddy viscosity and the two alternative forms for the mixing length.

As an alternative to (6.7.14), Karman suggested that the Prandtl mixing length l should depend only on the velocity profile in the immediate vicinity of the point considered and that the simplest relationship which is dimensionally correct would be

$$l=-k\frac{d\bar{u}/dy}{d^2\bar{u}/dy^2}. \tag{6.7.18}$$

If we substitute for l in (6.7.12) from (6.7.18), and assuming a constant shear stress with $\tau=\tau_0$ as before, we find that

$$\sqrt{\frac{\tau_0}{\rho}}=l\frac{d\bar{u}}{dy}=-\frac{k(d\bar{u}/dy)^2}{d^2u/dy^2}$$

i.e.

$$-\frac{d^2\bar{u}/dy^2}{(d\bar{u}/dy)^2}=\frac{k}{u_\tau}.$$

Hence, integrating, we have

$$\frac{1}{d\bar{u}/dy}=\frac{ky}{u_\tau}+\text{constant}.$$

We can take a value of zero for the constant, implying that if the velocity profile could be extended right to the wall the gradient would be infinite at $y=0$. It follows that

$$\frac{d\bar{u}}{dy}=\frac{u_\tau}{ky} \qquad (6.7.19)$$

which once again is identical with equation (6.5.14) leading to the logarithmic form of the law of the wall.

It might thus appear that almost any plausible hypothesis which is *dimensionally correct*, whether in the form of (6.7.7), (6.7.14), or (6.7.18) will lead to the same result in the shape of equation (6.7.19). It will be recalled that the derivation of the same equation (6.5.14) in §6.5 from the energy equation was essentially based on the method of dimensional analysis.

The mixing length model, as depicted in figure 51, also serves as a basis for *Reynolds analogy* between momentum transfer and heat transfer in turbulent flow. We will see in chapter 9, when we examine the complete energy equation, that heat transfer in steady flow problems can be expressed in terms of the change of *enthalpy* of the fluid, and that we can normally express an incremental enthalpy change ΔH as being equal to $c_p\,\Delta T$ where c_p is the specific heat at constant pressure. Thus in figure 51, if we consider a small fluid element of mass m moving from level y_1 to level y_2, the element will carry with it both the momentum and enthalpy appropriate to level y_1. At the same time, in order to satisfy the condition of continuity or conservation of mass, we must suppose that there is a balancing movement of another fluid element of mass m from level y_2 to level y_1 carrying with it the momentum and enthalpy appropriate to level y_2. The net transfer between levels y_1 and y_2 may then be expressed as follows:

$$\text{inward transfer of } x \text{ momentum} = m(\bar{u}_2 - \bar{u}_1)$$

$$\text{inward transfer of enthalpy} = mc_p(T_2 - T_1).$$

If we now assume that there are a large number of such movements and that the elements of fluid move on average between a level where the mean velocity is u_m and the bulk temperature of the fluid is T_m, and a level $y=0$ at the wall where the velocity is zero and where the wall temperature is T_0, we can say that

$$\frac{\text{rate of enthalpy transfer}}{\text{rate of momentum transfer}} = \frac{q_T}{\tau_0} = \frac{\rho c_p(T_m - T_0)}{\rho \bar{u}_m},$$

i.e.

$$\frac{q_T}{\rho \bar{u}_m c_p \, \Delta T} = \frac{\tau_0}{\rho \bar{u}_m^2}, \tag{6.7.20}$$

where q_T is the heat flux, and ΔT is the difference between the bulk fluid temperature T_m and the wall temperature T_0. Equation (6.7.20) may be expressed as

$$St = \tfrac{1}{2} c_f, \tag{6.7.21}$$

where St is the Stanton number defined by $q_T/\rho \bar{u}_m c_p \, \Delta T$.

We can obtain further insight into the implications of Reynolds analogy if we consider the gradient transport equation (6.7.3). Assuming that both momentum and enthalpy are transferred by the same turbulent mechanism, we would expect that a similar gradient transport equation (expressed in terms of the temperature gradient) would be applicable to heat transfer. We would also expect that a similar equation (expressed in terms of the concentration gradient) would apply to the mass transfer of a diffusing component. Thus we could assume that

for *momentum transfer* $\quad \tau = \rho l' \tilde{v}' \dfrac{d\bar{u}}{dy},$

for *heat transfer* $\quad q_T = \rho c_p l' \tilde{v}' \dfrac{dT}{dy},$

and for *mass transfer* $\quad N_A = l' \tilde{v}' \dfrac{dC_A}{dy}.$

If the kinematic eddy viscosity, or eddy diffusivity, $\nu_T = l' \tilde{v}'$ is the same for each of these three turbulent transfer processes, we can conclude that

$$\frac{\tau}{\rho\,\mathrm{d}\bar{u}/\mathrm{d}y}=\frac{q_T}{\rho c_p\,\mathrm{d}T/\mathrm{d}y}=\frac{N_A}{\mathrm{d}C_A/\mathrm{d}y}. \tag{6.7.22}$$

Provided we confine attention to a wall layer with constant shear stress $\tau=\tau_0$, it follows from (6.7.22) that

$$\frac{\mathrm{d}T}{\mathrm{d}y}=\frac{q_T}{\tau_0 c_p}\frac{\mathrm{d}\bar{u}}{\mathrm{d}y}. \tag{6.7.23}$$

Thus, if the heat flux is also constant within the wall layer, the temperature gradient will have a constant ratio to the velocity gradient for all values of y within the wall layer. This is equivalent to the statement that the temperature and velocity profiles will be similar. An identical argument may be applied to the concentration profile for the diffusing component.

The similarity of the velocity, temperature, and concentration profiles according to Reynolds analogy is illustrated in figure 52. If we integrate equation (6.7.23) through the wall layer from level $y=0$, where $u=0$ and $T=T_0$, to a level y where $u=\bar{u}_m$ and $T=T_m$, we get

$$T_m-T_0=\frac{q_T\bar{u}_m}{\tau_0 c_p}$$

or

$$\frac{q_T}{\rho\bar{u}_m c_p\,\Delta T}=\frac{\tau_0}{\rho\bar{u}_m^2}, \tag{6.7.24}$$

which is identical with (6.7.20) or (6.7.21).

It will be seen that the Reynolds analogy takes no account of the

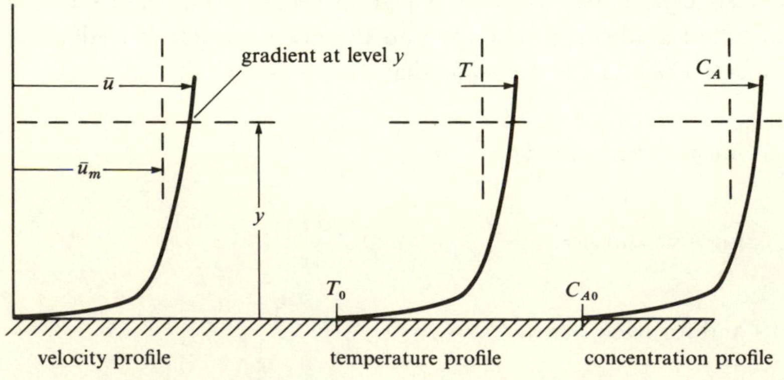

Fig. 52. Reynolds analogy.

existence of a viscous sub-layer adjacent to the wall. It must therefore be regarded as an over-simplification of the real physical situation. It is possible to extend the analogy, however, to allow for the existence of a laminar sub-layer as in the case of the Prandtl–Taylor analogy which is considered in chapter 17. A similar analysis applied to the multi-layer model of § 7.10 leads to the von Karman analogy which is also reviewed in chapter 17.

7

Boundary layers

7.1 The nature of flow in a boundary layer

The concept of a thin layer of fluid adjacent to a solid boundary, in which the fluid velocity changes spatially within a small distance measured normal to the surface from zero (the no-slip condition) at the surface to the full value of the main stream velocity at the outer edge of the layer, is of special importance in engineering applications of fluid mechanics. Boundary layers are formed wherever a fluid flows past a solid surface and the effects of viscosity are always important in a boundary-layer region.

It is observed in practice that the flow in a boundary layer may be either laminar or turbulent in character. In figure 53 we represent the simplest type of boundary layer formed on a flat plate which is held in position in a uniform fluid stream and located so that the plane of the surface is parallel to the velocity of the undisturbed stream.

At the leading edge of the plate there must be an abrupt change in the velocity of the oncoming fluid stream in a very thin layer close to the plate to satisfy the requirement of zero slip at the boundary. As we survey the situation further downstream, at a distance x measured from the leading edge, we will find that the thickness of the layer within which the velocity rises from $u=0$ at the plate to $u=u_1$ at the outer edge has increased as indicated in figure 53. It is generally found that the flow pattern in the boundary layer is laminar in character for a certain distance x_L measured from the leading edge, but that further downstream the flow in the boundary layer becomes turbulent. As a rough indication of what may be expected, we can say that if the Reynolds number defined by $Re=u_1x/\nu$ is less than about 1.2×10^5 the flow in the boundary layer will usually be laminar, but for greater values of the Reynolds number the flow may be expected to become turbulent. Transition from laminar to turbulent flow in a boundary layer, however, is a complex phenomenon and does not depend on a single parameter.

It is convenient to refer to the *thickness* of a boundary layer, represented

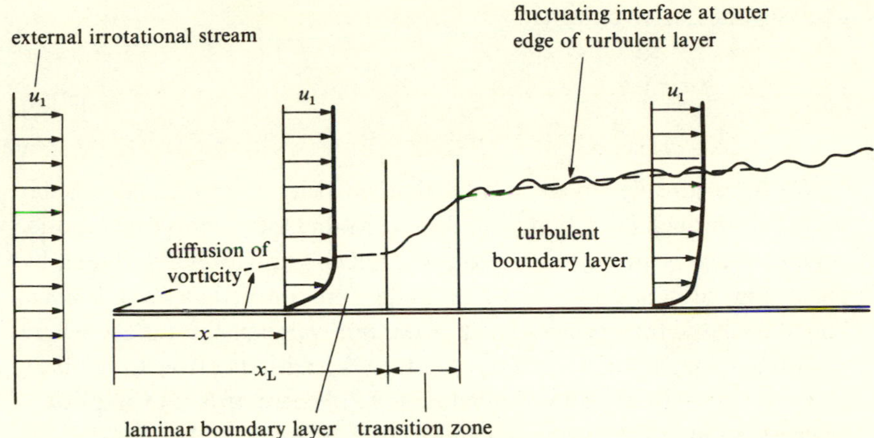

Fig. 53. Boundary layer on a flat plate.

by the symbol δ, although it is not altogether easy to provide a satisfactory definition of this quantity. In the case of a laminar boundary layer there is no distinct interface between the boundary-layer region and the undisturbed main stream and the velocity u at distance y from the wall approaches the value of the undisturbed stream velocity u_1 in an asymptotic manner as the distance y increases. One possibility is to define the thickness δ as the value of y for which the velocity ratio $u/u_1 = 0.99$, but other alternative definitions are discussed in §7.3. The thickness of a boundary layer δ (however we may choose to define it) formed on a flat plate in a steady uniform stream will generally be a function of the distance x measured from the leading edge, the stream velocity u_1, and the fluid properties ρ and μ. This relationship involving five physical quantities may be reduced by the normal method of dimensional analysis to a functional relationship between two dimensionless ratios, i.e. we can say that

$$\frac{\delta}{x} = f\left(\frac{\rho u_1 x}{\mu}\right) \quad \text{or} \quad f\left(\frac{u_1 x}{\nu}\right). \tag{7.1.1}$$

We will see in §7.4 that for a laminar boundary layer this relationship takes the form

$$\frac{\delta}{x} \sim \left(\frac{u_1 x}{\nu}\right)^{-\frac{1}{2}}, \tag{7.1.2}$$

while for a turbulent boundary layer, if we use the approximate method of analysis of § 7.8, we find that

$$\frac{\delta}{x} \sim \left(\frac{u_1 x}{\nu}\right)^{-\frac{1}{5}}. \tag{7.1.3}$$

We have already noted in § 3.8 that vorticity is created at a solid boundary through the action of fluid viscosity and the property of zero slip. Boundary layers are therefore regions in which vorticity is distributed by the combined action of viscous diffusion and convection. In a *laminar* boundary layer the vorticity has a maximum value at the surface where $y=0$ and its magnitude falls off progressively with distance from the surface, under steady-state conditions, tending towards zero at large values of y. There is no interface or line of demarcation between the boundary-layer region with its distribution of vorticity and the irrotational or nearly irrotational flow region of the main stream outside the boundary layer. The situation is different, however, with a *turbulent* boundary layer. While the vorticity in a turbulent boundary layer will again have a maximum value at the surface where $y=0$, the magnitude of the vorticity will be significant throughout the turbulent flow region since turbulence always involves a relatively high level of vorticity. We must therefore expect to observe a distinct interface between the flow in a turbulent boundary layer and the irrotational flow regime outside. This is illustrated diagrammatically in figure 53 but it must be emphasised that this outer layer or interface between the turbulent boundary layer and the undisturbed flow is highly irregular in shape and unsteady with time.

In this section we have presented the concept of boundary layer flow in its simplest form by taking the idealised case of steady uniform flow parallel to a thin flat plate. In engineering applications we have to deal with flow past curved surfaces more usually than flow past flat surfaces. Provided the curvature is not too great, however, we can treat the problem as being equivalent to flow past a flat surface with a longitudinal pressure gradient in the x direction and with an external stream velocity u_1 which also varies gradually with x. A good example is the case of flow past an aeroplane wing or aerofoil blade section where boundary layers are formed on both upper and lower surfaces extending from the nose or front stagnation point to the trailing edge.

Boundary layers are also observed in nature. The lower part of the earth's atmosphere, extending to a height of about 1000 metres, may be regarded as a boundary-layer region with the high-altitude winds forming the external stream. The flow pattern is naturally unsteady, however, since the high-altitude winds change in direction and themselves form part of the overall circulatory flow pattern around the globe.

7.2 The boundary-layer equations

We will take the basic case of an external stream with velocity u_1, which may, however, be a function of the distance x, flowing parallel to a flat surface as indicated in figure 54. We will assume that the flow pattern is essentially two-dimensional in character with no velocity component in the z direction. Figure 54 represents a plane section which is perpendicular to the z axis. The distance x is measured from the leading edge of the plate as indicated. The co-ordinate y is measured normal to the surface.

The essential equations of motion are the two components of the Navier–Stokes equation (5.4.3) for incompressible flow, together with the continuity equation (3.3.5) in its two-dimensional form, i.e.

$$\frac{\partial u}{\partial t}+u\frac{\partial u}{\partial x}+v\frac{\partial u}{\partial y}=-\frac{1}{\rho}\frac{\partial p}{\partial x}+\nu\left[\frac{\partial^2 u}{\partial x^2}+\frac{\partial^2 u}{\partial y^2}\right], \tag{7.2.1}$$

$$\frac{\partial v}{\partial t}+u\frac{\partial v}{\partial x}+v\frac{\partial v}{\partial y}=-\frac{1}{\rho}\frac{\partial p}{\partial y}+\nu\left[\frac{\partial^2 v}{\partial x^2}+\frac{\partial^2 v}{\partial y^2}\right], \tag{7.2.2}$$

and

$$\frac{\partial u}{\partial x}+\frac{\partial v}{\partial y}=0. \tag{7.2.3}$$

The boundary conditions which must be satisfied are that at the surface of the plate, where $y=0$, we have both $u=0$ (zero slip) and $v=0$. At the outer edge where $y\rightarrow\infty$ or where $y=\delta$ we have $u=u_1$ and $\partial u/\partial y=0$.

It is observed experimentally that the boundary layer region is relatively thin if the Reynolds number of the flow is large. In more precise terms, if we

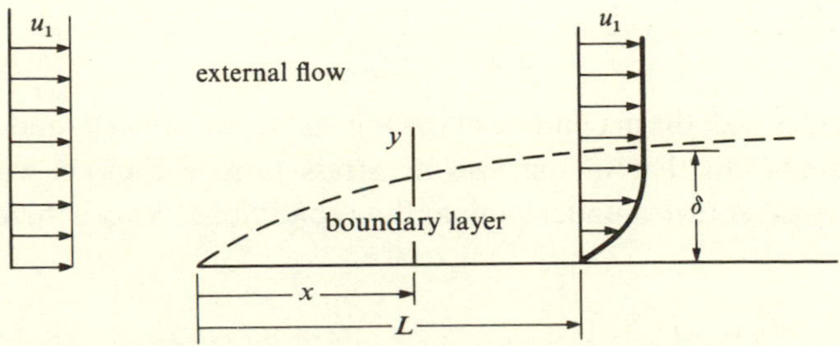

Fig. 54. Formation of a thin boundary layer.

examine the situation at a distance $x=L$ from the leading edge and if δ is a measure of the boundary-layer thickness at this point, we can say that δ/L will be small if the value of $Re=u_1L/\nu$ is large.

We can now examine the orders of magnitude of the terms which appear in equations (7.2.1) and (7.2.2). We note first of all that the velocity component u ranges from 0 to u_1 within the boundary layer, that the range for the co-ordinate x is from 0 to L, and for the co-ordinate y from 0 to δ. We can regard u_1 and L as the reference magnitudes for velocity and distance respectively. It is almost obvious from figure 54 that the velocity component v must be small in comparison with u or u_1, but this point can be proved by referring to the continuity equation (7.2.3). We may express the order of magnitude of the velocity component u and the velocity gradient $\partial u/\partial x$ in the form

$$\frac{u}{u_1}=0\left(\frac{x}{L}\right)=0(1) \quad \text{and} \quad \frac{\partial u}{\partial x}=0\left(\frac{u_1}{L}\right).$$

It will be evident from the continuity equation that the order of magnitude of the term $\partial v/\partial y$ cannot be greater than $0(u_1/L)$ and since the range of y is from 0 to δ we conclude that

$$\frac{v}{u_1}=0\left(\frac{y}{L}\right)=0\left(\frac{\delta}{L}\right).$$

It is easily verified that the two terms $u(\partial u/\partial x)$ and $v(\partial u/\partial y)$ which appear on the left-hand side of equation (7.2.1) are of similar order whose magnitude is represented by $0(u_1^2/L)$. The two second derivative terms $\partial^2u/\partial x^2$ and $\partial^2u/\partial y^2$, however, differ significantly. The term $\partial^2u/\partial x^2$ is $0(u_1/L^2)$ while $\partial^2u/\partial y^2$ is $0(u_1/\delta^2)$. We can thus ignore the term $\partial^2u/\partial x^2$ in comparison with the term $\partial^2u/\partial y^2$ within the square brackets on the right-hand side of (7.2.1). We may thus re-write equation (7.2.1) for a thin boundary layer as

$$\frac{\partial u}{\partial t}+u\frac{\partial u}{\partial x}+v\frac{\partial u}{\partial y}=-\frac{1}{\rho}\frac{\partial p}{\partial x}+\nu\frac{\partial^2 u}{\partial y^2}. \tag{7.2.4}$$

We would expect the magnitude of the inertial terms $u(\partial u/\partial x)$ and $v(\partial u/\partial y)$ to be similar to that of the viscous stress term $\nu(\partial^2u/\partial y)^2$ within the boundary-layer region under laminar flow conditions. We can therefore say that

$$0\left(\frac{u_1^2}{L}\right)=0\left(\nu\frac{u_1}{\delta^2}\right)$$

or, dividing through by u_1^2/L to make the statement dimensionless,

$$0(1)=0\left(\frac{\nu}{u_1 L}\cdot\frac{L^2}{\delta^2}\right).$$

It follows that for a *laminar boundary layer* $\delta^2/L^2=0(\nu/u_1 L)$ or

$$\frac{\delta}{L}=0(Re^{-\frac{1}{2}}), \tag{7.2.5}$$

where the Reynolds number is defined by $Re=u_1 L/\nu$. This confirms the experimental observation that the condition for the existence of a thin boundary layer is that the Reynolds number should be relatively large.

If we now look at equation (7.2.2) it will be seen that all the terms on the left-hand side are of order of magnitude $0(u_1^2\delta/L^2)$. For the viscous stress terms on the right-hand side of (7.2.2) we see that $\partial^2 v/\partial x^2$ is negligible in comparison with $\partial^2 v/\partial y^2$, and noting from (7.2.5) that $\nu/u_1 L$ is $0(\delta^2/L^2)$, we have $\nu(\partial^2 v/\partial y^2)=0(u_1^2\delta/L^2)$ which is the same order of magnitude as the terms on the left-hand side. We are left with the pressure gradient term $(-1/\rho)(\partial p/\partial y)$ which cannot therefore be greater than $0(u_1^2\delta/L^2)$ in magnitude. Hence for a thin boundary layer, where δ/L is small by definition, the pressure gradient $\partial p/\partial y$ normal to the surface must also be small and equation (7.2.2) consequently reduces to $\partial p/\partial y=0$. Thus the pressure in the fluid, which is established by the flow pattern in the external stream and which may be a function of the longitudinal distance x, is transmitted through the boundary layer without change in the y direction.

Summarising our conclusions at this stage, we can say that for steady flow past a flat or slightly curved surface at a value of the Reynolds number which is sufficiently large for the boundary layer to be thin, the basic equations (7.2.1), (7.2.2), and (7.2.3) for a *laminar boundary layer* take the form:

$$u\frac{\partial u}{\partial x}+v\frac{\partial u}{\partial y}=-\frac{1}{\rho}\frac{\partial p}{\partial x}+\nu\frac{\partial^2 u}{\partial y^2}, \tag{7.2.6}$$

$$\frac{\partial p}{\partial y}=0 \tag{7.2.7}$$

and

$$\frac{\partial u}{\partial x}+\frac{\partial v}{\partial y}=0. \tag{7.2.8}$$

The manner in which these equations have to be modified or extended to cover flow in a *turbulent boundary layer* will be considered in §7.9.

It will be seen from (7.2.6) that outside the boundary layer, where viscous effects are negligible and where $u = u_1$ which is assumed to be a function of x only, the equation reduces to

$$u_1 \frac{du_1}{dx} = -\frac{1}{\rho}\frac{dp}{dx}, \tag{7.2.9}$$

which is simply the Bernoulli equation in differential form.

7.3 Definition of boundary-layer thickness

We have already noted that a difficulty arises in specifying the boundary-layer thickness δ because the velocity profile must merge imperceptibly into the external stream, i.e. as $y \to \delta$, $u \to u_1$ and $\partial u/\partial y \to 0$.

A quantity known as the *displacement thickness* δ^* can be defined in a precise way, however, as follows. Referring to figure 55, if the volume flow in the boundary layer is Q, we can say that

$$Q = \int_0^\delta u \, dy = u_1(\delta - \delta^*), \tag{7.3.1}$$

i.e.

$$\delta^* = \int_0^\delta \left(1 - \frac{u}{u_1}\right) dy. \tag{7.3.2}$$

The physical meaning of this definition is that δ^* represents the distance by which an equivalent uniform stream would have to be displaced from the surface as indicated in figure 55 to give the same total volume flow.

A similar picture may be drawn for the *momentum flow* in the boundary layer. Referring to figure 56, if the momentum flow in the boundary layer is

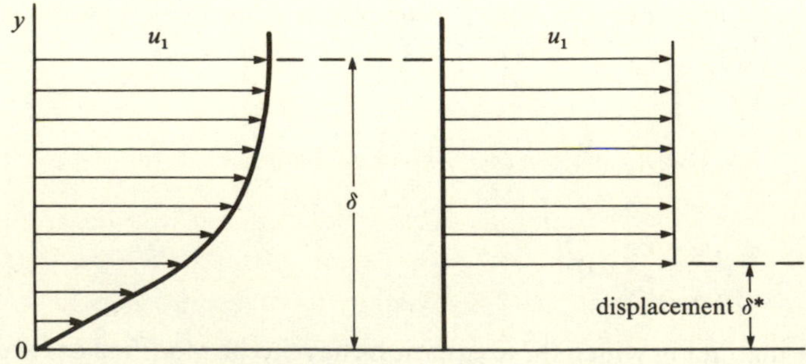

Fig. 55. Displacement thickness.

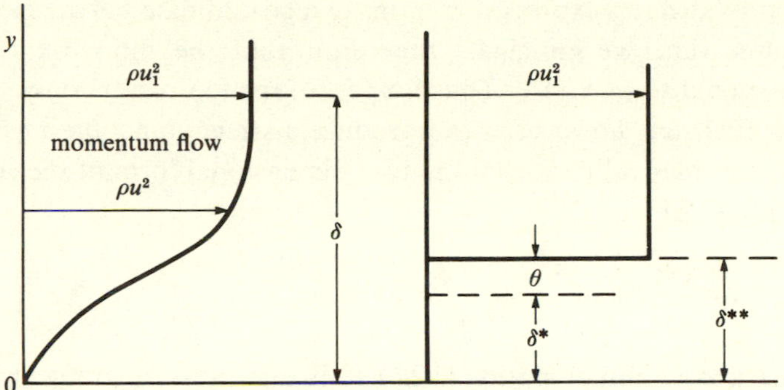

Fig. 56. Momentum thickness.

M, we can say that

$$M = \int_0^\delta \rho u^2 \, dy = \rho u_1^2(\delta - \delta^{**}) \tag{7.3.3}$$

and, assuming constant density, we therefore have

$$\delta^{**} = \int_0^\delta \left(1 - \frac{u^2}{u_1^2}\right) dy. \tag{7.3.4}$$

It would be logical to describe the quantity δ^{**} as the momentum thickness or the momentum-displacement thickness. By a long-established custom, however, the name *momentum thickness* is reserved for another quantity θ as indicated in figure 56 and defined by $\theta = \delta^{**} - \delta^*$. From (7.3.2) and (7.3.4) we have

$$\theta = \delta^{**} - \delta^* = \int_0^\delta \frac{u}{u_1}\left(1 - \frac{u}{u_1}\right) dy. \tag{7.3.5}$$

It will be seen in §7.6 that the momentum thickness θ, defined by (7.3.5), plays an important role in the momentum equation for boundary-layer flow.

7.4 Exact solution for a laminar boundary layer

It is possible to obtain an exact solution to the laminar boundary-layer equations (7.2.6) and (7.2.8) in the special case when the longitudinal pressure gradient $\partial p/\partial x$ is zero. The key to this solution, which was first obtained by Blasius in 1908, is to introduce the hypothesis that the velocity profile in the boundary layer is similar from section to section along the plate provided it is expressed in terms of a co-ordinate y/δ in place of y. At the same time we anticipate the result that the ratio δ/x is directly proportional to $(u_1x/\nu)^{-\frac{1}{2}}$. Thus u/u_1 is a function of $y/x^{\frac{1}{2}}$ only.

The first step, however, is to introduce a stream function ψ as in (4.4.4) which automatically satisfies the two-dimensional form of the continuity equation (7.2.8), i.e.

$$u=\frac{\partial\psi}{\partial y} \quad \text{and} \quad v=-\frac{\partial\psi}{\partial x}. \tag{7.4.1}$$

The basic equation of motion (7.2.6) with zero pressure gradient takes the form

$$u\frac{\partial u}{\partial x}+v\frac{\partial v}{\partial y}=\nu\frac{\partial^2 u}{\partial y^2} \tag{7.4.2}$$

and, substituting from (7.4.1), this becomes

$$\frac{\partial\psi}{\partial y}\frac{\partial^2\psi}{\partial x\,\partial y}-\frac{\partial\psi}{\partial x}\frac{\partial^2\psi}{\partial y^2}=\nu\frac{\partial^3\psi}{\partial y^3}. \tag{7.4.3}$$

If we now introduce new dimensionless variables defined by

$$\eta=\frac{1}{2}\left(\frac{u_1x}{\nu}\right)^{\frac{1}{2}}\frac{y}{x} \quad \text{and} \quad f=\frac{\psi}{u_1x}\left(\frac{u_1x}{\nu}\right)^{\frac{1}{2}}, \tag{7.4.4}$$

we find that the partial differential equation (7.4.3) can be reduced to an ordinary differential equation of the form

$$\frac{\mathrm{d}^3f}{\mathrm{d}\eta^3}+f\frac{\mathrm{d}^2f}{\mathrm{d}\eta^2}=0. \tag{7.4.5}$$

The fact that we now have an ordinary differential equation with only two variables proves the validity of the hypothesis that u/u_1 is a function of $y/x^{\frac{1}{2}}$ only, and that velocity profiles are similar when expressed in terms of the new variable $\eta=\frac{1}{2}(u_1/\nu x)^{\frac{1}{2}}y$.

The third-order differential equation (7.4.5) may be solved by expressing the variable f as a series in powers of η. A solution which satisfies the boundary condition that $u=0$ and $v=0$ at the wall where $\eta=0$ takes the form

$$f=\frac{\alpha\eta^2}{2!}-\frac{\alpha^2\eta^5}{5!}+11\frac{\alpha^3\eta^8}{8!}-375\frac{\alpha^4\eta^{11}}{11!}+\cdots. \tag{7.4.6}$$

where α is a numerical constant which has to be determined so that the remaining boundary condition, $u \to u_1$ as $y \to \infty$ is also satisfied. The numerical value of α which meets this requirement is $\alpha = 1.3282$.

It will be seen from (7.4.1) and (7.4.4) that $u/u_1 = \frac{1}{2}(\mathrm{d}f/\mathrm{d}\eta)$ and the velocity profile may thus be expressed in terms of η by

$$\frac{u}{u_1} = \frac{1}{2}\left[\alpha\eta - \frac{\alpha^2\eta^4}{4!} + 11\frac{\alpha^3\eta^7}{7!} - 375\frac{\alpha^4\eta^{10}}{10!} + \cdots\right]. \tag{7.4.7}$$

The skin friction at the wall is given by

$$\tau_0 = \mu\left(\frac{\partial u}{\partial y}\right)_{y=0} = \frac{\mu}{2}\left(\frac{u_1}{\nu x}\right)^{\frac{1}{2}}\left(\frac{\partial u}{\partial \eta}\right)_{\eta=0} = \frac{\mu u_1 \alpha}{4}\left(\frac{u_1}{\nu x}\right)^{\frac{1}{2}} \tag{7.4.8}$$

and hence

$$c_f = \frac{\tau_0}{\frac{1}{2}\rho u_1^2} = \frac{\alpha}{2}\left(\frac{u_1 x}{\nu}\right)^{-\frac{1}{2}} = 0.664\left(\frac{u_1 x}{\nu}\right)^{-\frac{1}{2}}. \tag{7.4.9}$$

The displacement thickness δ^* is given by

$$\frac{\delta^*}{x} = 1.7208\left(\frac{u_1 x}{\nu}\right)^{-\frac{1}{2}} \tag{7.4.10}$$

and the momentum thickness θ is given by

$$\frac{\theta}{x} = 0.664\left(\frac{u_1 x}{\nu}\right)^{-\frac{1}{2}}. \tag{7.4.11}$$

7.5 Limits of laminar flow and the effects of pressure gradient

We have already noted in § 7.1 that laminar flow is likely to prevail in a boundary layer at values of the Reynolds number $Re = u_1 x/\nu$ up to about 1.2×10^5. Alternatively we can define a Reynolds number in terms of the boundary-layer thickness δ or the displacement thickness δ^*. If we use the latter definition we can conclude from the result (7.4.10) that, for a laminar boundary layer formed on a flat plate with zero longitudinal pressure gradient,

$$\frac{u_1\delta^*}{\nu} = 1.7208\left(\frac{u_1 x}{\nu}\right)^{\frac{1}{2}}. \tag{7.5.1}$$

If we put $u_1 x/\nu = 1.2 \times 10^5$ the equivalent value of $u_1\delta^*/\nu$ is approximately 600 and this may be taken as a critical value for the Reynolds number defined in terms of the displacement thickness.

Mathematical analysis of the stability of two-dimensional flow in a laminar boundary layer suggests that the flow pattern may become unstable with respect to disturbances, whose wavelengths fall within a certain range, at values of the Reynolds number $u_1\delta^*/\nu$ greater than 600. This analysis supports the experimental observation noted above, although it is not by any means certain that the amplification of small two-dimensional disturbances is the main cause of transition in a boundary layer. It has to be recalled that turbulence is essentially three-dimensional in character and it is possible that transition to turbulent flow in a boundary layer is more frequently caused either by the formation of three-dimensional eddies within the boundary-layer region due to the presence of small projections or roughness particles on the surface, or alternatively through the agency of small eddies or vortices which may be present in the external flow pattern and which penetrate the boundary layer.

It is important also to appreciate that if a spanwise view is taken of the flow past the plate, transition to turbulence does not take place precisely along a line at the same value of x at every point. In practice we find that patches of turbulence develop at different points along the span of the plate, and that these subsequently spread out and merge downstream into a completely turbulent regime for the boundary layer. Transition from laminar to turbulent flow in a boundary layer is in fact an extremely complex phenomenon and is not controlled by a single parameter.

With these provisos and reservations, however, the concept of a critical distance x_L is useful in providing a guide as to the probable extent of the stable laminar flow regime which may be expected to prevail for a limited distance downstream from the leading edge. Just how limited this distance is will be apparent from the following table which gives the critical distance x_L corresponding to a value of 1.2×10^5 for the Reynolds number for (a) air, and (b) water flowing at different velocities past a flat plate.

Air at 20 °C

stream velocity u_1	5	10	20	50	100	m s^{-1}
critical length x_L	0.36	0.18	0.09	0.036	0.018	m;

Water at 20 °C

stream velocity u_1	0.5	1	2	5	10	m s^{-1}
critical length x_L	0.242	0.121	0.061	0.024	0.012	m.

The exact solution of the laminar boundary-layer equations outlined above in § 7.4 is only applicable to the case of flow parallel to a flat plate with zero external pressure gradient $\partial p/\partial x$. In many important applications, however, we have to consider the flow in a boundary layer where there is either a favourable (falling) or adverse (rising) pressure gradient in the direction of the external stream. In the case of flow past an aerofoil section, for example, we have a falling pressure gradient over the upper surface from the front stagnation point to a point near the quarter-chord position (see §4.7) followed by a rising pressure gradient over the remaining part of the upper surface as far as the trailing edge. It is found experimentally that a falling pressure gradient has a stabilising effect on laminar flow in the boundary layer and tends to prolong the extent of the laminar region. Conversely, a rising pressure gradient has a destabilising effect and tends to promote transition to turbulent flow in the boundary layer.

It is important to note also that we no longer have velocity profile similarity from section to section in a boundary layer when there is a variable external longitudinal pressure gradient. The influence of the pressure gradient on the velocity profile will be evident if we take the laminar boundary-layer equation (7.2.6) and put $y=0$. Since both velocity components u and v are zero at the wall where $y=0$, the terms on the left-hand side of (7.2.6) vanish and we are left with the condition that, at $y=0$,

$$\mu\left(\frac{\partial^2 u}{\partial y^2}\right)_{y=0}=\frac{\partial p}{\partial x}. \tag{7.5.2}$$

If $\partial p/\partial x$ is zero, it follows from (7.5.2) that $(\partial^2 u/\partial y^2)_{y=0}=0$ and the velocity gradient close to the wall is a constant $=\tau_0/\mu$. In this case the velocity gradient falls off progressively with increasing distance from the wall as indicated in figure 57(*a*).

If $\partial p/\partial x$ is negative, representing a falling or favourable pressure gradient, it will be seen from (7.5.2) that $\partial^2 u/\partial y^2$ is negative at $y=0$ and that the velocity profile therefore takes the form indicated in figure 57(*b*). If, on the other hand, $\partial p/\partial x$ is positive, it follows from (7.5.2) that $\partial^2 u/\partial y^2$ must be positive at $y=0$ and the velocity profile takes the form shown in figure 57(*c*),

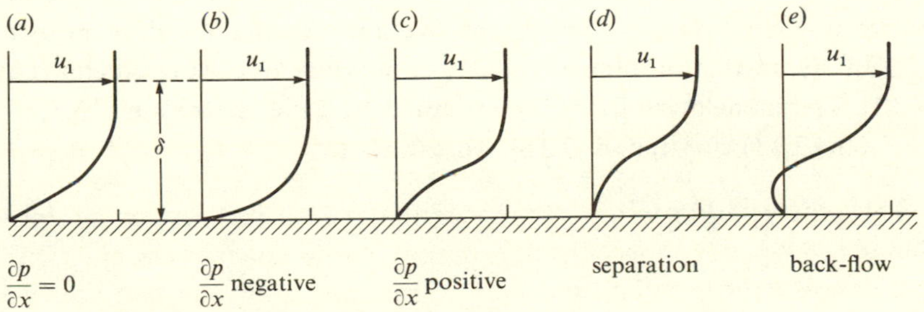

Fig. 57. Effect of pressure gradient on velocity profile.

i.e. the velocity gradient $\partial u/\partial y$ increases initially with increasing distance from the wall before reaching a maximum value at some intermediate value of y and subsequently falling off to zero at the outer edge of the boundary layer. Under extreme conditions with a severe adverse pressure gradient the S-shaped profile of figure 57(*d*) may reach the point where the velocity gradient $\partial u/\partial y$ is zero at the wall. Under these conditions *separation of flow* occurs and there will be dead-water or back-flow region adjacent to the wall.

While separation of a laminar boundary layer is observed in certain flow patterns, it frequently happens that transition to turbulent flow takes place in the boundary layer before laminar separation can occur. The boundary layer is then able to adhere to the surface for a greater distance in face of the adverse pressure gradient before separation of the turbulent boundary layer occurs.

7.6 The momentum equation for the boundary layer

We have seen in §7.4 that, although an exact solution of the laminar boundary-layer equations can be obtained in the simplest case when $\partial p/\partial x = 0$, the mathematical solution of the equations of motion is not altogether easy. An alternative approach is to formulate the momentum equation for steady flow in a boundary-layer region on the basis of the principles outlined in §3.7. This approach makes it possible to derive simple approximate solutions for the laminar boundary layer which are very close to the exact results. The method may also be used to obtain useful approximations for the boundary-layer thickness and skin-friction coefficient for a turbulent boundary layer.

We start by considering a short length dx of the boundary layer as shown in figure 58. Let the nominal thickness of the boundary layer at distance x from the leading edge be represented by δ. We assume for the purposes of the following discussion that δ is a finite and measurable quantity although its precise definition is not important. We use the same terminology as in §7.3 and represent the volume flow in the boundary layer by Q as defined by (7.3.1). The momentum flow is represented by M as defined in (7.3.3).

From (7.3.1), (7.3.3) and (7.3.5), we can write:

$$\rho Q = \rho u_1(\delta - \delta^*) \tag{7.6.1}$$

and

$$M = \rho u_1^2(\delta - \delta^* - \theta). \tag{7.6.2}$$

If the volume flow Q included within the boundary-layer region increases

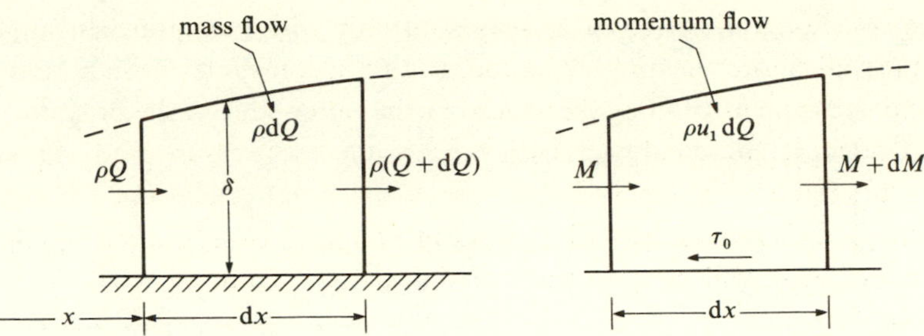

Fig. 58. Mass and momentum balances on an elementary length of boundary layer.

by the amount $\mathrm{d}Q$ in distance $\mathrm{d}x$, the principle of continuity requires that an inward flow $\mathrm{d}Q$ should occur across the outer edge of the boundary layer as shown in figure 58. This is simply another way of expressing the fact that the boundary-layer thickness δ is increasing steadily as the flow proceeds parallel to the surface in the x direction. The inward flow of mass $\rho\, \mathrm{d}Q$ involves a corresponding inward flow of momentum $\rho u_1\, \mathrm{d}Q$. If we now take a *control surface* extending from the wall at $y=0$ to the outer edge of the boundary layer, and from distance x to distance $x+\mathrm{d}x$, and if we consider unit width of flow in the z direction, we can express the steady-flow momentum equation as follows:

$$-\tau_0\, \mathrm{d}x - \delta \frac{\partial p}{\partial x}\, \mathrm{d}x = \text{net outflow of } x \text{ momentum}$$

$$= \frac{\mathrm{d}M}{\mathrm{d}x}\, \mathrm{d}x - \rho u_1 \frac{\mathrm{d}Q}{\mathrm{d}x}\, \mathrm{d}x.$$

Substituting from (7.6.1) and (7.6.2) gives

$$-\frac{\tau_0}{\rho} - \frac{\delta}{\rho}\frac{\partial p}{\partial x} = \frac{\mathrm{d}}{\mathrm{d}x}\left[u_1^2(\delta - \delta^* - \theta)\right] - u_1 \frac{\mathrm{d}}{\mathrm{d}x}\left[u_1(\delta - \delta^*)\right],$$

which simplifies to

$$-\frac{\tau_0}{\rho} - \frac{\delta}{\rho}\frac{\partial p}{\partial x} = -u_1^2 \frac{\mathrm{d}\theta}{\mathrm{d}x} + (\delta - \delta^* - 2\theta) u_1 \frac{\mathrm{d}u_1}{\mathrm{d}x}. \qquad (7.6.3)$$

But from (7.2.9) at the outer edge of the boundary layer

$$u_1 \frac{\mathrm{d}u_1}{\mathrm{d}x} = -\frac{1}{\rho}\frac{\mathrm{d}p}{\mathrm{d}x} \qquad (7.6.4)$$

and since the pressure is transmitted without change in the y direction through the boundary layer $\partial p/\partial y=0$ and $\partial p/\partial x=\mathrm{d}p/\mathrm{d}x$. Thus from (7.6.3) and (7.6.4) we have

$$\frac{\tau_0}{\rho}=u_1^2\frac{\mathrm{d}\theta}{\mathrm{d}x}+(\delta^*+2\theta)u_1\frac{\mathrm{d}u_1}{\mathrm{d}x} \tag{7.6.5}$$

and it will be seen that the two terms in (7.6.3) involving the nominal thickness δ conveniently cancel out by virtue of (7.6.4).

Dividing through by u_1^2, we can express the momentum equation (7.6.5) in dimensionless form as

$$\frac{\tau_0}{\rho u_1^2}=\frac{\mathrm{d}\theta}{\mathrm{d}x}+(\delta^*+2\theta)\frac{1}{u_1}\frac{\mathrm{d}u_1}{\mathrm{d}x}. \tag{7.6.6}$$

In the special case when the pressure is constant outside the boundary layer and $\mathrm{d}u_1/\mathrm{d}x=0$, the momentum equation reduces to the simple form

$$\frac{\tau_0}{\rho u_1^2}=\frac{\mathrm{d}\theta}{\mathrm{d}x}. \tag{7.6.7}$$

It will be seen from (7.6.6) and (7.6.7) that it does not matter if the nominal thickness δ of the boundary layer, as used in figure 58, cannot be defined precisely. All that is required is that we have precise definitions for δ^* and θ as given by (7.3.2) and (7.3.5).

7.7 Approximate solution for laminar boundary layer using the momentum equation

We can use the momentum equation to derive a very simple approximate solution for a laminar boundary layer in the case where the external pressure gradient $\mathrm{d}p/\mathrm{d}x$ is zero. In this situation we may assume similarity of the velocity profile from section to section as in the exact solution of §7.4, i.e. we may say that

$$\frac{u}{u_1}=f\left(\frac{y}{\delta}\right), \tag{7.7.1}$$

where δ is the nominal thickness of the boundary layer.

The procedure is to choose a particular form for the velocity profile (7.7.1) which will resemble the experimental profile of figure 55 and which satisfies as many of the boundary conditions as possible. We can then evaluate the momentum thickness θ for this profile in terms of δ, and we can also express τ_0 in terms of δ by making use of the fact that the shear stress at the wall is given by $\tau_0 = \mu(\partial u/\partial y)_{y=0}$.

The boundary conditions which should be satisfied by the velocity profile are as follows:

$$\text{at} \quad y=0, \quad u=0, \quad \frac{\partial u}{\partial y} \text{ must be finite} \quad \text{and} \quad \frac{\partial^2 u}{\partial y^2}=0,$$

$$y=\delta, \quad u=u_1, \quad \frac{\partial u}{\partial y}=0 \qquad \text{and} \quad \frac{\partial^2 u}{\partial y^2}=0.$$

The simplest mathematical form for (7.7.1) which will meet all these requirements is

$$\frac{u}{u_1}=2\frac{y}{\delta}-2\left(\frac{y}{\delta}\right)^3+\left(\frac{y}{\delta}\right)^4. \tag{7.7.2}$$

With this form for the profile, the displacement and momentum thicknesses will then be given by

$$\frac{\delta^*}{\delta}=\int_0^1\left(1-\frac{u}{u_1}\right)\mathrm{d}\left(\frac{y}{\delta}\right)=\frac{3}{10} \tag{7.7.3}$$

and

$$\frac{\theta}{\delta}=\int_0^1\frac{u}{u_1}\left(1-\frac{u}{u_1}\right)\mathrm{d}\left(\frac{y}{\delta}\right)=\frac{37}{315} \tag{7.7.4}$$

and the shear stress at the wall is given by

$$\tau_0=\mu\left(\frac{\partial u}{\partial y}\right)_{y=0}=\frac{2\mu u_1}{\delta}. \tag{7.7.5}$$

If we now substitute from (7.7.4) and (7.7.5) in the momentum equation (7.6.7), we have the following result:

$$\frac{2\nu}{u_1\delta}=\frac{37}{315}\frac{\mathrm{d}\delta}{\mathrm{d}x} \tag{7.7.6}$$

and hence integrating from 0 to x, noting that δ is zero at $x=0$,

$$\delta^2=34.05\,\frac{\nu x}{u_1}$$

i.e.

$$\delta=5.836\left(\frac{\nu x}{u_1}\right)^{\frac{1}{2}} \tag{7.7.7}$$

or

$$\frac{\delta}{x} = 5.836\left(\frac{\nu}{u_1 x}\right)^{\frac{1}{2}} = 5.836\, Re^{-\frac{1}{2}}, \tag{7.7.8}$$

which confirms equations (7.1.2) and (7.2.5).

Bearing in mind that the nominal thickness δ is a rather arbitrary quantity, it is of more interest to obtain expressions for the displacement thickness δ^* and the momentum thickness θ which may then be compared directly with the exact solution obtained in §7.4. For the displacement thickness we have from (7.7.3)

$$\frac{\delta^*}{x} = 1.751\left(\frac{u_1 x}{\nu}\right)^{-\frac{1}{2}}, \tag{7.7.9}$$

which is seen to be in good agreement with the exact result of (7.4.10). For the momentum thickness we have from (7.7.4)

$$\frac{\theta}{x} = 0.686\left(\frac{u_1 x}{\nu}\right)^{-\frac{1}{2}}, \tag{7.7.10}$$

which is also seen to be in good agreement with the exact result of (7.4.11).

The skin-friction coefficient c_f may be obtained either from (7.7.5) or directly from the momentum equation (7.6.7) with the result

$$c_f = \frac{\tau_0}{\frac{1}{2}\rho u_1^2} = 0.686\left(\frac{u_1 x}{\nu}\right)^{-\frac{1}{2}}, \tag{7.7.11}$$

which compares with (7.4.9).

Other simple mathematical forms which may be used to represent the velocity profile (7.7.1) are

$$\frac{u}{u_1} = 2\frac{y}{\delta} - \left(\frac{y}{\delta}\right)^2 \tag{7.7.12}$$

and

$$\frac{u}{u_1} = \sin\left(\frac{\pi}{2}\frac{y}{\delta}\right). \tag{7.7.13}$$

These also give reasonably good agreement with the exact solution although they do not satisfy the boundary condition that $\partial^2 u/\partial y^2$ should be zero when $y=\delta$, and the quadratic form (7.7.12) does not satisfy the boundary condition that $\partial^2 u/\partial y^2 = 0$ at $y=0$.

The solution of the momentum equation in the form (7.6.6), in the case where there is an external pressure gradient $\partial p/\partial x$ in the x direction, is more difficult, since in these circumstances we cannot invoke the principle of profile similarity from section to section. Analysis is possible, however, by compounding two different basic types of velocity profile in varying proportion with the distance x to take account of the changing form of the actual velocity distribution from section to section.

7.8 Approximate method for turbulent boundary layer using the momentum equation

The momentum equation (7.6.7) may also be used to investigate the growth in the thickness of a turbulent boundary layer formed on a flat plate. In the case of turbulent flow, however, we cannot expect to calculate the shear stress at the wall from the expression $\mu(\partial\bar{u}/\partial y)_{y=0}$ when using a very approximate form for the velocity profile which is applicable only in the turbulent zone outside the viscous sub-layer. We therefore make use of empirical results for the shear stress τ_0 based on measurements of turbulent flow in pipes such as those quoted in §6.6.

The simplest mathematical representation of the velocity distribution in a turbulent boundary layer is the one-seventh power law (6.6.17) derived for flow in pipes. Taking the nominal thickness δ of the boundary layer in place of the radius a of the pipe, we could assume that

$$\frac{\bar{u}}{u_1}=\left(\frac{y}{\delta}\right)^{\frac{1}{7}}. \tag{7.8.1}$$

It should be noted that the same reservations apply to this velocity profile as in the case of flow through a pipe discussed in §6.6. The velocity gradient $\mathrm{d}\bar{u}/\mathrm{d}y$ is given by (6.6.7), putting $a=\delta$, and is infinite at $y=0$. At the outer edge of the boundary layer, where $y=\delta$, the velocity gradient $\mathrm{d}\bar{u}/\mathrm{d}y=u_1/7\delta$ whereas it should logically be zero at this point. We can overcome the first objection, however, by setting an inner limit of $y=\delta_L$ for the turbulent profile (7.8.1) and by assuming that we have a viscous sub-layer adjacent to the wall extending from $y=0$ to $y=\delta_L$ as indicated in figure 59.

We can evaluate the momentum thickness θ for this simple two-layer model of the turbulent boundary layer using (7.3.5) and (7.8.1). The departure of the velocity profile from (7.8.1) in the viscous sub-layer for values of y less than δ_L will not significantly affect the evaluation and need not be taken into account. The result is

$$\frac{\theta}{\delta}=\frac{7}{72}. \tag{7.8.2}$$

For the turbulent shear stress τ_0 we can take the experimental result for

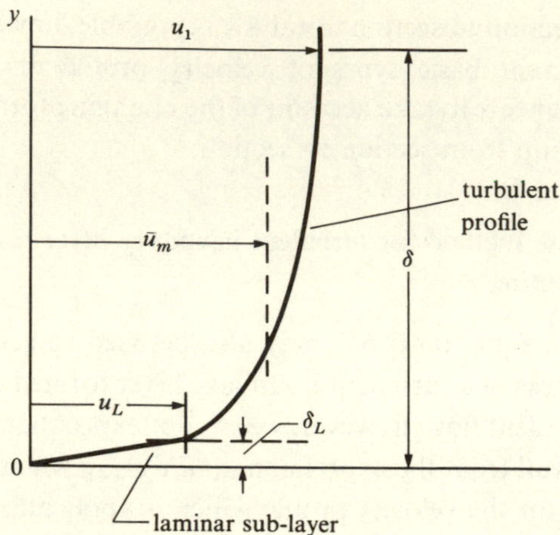

Fig. 59. Turbulent profile with laminar sub-layer.

flow in pipes which was quoted as equation (6.6.11) in chapter 6.

$$\frac{\tau_0}{\frac{1}{2}\rho\bar{u}_m^2}=0.0664\left(\frac{\bar{u}_m a}{\nu}\right)^{-\frac{1}{4}}. \tag{7.8.3}$$

If this expression for turbulent flow in pipes is re-stated in terms of the maximum velocity u_1 measured at the centre-line of the pipe, instead of the mean velocity u_m averaged over the cross-section of the pipe, we can say that

$$\frac{\tau_0}{\rho u_1^2}=0.0332\left(\frac{\bar{u}_m}{u_1}\right)^{\frac{7}{4}}\left(\frac{u_1 a}{\nu}\right)^{-\frac{1}{4}}$$

and taking the value $\bar{u}_m/u_1=0.817$ for the one-seventh power law profile (6.6.17) as calculated in §6.6, we have

$$\frac{\tau_0}{\rho u_1^2}=0.0233\left(\frac{u_1 a}{\nu}\right)^{-\frac{1}{4}}. \tag{7.8.4}$$

This result may be applied to the boundary layer of figure 59 by substituting δ for a.

If we now take the momentum equation (7.6.7) for boundary-layer flow with zero external pressure gradient, we see from (7.8.2) and (7.8.4) that

$$0.0233\left(\frac{u_1\delta}{\nu}\right)^{-\frac{1}{4}}=\frac{7}{72}\frac{\mathrm{d}\delta}{\mathrm{d}x}$$

or

$$\frac{\mathrm{d}\delta}{\mathrm{d}x}=0.239\left(\frac{\nu}{u_1\delta}\right)^{\frac{1}{4}}. \tag{7.8.5}$$

Integrating from 0 to x, we find that

$$\frac{\delta}{x}=0.379\left(\frac{\nu}{u_1 x}\right)^{\frac{1}{5}}=0.379\ Re^{-\frac{1}{5}}, \tag{7.8.6}$$

which may be compared with the corresponding result for a laminar boundary layer obtained in (7.7.8).

For the turbulent skin-friction coefficient c_f we have

$$c_f=\frac{\tau_0}{\frac{1}{2}\rho u_1^2}=0.0588\left(\frac{u_1 x}{\nu}\right)^{-\frac{1}{5}}, \tag{7.8.7}$$

which may be compared with (7.7.11) for the laminar boundary layer.

It should be noted that, in carrying out the integration of the momentum equation (7.8.5) from 0 to x, we have implied that the turbulent boundary layer extends from the leading edge and we have ignored the possible existence of a laminar boundary layer over the front part of the plate. If the overall length of the plate is large compared with the critical length x_L for a laminar boundary layer, however, the effect of ignoring the laminar flow portion will not be very significant. Subject to this proviso, the expression (7.8.7) for the turbulent skin-friction coefficient is in good agreement with experimental measurements. For the *drag coefficient*, or average skin-friction coefficient for one side of a plate of length x, it follows from (7.8.7) that

$$c_D=\frac{D}{\frac{1}{2}\rho u_1^2 x}=0.073\left(\frac{u_1 x}{\nu}\right)^{-\frac{1}{5}}, \tag{7.8.8}$$

where D is the drag exerted on one side of the plate per unit width. Note that $D=\int_0^x \tau_0\, \mathrm{d}x$.

It will be appreciated that the two-layer model of the turbulent boundary layer, as represented in figure 59, is an over-simplification of the complex nature of flow in a boundary-layer region. We will consider a slightly more sophisticated multi-layer model in § 7.10 using the logarithmic form for the law of the wall derived in § 6.5. In the meantime, however, it will be of interest to obtain a preliminary estimate of the thickness of the viscous sub-layer, or laminar sub-layer, for the two-layer model of figure 59. Assuming a linear velocity distribution in the sub-layer, corresponding to constant shear stress, we will have for values of y less than δ_L

$$\frac{u}{u_L}=\frac{y}{\delta_L} \tag{7.8.9}$$

and the shear stress τ_0 will be given by

$$\tau_0 = \mu \frac{u_L}{\delta_L}. \tag{7.8.10}$$

But from (7.8.4) the shear stress in the turbulent layer is

$$\tau_0 = 0.0233\, \rho u_1^2 \left(\frac{u_1 \delta}{\nu}\right)^{-\frac{1}{4}}. \tag{7.8.11}$$

Hence for equality of the shear stress we must have

$$\mu \frac{u_L}{\rho_L} = 0.0233\, \rho u_1^2 \left(\frac{u_1 \delta}{\nu}\right)^{-\frac{1}{4}}$$

or

$$\frac{u_L}{u_1} = 0.0233 \frac{\delta_L}{\delta} \left(\frac{u_1 \delta}{\nu}\right)^{\frac{3}{4}}. \tag{7.8.12}$$

For the turbulent profile (7.8.1), however, the velocity at $y = \delta_L$ is given by

$$\frac{u_L}{u_1} = \left(\frac{\delta_L}{\delta}\right)^{\frac{1}{7}}. \tag{7.8.13}$$

Hence, eliminating u_L/u_1 between (7.8.12) and (7.8.13), we find

$$\frac{\delta_L}{\delta} = 80 \left(\frac{u_1 \delta}{\nu}\right)^{-\frac{7}{8}}. \tag{7.8.14}$$

From (7.8.11) we can say that

$$\sqrt{\frac{\tau_0}{\rho}} = 0.153\, u_1 \left(\frac{u_1 \delta}{\nu}\right)^{-\frac{1}{8}}. \tag{7.8.15}$$

Hence from (7.8.14) and (7.8.15) we have

$$\frac{\delta_L \sqrt{(\tau_0/\rho)}}{\nu} = 12 \tag{7.8.16}$$

or, using the notation of (6.5.9), we see that the thickness of the laminar sub-layer according to the simple two-layer model corresponds to a value of $y^+ = y\sqrt{(\tau_0/\rho)}/\nu = 12$. This result is very close to the value $y^+ = 11.6$ which represents the intersection point between the linear expression $u^+ = y^+$ for the laminar sub-layer (6.5.10) and the logarithmic form for the law of the wall (6.5.17).

7.9 The turbulent boundary-layer equations

In §7.2 we derived the laminar boundary-layer equations for two-dimensional flow past a flat or slightly curved surface. It will be of interest to go back to the basic Navier–Stokes equations (7.2.1) and (7.2.2) and the continuity equation in order to investigate the corresponding problem of turbulent flow in a boundary layer. We have to remember that, although we can have two-dimensional flow conditions for the *mean flow* pattern, there is no such thing as two-dimensional turbulence. Turbulence is intrinsically three-dimensional in character as we have noted in chapter 6. In the case of a turbulent boundary layer, the mean velocity component $\bar{w}$ in the z direction may be zero but the fluctuating turbulent velocity component w' will exist and will be comparable in magnitude with the other components u' and v'.

Using the notation of (6.2.1), the x component of the equation of motion (7.2.1) takes the form

$$\frac{\partial}{\partial t}(\bar{u}+u')+(\bar{u}+u')\frac{\partial}{\partial x}(\bar{u}+u')+(\bar{v}+v')\frac{\partial}{\partial y}(\bar{u}+u')+w'\frac{\partial}{\partial z}(\bar{u}+u')$$

$$=-\frac{1}{\rho}\frac{\partial}{\partial x}(\bar{p}+p')+\nu\left[\frac{\partial^2}{\partial x^2}(\bar{u}+u')+\frac{\partial^2}{\partial y^2}(\bar{u}+u')\right]. \tag{7.9.1}$$

If we now assume a steady mean flow pattern, and if we take a time or ensemble average of equation (7.9.1), we are left with

$$\bar{u}\frac{\partial \bar{u}}{\partial x}+\bar{v}\frac{\partial \bar{u}}{\partial y}+\left(\overline{u'\frac{\partial u'}{\partial x}}+\overline{v'\frac{\partial u'}{\partial y}}+\overline{w'\frac{\partial u'}{\partial z}}\right)=-\frac{1}{\rho}\frac{\partial \bar{p}}{\partial x}+\nu\left[\frac{\partial^2 \bar{u}}{\partial x^2}+\frac{\partial^2 \bar{u}}{\partial y^2}\right]. \tag{7.9.2}$$

As in the case of the laminar boundary layer, we can say that $\bar{v}$ is small compared with $\bar{u}$, $\partial\bar{u}/\partial x$ is small compared with $\partial\bar{u}/\partial y$, and $\partial^2\bar{u}/\partial x^2$ is small compared with $\partial^2\bar{u}/\partial y^2$. We can then ignore the term $\partial^2\bar{u}/\partial x^2$ in comparison with $\partial^2\bar{u}/\partial y^2$ on the right-hand side, and the equation reduces to

$$\bar{u}\frac{\partial \bar{u}}{\partial x}+\bar{v}\frac{\partial \bar{u}}{\partial y}+\left(\overline{u'\frac{\partial u'}{\partial x}}+\overline{v'\frac{\partial u'}{\partial y}}+\overline{w'\frac{\partial u'}{\partial z}}\right)=-\frac{1}{\rho}\frac{\partial \bar{p}}{\partial x}+\nu\frac{\partial^2 \bar{u}}{\partial y^2}. \tag{7.9.3}$$

If we apply a similar procedure to the y component of the equation of motion (7.2.2) for the case of turbulent flow, and take note of the relative magnitude of terms in the equation as in §7.2, we find that we are left with

$$\left(\overline{u'\frac{\partial v'}{\partial x}}+\overline{v'\frac{\partial v'}{\partial y}}+\overline{w'\frac{\partial v'}{\partial z}}\right)=-\frac{1}{\rho}\frac{\partial\bar{p}}{\partial y}. \tag{7.9.4}$$

If we now turn attention to the continuity equation and recall from (6.3.3) and (6.3.4) that the continuity equation for incompressible turbulent flow applies separately to both the mean velocity components and to the turbulent components, we have

$$\frac{\partial\bar{u}}{\partial x}+\frac{\partial\bar{v}}{\partial y}=0 \tag{7.9.5}$$

and

$$\frac{\partial u'}{\partial x}+\frac{\partial v'}{\partial y}+\frac{\partial w'}{\partial z}=0. \tag{7.9.6}$$

Note again that, although the mean flow is two-dimensional with $\bar{w}=0$, the turbulent motion is three-dimensional.

It will be seen that the three terms on the left-hand side of equation (7.9.3) involving the fluctuating turbulent velocity components may be re-written with the aid of the continuity equation (7.9.6) as follows

$$\begin{aligned}\overline{u'\frac{\partial u'}{\partial x}}+\overline{v'\frac{\partial u'}{\partial y}}+\overline{w'\frac{\partial u'}{\partial z}}&=\frac{\partial}{\partial x}(\overline{u'^2})+\frac{\partial}{\partial y}(\overline{u'v})+\frac{\partial}{\partial z}(\overline{u'w'})\\&\quad-\left(\overline{u'\frac{\partial u'}{\partial x}}+\overline{u'\frac{\partial v'}{\partial y}}+\overline{u'\frac{\partial w'}{\partial z}}\right)\\&=\frac{\partial}{\partial x}(\overline{u'^2})+\frac{\partial}{\partial y}(\overline{u'v'})+\frac{\partial}{\partial z}(\overline{u'w'}).\end{aligned} \tag{7.9.7}$$

Thus equation (7.9.3) may be expressed in the form:

$$\bar{u}\frac{\partial\bar{u}}{\partial x}+\bar{v}\frac{\partial\bar{u}}{\partial y}+\frac{\partial}{\partial x}(\overline{u'^2})+\frac{\partial}{\partial y}(\overline{u'v'})+\frac{\partial}{\partial z}(\overline{u'w'})=-\frac{1}{\rho}\frac{\partial\bar{p}}{\partial x}+\nu\frac{\partial^2\bar{u}}{\partial y^2}. \tag{7.9.8}$$

If this is compared with the equivalent equation (7.2.6) for a laminar boundary layer, it will be seen that the form of the equation is similar but that we have additional terms, as might be expected, involving the spatial derivatives of the Reynolds stresses. Although the magnitudes of the mean values of the velocity products $\overline{u'^2}$, $\overline{u'v'}$, $\overline{u'w'}$, will generally be comparable, the spatial gradients of these quantities will differ. For example, the gradient $(\partial/\partial x)(\overline{u'^2})$ will usually be small compared with the transverse gradient $(\partial/\partial y)(\overline{u'v'})$ in a thin boundary layer. We can also say that, provided the *mean flow pattern* is two-dimensional in character, the gradient in the z direction $(\partial/\partial z)(\overline{u'w'})$ may be ignored.

If we apply the same procedure as in (7.9.7) to the three mean product terms on the left-hand side of equation (7.9.4), the y component of the equation of motion takes the form:

$$\frac{\partial}{\partial x}(\overline{u'v'})+\frac{\partial}{\partial y}(\overline{v'^2})+\frac{\partial}{\partial z}(\overline{v'w'})=-\frac{1}{\rho}\frac{\partial \bar{p}}{\partial y}. \tag{7.9.9}$$

For a thin boundary layer with only gradual change of velocity in the x direction, we can usually say that the longitudinal gradient $(\partial/\partial x)(\overline{u'v'})$ in this equation will be small compared with the transverse gradient term $(\partial/\partial y)(\overline{v'^2})$. Also, for a two-dimensional mean flow pattern, we can ignore the term $(\partial/\partial z)(\overline{v'w'})$. Hence the transverse component of the turbulent boundary-layer equation can be expressed as

$$\frac{\partial}{\partial y}(\bar{p}+\rho\, \overline{v'^2})=0, \tag{7.9.10}$$

which may be compared with (7.2.7) for the laminar boundary layer. We can integrate (7.9.10) from the outside of the boundary layer, where $\overline{v'^2}=0$, to a level y within the layer, with the result:

$$\bar{p}+\rho\, \overline{v'^2}=p_1, \tag{7.9.11}$$

where p_1 is the pressure just outside the boundary layer.

Substituting for p from (7.9.11) in equation (7.9.8), we can express the basic equation of motion for a two-dimensional turbulent boundary layer in the form:

$$\bar{u}\frac{\partial \bar{u}}{\partial x}+\bar{v}\frac{\partial \bar{u}}{\partial y}+\frac{\partial}{\partial x}(\overline{u'^2}-\overline{v'^2})+\frac{\partial}{\partial y}(\overline{u'v'})=-\frac{1}{\rho}\frac{\partial p_1}{\partial x}+\nu\frac{\partial^2 \bar{u}}{\partial y^2}. \tag{7.9.12}$$

If we neglect the longitudinal gradient term $(\partial/\partial x)(\overline{u'^2}-\overline{v'^2})$ this equation reduces to:

$$\bar{u}\frac{\partial \bar{u}}{\partial x}+\bar{v}\frac{\partial \bar{u}}{\partial y}+\frac{\partial}{\partial y}(\overline{u'v'})=-\frac{1}{\rho}\frac{\partial p_1}{\partial x}+\nu\frac{\partial^2 \bar{u}}{\partial y^2}. \tag{7.9.13}$$

Outside the viscous sub-layer we can ignore the last term on the right-hand side involving the viscous stresses associated with the mean velocity gradient. Thus the simplest form of the turbulent boundary-layer equation may be expressed as:

$$\bar{u}\frac{\partial \bar{u}}{\partial x}+\bar{v}\frac{\partial \bar{u}}{\partial y}+\frac{\partial}{\partial y}(\overline{u'v'})=-\frac{1}{\rho}\frac{\partial p_1}{\partial x}. \tag{7.9.14}$$

7.10 Multi-layer model for turbulent boundary-layer flow

The approximate method of analysis for a turbulent boundary layer outlined in §7.8, using the one-seventh power law (7.8.1) to represent the mean velocity distribution, in conjunction with the momentum equation, yields some useful results such as (7.8.7) and (7.8.8), but inevitably it over-simplifies the problem. It may be noted that for a turbulent boundary layer (even with zero pressure gradient in the x direction) there is no obvious *a priori* reason why we should have profile similarity from section to section in a developing flow. If equations (7.2.6) and (7.9.13) are compared it will be seen that the presence of the additional term $(\partial/\partial y)(\overline{u'v'})$ in the latter case introduces a complicating factor which precludes the possibility of any simple mathematical solution. The concept of *self-preservation* of developing flows, formulated by A. A. Townsend, represents an extension of the idea of profile similarity in laminar flow and requires that the variation of any mean quantity, over a plane $x=$constant for instance, should be capable of being expressed non-dimensionally through suitable scales of length and velocity, L_0 and U_0, as a universal function of y/L_0. In addition it is necessary that the dimensionless ratios associated with the turbulence, for example, $\overline{u'v'}/U_0^2$, should be functions of y/L_0 only.

Experimental study of the flow in turbulent boundary layers suggests that it is appropriate to divide the flow into two main zones. The *inner layer* may be taken as extending from the wall to a level defined very approximately by $y=0.2\delta$. The method of analysis outlined in §6.5 is applicable in this zone. The *outer layer* is assumed to extend from $y=0.2\delta$ to $y=\delta$ and the structure of the turbulence in this zone is even more complex owing to the larger size of the eddies.

The inner layer, as indicated in figure 60, may be regarded as a region of nearly constant shear stress and the analysis outlined in §6.5 is therefore applicable. We can sub-divide the inner layer into a *viscous sub-layer* very close to the wall where viscous stresses are significant, and an *equilibrium turbulent layer* where viscous stresses are negligible and where the 'law of the wall' applies. The viscous sub-layer may itself be further sub-divided into an even thinner *laminar sub-layer* (sometimes referred to as the linear

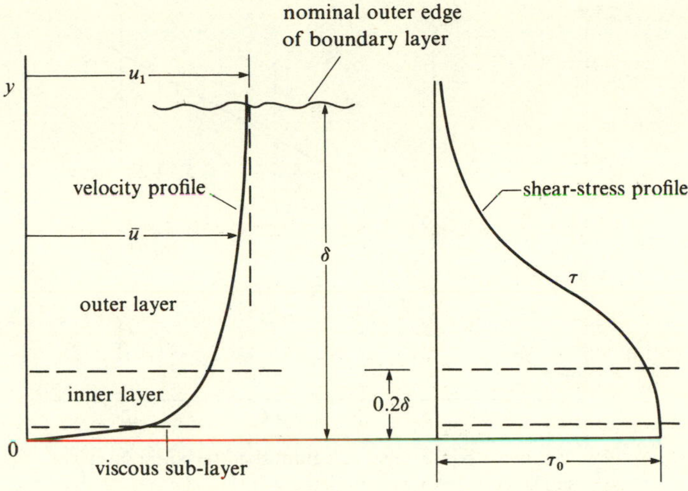

Fig. 60. Velocity and shear stress profiles in a turbulent boundary layer.

sub-layer) where the Reynolds stresses are negligible, and a *buffer layer* where viscous stresses and Reynolds stresses are of comparable magnitude.

It is convenient to use the dimensionless ratios for velocity $\bar{u}^+$, and distance measured from the plane of the wall y^+, as defined by (6.5.7) and (6.5.8) when we are discussing the inner layer. In figure 61 the velocity ratio $\bar{u}^+$ is plotted against log y^+. Thus the logarithmic form of the law of the wall (6.5.17) appears as a straight line in this diagram.

The structure of the flow in the *inner layer* may be summarised as follows:

$y^+ <$ about 5	laminar sub-layer	$\bar{u}^+ = y^+$ as in (6.5.10)
$5 < y^+ < 30$	buffer zone suggested profile	$\bar{u}^+ = -3.05 + 5.0 \log y^+$,
$30 < y^+$	equilibrium turbulent layer	$\bar{u}^+ = 5.5 + 2.5 \log y^+$,
and $y < 0.2\delta$	(law of the wall as in (6.5.17)).	

Although we have indicated the thickness of the inner layer as approximately 0.2δ, this is a rather arbitrary figure and there is in fact no clear demarcation line between the inner and outer zones. It will be evident from figure 60 that, as the value of y increases, we depart progressively from the constant shear-stress condition to which the law of the wall (6.5.17) relates. It is perhaps preferable to say that the thickness of the inner layer is somewhere between 0.1δ and 0.2δ, bearing in mind that the nominal

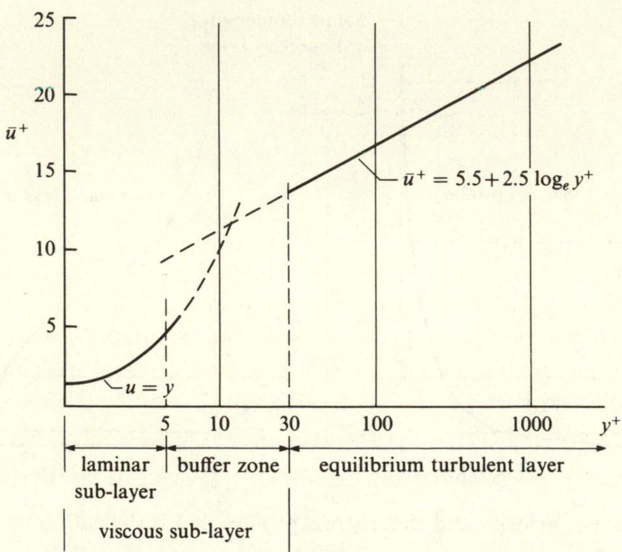

Fig. 61. Variation of dimensionless velocity with dimensionless distance in a turbulent boundary layer.

thickness δ of the complete boundary layer cannot itself be defined precisely.

Although the range of variation of the mean flow velocity $\bar{u}$ becomes smaller as we move into the outer layer, with increasing value of y, and the velocity profile becomes flatter as indicated in figure 60, the turbulent eddies become progressively larger. It is thus more difficult to present an analytical description of the flow pattern. The magnitude of the transverse velocity components may become comparable with the mean velocity $\bar{u}$ in the outer regions of the boundary layer and very large transient movements may be observed in the fluctuating flow pattern. The linear scale of the eddies in this region is roughly of the same magnitude as the boundary-layer thickness δ.

We have already noted in § 7.1 that an interface must exist in the outer region of the boundary layer dividing the turbulent motion with its distributed vorticity from the irrotational motion of the external stream. A sharp change in the level of vorticity in the fluid can only be brought about through the action of viscosity and this consideration has led to the hypothesis that there must be a *viscous super-layer* at the outer edge of the turbulent boundary layer. It should be appreciated, however, that this interface is highly irregular and unsteady in character and a hot wire probe located in this region will give an intermittent signal. The thickness of the viscous super-layer is believed to be of the same order of magnitude as the Kolmogorov length l_ε. Outside this interface there will be a zone of unsteady irrotational flow with relatively large irrotational eddies penetrating into the main external stream.

7.11 Turbulent mixing layers and turbulent jets

Although we are most frequently concerned in engineering applications with turbulent shear flow in the vicinity of solid boundaries, there is another class of flow problem where solid surfaces are not directly involved and which may be described under the general heading of *free turbulent shear flow*. An important example is the case of a turbulent mixing layer illustrated diagrammatically in figure 62. We assume in this example that a uniform irrotational stream emerges from a wide nozzle or duct into a region where the surrounding fluid is at rest. We will simplify the problem by ignoring the presence of a thin boundary layer at the wall of the duct or nozzle immediately upstream from section (1) in the diagram, and we will treat the flow as being effectively two-dimensional. A sheet of vorticity will be generated at the interface between the moving fluid and the stationary fluid and this leads to the formation of a wedge-shaped turbulent mixing layer. Fluid will be steadily entrained from the stagnant region at the outer edge of the mixing layer. It is observed experimentally that the turbulent mixing layer spreads outwards at a relatively small angle as indicated in the diagram, although it should be noted that the edges of the turbulent zone are intermittent in character like the outer edge of a turbulent boundary

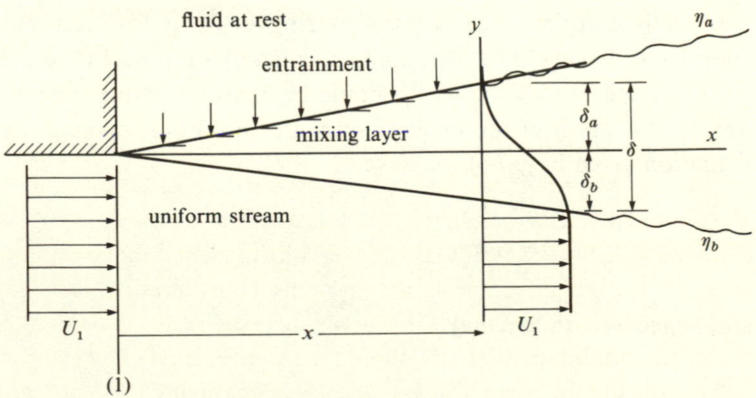

Fig. 62. Turbulent mixing layer.

layer. We will make the assumption that the thickness δ is proportional to the longitudinal distance x, i.e. $\delta=\beta x$, where β is a constant whose experimental value is in the range 0.225–0.25.

If we apply the turbulent boundary layer equation (7.9.14) to the mixing layer, and if we take the pressure gradient in the x direction as zero, we have

$$\bar{u}\frac{\partial \bar{u}}{\partial x}+\bar{v}\frac{\partial \bar{u}}{\partial y}=-\frac{\partial}{\partial y}(\overline{u'v'}). \tag{7.11.1}$$

We can satisfy the equation of continuity for the mean flow by introducing a stream function ψ such that

$$\bar{u}=\frac{\partial \psi}{\partial y} \quad \text{and} \quad \bar{v}=-\frac{\partial \psi}{\partial x}. \tag{7.11.2}$$

We can now introduce the hypothesis of profile similarity, i.e. that the velocity ratio u/U_1 can be expressed as a function of y/δ or y/x, by defining the new variable $\eta=y/x$ and by taking a stream function in the form

$$\psi=U_1 xF(\eta), \tag{7.11.3}$$

where $F(\eta)$ is a function of η only. From (7.11.2) and (7.11.3) we have:

$$\bar{u}=U_1\frac{\mathrm{d}F}{\mathrm{d}\eta} \quad \text{and} \quad \bar{v}=U_1\left(\eta\frac{\mathrm{d}F}{\mathrm{d}\eta}-F\right) \tag{7.11.4}$$

and it is easily verified that the left-hand side of (7.11.1) may be expressed in terms of the function F in the form:

$$\bar{u}\frac{\partial \bar{u}}{\partial x}+\bar{v}\frac{\partial \bar{u}}{\partial y}=-\frac{U_1^2}{x}F\frac{\mathrm{d}^2F}{\mathrm{d}\eta^2}. \tag{7.11.5}$$

Various assumptions are possible regarding the Reynolds stress term on the right-hand side of (7.11.1). We might reasonably suppose, for example, that the kinematic eddy viscosity ν_T should be constant with respect to y over any cross-section of the mixing layer. On this basis, making use of the definition given in (6.7.5), we have

$$-\frac{\partial}{\partial y}(\overline{u'v'})=\nu_T\frac{\partial^2 \bar{u}}{\partial y^2} \tag{7.11.6}$$

and hence we can say that

$$-\frac{\partial}{\partial y}(\overline{u'v'})=\frac{\nu_T U_1}{x^2}\frac{\mathrm{d}^3F}{\mathrm{d}\eta^3}. \tag{7.11.7}$$

Substituting from (7.11.5) and (7.11.7) in equation (7.11.1) gives:

$$\frac{\mathrm{d}^3F}{\mathrm{d}\eta^3}+\frac{U_1x}{\nu_T}F\frac{\mathrm{d}^2F}{\mathrm{d}\eta^2}=0. \tag{7.11.8}$$

If we now make the further assumption that the eddy viscosity ν_T is proportional to the thickness δ of the mixing layer multiplied by the overall velocity difference U_1, we can say that

$$\nu_T=cU_1x \tag{7.11.9}$$

and, substituting in (7.11.8), we have the result:

$$\frac{\mathrm{d}^3F}{\mathrm{d}\eta^3}+\frac{1}{c}F\frac{\mathrm{d}^3F}{\mathrm{d}\eta^2}=0. \tag{7.11.10}$$

We have thus reduced the partial differential equation (7.11.1) to the ordinary differential equation (7.11.10) which confirms that the hypothesis of profile similarity, or self-preservation, in a two-dimensional mixing layer is consistent with the assumptions made in (7.11.6) and (7.11.9) regarding the eddy viscosity.

Finally, if we substitute a new variable defined by $\eta'=\eta/c$ equation (7.11.10) reduces to the standard form:

$$\frac{\mathrm{d}^3F}{\mathrm{d}\eta'^3}+F\frac{\mathrm{d}^2F}{\mathrm{d}\eta'^2}=0, \tag{7.11.11}$$

which is similar to (7.4.5) which appeared in the solution of the laminar boundary-layer equation. The boundary conditions, however, are different for the turbulent mixing layer.

It is slightly more convenient, when applying the boundary conditions, if we take the edge of the uniform stream (defined by the value η_b in figure 62) as the datum level rather than the x axis. For this purpose we can re-define the parameter η' as $(\eta-\eta_b)/c$, in place of η/c, without affecting equation (7.11.11). We then have the requirement that at $\eta=\eta_b$, or $\eta'=0$,

$$\bar{u}=U_1,\quad \bar{v}=0,\quad \text{and}\quad \frac{\partial\bar{u}}{\partial y}=0.$$

Hence, from (7.11.4), we require:

$$\frac{\mathrm{d}F}{\mathrm{d}\eta'}=c,\quad F=\eta_b,\quad \text{and}\quad \frac{\mathrm{d}^2F}{\mathrm{d}\eta'^2}=0.$$

At the outer edge of the mixing layer where $\eta = \delta_a/x$, or $\eta' = \delta/cx$,

$$\bar{u} = 0 \quad \text{and} \quad \frac{\partial \bar{u}}{\partial y} = 0,$$

i.e. we require $\mathrm{d}F/\mathrm{d}\eta' = 0$ and $\mathrm{d}^2F/\mathrm{d}\eta'^2 = 0$. In principle, therefore, we can determine the velocity distribution in the mixing layer, and also evaluate the growth of the layer thickness, by solving equation (7.11.11) and applying the specified boundary conditions. In actual fact, the procedure involves some mathematical complications since there is no simple analytical solution to equation (7.11.11). It must also be noted that the velocity profile derived from (7.11.11) gives an asymptotic approach of $\bar{u}$ to zero as $\eta' \to \infty$ so that we cannot obtain a precise value for the thickness δ.

The case of a relatively narrow *two-dimensional jet* is illustrated in figure 63. It will be seen that a mixing layer is formed initially at the edge of the jet as it emerges from the nozzle into the region occupied by stagnant fluid. When the mixing layers meet at the centre-line a symmetrical turbulent jet is formed which continues to expand and to entrain fluid from the surrounding region.

Since the entrained fluid enters the jet with zero x momentum we can say that, although the mass-flow rate increases, the rate of flow of x momentum in the jet must remain constant, i.e.

$$\int_{-\delta}^{+\delta} \rho \bar{u}^2 \, \mathrm{d}y = \text{constant} \tag{7.11.12}$$

or

$$\rho U^2 \delta \int_{-1}^{+1} \left(\frac{u}{U}\right)^2 \mathrm{d}\left(\frac{y}{\delta}\right) = \text{constant} = M, \tag{7.11.13}$$

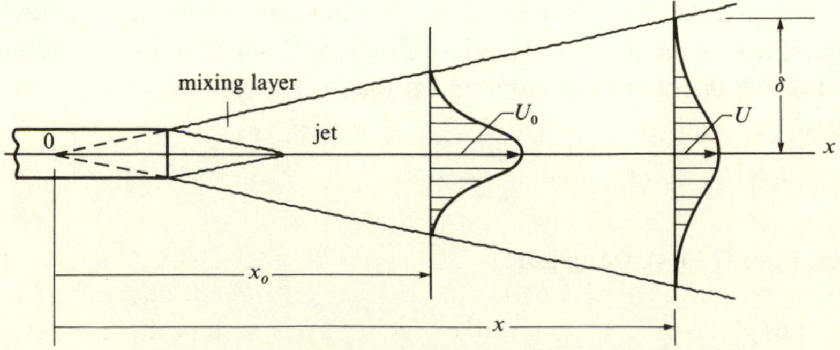

Fig. 63. Two-dimensional turbulent jet.

where U is the mean velocity at the centre-line of the jet. It will be appreciated that U is a function of the axial distance x, and it follows from (7.11.13) under conditions of profile similarity that $U^2\delta$ must be constant. Hence if $\delta=\beta x$, the velocity U at the centre-line must be proportional to $1/x^{\frac{1}{2}}$.

It is convenient to choose a fictitious origin, as shown in figure 63, and to measure axial distances from this point. It is also convenient to select a reference length x_0 and a corresponding velocity U_0 at the centre-line for this section. The velocity U at another section defined by x will then be given by

$$U=U_0\left(\frac{x_0}{x}\right)^{\frac{1}{2}}. \tag{7.11.14}$$

We can analyse the mean flow pattern for the symmetrical turbulent jet under two-dimensional flow conditions using exactly the same procedure as we employed for the mixing layer. We start with equation (7.11.1) as before, and with a stream function defined by (7.11.2). In view of the variation of the centre-line velocity U with $x^{-\frac{1}{2}}$, however, we must take the following form for the stream function:

$$\psi=U_0x_0^{\frac{1}{2}}x^{\frac{1}{2}}F(\eta). \tag{7.11.15}$$

It follows that

$$\bar{u}=U_0\left(\frac{x_0}{x}\right)^{\frac{1}{2}}\frac{\mathrm{d}F}{\mathrm{d}\eta} \quad \text{and} \quad \bar{v}=U_0\left(\frac{x_0}{x}\right)^{\frac{1}{2}}\left(\eta\frac{\mathrm{d}F}{\mathrm{d}\eta}-\frac{1}{2}F\right). \tag{7.11.16}$$

We can make use of (7.11.6) for the Reynolds stress term with the same assumption as before that the eddy viscosity ν_T will be constant over any cross-section of the jet. Hence we find that

$$-\frac{\partial}{\partial y}(\overline{u'v'})=\frac{\nu_T U_0}{x}\left(\frac{x_0}{x}\right)^{\frac{1}{2}}\frac{\mathrm{d}^3F}{\mathrm{d}\eta^3}. \tag{7.11.17}$$

It follows from (7.11.16) and (7.11.17) that equation (7.11.1) may be expressed in terms of the function F in the form:

$$\frac{\mathrm{d}^3F}{\mathrm{d}\eta^3}+\frac{U_0x}{\nu_T}\frac{1}{2}\left(\frac{x_0}{x}\right)^{\frac{1}{2}}\left[\left(\frac{\mathrm{d}F}{\mathrm{d}\eta}\right)^2+F\frac{\mathrm{d}^2F}{\mathrm{d}\eta^2}\right]=0, \tag{7.11.18}$$

which is the equivalent of equation (7.11.8), but it should be noted that the function F will be different in the two cases. If we now assume that the value of the eddy viscosity at any section x is proportional to the width of the jet at that section (which itself is proportional to x), multiplied by the velocity U at the centre-line, we can say that

$$v_T = cUx, \tag{7.11.19}$$

where c is a constant. Hence from (7.11.14) we have

$$v_T = cU_0 x\left(\frac{x_0}{x}\right)^{\frac{1}{2}}. \tag{7.11.20}$$

Substituting for v_T in (7.11.18) gives

$$\frac{d^3F}{d\eta^3} + \frac{1}{2c}\left[\left(\frac{dF}{d\eta}\right)^2 + F\frac{d^2F}{d\eta^2}\right] = 0 \tag{7.11.21}$$

and this may be written as

$$\frac{d^3F}{d\eta^3} + \frac{1}{2c}\frac{d}{d\eta}\left[F\frac{dF}{d\eta}\right] = 0,$$

which may be integrated immediately to give

$$\frac{d^2F}{d\eta^2} + \frac{1}{2c}F\frac{dF}{d\eta} = A, \tag{7.11.22}$$

where A is a constant of integration. It is easily verified, however, that the value of A must be zero in order to satisfy the boundary conditions at the centre-line of the jet where $\eta = 0$ and where $\bar{u} = U$, $\partial\bar{u}/\partial y = 0$, and $\bar{v} = 0$. Hence we can re-write (7.11.22) in the form

$$\frac{d^2F}{d\eta^2} + \frac{1}{4c}\frac{d}{d\eta}(F^2) = 0.$$

This may be integrated to give

$$\frac{dF}{d\eta} + \frac{F^2}{4c} = B, \tag{7.11.23}$$

where B is another constant of integration. If we now put $a^2 = 4cB$, equation (7.11.23) may be expressed as

$$4c\frac{dF}{d\eta} = a^2 - F^2$$

and hence, integrating from 0 to η and noting that $F = 0$ at $\eta = 0$,

$$\frac{\eta}{4c} = \int_0^F \frac{dF}{a^2 - F^2} = \frac{1}{a}\tanh^{-1}\left(\frac{F}{a}\right)$$

or

$$\frac{F}{a} = \tanh\left(\frac{a\eta}{4c}\right). \tag{7.11.24}$$

We thus have for the velocity profile

$$\frac{\bar{u}}{U} = \frac{\mathrm{d}F}{\mathrm{d}\eta} = \frac{a^2}{4c}\,\mathrm{sech}^2\left(\frac{a\eta}{4c}\right). \tag{7.11.25}$$

Since $\bar{u} = U$ at the centre-line of the jet where $\eta = 0$, and since sech $0 = 1$, we must have $a^2 = 4c$ and therefore $B = 1$. The final expression for the velocity profile is therefore simply

$$\frac{\bar{u}}{U} = \mathrm{sech}^2\,(\eta/a), \tag{7.11.26}$$

where $a = 2\sqrt{c}$. The form of the profile is shown in figure 64.

We cannot obtain a precise value for the width of the jet since the function $\mathrm{sech}^2\,(\eta/a)$ is asymptotic, tending to zero as η tends to infinity. However, we can conveniently obtain the value of the parameter η corresponding to a velocity ratio $\bar{u}/U = \frac{1}{2}$. Putting $\bar{u}/U = \frac{1}{2}$ in (7.11.26) gives $\eta/a = 0.881$.

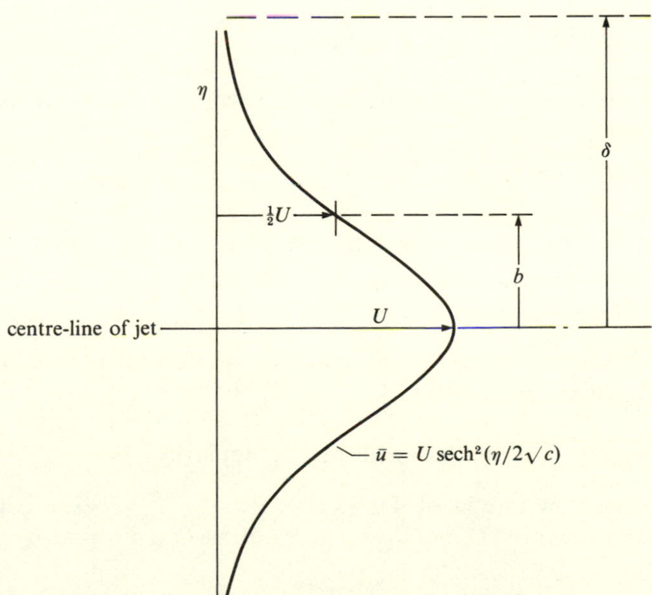

Fig. 64. Velocity profile in two-dimensional turbulent jet.

Thus if the dimension b in figure 64 measures the distance in the y direction from the centre-line to the 'half-velocity point', we have at $y=b$:

$$\eta=\frac{b}{x}=0.881\ a=1.762\ \sqrt{c}. \tag{7.11.27}$$

It is found experimentally that $b/x\approx 0.115$ which implies that the constant $c=0.004\,26$ and that $a=0.130$ approximately. The expression (7.11.19) for ν_T may then be written as

$$\nu_T=0.004\,26\ Ux=0.037\ Ub. \tag{7.11.28}$$

Inevitably a *three-dimensional jet*, emerging from an aperture or nozzle having a circular cross-section, is more difficult to analyse than the two-dimensional case which we have just considered. The essential features of the flow in an axisymmetric turbulent jet, however, are similar to those of the two-dimensional jet. It is again observed experimentally that the edge of the jet expands in a linear manner with the effective distance from the origin. The rate of flow of x momentum in an axisymmetric jet is given by

$$M=\int_0^{r_a}\rho\bar{u}^2 2\pi r\ \mathrm{d}r=\text{constant}$$

or

$$M=2\pi\rho r_a^2U^2\int_0^1\left(\frac{\bar{u}}{U}\right)^2\frac{r}{r_a}\mathrm{d}\left(\frac{r}{r_a}\right). \tag{7.11.29}$$

Hence for a self-preserving flow pattern with profile similarity we must have

$$U\sim\frac{1}{r_a}\sim\frac{1}{x}. \tag{7.11.30}$$

The mass-flow rate in the jet will not be constant from section to section because of entrainment at the edge from the surrounding fluid. The mass-flow rate at any section x will be given by

$$\int_0^{r_a}\rho\bar{u}2\pi r\ \mathrm{d}r=2\pi\rho r_a^2U\int_0^1\frac{\bar{u}}{U}\frac{r}{r_a}\mathrm{d}\left(\frac{r}{r_a}\right).$$

It follows from (7.11.30) that, with profile similarity,

$$\text{mass-flow rate in jet}\sim r_a\sim x. \tag{7.11.31}$$

8

Flow in pipes and ducts

8.1 Flow in the inlet region of a pipe or duct

We normally use the term 'pipe' to describe a tubular duct having a circular cross-section. The term 'duct' is a more general one and may be used to describe an enclosed passage of any cross-sectional shape, although it is most frequently applied to a passage of rectangular shape. The distinctive feature of fluid flow through a pipe or duct is that the fluid is completely enclosed, except at the inlet and outlet, by the walls of the duct. The influence of viscosity will be significant in these circumstances and, because of the property of zero slip at a solid boundary, the flow pattern will be characterised by a high level of vorticity. In this situation we must expect the normal type of fluid motion to be turbulent rather than laminar, although at sufficiently low values of the Reynolds number the damping effect of viscosity may be sufficient to suppress the development of turbulence.

It is appropriate to start by considering the nature of the flow in the inlet region of a duct as indicated in figure 65.

We assume in figure 65 that the duct entry takes the form of a short convergent nozzle. We may therefore expect that at section (0), which represents the start of the parallel-sided duct, the velocity will be effectively uniform over the cross-section. A thin boundary layer will be formed at the wall of the duct, however, as a result of the zero slip property. The thickness of the boundary-layer region will increase in the normal way, as described in chapter 7, with distance measured axially along the duct. The fluid in the central core, outside the boundary-layer region, will at first be unaffected by the presence of the duct walls and the flow in the central region will be nearly irrotational in character. It should be noted, however, that the fluid in the central core must experience a gradual acceleration as the boundary layer thickens with distance along the duct.

It will be clear from figure 65 that, at a certain distance along the duct, the boundary layers will meet at the centre-line and the size of the central core

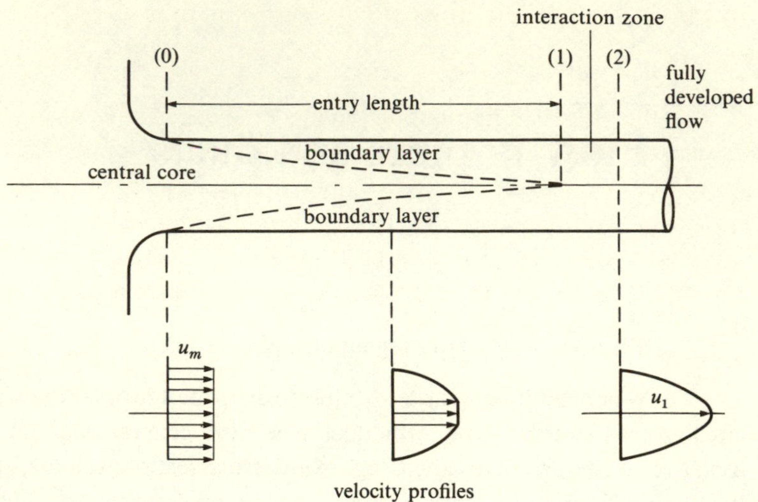

Fig. 65. Entry region of a pipe.

will be reduced to zero. A zone of boundary-layer interaction will then be observed which leads eventually to the establishment of *fully developed internal flow*.

We can follow the course of events in the inlet region quite easily in the case of laminar flow. The growth of the laminar boundary layer can be calculated using the approximate methods of §§ 7.6 and 7.7. It should be noted, however, that the fluid velocity in the central core or 'external flow' is increasing with distance along the duct and that there is a negative pressure gradient in the direction of flow. We therefore have to use the momentum equation in the form of (7.6.6).

The velocity profile in the fully developed laminar flow region downstream from section (1) is given by (5.8.12) or (5.8.14) and this may be expressed conveniently in the form:

$$\frac{u}{u_1} = 1 - \frac{r^2}{a^2}. \tag{8.1.1}$$

It will also be noted from (5.8.16) that the maximum velocity u_1 at the centre-line is equal to twice the mean velocity u_m averaged over the cross-section. If we apply Bernoulli's equation to the irrotational flow in the central core from section (0) to section (1) in figure 65, we can say that

$$p_0 + \tfrac{1}{2}\rho u_m^2 = p_1 + \tfrac{1}{2}\rho u_1^2$$

and hence the pressure drop over the entry length is given by

$$p_0 - p_1 = \tfrac{3}{2}\rho u_m^2. \tag{8.1.2}$$

For the velocity distribution in the boundary-layer region we can assume a simple parabolic profile of the form (7.7.12). Noting that $y = a - r$, it will be seen that (7.7.12) becomes identical with (8.1.1) when the boundary-layer region extends to the centre-line and δ in equation (7.7.12) becomes equal to the pipe radius a at section (1). Using the continuity and momentum equations the entry length x_e may be calculated and the result may be expressed in the form

$$\frac{x_e}{a} = 0.0575\, Re, \tag{8.1.3}$$

where $Re = 2au_m/\nu$. Thus, if $Re = 100$ we have $x_e/a = 5.75$, but if $Re = 1000$ we have $x_e/a = 57.5$.

It is of interest to examine the conditions under which the boundary-layer flow in the inlet region is likely to remain laminar. We noted in § 7.5 that for a boundary layer formed on a flat plate the flow will normally remain laminar provided $u_1\delta^*/\nu$ is less than about 600. The displacement thickness δ^* for a laminar boundary layer is typically about one-third of the nominal thickness δ. For example from (7.7.8) and (7.7.9) we have $\delta^*/\delta = 0.300$ while for the profile of (7.7.12) the ratio is 0.333. We may therefore express the normal condition for laminar flow in the boundary layer as

$$\frac{u_1\delta}{\nu} < 2000 \quad \text{approximately.} \tag{8.1.4}$$

At section (1) in figure 65, where the boundary-layer region reaches the centre-line of the pipe and the thickness δ becomes equal to the pipe radius a, we have the velocity $u_1 = 2u_m$ from (5.8.16). Hence the condition (8.1.4) that laminar flow will be maintained throughout the entry length up to section (1) may be expressed as

$$\frac{2u_m a}{\nu} < 2000 \quad \text{approximately,} \tag{8.1.5}$$

which is the same as the normal experimental observation that laminar flow prevails for flow in pipes when the value of the Reynolds number is less than about 2000.

It will be appreciated that this is only a very rough calculation. Flow in a boundary layer formed on the curved inside surface of a pipe is not quite the same as two-dimensional flow in a boundary layer formed on a flat plate.

Furthermore there is a longitudinal pressure gradient, as we have noted, in the case of the pipe inlet region. The coincidence of the critical value of the Reynolds number must therefore be regarded to some extent as fortuitous.

With values of the Reynolds number greater than about 2000 the flow pattern in the region of the pipe inlet becomes more complex. Small disturbances are no longer damped out through the action of viscosity and the flow in the inlet region may be regarded as becoming unstable. The amplification of small disturbances leads ultimately to the establishment of turbulent flow in the pipe. Because of the complex and unsteady nature of the flow in the entry region of the pipe it is not possible to give any precise analysis for the calculation of the entry length. However, we can make a very rough estimate of the order of magnitude of the entry length with turbulent flow if we assume that a turbulent boundary layer is formed at the wall of the pipe all the way from section (0) in figure 65.

Using the approximate method of § 7.8, and assuming a simple power law expression for the velocity distribution in the boundary layer as in (7.8.1), we can express the growth of the boundary-layer thickness by means of (7.8.6), i.e.

$$\frac{\delta}{x}=0.379\left(\frac{\bar{u}_1 x}{\nu}\right)^{-\frac{1}{5}}=0.379\left(\frac{\bar{u}_1 \delta}{\nu}\right)^{-\frac{1}{5}}\left(\frac{x}{\delta}\right)^{-\frac{1}{5}}$$

and hence

$$\frac{\delta}{x}=0.297\left(\frac{\bar{u}_1 \delta}{\nu}\right)^{-\frac{1}{4}}$$

or

$$\frac{x}{\delta}=3.363\left(\frac{\bar{u}_1 \delta}{\nu}\right)^{\frac{1}{4}}.$$

If we now put $\delta=a$, and noting that $\bar{u}_m/\bar{u}_1=0.817$ from § 6.8, we get

$$\frac{x_e}{a}=2.975\left(\frac{2u_m a}{\nu}\right)^{\frac{1}{4}}. \tag{8.1.6}$$

Thus if

$$Re=10^4 \quad \frac{x_e}{a}=30 \quad \text{approximately,}$$

and if

$$Re=10^5 \quad \frac{x_e}{a}=53 \quad \text{approximately.}$$

As a general rule we must assume that there will be an entry length of the order of 50 to 100 times the pipe radius before a fully developed turbulent profile is established.

8.2 Fully developed internal flow in a pipe or duct

As we have seen in § 8.1, at a sufficient distance from the pipe entry, the boundary-layer region formed at the wall of the pipe in the inlet region extends to the centre and effectively fills the entire cross-section of the pipe. We can describe this situation as enclosed or *internal flow* to distinguish it from the external flow which is characteristic of a wide fluid stream bounded by only one solid surface or wall.

When fully developed internal flow is finally established in a long pipe the velocity profile will be the same from section to section provided the fluid is incompressible, and it follows that the shear stress at the wall τ_0 will be independent of the distance measured along the length of the pipe. The pressure p will be uniform over any cross-section and the pressure gradient $\mathrm{d}p/\mathrm{d}x$ (measured in the direction of flow) will be constant.

The skin-friction coefficient for fully developed flow in a long pipe is defined by

$$c_f = \frac{\tau_0}{\frac{1}{2}\rho \bar{u}_m^2}, \tag{8.2.1}$$

where $\bar{u}_m$ is the mean flow velocity, i.e. the volume flow rate divided by the cross-sectional flow area of the pipe. Note that the definition of c_f for flow in a pipe differs from the definition for flow in a boundary layer. In chapter 7, when discussing boundary-layer flow we defined the skin-friction coefficient in terms of the external undisturbed stream velocity u_1. When dealing with internal flow in a pipe, however, it is more convenient to base the definition on the mean velocity u_m as in (8.2.1).

There is a very simple relationship between the skin-friction coefficient c_f and the pressure drop Δp occurring over a length L of a long pipe under fully developed steady-flow conditions. Referring to figure 66, it is evident that with no acceleration of the fluid the equilibrium condition must be

$$\pi a^2 \, \Delta p = 2\pi a L \tau_0$$

or

$$\Delta p = 2\frac{L}{a}\tau_0, \tag{8.2.2}$$

where a is the internal radius. Substituting for τ_0 from (8.2.1), the pressure

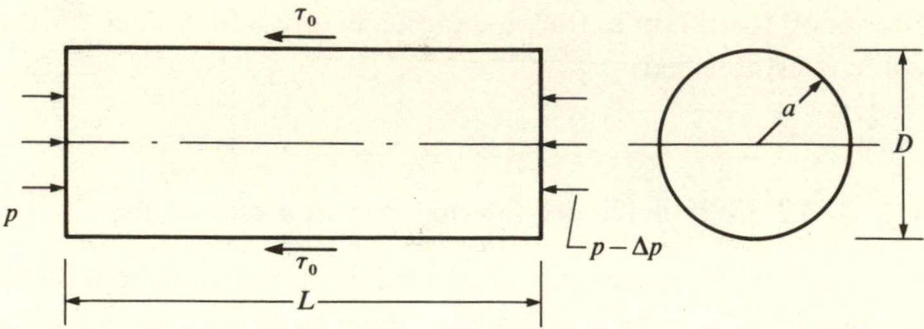

Fig. 66. Force balance for steady flow in a pipe.

drop is given by

$$\Delta p = c_f \frac{L}{a} \rho u_m^2$$

or

$$\Delta p = 2c_f \frac{L}{D} \rho u_m^2, \tag{8.2.3}$$

which is a frequently quoted result known as Fanning's equation.

We have seen in §5.8 that for the special case of fully developed *laminar flow* in a pipe the pressure drop is given by equation (5.8.17). Comparing (5.8.17) and (8.2.3), it will be evident that the skin-friction coefficient for laminar flow in a long pipe can be expressed as

$$c_f = \frac{16\mu}{\rho u_m D} = \frac{16}{Re}. \tag{8.2.4}$$

In the more normal case of *turbulent flow*, which is usually observed at values of the Reynolds number greater than about 2000, we have to use an empirical expression for the skin-friction coefficient such as the Blasius formula (6.6.10), or a semi-empirical expression in the form of the logarithmic resistance formula derived in §8.4. It will be noted, however, that the skin-friction coefficient c_f is always a function of the Reynolds number.

An important result regarding the variation of the shear stress τ with radius, under fully developed steady-flow conditions, can easily be deduced by considering the equilibrium of a cylindrical element of fluid of radius r as indicated in figure 67. The equilibrium condition may be expressed as

$$\pi r^2 \Delta p = 2\pi r L \tau$$

or

$$\Delta p = 2\frac{L}{r}\tau, \tag{8.2.5}$$

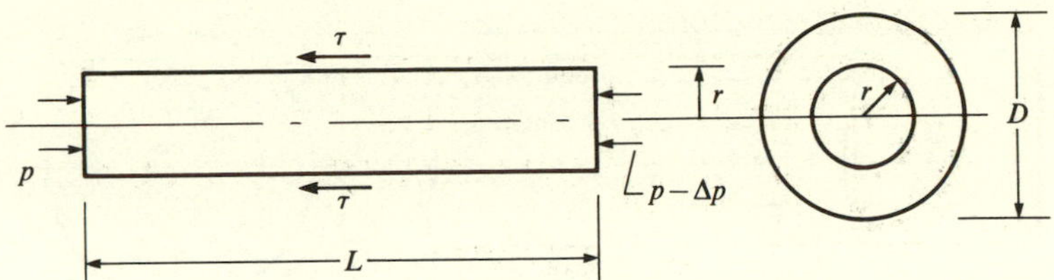

Fig. 67. Force balance on a cylindrical element of fluid.

which is the equivalent of (8.2.2). It follows from (8.2.2) and (8.2.5) that

$$\frac{\tau}{\tau_0}=\frac{r}{a}. \tag{8.2.6}$$

Thus the shear stress varies in a linear manner with the radius, falling from its maximum value τ_0 at the pipe wall to zero at the centre-line.

8.3 The velocity profile for fully developed turbulent flow in a long pipe or duct

In most practical examples of fluid flow through long pipes or ducts we encounter fairly large values for the Reynolds number and we must regard turbulence as the normal state of fluid motion. We have seen in § 8.1 that the final steady-flow regime which is established at some distance from the entry results from the merging of the wall boundary-layer region so that it fills the entire cross-section. A natural starting point for the investigation of fully developed turbulent flow in a pipe or duct, therefore, is to take the analysis of turbulent shear flow outlined in §§ 6.5 and 7.10 and to apply this to the wall region. As we have already noted in § 6.6, however, there is one significant difference between the case of *external flow* in a two-dimensional boundary layer where there is only one wall influencing the flow pattern, and the case of fully developed *internal flow* in a duct where the fluid is totally enclosed by the walls of the duct. While it is possible for the shear stress to be nearly constant in the wall region of a single-wall boundary layer provided the longitudinal pressure gradient is zero, the shear stress which is established in the case of fully developed internal flow in a parallel-walled duct or pipe must vary in a linear manner from zero at the centre-line to a maximum value of τ_0 at the wall as in (8.2.6). There can thus be no region of constant shear stress in fully developed internal flow.

Figure 68 represents a longitudinal cross-section through a long parallel-walled duct. The diagram is applicable equally to a pipe of radius a

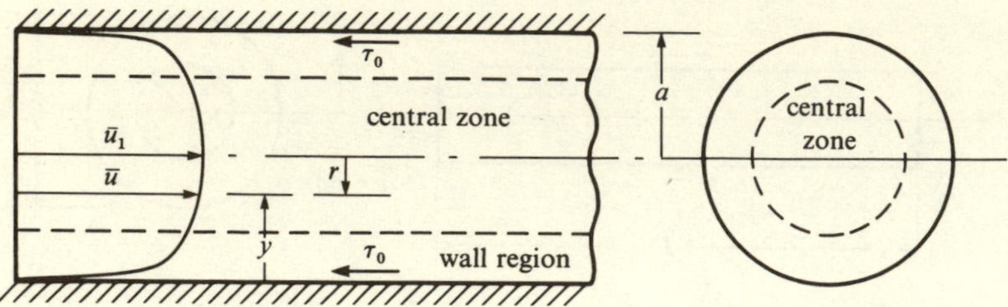

Fig. 68. Velocity distribution in a duct.

and circular cross-section, or to a broad rectangular duct where the flow in the central region can be treated as being equivalent to two-dimensional flow between parallel walls. We can argue in the most general way that the velocity $\bar{u}$ at distance y from the wall, or distance r from the centre-line, may be expressed as a function of five independent variables, i.e.

$$\bar{u}=f(\tau_0, \rho, \nu, y, a).$$

Using the normal method of dimensional analysis we can reduce this statement to a functional relationship between three dimensionless ratios, i.e.

$$\frac{\bar{u}}{\sqrt{(\tau_0/\rho)}}=f\left(\frac{y\sqrt{(\tau_0/\rho)}}{\nu}, \frac{a\sqrt{(\tau_0/\rho)}}{\nu}\right). \tag{8.3.1}$$

It is convenient to make use of the shorthand notation introduced in (6.5.6), (6.5.7), and (6.5.8), i.e. we define:

$$u_\tau=\sqrt{\frac{\tau_0}{\rho}}, \quad u^+=\frac{\bar{u}}{u_\tau}, \quad \text{and} \quad y^+=y\frac{\sqrt{(\tau_0/\rho)}}{\nu}.$$

We can thus re-write (8.3.1) in the form

$$u^+=f(y^+, a^+), \tag{8.3.2}$$

where a^+ is the value of y^+ at the centre-line where $y=a$. Equation (8.3.2) is of general validity and does not in any way imply that the shear stress τ at distance y from the wall is equal to the shear stress τ_0 at the wall.

In the wall layer, extending from $y=0$ to about $y=0.2\,a$, we can argue as in §§ 6.5 and 6.6 that the velocity distribution should be independent of the duct width or pipe radius. With this assumption the general functional relationship (8.3.2) reduces to

$$u^+=f(y^+), \tag{8.3.3}$$

which is identical to (6.6.2) and represents the general form of the *law of the wall.*

At the centre-line of the duct or pipe, where $y = a$, we have $\bar{u} = \bar{u}_1$ and it follows from (8.3.1) or (8.3.2) that

$$\frac{\bar{u}_1}{u_\tau} = f\left(\frac{au_\tau}{\nu}\right) \quad \text{or} \quad u_1^+ = f(a^+). \tag{8.3.4}$$

This result also follows directly from the definition of the skin-friction coefficient c_f. From (8.2.1) we have, as in (6.6.9),

$$\frac{\bar{u}_m}{u_\tau} = \sqrt{\frac{2}{c_f}}. \tag{8.3.5}$$

We also see that

$$a^+ = \frac{au_\tau}{\nu} = \frac{a\bar{u}_m}{\nu}\frac{u_\tau}{\bar{u}_m} = \tfrac{1}{2}\,Re\sqrt{(\tfrac{1}{2}c_f)} \tag{8.3.6}$$

and since c_f is a function of Re only, a^+ is also a function of Re only.

The velocity ratio u_1^+ may be expressed as:

$$u_1^+ = \frac{\bar{u}_1}{u_\tau} = \frac{\bar{u}_1}{\bar{u}_m}\frac{\bar{u}_m}{u_\tau} = \frac{\bar{u}_1}{\bar{u}_m}\sqrt{\frac{2}{c_f}}. \tag{8.3.7}$$

The ratio $\bar{u}_1/\bar{u}_m$, however, is a function of the Reynolds number and we can therefore say, from (8.3.7), that u_1^+ is also a function of Re only. Hence we conclude that

$$u_1^+ = f(a^+)$$

as in (8.3.4).

If we now adopt the point of view of an imaginary observer who is located at the centre-line of the duct or pipe, and who is able to measure the flow velocity over a short range of radial distance r on either side of the centre-line, we could argue that the velocity defect $\bar{u}_1 - \bar{u}$ observed at position r could be expressed as a function of only four independent variables, i.e.

$$\bar{u}_1 - \bar{u} = f(\tau_0, \rho, r, a)$$

or

$$\bar{u}_1 - \bar{u} = f(\tau_0, \rho, y, a). \tag{8.3.8}$$

Note that we have omitted the viscosity ν from this functional expression on the grounds that, while viscosity influences the overall flow pattern through its action in the wall region which is duly allowed for by the inclusion of the variable τ_0 in the functional relationship, the viscous stresses play a negligible role in comparison with the turbulent Reynolds stresses *in the central zone.* In other words, relative flow velocities, or velocity differences, are not influenced by viscosity in the central zone near the centre-line. If we now apply the normal method of dimensional analysis to the above relationship (8.3.8) we can reduce the expression to a functional relationship between two dimensionless ratios, i.e.

$$\frac{\bar{u}_1 - \bar{u}}{\sqrt{(\tau_0/\rho)}} = f\left(\frac{y}{a}\right) \tag{8.3.9}$$

or

$$u_1^+ - u^+ = f\left(\frac{y}{a}\right). \tag{8.3.10}$$

This is known as the *velocity-defect law*. It is applicable only in the central zone of fully developed turbulent flow and should be regarded as a statement of the limiting case where local viscous stresses become negligible.

8.4 Review of different formulae to represent the velocity profile with fully developed turbulent flow in a pipe or duct

In view of the different requirements and formulation of the law of the wall (8.3.3), on the one hand, and the velocity-defect law (8.3.10) on the other, we cannot expect to find a single mathematical expression which will give a satisfactory representation of the velocity profile across the entire turbulent region extending from the outer edge of the viscous sub-layer to the centre-line of the pipe or duct. Various formulae of a semi-empirical nature have been proposed, however, in an attempt to provide an approximate representation of the velocity profile. In this section we will examine three of these formulae and investigate how far they comply with the basic requirements specified in §8.3.

We have already quoted one such formula in §6.6. The *one-seventh power law* expression (6.6.15) was derived from the experimental relationship (6.6.10) for the skin-friction coefficient in such a way as to satisfy the requirements of the law of the wall (8.3.3). For convenience we will quote it again at this point:

$$u^+ = 8.56(y^+)^{\frac{1}{7}}. \tag{8.4.1}$$

We would expect this formula to give good correlation with the observed velocity distribution in the wall layer, up to a value of $y=0.2\,a$ or thereabouts, but to show an increasing divergence from measured velocities near the centre of the pipe. In fact, with a slight adjustment of the constant to a value of 8.7, equation (8.4.1) gives reasonably good agreement with measured velocities over the entire cross-section of the pipe. As already noted in § 6.6, however, the equation cannot satisfy the condition that the velocity gradient $d\bar{u}/dy$ should be zero at the centre-line where $y=a$.

If we attempt to apply (8.4.1) in the central zone, using the modified value of 8.7 in place of 8.56 for the numerical constant, we obtain for the value u_1^+ at the centre-line

$$u_1^+ = 8.7\,a^{+\frac{1}{7}} \tag{8.4.2}$$

and for the velocity defect

$$u_1^+ - u^+ = 8.7(a^{+\frac{1}{7}} - y^{+\frac{1}{7}}) = 8.7\,a^{+\frac{1}{7}}\left[1-\left(\frac{y}{a}\right)^{\frac{1}{7}}\right]$$

or

$$u_1^+ - u^+ = u_1^+\left[1-\left(\frac{y}{a}\right)^{\frac{1}{7}}\right]. \tag{8.4.3}$$

It will be noted that this does not comply with the correct form of the velocity defect law (8.3.10) owing to the presence of the factor $a^{+\frac{1}{7}}$ or u_1^+ (involving the viscosity ν) on the right-hand side.

The *logarithmic form* of the law of the wall (6.5.16) or (6.5.17) was derived for a turbulent shear layer with constant shear stress τ. We have noted in §§ 8.2 and 8.3 that a constant value for the shear stress is not possible in the case of fully developed internal flow in a pipe, and it is therefore illogical to expect this formula to be applicable to the velocity profile in a pipe or duct as distinct from a one-wall boundary layer. Despite this anomaly the logarithmic formula is widely used as a semi-empirical expression to represent the velocity distribution in a long pipe or duct. We have seen in § 6.6 that the numerical difference between the values of u^+ as calculated from (6.5.17) and (6.6.16) is not very great and, given the fact that the Karman constant k has been evaluated mainly from experimental measurements of the velocity distribution in pipes rather than in one-wall boundary layers, it is perhaps not surprising that a logarithmic formula as in (6.5.16) gives reasonably good correlation with measured velocity profiles in pipe flow.

We may therefore take as our second example of a semi-empirical formula for the velocity distribution:

$$u^+ = \frac{1}{k} \log_e y^+ + B, \tag{8.4.4}$$

where k is the Karman constant whose experimental value may be taken as approximately 0.4. As in the case of (8.4.1), this is simply another example of the law of the wall (8.3.3) and might be expected to apply only within the wall region. It is found to give good agreement with experimental measurements in pipe flow for a range of y^+ extending from the outer edge of the viscous sub-layer up to about $y = 0.5\,a$, if we take a value of about 5.5 for the numerical constant B.

If we attempt to extend the application of this formula to the central zone of the pipe, with a suitable adjustment to the numerical value of the constant B to fit the observed velocity at $y = a$, we could express the value of u_1 at the centre-line by

$$u_1^+ = \frac{1}{k} \log_e a^+ + B'. \tag{8.4.5}$$

From (8.4.4) and (8.4.5) the velocity defect is then given by

$$u_1^+ - u^+ = \frac{1}{k} \log_e \frac{a}{y} + (B' - B) \tag{8.4.6}$$

and it will be seen that this is consistent with the general form of the velocity-defect law (8.3.10) since the terms involving the coefficient of viscosity ν have cancelled out on the right-hand side.

The fact that the logarithmic formula (8.4.4) is able to satisfy *both* the law of the wall (8.3.3) and the velocity defect law (8.3.10) is probably no more than a mathematical coincidence and has little physical significance, but the formula has tended to acquire an enhanced status on this account. It must be emphasised, however, that no matter how the numerical value of the constant B or B' is manipulated in equation (8.4.4) or (8.4.6) the logarithmic formula cannot meet the requirement that the velocity gradient should be zero at the centre-line where $y = a$.

We have noted that equation (8.4.4) gives quite a good representation of the velocity profile in a pipe, from the outer edge of the viscous sub-layer up to a level of y equal to about 0.5 a, provided we take values of $k = 0.40$ and $B = 5.5$, i.e.

$$u^+ = 2.5 \log_e y^+ + 5.5. \tag{8.4.7}$$

This profile intersects the line $u^+ = y^+$ representing the laminar sub-layer at

$y^+=11.6$. A better picture of the profile in the wall layer is obtained, however, if we regard the true laminar sub-layer (sometimes referred to as the linear sub-layer) as extending from $y^+=0$ to $y^+=5$ with an intermediate zone representing the outer part of the viscous sub-layer extending from $y^+=5$ to $y^+=30$ approximately. The empirical equation proposed by von Karman to cover the range from $y^+=5$ to $y^+=30$ is

$$u^+=5.0\log_e y^+-3.05. \tag{8.4.8}$$

This expression is tangential to $u^+=y^+$ at $y^+=5$ and intersects equation (8.4.7) at $y^+=30$ where $u^+=14$ approximately. The combination of these three expressions is sometimes referred to as the *universal velocity profile.* This is a misleading title, however, since it implies a fundamental theoretical basis which does not exist in reality.

A reasonably accurate numerical value for the velocity at the centre-line may be obtained by taking a value of about 6.15 for the constant B' in equation (8.4.5), i.e. at $y=a$ we have

$$u_1^+=2.5\log_e a^+ +6.15. \tag{8.4.9}$$

If we are to use the logarithmic formula to represent the velocity distribution near the centre-line in the central zone of the flow we must take the same value for the constant by putting $B=B'$ since there cannot be a step change in the velocity at $y=a$. Thus the velocity-defect law (8.4.6) takes the more plausible form

$$u_1^+-u^+=\frac{1}{k}\log_e\frac{a}{y}=-\frac{1}{k}\log_e\frac{y}{a}$$

or, with $k=0.4$ as before,

$$u_1^+-u^+=-2.5\log_e\frac{y}{a}. \tag{8.4.10}$$

We still have the unsatisfactory feature, which is basic to the logarithmic formula, that the velocity gradient at the centre-line is given by $\mathrm{d}u/\mathrm{d}y=u_\tau/ky$ which is clearly incorrect. We cannot therefore regard equation (8.4.10) as an acceptable statement of the velocity profile near the centre-line.

In order to obtain a more satisfactory expression for the velocity distribution in the central zone it is necessary to go back to equation (6.5.14) or (6.7.16) and to introduce the fact that the shear stress is *not* constant and equal to τ_0, but is given correctly by (8.2.6). In place of $\sqrt{(\tau_0/\rho)}$ in (6.7.15), for instance, we should have $\sqrt{(\tau/\rho)}$, and from (8.2.6) we can say

$$\frac{\tau}{\rho}=\frac{\tau_0}{\rho}\frac{r}{a}=u_\tau^2\frac{r}{a}=u_\tau^2\left(1-\frac{y}{a}\right). \tag{8.4.11}$$

Hence equation (6.5.14) or (6.7.16) for the velocity gradient should take the modified form

$$\frac{\mathrm{d}\bar{u}}{\mathrm{d}y}=\frac{u_\tau}{ky}\left(1-\frac{y}{a}\right)^{\frac{1}{2}}, \tag{8.4.12}$$

which automatically satisfies the condition that the velocity gradient is zero at $y=a$.

We can integrate equation (8.4.12) from the centre-line, where $y=a$ and $u=u_1$, to an intermediate level y, i.e.

$$\frac{\bar{u}_1-\bar{u}}{u_\tau}=\frac{1}{k}\int_a^y\frac{[1-(y/a)]^{\frac{1}{2}}}{y}\,\mathrm{d}y. \tag{8.4.13}$$

It is slightly easier to evaluate the integral on the right-hand side if we express it in terms of the variable $r=a-y$. The result is

$$\bar{u}_1^+-u^+=\frac{1}{k}\left[\log_e\left\{\frac{1+(r/a)^{\frac{1}{2}}}{1-(r/a)^{\frac{1}{2}}}\right\}-2\left(\frac{r}{a}\right)^{\frac{1}{2}}\right]. \tag{8.4.14}$$

Noting that $r/a=1-(y/a)$, it will be seen that equation (8.4.14) complies with the velocity defect law (8.3.10). As might be expected, this equation gives good agreement with the measured velocity profile in the central zone of a pipe or duct but there is an increasing divergence between calculated and experimental values of the velocity as we approach the wall of the pipe.

We can summarise the situation at this point by noting that expressions (8.4.1) and (8.4.7) are tailored to the requirements of the wall layer but are unsatisfactory in the central zone. The third expression (8.4.14) is tailored to the requirements of the central zone but is less satisfactory in the wall layer. We have to recognise the fact that there is no single analytical expression which can give an accurate representation of the velocity distribution across the entire section of the pipe. This is perhaps not surprising when one considers the extremely complex nature of the motion in turbulent flow.

8.5 The skin-friction coefficient for turbulent flow in a pipe

We have already noted in (6.6.8) and (8.2.1) that the skin-friction coefficient c_f is defined in terms of the shear stress at the wall and the mean flow velocity $\bar{u}_m$, i.e.

$$c_f = \frac{2\tau_0}{\rho \bar{u}_m^2} \tag{8.5.1}$$

and this may be expressed in the alternative forms:

$$c_f = 2\left(\frac{u_\tau}{\bar{u}_m}\right)^2 \tag{8.5.2}$$

or

$$\frac{\bar{u}_m}{u_\tau} = \sqrt{\frac{2}{c_f}}. \tag{8.5.3}$$

Thus if we know the relationship governing the skin-friction coefficient c_f we can determine $\bar{u}_m/u_\tau$ and vice versa.

In § 6.6 we started with the experimental formula of Blasius for the skin friction with turbulent flow in smooth-walled pipes:

$$c_f = 0.079\, Re^{-\frac{1}{4}} \tag{8.5.4}$$

and we proceeded to derive the corresponding expression for $\bar{u}_m/u_\tau$. By taking into account the law of the wall (6.6.2) we then arrived at an expression for the velocity profile (6.6.15) in the wall region. In this section we will carry out the reverse process, taking the logarithmic formula (8.4.7) to represent the velocity distribution in the pipe, and we will derive the corresponding expression for the skin-friction coefficient c_f.

The first step is to evaluate the spatial mean-flow velocity $\bar{u}_m$ which is defined by

$$\text{volume flow} = \pi a^2 \bar{u}_m = \int_0^a 2\pi r \bar{u}\, \mathrm{d}r = \int_0^a 2\pi(a-y)\bar{u}\, \mathrm{d}y.$$

or

$$a^2 \bar{u}_m = \int_0^a 2(a-y)\bar{u}\, \mathrm{d}y. \tag{8.5.5}$$

If we multiply each side by u_τ and divide through by ν^2 we get the dimensionless expression

$$a^{+2}\frac{\bar{u}_m}{u_\tau} = \int_0^{a^+} 2(a^+ - y^+)u^+\, \mathrm{d}y^+. \tag{8.5.6}$$

We can now insert the logarithmic form of the velocity profile

$$u^+ = A \log_e y^+ + B, \tag{8.5.7}$$

so that equation (8.5.6) becomes

$$a^{+2}\frac{\bar{u}_m}{u_\tau}=2\int_0^{a^+}(a^+-y^+)(A\log_e y^+ +B)\,\mathrm{d}y^+. \tag{8.5.8}$$

There is a slight problem at this point since the logarithmic profile does not apply at very small values of y within the viscous sub-layer and would give a value of $-\infty$ for the velocity $\bar{u}$ if taken all the way to $y=0$. We can by-pass this difficulty, however, by taking $y=\delta$ or $y^+=\delta^+=\delta u_\tau/\nu$, where δ is the thickness of the viscous sub-layer, in place of $y=0$ for the lower limit in the integration.

Hence from (8.5.8), noting that $\int \log_e x\,\mathrm{d}x=x\log_e x-x$ and that $\int x\log_e \mathrm{d}x=\frac{1}{2}x^2\log_e x-\frac{1}{4}x^2$, we get

$$a^{+2}\frac{\bar{u}_m}{u_\tau}=2\left[Aa^+y^+(\log_e y^+-1)-\frac{A}{2}y^{+2}(\log_e y^+-\tfrac{1}{2})+Ba^+y^+-\frac{B}{2}y^{+2}\right]_{\delta^+}^{a^+}.$$

We find in fact that there is no problem with the lower limit since both $y^+\log_e y^+$ and $y^{+2}\log_e y^+$ tend to zero as $y^+\to 0$. We thus have the result, provided δ^+ is small, that

$$\frac{\bar{u}_m}{u_\tau}=A\log_e a^+-\tfrac{3}{2}A+B. \tag{8.5.9}$$

If we take $A=2.5$ and $B=5.5$ as in (8.4.7) we get

$$\frac{\bar{u}_m}{u_\tau}=2.5\log_e a^+ +1.75. \tag{8.5.10}$$

The dimensionless ratio a^+ is related to the Reynolds number and the skin-friction coefficient in the following way:

$$a^+=\frac{au_\tau}{\nu}=\frac{a\bar{u}_m}{\nu}\frac{u_\tau}{\bar{u}_m}=\frac{Re}{2}\sqrt{\frac{c_f}{2}}. \tag{8.5.11}$$

Hence from (8.5.3), (8.5.10), and (8.5.11) we have

$$\sqrt{\frac{2}{c_f}}=2.5\log_e\left(\frac{Re}{2}\sqrt{\frac{c_f}{2}}\right)+1.75$$

$$=2.5\log_e(Re\sqrt{c_f})-0.85$$

or

$$\sqrt{\frac{1}{c_f}}=1.77\log_e(Re\sqrt{c_f})-0.60. \tag{8.5.12}$$

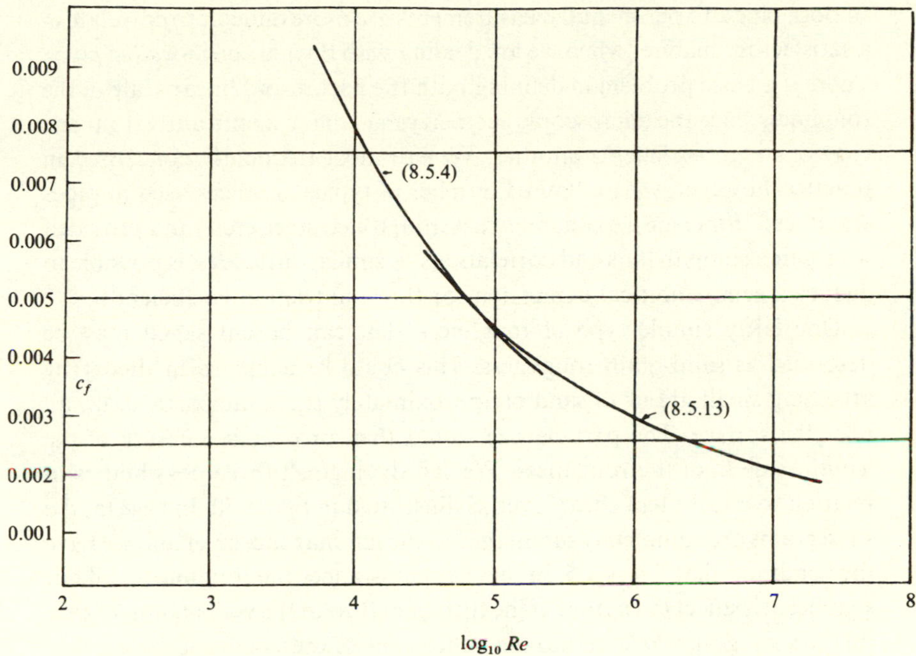

Fig. 69. Skin friction coefficient as a function of Reynolds number.

If the values of the constants in this expression are slightly adjusted in order to give the best correlation with measured values of c_f over a fairly wide range of the Reynolds number, we get the *Karman–Nikuradse* formula for the skin-friction coefficient:

$$\sqrt{\frac{1}{c_f}} = 1.74 \log_e (Re\sqrt{c_f}) - 0.40. \tag{8.5.13}$$

Although this formula is less convenient to use for calculation purposes than (8.5.4) it covers a wider range of the Reynolds number. For values of Re up to about 10^5 the Blasius formula (8.5.4) is perfectly adequate, but for values of Re above 10^5 the Karman–Nikuradse formula (8.5.13) should be used.

The two basic formulae for the skin-friction coefficient (8.5.4) and (8.5.13) are plotted graphically in figure 69.

8.6 The velocity profile and friction coefficient for flow in rough pipes

The investigation of the velocity profile and skin-friction coefficient which we have carried out in §§ 8.4 and 8.5 is only applicable to flow in smooth pipes. Experimental measurements are more difficult to correlate in a satisfactory manner when we are dealing with flow in *rough-walled pipes.* There is a basic problem in defining both the nature and linear scale of the roughness since the microscopic geometry may differ significantly from one type of rough surface to another. We are most frequently concerned in practice, however, with a limited number of typical surfaces used in pipes and ducts – for example concrete, cast iron, rolled steel, etc. – and provided we confine comparisons and correlations to similar surfaces it is possible to deduce a semi-empirical expression for the skin-friction coefficient.

One fairly simple type of roughness that can be envisaged may be described as sand-grain roughness. This could be achieved in theory by attaching small grains of sand of approximately the same particle size all over the surface. The particle size would then provide a measure of the length scale k_r of the roughness. We can distinguish three possibilities in relation to a turbulent shear layer as illustrated in figure 70. In case (a) the sand grains are completely submerged in the laminar sub-layer and we have the condition that $k_r u_\tau/\nu < 5$. In these circumstances the roughness will not significantly affect the nature of the turbulent flow in the wall region. In case (b) the sand grains extend into the buffer zone of the viscous layer, i.e. into the region between $y^+ = 5$ and $y^+ = 30$, and will consequently introduce another parameter into the law of the wall which may now be expected to take the form

$$u^+ = f\left(y^+, \frac{k_r u_\tau}{\nu}\right). \tag{8.6.1}$$

In case (c) the sand grains penetrate into the main turbulent wall layer, i.e. $k_r u_\tau/\nu > 30$ approximately. In this situation the eddies caused by the roughness particles will dominate the turbulent motion in the wall layer to

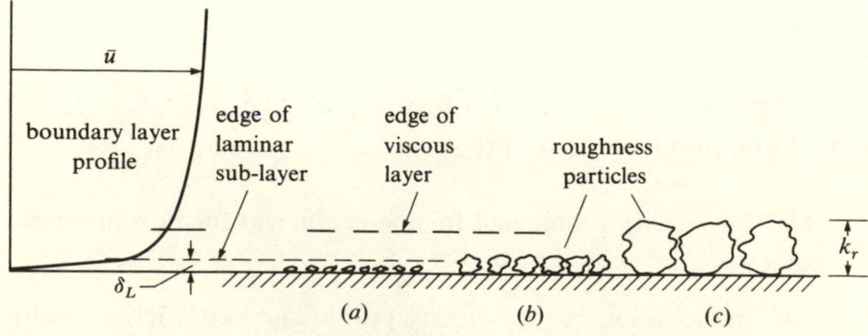

Fig. 70. Effect of wall roughness.

the exclusion of viscous effects. We could thus say that $u=f(\tau_0, \rho, y, k_r)$ in the wall region and, using the normal method of dimensional analysis, this reduces to

$$\frac{u}{\sqrt{(\tau_0/\rho)}}=f\left(\frac{y}{k_r}\right). \tag{8.6.2}$$

If we now make the further assumption of a constant shear stress region in the wall layer, although not strictly correct for flow in a pipe as we have already observed, we might expect equation (6.5.14) or (6.7.16) to be applicable over a limited range, i.e.

$$\frac{\mathrm{d}u}{\mathrm{d}y}=\frac{u_\tau}{ky},$$

which can be integrated as before to give the logarithmic form

$$\frac{\bar{u}}{u_\tau}=\frac{1}{k}\log_e y+\text{constant}. \tag{8.6.3}$$

We can therefore satisfy both (8.6.2) and (8.6.3) by expressing the velocity distribution in the wall layer in the form

$$u^+=\frac{1}{k}\log_e\frac{y}{k_r}+C \tag{8.6.4}$$

or

$$u^+=A\log_e\frac{y}{k_r}+C, \tag{8.6.5}$$

where $A=1/k=2.5$ if we have the same value as before for the Karman constant k. Equation (8.6.4) is the counterpart for flow in a rough pipe, with a linear roughness scale $k_r u_\tau/\nu>30$, of equation (8.4.4) for the velocity distribution in a smooth-walled pipe.

Although it is not strictly justifiable to extend the application of equation (8.6.4) or (8.6.5) into the central zone of the pipe, we can carry out exactly the same operation on (8.6.5) that we did in § 8.5 on the corresponding equation (8.5.7) in order to evaluate $\bar{u}_m/u_\tau$ and hence obtain an expression for the skin-friction coefficient. From (8.5.5), dividing through by u_τ we get

$$a^2\frac{\bar{u}_m}{u_\tau}=\int_0^a 2(a-y)u^+\,\mathrm{d}y$$

$$=2\int_0^a (a-y)[A\log_e y-A\log_e k_r+C]\,\mathrm{d}y. \tag{8.6.6}$$

Hence, carrying out the integration in a similar manner as in §8.5 and inserting the limits, we find

$$\frac{\bar{u}_m}{u_\tau} = A \log_e \frac{a}{k_r} - \tfrac{3}{2}A + C, \tag{8.6.7}$$

which is the counterpart of equation (8.5.9) for a smooth pipe. Taking $A = 2.5$ as before, we get

$$\frac{\bar{u}_m}{u_\tau} = \sqrt{\frac{2}{c_f}} = 2.5 \log_e \frac{a}{k_r} + (C - 3.75)$$

or

$$\sqrt{\frac{1}{c_f}} = 1.77 \log_e \frac{a}{k_r} + \frac{C}{\sqrt{2}} - 2.65. \tag{8.6.8}$$

This is the equivalent equation to (8.5.12). With adjusted values for the constants it becomes

$$\sqrt{\frac{1}{c_f}} = 1.74 \log_e \frac{a}{k_r} + 3.46. \tag{8.6.9}$$

8.7 Secondary flow effects

We have seen in §8.1 that quite a long entry length is required before a fully developed self-preserving turbulent flow velocity profile is established in a straight pipe. A long straight run of pipe, however, is the exception rather than the rule in the majority of engineering installations. When we encounter pipe bends or sharp corners there will be significant effects on the internal flow pattern. We also have to deal on occasions with the special problems of flow in pipes or ducts of non-circular cross-section, for example with the case of flow in a rectangular duct where complicated three-dimensional flow patterns are encountered. We can describe such effects under the general heading of *secondary flow* but the possibilities for detailed analysis are inevitably limited by the complexity of the three-dimensional turbulent flow patterns. By way of example, however, we will consider very briefly two illustrations of secondary flow effects.

Figure 71 illustrates the case of flow round a right-angle bend in a pipe of circular cross-section. We assume that, at entry to the bend, we have an axisymmetrical mean velocity profile corresponding to fully developed internal flow. Associated with this mean velocity profile will be a system of vortex lines in the form of concentric circles. We have already seen in §3.8 that vortex lines move with the fluid in the absence of significant viscous

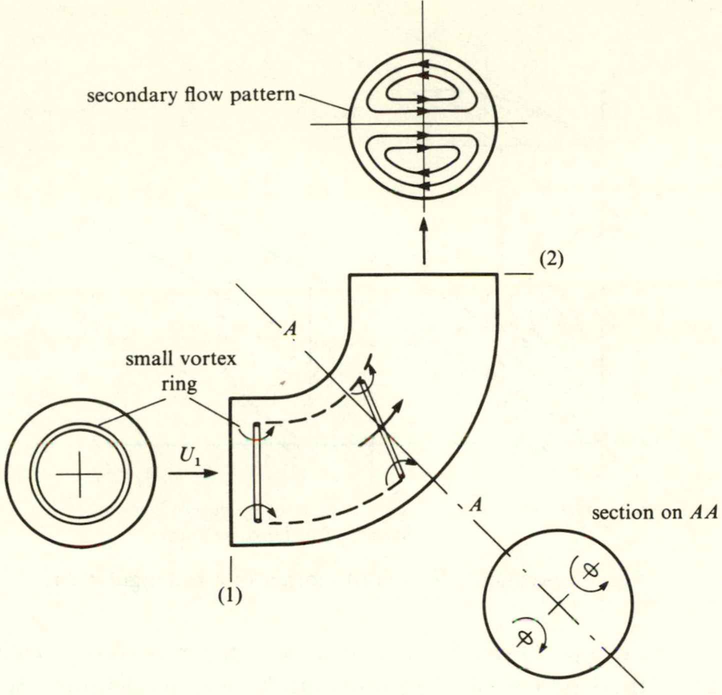

Fig. 71. Secondary flow in a pipe bend.

effects. As the fluid in the pipe flows round the bend the vortex lines, or vortex rings, become progressively distorted and tilted relative to the direction of the mean flow. If we ignore the effect of the diffusion of vorticity through the action of viscosity over the relatively short distances involved in the region of the pipe bend, it will be seen from figure 71 that the circumferential component of vorticity at section (1) produces axial components at section (2). This is illustrated in the figure where a small vortex ring in the axisymmetric flow pattern at section (1) is shown subsequently at the 45° position during its passage with the fluid round the bend. The consequence of this redistribution of vorticity components is the secondary flow pattern shown in the diagram at section (2). This process may be regarded as an example of the effect of the term $\boldsymbol{\omega} \cdot \nabla \mathbf{u}$ which appeared in the vorticity equation (5.6.8) and which represents the interaction between the components of the vorticity and the velocity gradient. It applies to both laminar and turbulent flow.

We can also explain the resulting secondary flow pattern, shown in figure 71, by considering the forces acting on a fluid element as it passes round the

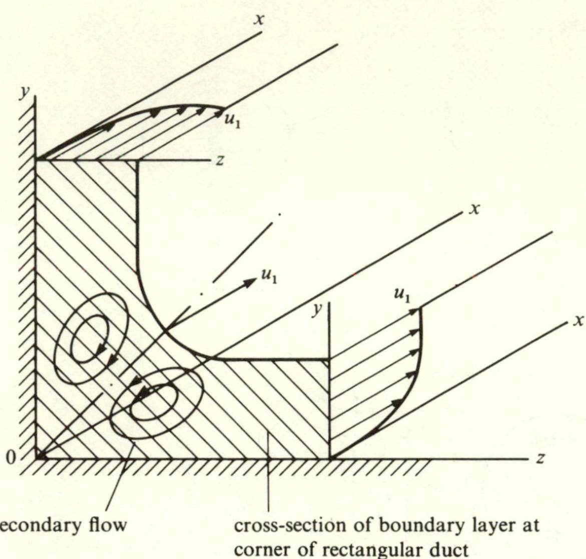

Fig. 72. Secondary flow in the corner of a rectangular duct.

bend in the pipe. Owing to the curvature of the mean flow velocity in the plane of the diagram, a fluid element must have an acceleration towards the centre of curvature and a radial transverse pressure gradient must therefore be set up across the pipe to balance the centrifugal force. Fluid which is located near a transverse centre-line (lying in the plane of the pipe bend), however, will have a higher velocity than fluid which is located close to the wall of the pipe. Thus the transverse pressure gradient necessary for equilibrium is smaller in the case of the fluid which is close to the wall. The net result is that there must be an outward movement of fluid in the vicinity of the transverse centre-line (drawn through the centre of curvature) and a corresponding inward flow in the region of the pipe wall.

A different type of secondary flow is encountered when we have turbulent flow in a straight duct of non-circular cross-section, for example a square- or rectangular-shaped duct as illustrated in figure 72. Near the corners of such a duct we have interaction between two boundary layers as indicated in the diagram. Owing to the transverse gradients of the Reynolds stress terms, $\rho\,\overline{u'^2}$, $\rho\,\overline{u'v'}$, $\rho\,\overline{u'w'}$, etc., in the interacting flow region, corner vortices are established which give rise to a secondary circulatory-type flow in the yz plane as shown diagrammatically in figure 72.

9

The energy equation

9.1 The complete energy equation for steady flow along a stream tube

In chapter 3 we used the concept of a control surface to enable us to derive the mechanical energy equation for the special case of the steady flow of a frictionless fluid along a stream tube. We will now extend the method of § 3.4 to the case of a real fluid with viscous stresses, and we will obtain the complete energy equation for steady flow along a stream tube by invoking the principle of the overall conservation of energy.

We select a fixed control surface represented by the dotted outline in figure 73, extending from section (1) to section (2), and coinciding with the stream tube between these points. We will assume that shear stresses exist within the fluid as a result of transverse velocity gradients, and that heat transfer is taking place within the fluid as a result of transverse temperature gradients. As in the case of § 3.3 (figure 16) and § 3.4 (figure 17), we picture a material volume of fluid moving along the stream tube with fluid crossing the control surface only at sections (1) and (2).

We can calculate the rate at which *flow work* is being done at sections (1) and (2) using exactly the same method of analysis as in § 3.4. From (3.4.2) we have:

$$\text{flow work per second} = p_1 a_1 u_1 - p_2 a_2 u_2 = \left(\frac{p_1}{\rho_1} - \frac{p_2}{\rho_2}\right) m, \tag{9.1.1}$$

where m is the mass-flow rate $= \rho_1 a_1 u_1 = \rho_2 a_2 u_2$. It will be seen from (9.1.1) that the rate at which flow work is being done at the boundary of the control surface is equal to the surface integral of the product $p\,\mathbf{u}$ evaluated over the entire boundary.

We can obtain the rate of flow work *per unit mass-flow rate* by dividing equation (9.1.1) by m. This is the same as the flow work per unit mass of fluid

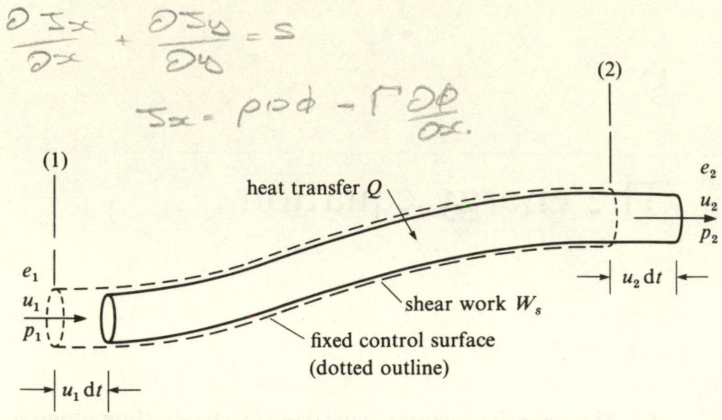

Fig. 73. Energy balance on a stream tube.

passing through the control surface, i.e.

$$\text{flow work per unit mass} = \frac{p_1}{\rho_1} - \frac{p_2}{\rho_2}, \tag{9.1.2}$$

which is identical with (3.4.3). It will be noted that this expression represents the work done *by* the surrounding fluid *on* the fluid which is inside the surface.

If there is a tangential shear stress acting at the wall of the stream tube at any point, due to the presence of a transverse velocity gradient, work will also be done at the wall of the stream tube through the agency of this stress acting in conjunction with the absolute velocity of the fluid. We can describe this as the *shear work*, using the same terminology as in § 5.5. The shear work is analogous to the flow work and is equal to the surface integral of the scalar product of the vector shear stress $\boldsymbol{\tau}$ and the absolute velocity $\mathbf{u}$ of the fluid evaluated over the boundary of the material volume.

It will be evident that the shear work, like the flow work, may be either positive or negative. In other words, the fluid contained within the material volume may be either gaining energy from, or losing energy to, the surrounding fluid as it passes through the control surface. By analogy with (9.1.2) we will define the shear work per unit mass W_s as the work which is done *by* the surrounding fluid *on* the fluid which is inside the surface.

The total amount of work done by the surrounding fluid on the fluid inside the material volume, per unit mass passing through the control surface, is therefore given by

$$\text{surface work per unit mass} = \left(\frac{p_1}{\rho_1} - \frac{p_2}{\rho_2}\right) + W_s. \tag{9.1.3}$$

If there are temperature gradients within the fluid, heat will be transferred by conduction across the boundary of the material volume as it passes through the control surface. We will represent the net rate of heat supply, per unit mass-flow rate, by the symbol Q. This is the same as the net amount of heat supplied per unit mass of fluid passing through the control surface.

We can now carry out an overall energy balance on the control surface and state that the total energy per unit mass of the fluid leaving the control surface at section (2) minus the total energy per unit mass of the fluid entering at section (1) must be equal to the sum of the heat supplied across the boundary plus the total amount of work done by the surrounding fluid at the boundary per unit mass of fluid passing through the control surface. The total energy per unit mass of fluid is given by the sum of the *internal energy* e, the *kinetic energy* $\frac{1}{2}u^2$, and the *potential energy* gz. It should be noted that each of these energy quantities is expressed *per unit mass* of fluid. Thus the overall energy balance may be expressed as:

$$(e_2+\tfrac{1}{2}u_2^2+gz_2)-(e_1+\tfrac{1}{2}u_1^2+gz_1)=Q+\left(\frac{p_1}{\rho_1}-\frac{p_2}{\rho_2}\right)+W_s. \tag{9.1.4}$$

This may be re-arranged as follows:

$$e_2+\frac{p_2}{\rho_2}+\tfrac{1}{2}u_2^2+gz_2=e_1+\frac{p_1}{\rho_1}+\tfrac{1}{2}u_1^2+gz_1+Q+W_s. \tag{9.1.5}$$

It is convenient, when dealing with cases of steady flow, to make use of the thermodynamics property known as the *enthalpy*, H. The enthalpy of a given mass of substance is defined by $H=E+pV$, where E is the internal energy and V is the volume. When expressed *per unit mass* of fluid, however, it is customary to use the lower-case letters h and e and to define the specific enthalpy by

$$h=e+\frac{p}{\rho}. \tag{9.1.6}$$

If we adopt this convention, we can re-write the steady-flow energy equation (9.1.5) for a stream tube in the form:

$$h_2+\tfrac{1}{2}u_2^2+gz_2=h_1+\tfrac{1}{2}u_1^2+gz_1+Q+W_s. \tag{9.1.7}$$

There is one disadvantage, however, in using h rather than H for the enthalpy per unit mass, since it is also customary in heat transfer calculations to use the same symbol h for the heat transfer coefficient. Later in this chapter, for example, in §§ 9.6 and 9.7, we will use the symbol h in this

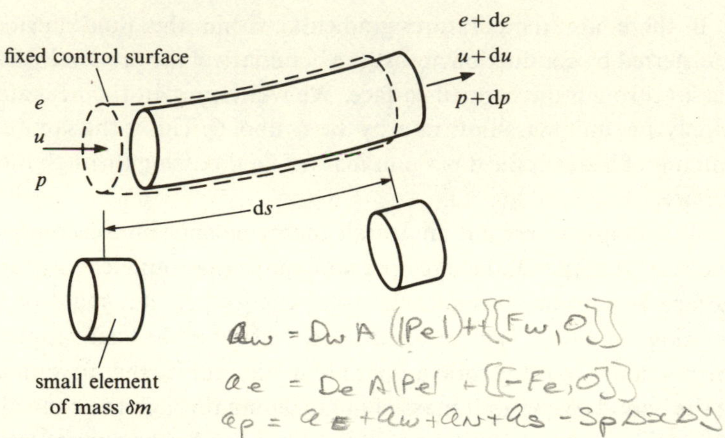

Fig. 74. Energy balance on a short length of stream tube.

latter sense. Care must therefore be taken to avoid any misunderstanding on this point.

It is unusual in engineering applications to encounter both large thermal effects and significant changes of potential energy in the same problem. Thus if we ignore the effects of changes in level we can express the steady-flow energy equation as:

$$h_2+\tfrac{1}{2}u_2^2=h_1+\tfrac{1}{2}u_1^2+Q+W_s. \tag{9.1.8}$$

We can obtain the steady-flow energy equation (9.1.4) in differential form by considering a short length of the stream tube as indicated in figure 74. If we ignore the effect of change of level we can say from equation (9.1.4) that the total energy gain, per unit mass of fluid, experienced over a short length ds of the stream tube is given by

$$\mathrm{d}\left(e+\frac{u^2}{2}\right)=\mathrm{d}Q+\mathrm{d}W_s-\mathrm{d}\left(\frac{p}{\rho}\right) \tag{9.1.9}$$

or

$$\mathrm{d}\left(h+\frac{u^2}{2}\right)=\mathrm{d}Q+\mathrm{d}W_s, \tag{9.1.10}$$

where dQ is the net rate of heat supply per unit mass-flow rate, across the short control surface of length ds indicated in figure 74, and dW_s is the amount of shear work done on the fluid per unit mass passing through the short control surface.

We will now anticipate the more exact analysis of §9.3 by considering the application of the first law of thermodynamics to a small element of fluid of mass δm, as indicated in the diagram, as it moves along the short length ds of the stream tube. For this purpose we mentally isolate the small element of mass δm by supposing it to be enclosed by an imaginary flexible boundary which moves with the fluid and always encloses the same quantity of material. In order to apply the first law of thermodynamics to this material volume we have to adopt the standpoint of an observer who moves with the fluid. We therefore ignore the absolute motion and concern ourselves only with the relative motion involving expansion or contraction and distortion in the shape of the element.

As we have already seen in (3.4.12), the *compression work* per unit mass is given by $(p/\rho^2)\,\mathrm{d}\rho$ where $\mathrm{d}\rho$ is the change of density which is experienced by the element during the movement along the length ds of the stream tube. We will express the *work of distortion* which is done on the element, per unit mass passing through the control surface of length ds in figure 74 by the term $\mathrm{d}W_\phi$. The heat received by the element of mass δm as it passes through the control surface of length ds is simply $\delta m\,\mathrm{d}Q$. We can thus express the first law of thermodynamics by saying that the change of internal energy of the fluid element of mass δm as it passes along the length ds of the stream tube is given by:

$$\delta m\,\mathrm{d}e = \delta m\left(\mathrm{d}Q + \mathrm{d}W_\phi + \frac{p}{\rho^2}\,\mathrm{d}\rho\right)$$

or

$$\mathrm{d}e = \mathrm{d}Q + \mathrm{d}W_\phi + \frac{p}{\rho^2}\,\mathrm{d}\rho. \tag{9.1.11}$$

If we now subtract the first law of thermodynamics, equation (9.1.11), from the complete steady-flow energy equation (9.1.9) we get:

$$\mathrm{d}(\tfrac{1}{2}u^2) = -\mathrm{d}\left(\frac{p}{\rho}\right) - \frac{p}{\rho^2}\,\mathrm{d}\rho + \mathrm{d}W_s - \mathrm{d}W_\phi, \tag{9.1.12}$$

i.e.

gain of kinetic energy = flow work − compression work

+ shear work − work of distortion.

This is the *mechanical energy equation* for steady flow along a stream tube. Thus the complete steady-flow energy equation (9.1.9) is simply the sum of the first law of thermodynamics (9.1.11) applied to a moving fluid element plus the mechanical energy equation for steady flow (9.1.12).

If equation (9.1.12) is compared with (5.5.12) it will be seen that the two equations are exactly similar in form. In the general form of the mechanical energy equation for steady flow, (5.5.12), the terms representing the shear work W_τ and the work of distortion Φ are expressed in each case as a *rate of work per unit material volume.* In equation (9.1.12), however, the terms W_s and W_ϕ represent the *amount* of work done *per unit mass of fluid* passing through the control surface of figure 74.

If equation (9.1.12) is compared with (3.4.11) it will be seen that (9.1.12) is simply an extension of the differential form of Bernoulli's equation with the additional terms representing the effect of the shear stresses. If we rearrange the first two terms on the right-hand side of (9.1.12) the equation may be written in the more concise form:

$$\mathrm{d}(\tfrac{1}{2}u^2) = -\frac{1}{\rho}\,\mathrm{d}p + \mathrm{d}W_s - \mathrm{d}W_\phi \tag{9.1.13}$$

and this may be seen as an extension of Euler's equation for steady flow in a stream tube as expressed by (3.4.7).

9.2 The energy equation for steady flow through a duct

We can picture a finite fluid stream being formed by adding together a number of individual stream tubes in a bundle. Strictly speaking, the stream tubes would have to be of suitable shape, for example with a square or hexagonal cross-section, in order to fit together, but the actual cross-sectional shape does not influence the argument. If we form a bundle of stream tubes in this manner, the shear work W_s will cancel out from one stream tube to another since the work done *on* one part of the fluid is work done *by* another part. However, we still have to evaluate the shear work for the complete stream *at the edge of the bundle.*

If the bundle of stream tubes amounts in total to flow through a *fixed duct*, as indicated in figure 75, there can be no shear work at the outer boundary of the stream since, with the no-slip condition, the absolute velocity of the fluid must be zero at the wall of the duct.

For flow through a pipe or duct which is fixed in space, therefore, the steady-flow energy equation (9.1.7) reduces to

$$h_2 + \frac{u_2^2}{2} + gz_2 = h_1 + \frac{u_1^2}{2} + gz_1 + Q, \tag{9.2.1}$$

where Q is the heat supplied across the wall of the duct to the fluid, expressed per unit mass of fluid passing through the duct.

It is important to appreciate that this form of the steady flow energy equation is quite general in its application, provided we have a fixed duct,

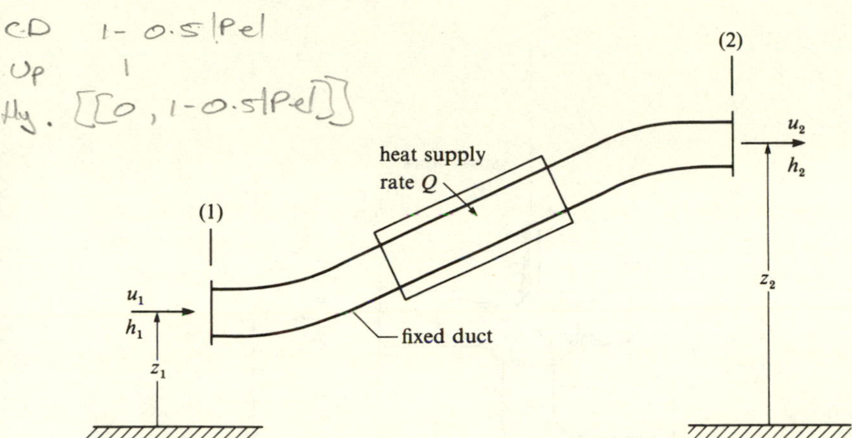

Fig. 75. Energy balance on a fixed duct with external heat supply.

and does *not* involve any assumption of frictionless flow. It must be noted, however, that we have over-simplified the problem in one respect. Since we are dealing with a finite flow area, and with a non-uniform velocity profile, it is necessary to take appropriate mean values, averaged over the cross-section of the duct, for the enthalpy h and the kinetic energy of the stream $\frac{1}{2}u^2$ per unit mass, when using equation (9.2.1).

If the effects of changes in level are negligible, as in the case of equation (9.1.8), we can express the steady-flow energy equation for flow through a fixed duct with heat transfer in the form

$$h_2+\frac{u_2^2}{2}=h_1+\frac{u_1^2}{2}+Q. \tag{9.2.2}$$

If, in addition, the kinetic energy term is small compared with the enthalpy term, or if $u_1=u_2$, equation (9.2.2) reduces to

$$h_2-h_1=Q, \tag{9.2.3}$$

i.e. the heat transferred is equal to the change of enthalpy of the fluid. This may be regarded as the basic heat transfer equation for a flow system where we have steady flow through a fixed duct at constant velocity and constant level.

If the fluid is a perfect gas, the enthalpy h may be expressed as a function of temperature by the relationship $\mathrm{d}h=c_p\,\mathrm{d}T$, and if the specific heat c_p is effectively constant over the range of temperature considered, equation (9.2.3) takes the form

$$c_p(T_2-T_1)=Q. \tag{9.2.4}$$

This form is also applicable to most liquids.

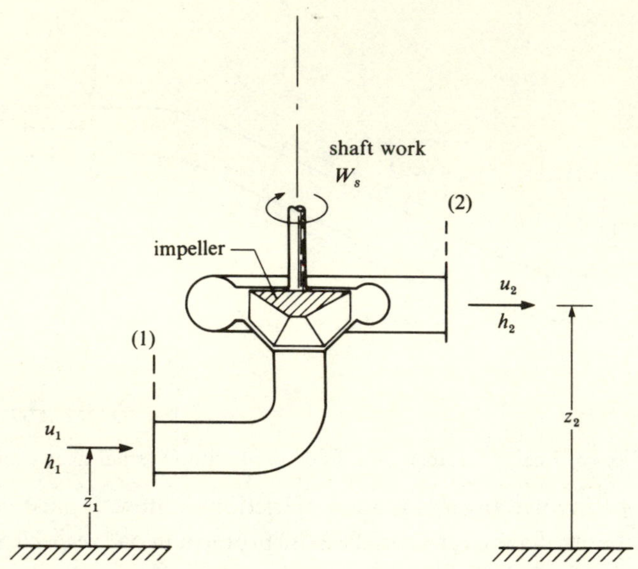

Fig. 76. Energy balance on a pump or compressor.

A different situation prevails if the fluid stream is bounded, either entirely or in part, by a *moving duct*. Figure 76 illustrates the case of a radial-flow pump or compressor where the rotating impeller forms part of the boundary of the fluid stream. Another example would be the case of flow through a row of moving blades in an axial flow compressor or turbine. When the fluid at the outer edge of the stream has an absolute velocity, by virtue of the fact that the wall of the duct itself is moving, it follows that shear work will be done on or by the fluid at the duct wall. When we are dealing with flow through turbo-machines it is convenient to describe this as the *shaft work*, W_s.

It follows from (9.1.7) that the steady-flow energy equation for a finite stream flowing through a moving duct, or through a system in which shaft work is being done on the fluid, takes the general form:

$$h_2+\tfrac{1}{2}u_2^2+gz_2=h_1+\tfrac{1}{2}u_1^2+gz_1+Q+W_s, \tag{9.2.5}$$

where W_s is the shaft work done on the fluid per unit mass, and Q is the heat transferred to the fluid per unit mass. If there is no heat transfer involved, the energy equation with shaft work is simply:

$$h_2+\tfrac{1}{2}u_2^2+gz_2=h_1+\tfrac{1}{2}u_1^2+gz_1+W_s. \tag{9.2.6}$$

If there is no significant change of level between inlet and outlet, this reduces to:

$$h_2+\tfrac{1}{2}u_2^2=h_1+\tfrac{1}{2}u_1^2+W_s. \tag{9.2.7}$$

If, in addition, the kinetic energy term is small compared with the enthalpy term, or if $u_1=u_2$, the equation reduces further to:

$$h_2-h_1=W_s \tag{9.2.8}$$

or for a perfect gas with constant specific heat c_p,

$$c_p(T_2-T_1)=W_s. \tag{9.2.9}$$

It is important to appreciate that the specific heat at constant pressure, c_p, only appears in equations (9.2.4) and (9.2.9) by virtue of the fact that the enthalpy of a perfect gas is related to the thermodynamic temperature by the statement $\mathrm{d}h=c_p\,\mathrm{d}T$. The equations quoted above are quite general in their application to *steady-flow* processes and are not in any way restricted to constant pressure applications.

9.3 The first law of thermodynamics applied to a moving fluid

We have seen in §9.1 that the complete energy equation for the special case of steady 'one-dimensional flow' along a stream tube may be regarded as a combination of the *mechanical energy equation* for flow in the stream tube together with the *first law of thermodynamics applied to a material volume of fluid* which moves along the stream tube. It will be appropriate at this stage to clarify the application of the first law of thermodynamics to a moving fluid and to obtain a general statement of the law as it applies to a three-dimensional flow system.

The first law of thermodynamics expresses the relationship between internal energy, heat, and work. In its usual formulation, however, it is applicable only to a *closed system* containing a fixed quantity of matter. For any particular equilibrium state, specified by the temperature, pressure, and composition, etc., a closed system will have a definite value for its *internal energy E*. The first law states that, for any change from one equilibrium state to another, the increase in the internal energy of the system will be equal to the heat supplied to the system plus the work done *on* the system, i.e.

$$\Delta E=\Delta Q+\Delta W. \tag{9.3.1}$$

It is important to note that the heat supplied ΔQ, and the work done ΔW, are evaluated at the boundary of the system. The law is formulated, in effect, from the point of view of an observer who is outside the system but who is

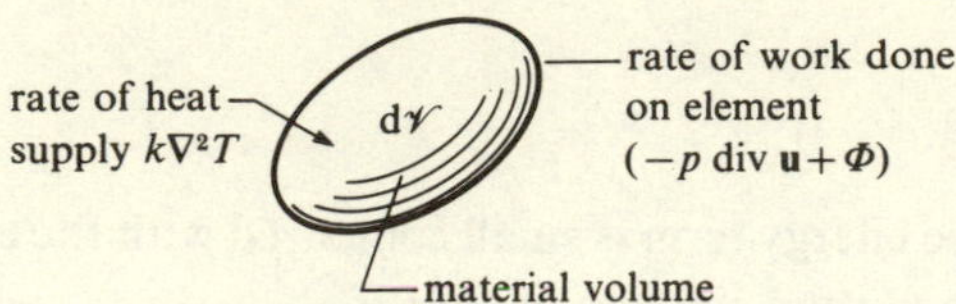

Fig. 77. Energy balance on a small material volume.

able to patrol the boundary. It is also important to appreciate that, whereas the internal energy E is a function of the state of the system as defined by its temperature, pressure, etc., the heat and work quantities in (9.3.1) depend on the particular process which is followed and not simply on the end points. We can express (9.3.1) in differential form, i.e.

$$dE = dQ + dW, \tag{9.3.2}$$

but we have to be careful when manipulating the differential quantities dQ and dW since these only have a meaning in thermodynamics when we can define the precise nature of the changes which are taking place.

We can meet the requirements of a closed system in a moving fluid by introducing the concept of an imaginary flexible membrane which encloses a definite mass of fluid and which follows the motion of the fluid with time. If we select a small material volume $d\mathscr{V}$ in this manner, as indicated in figure 77, we can say from (2.4.3) that the rate of change of the internal energy of the fluid contained within this material volume is $\rho(De/Dt)\, d\mathscr{V}$, where e is the internal energy of the fluid expressed *per unit mass.*

The rate at which heat is being supplied by conduction per unit material volume is given by (2.9.6), i.e. by $k\,\nabla^2 T$, where k is the thermal conductivity of the fluid (replacing the symbol k_h which was used in § 2.9). Hence the rate of heat supply to the material volume $d\mathscr{V}$ is given by $k\,\nabla^2 T\, d\mathscr{V}$.

The rate at which work is being done on the fluid in the material volume by the surrounding fluid is represented by the sum of two terms describing the work of compression and the work of distortion respectively. We have already seen in § 5.5 that the rate of compression work per unit volume is simply $-p$ div $\mathbf{u}$. The rate of work of distortion per unit volume is given by the dissipation function Φ defined by (5.5.11). In the case of the laminar flow of a Newtonian fluid the dissipation function Φ is represented by (5.5.14) or (5.5.15). This may be expressed in the concise form of (5.5.16), using tensor notation, as

$$\Phi = \frac{\mu}{2}\left(\frac{\partial u_i}{\partial x_j} + \frac{\partial u_j}{\partial x_i}\right)^2 - \tfrac{2}{3}\mu(\text{div } \mathbf{u})^2. \tag{9.3.3}$$

Thus the total rate at which work is being done on the fluid contained within the material volume $d\mathscr{V}$ is:

$$\text{rate of work} = (-p \operatorname{div} \mathbf{u} + \Phi)\, d\mathscr{V}. \tag{9.3.4}$$

We can now apply the first law of thermodynamics to the material volume $d\mathscr{V}$ and say that

$$\rho \frac{De}{Dt} d\mathscr{V} = (k\, \nabla^2 T - p \operatorname{div} \mathbf{u} + \Phi)\, d\mathscr{V}$$

or

$$\rho \frac{De}{Dt} = k\, \nabla^2 T - p \operatorname{div} \mathbf{u} + \Phi. \tag{9.3.5}$$

Thus the dissipation function Φ, which represented a loss of kinetic plus potential energy in the mechanical energy equation (5.5.12), reappears in equation (9.3.5) to represent a gain in the internal energy of the fluid. It thus describes an irreversible internal transfer process by which mechanical energy is transformed into internal energy.

The dissipation of mechanical energy in a moving fluid is much greater with turbulent flow than with laminar flow, and we have already observed that the great majority of actual engineering flow situations, for example in pipes, ducts, and boundary layers, involve turbulent motion. As we have seen in chapter 6, energy is transferred in turbulent flow from the mean motion of the fluid to large eddies by the working of the mean motion against the Reynolds stresses. This energy is then transferred from large eddies to smaller eddies and hence to the small-scale micro-turbulence. Finally the energy is dissipated through the action of the small-scale isotropic turbulence against the local fluctuating viscous stresses. In § 6.4 we introduced the term ε to represent the turbulent energy dissipation rate *per unit mass* of fluid. It is thus appropriate when considering turbulent flow to replace the dissipation function Φ in equation (9.3.5) by the turbulent energy dissipation rate *per unit volume*, $\rho\varepsilon$.

It was shown in § 6.4 that, for an incompressible fluid, the energy dissipation rate in turbulent flow is given by (6.4.8) or (6.4.9). Hence we can say that the rate per unit volume is

$$\rho\varepsilon = \tfrac{1}{2}\mu \overline{\left(\frac{\partial u_i'}{\partial x_j} + \frac{\partial u_j'}{\partial x_i}\right)^2} \approx \mu \overline{\left(\frac{\partial u_i'}{\partial x_j}\right)^2}. \tag{9.3.6}$$

Equation (9.3.5) may then be written for incompressible turbulent motion in the form

$$\rho \frac{\mathrm{D}e}{\mathrm{D}t} = k\,\nabla^2 T + \rho\varepsilon. \tag{9.3.7}$$

It will be seen that the turbulent energy dissipation term $\rho\varepsilon$ could equally well be regarded as a thermal energy 'source' supplying internal energy to the fluid.

9.4 The complete energy equation for a three-dimensional flow system

We have already obtained the mechanical energy equation for a fluid with viscous stresses in §5.5. We have also derived the first law of thermodynamics applied to a moving fluid in §9.3. By combining these two basic equations we can obtain the complete energy equation for a three-dimensional flow system.

From (5.5.12), the mechanical energy equation for steady three-dimensional flow may be expressed as

$$\rho \frac{\mathrm{D}}{\mathrm{D}t}(\tfrac{1}{2}\mathbf{u}^2 + \Psi) = -\mathrm{div}\,(p\mathbf{u}) + p\,\mathrm{div}\,\mathbf{u} + W_\tau - \Phi, \tag{9.4.1}$$

where W_τ is the rate at which shear work is being done on the fluid per unit volume as defined by (5.5.8), and Φ is the dissipation function defined by (5.5.11).

From (9.3.5), the first law of thermodynamics applied to a moving fluid may be expressed as

$$\rho \frac{\mathrm{D}e}{\mathrm{D}t} = k\,\nabla^2 T - p\,\mathrm{div}\,\mathbf{u} + \Phi. \tag{9.4.2}$$

If we now add equations (9.4.1) and (9.4.2) we obtain the complete energy equation for steady three-dimensional flow:

$$\rho \frac{\mathrm{D}}{\mathrm{D}t}(e + \tfrac{1}{2}\mathbf{u}^2 + \Psi) = k\,\nabla^2 T - \mathrm{div}\,(p\mathbf{u}) + W_\tau. \tag{9.4.3}$$

This simply states that the rate of gain of the total energy of the fluid per unit volume is equal to the rate of supply of heat by conduction plus the rate at which flow work is being done on the fluid plus the rate at which shear work is being done on the fluid. It will be noted that the terms representing compression work and distortion work have disappeared from the equation. These terms describe internal transfer processes and therefore cancel out between equations (9.4.1) and (9.4.2).

It will be appreciated that equations (9.4.2) and (9.4.3) refer to the normal case where there is no internal release of heat within the fluid as a result of chemical or nuclear reactions. If, however, there is an internal energy source of this type an extra term must be added to the right-hand side of equations (9.4.2) and (9.4.3) expressing the rate at which this energy is being supplied to the fluid per unit volume.

It will be noted that the procedure of adding equations (9.4.1) and (9.4.2) to obtain the complete energy equation (9.4.3) is exactly equivalent to the addition of (9.1.12) and (9.1.11) to give the differential form of the one-dimensional steady-flow energy equation (9.1.9) for a stream tube.

From the definition of the enthalpy given by (9.1.6) we have

$$h=e+\frac{p}{\rho}$$

and therefore

$$\frac{\mathrm{D}h}{\mathrm{D}t}=\frac{\mathrm{D}e}{\mathrm{D}t}+\frac{1}{\rho}\frac{\mathrm{D}p}{\mathrm{D}t}-\frac{p}{\rho^2}\frac{\mathrm{D}\rho}{\mathrm{D}t}. \tag{9.4.4}$$

From the equation of continuity, in the form of (2.3.7), however, we have:

$$\frac{1}{\rho}\frac{\mathrm{D}\rho}{\mathrm{D}t}=-\operatorname{div}\mathbf{u}$$

and hence (9.4.4) may be written as:

$$\rho\frac{\mathrm{D}h}{\mathrm{D}t}=\rho\frac{\mathrm{D}e}{\mathrm{D}t}+\frac{\mathrm{D}p}{\mathrm{D}t}+p\operatorname{div}\mathbf{u}. \tag{9.4.5}$$

This is a general mathematical statement relating the rate of change of the enthalpy of a moving fluid element to the rate of change of its internal energy.

If we now make use once again of the mathematical relationship:

$$\operatorname{div}(p\mathbf{u})=\mathbf{u}\cdot\nabla p+p\operatorname{div}\mathbf{u}$$

and noting that for *steady flow* $\mathbf{u}\cdot\nabla p=\mathrm{D}p/\mathrm{D}t$, we can write:

$$\operatorname{div}(p\mathbf{u})=\frac{\mathrm{D}p}{\mathrm{D}t}+p\operatorname{div}\mathbf{u}. \tag{9.4.6}$$

Hence under *steady-flow conditions*, from (9.4.5) and (9.4.6),

$$\rho \frac{\mathrm{D}h}{\mathrm{D}t} = \rho \frac{\mathrm{D}e}{\mathrm{D}t} + \operatorname{div}(p\mathbf{u}). \tag{9.4.7}$$

We can now substitute from (9.4.7) in (9.4.3) and thus express the steady-flow energy equation in terms of the enthalpy h instead of the internal energy e, hence

$$\frac{\mathrm{D}}{\mathrm{D}t}(h + \tfrac{1}{2}\mathbf{u}^2 + \Psi) = k\,\nabla^2 T + W_\tau. \tag{9.4.8}$$

This simply states that the rate of change of the enthalpy plus the kinetic energy plus the potential energy of the fluid, per unit volume, is equal to the rate of supply of heat by conduction plus the rate at which shear work is being done on the fluid per unit volume. If we compare equations (9.4.3) and (9.4.8) it will be seen that we have simply transferred the flow work term $-\operatorname{div}(p\mathbf{u})$ from the right-hand side of equation (9.4.3) to the left-hand side of equation (9.4.8) where it is included in the term $\rho(\mathrm{D}h/\mathrm{D}t)$ as indicated by (9.4.7).

In most engineering problems where thermal effects are significant we can usually ignore potential energy changes, and the steady-flow energy equation then takes the form

$$\rho \frac{\mathrm{D}}{\mathrm{D}t}(h + \tfrac{1}{2}\mathbf{u}^2) = k\,\nabla^2 T + W_\tau. \tag{9.4.9}$$

We are normally only concerned with the energy equation under steady mean flow conditions. However, provided we can omit the potential energy term on the left-hand side of equations (9.4.1) and (9.4.3), the basic energy equation in the form of (9.4.3) is also applicable to unsteady flow. We can still make use of equation (9.4.5), which is of general validity, but in place of (9.4.7) we have to write:

$$\rho \frac{\mathrm{D}h}{\mathrm{D}t} = \rho \frac{\mathrm{D}e}{\mathrm{D}t} + \frac{\partial p}{\partial t} + \operatorname{div}(p\mathbf{u}). \tag{9.4.10}$$

Thus the energy equation, expressed in terms of the enthalpy h, takes the following form for unsteady flow:

$$\rho \frac{\mathrm{D}}{\mathrm{D}t}(h + \tfrac{1}{2}\mathbf{u}^2) = \frac{\partial p}{\partial t} + k\,\nabla^2 T + W_\tau, \tag{9.4.11}$$

which replaces (9.4.9).

When we are dealing with a fluid, such as a perfect gas, where the enthalpy h can be expressed in terms of the thermodynamic temperature by the relationship $\mathrm{d}h = c_p\,\mathrm{d}T$, and if the specific heat c_p is effectively constant, we can then express the steady-flow energy equation (9.4.9) in the form:

$$\rho \frac{\mathrm{D}}{\mathrm{D}t}(c_p T + \tfrac{1}{2}\mathbf{u}^2) = k\,\nabla^2 T + W_\tau. \tag{9.4.12}$$

In the *special case* where kinetic energy changes are small, and where the shear work term can be neglected, the equation reduces to

$$\frac{\mathrm{D}T}{\mathrm{D}t} = \frac{k}{\rho c_p}\nabla^2 T, \tag{9.4.13}$$

which is similar to the simple form (2.9.7) of the transport equation derived in §2.9, but it should be noted that we now have c_p in place of c_v.

9.5 Similarity for heat transfer in fluid flow

We obtained the basic conditions for dynamical similarity in §5.7 by expressing the equation of motion for a viscous fluid in dimensionless form. We can apply the same technique to the energy equation in order to determine the requirements for similarity of the heat transfer pattern between two cases of steady flow with geometrically similar boundaries and with similar thermal boundary conditions.

If we ignore potential energy changes, and if we are dealing with a fluid such as a perfect gas where we can express the enthalpy as a function of the absolute temperature with a constant value for the specific heat c_p, we can regard (9.4.12) as the basic form of the steady-flow energy equation for our present purposes. We thus have

$$\rho \frac{\mathrm{D}}{\mathrm{D}t}\left(c_p T + \frac{\mathbf{u}^2}{2}\right) = k\,\nabla^2 T + W_\tau \tag{9.5.1}$$

and from (5.5.8), where

$$W_\tau = \frac{\partial}{\partial x}(\mathbf{u}\cdot\boldsymbol{\tau}_x) + \frac{\partial}{\partial y}(\mathbf{u}\cdot\boldsymbol{\tau}_y) + \frac{\partial}{\partial z}(\mathbf{u}\cdot\boldsymbol{\tau}_z). \tag{9.5.2}$$

We start by defining a reference length L and a reference velocity U_0 as in §5.7. The development of the argument is clarified if we think in terms of a specific case, for example the steady flow of a fluid past a heated sphere of diameter L whose surface is maintained at a constant temperature T_s as

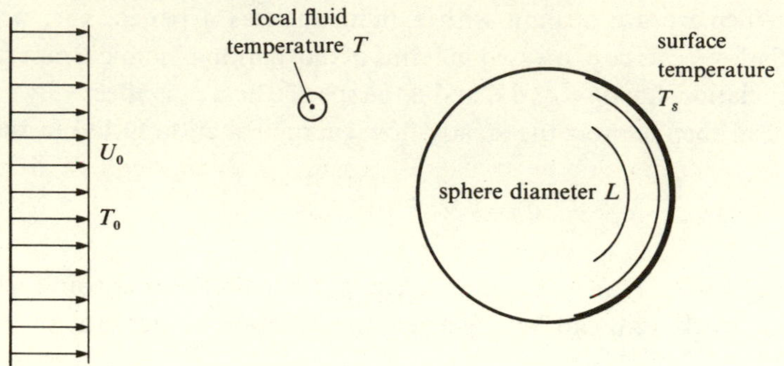

Fig. 78. Heat transfer from a sphere in a moving stream.

pictured in figure 78. The method of analysis, however, is entirely general in its range of application.

Heat transfer by conduction and convection depends on the existence of temperature differences. It is therefore appropriate to define both a *reference temperature* and a *reference temperature difference*.

In the particular example illustrated in figure 78 the fluid stream is assumed to approach the sphere with a uniform velocity U_0 and a uniform temperature T_0. We may therefore take T_0 as the reference temperature and define the reference temperature difference by $\Delta T_s = (T_s - T_0)$. If the local temperature of the fluid at any point in the flow system is T, we can then introduce a dimensionless temperature ratio T' defined by

$$T' = \frac{T - T_0}{T_s - T_0} = \frac{T - T_0}{\Delta T_s}. \tag{9.5.3}$$

For the remaining variables which appear in (9.5.1) and (9.5.2) we can introduce the following dimensionless ratios which are similar to those used in connection with the equation of motion in §5.7:

$$\left.\begin{array}{lll} u' = \dfrac{u}{U_0} & v' = \dfrac{v}{U_0} & w' = \dfrac{w}{U_0} \\[2ex] x' = \dfrac{x}{L} & y' = \dfrac{y}{L} & z' = \dfrac{z}{L} \\[2ex] t' = \dfrac{U_0 t}{L} & \tau'_{xy} = \dfrac{\tau_{xy}}{\rho U_0^2}, & \text{etc.} \end{array}\right\} \tag{9.5.4}$$

Noting the definition of the viscous stress vector $\boldsymbol{\tau}_x$ as given by (5.2.8), we can write:

$$\boldsymbol{\tau}'_x = \tau'_{xx}\,\mathbf{i} + \tau'_{xy}\,\mathbf{j} + \tau'_{xz}\,\mathbf{k}, \tag{9.5.5}$$

with similar expressions for $\boldsymbol{\tau}'_y$ and $\boldsymbol{\tau}'_z$. Thus the rate of shear work W_τ as given by (9.5.2) may be expressed in terms of dimensionless variables by writing

$$W_\tau = \frac{\rho U_0^3}{L} W'_\tau = \frac{\rho U_0^3}{L}\left[\frac{\partial}{\partial x'}(\mathbf{u}'\cdot\boldsymbol{\tau}'_x) + \frac{\partial}{\partial y'}(\mathbf{u}'\cdot\boldsymbol{\tau}'_y) + \frac{\partial}{\partial z'}(\mathbf{u}'\cdot\boldsymbol{\tau}'_z)\right]. \tag{9.5.6}$$

The energy equation (9.5.1) may now be re-written in terms of the dimensionless variables in the form

$$\frac{\rho U_0}{L}\frac{\mathrm{D}}{\mathrm{D}t'}[c_p T'\,\Delta T_s + \tfrac{1}{2}\mathbf{u}'^2 U_0^2] = \frac{k\,\Delta T_s}{L^2}\nabla'^2 T' + \frac{\rho U_0^3}{L} W'_\tau,$$

where the symbol ∇'^2 has the same meaning as in § 5.7. Dividing through by $(\rho U_0 c_p \Delta T_s)/L$, we get

$$\frac{\mathrm{D}}{\mathrm{D}t'}\left[T' + \frac{U_0^2}{c_p \Delta T_s}\frac{\mathbf{u}'^2}{2}\right] = \frac{k}{\rho U_0 c_p L}\nabla'^2 T' + \frac{U_0^2}{c_p \Delta T_s} W'_\tau, \tag{9.5.7}$$

which is the dimensionless form of the energy equation (9.5.1).

It will be seen that, for given boundary conditions and with geometric similarity, the solution of equation (9.5.7) depends only on the numerical values of the two dimensionless groups $U_0^2/(c_p \Delta T_s)$ and $k/\rho U_0 c_p L$.

The first of these dimensionless ratios is known as the *Eckert number*, *Ec*. It will be seen that it provides a measure of the relative magnitude of the kinetic energy term compared with the temperature or enthalpy term within the square bracket of the convective differential on the left-hand side of equation (9.5.7). It equally expresses the relative magnitude of the last term on the right-hand side of (9.5.7), representing the rate of shear work, compared with the convective rate of change of total energy on the left-hand side of the equation. The Eckert number is closely related to the *Mach number*, *Ma*, which is used in the analysis of high-speed gas flow. We will see in chapter 11 that for a perfect gas the relationship between these two ratios is given by

$$Ec = (\gamma - 1)\frac{T_0}{\Delta T_s} Ma^2. \tag{9.5.8}$$

The other dimensionless ratio which appears in (9.5.7), $k/\rho U_0 c_p L$, evidently provides a measure of the relative magnitude of heat transfer by conduction within the fluid to the rate of change of energy by convection. The inverse of this ratio is known as the *Péclet number*, *Pé*, i.e. $Pé = \rho U_0 c_p L/k$. The Péclet number is related to the *Reynolds number*, *Re*, by

$$Pé = Re \times \frac{\mu c_p}{k} = Re \times Pr, \tag{9.5.9}$$

where $Pr = \mu c_p/k$ is the *Prandtl number*. Since the Reynolds number is required independently as a condition for dynamical similarity in connection with the equation of motion, as shown in § 5.7, it is customary when analysing heat transfer in forced convection to work in terms of the Reynolds and Prandtl numbers rather than the Reynolds and Péclet numbers. The Prandtl number is convenient to evaluate since it involves only the physical properties of the fluid μ, c_p, and k.

9.6 Heat transfer coefficients – the Nusselt and Stanton numbers

The discussion of the energy equation for a moving fluid in §§ 9.1–9.5 has necessarily involved consideration of heat conduction within the fluid and of heat transfer to or from the fluid across a boundary surface such as the wall of a duct. The detailed analysis of heat transfer is the main subject of chapters 13, 16, and 17. It will be appropriate, however, to take a preliminary view at this stage of the concept of a heat transfer coefficient and to consider some of the basic relationships involving the energy equation which are applicable to flow in a stream tube or duct.

Heat transfer across a boundary surface, such as a tube wall, may be expressed in terms of the *heat flux*, *q*, per unit area. When we are dealing with transfer by conduction and convection, but excluding transfer by radiation, the heat flux is generally proportional to the overall temperature difference. It is therefore convenient to introduce a heat transfer factor or coefficient *h* defined by

$$q = h\,\Delta T, \tag{9.6.1}$$

where ΔT is the difference between the wall temperature and the bulk temperature of the fluid. The SI units for heat flux are $\mathrm{W\,m^{-2}}$ and the units for the heat transfer coefficient *h* are therefore $\mathrm{W\,m^{-2}\,K^{-1}}$.

As already noted in § 9.1, care should be taken not to confuse the use of the lower-case letter *h* in §§ 9.1, 9.2, and 9.4, where it represents the enthalpy of the fluid per unit mass, with the use of the same letter in §§ 9.6 and 9.7, where it represents the heat transfer factor as defined by (9.6.1).

Anticipating the more-detailed analysis of heat conduction in chapter 13, we may simply note at this stage that the Fourier law of conduction, which we have already used in §9.3, takes the following form for the one-dimensional flow of heat in the y direction:

$$q = -k\frac{\mathrm{d}T}{\mathrm{d}y}, \tag{9.6.2}$$

where k is the thermal conductivity of the fluid. If we now introduce the same dimensionless ratios as in §9.5, i.e. if we define $y' = y/L$ and $T' = (T - T_0)/\Delta T$, where L is a reference length and ΔT a reference temperature difference, equation (9.6.2) can be expressed in dimensionless form as

$$\frac{qL}{k\,\Delta T} = -\frac{\mathrm{d}T'}{\mathrm{d}y'}. \tag{9.6.3}$$

Provided the reference temperature difference is chosen to be identical with that used in the definition of h in (9.6.1), the dimensionless form of the equation for heat conduction within the fluid may be written as

$$\frac{hL}{k} = -\frac{\mathrm{d}T'}{\mathrm{d}y'}. \tag{9.6.4}$$

The dimensionless quantity hL/k is known as the *Nusselt number*, *Nu*.

It is equally possible to relate the heat flux q across a boundary surface to the transport of heat by convection within the fluid. If we consider the simple case of 'one-dimensional' flow in a stream tube or duct, as indicated in figure 79, and if we apply the energy equation in the form of (9.2.4) to a short length $\mathrm{d}x$ of the tube, we have

$$\rho u_m \pi \frac{D^2}{4} c_p \,\mathrm{d}T = q\pi D\,\mathrm{d}x$$

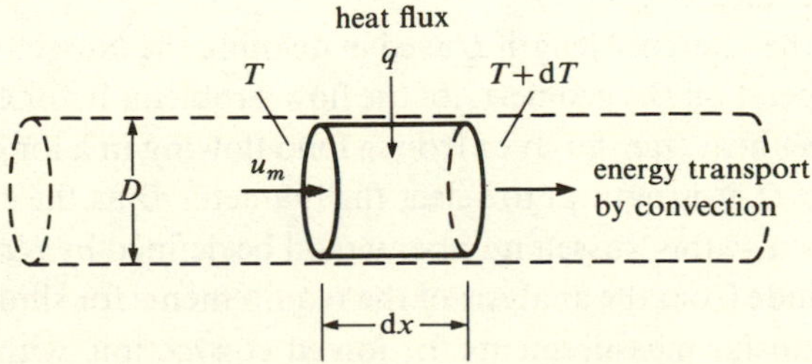

Fig. 79. Heat balance on an elemental length of tube.

or

$$q=\rho u_m c_p \frac{D}{4}\frac{\mathrm{d}T}{\mathrm{d}x}, \tag{9.6.5}$$

where D is the diameter of the tube, and u_m is the fluid velocity. We can express the energy equation (9.6.5) in dimensionless form by making the same substitutions as before, i.e. we define

$$x'=\frac{x}{L} \quad \text{and} \quad T'=\frac{T-T_0}{\Delta T},$$

and hence

$$\frac{q}{\rho u_m c_p \Delta T}=\frac{D}{4L}\frac{\mathrm{d}T'}{\mathrm{d}x'}. \tag{9.6.6}$$

Provided the reference temperature difference is again chosen to be identical with that used in the definition of h, the last result may be written as

$$\frac{h}{\rho u_m c_p}=\frac{D}{4L}\frac{\mathrm{d}T'}{\mathrm{d}x'}. \tag{9.6.7}$$

The dimensionless quantity $h/\rho u_m c_p$ is known as the *Stanton number*, *St*.

The Nusselt and Stanton numbers are alternative forms of dimensionless heat transfer coefficient. It will be clear from the above derivation that the Nusselt number provides a measure of the actual overall rate of heat transfer from surface to fluid compared with a theoretical rate which would be achieved by conduction alone within a static fluid with the same reference temperature difference applied over a conduction path of length L. Similarly the Stanton number gives a measure of the actual rate of heat transfer from surface to fluid compared with a theoretical rate of heat transport by convection which would be achieved by a fluid with mass velocity ρu_m experiencing a temperature rise equal to the reference temperature difference ΔT.

The choice of the reference length L used in defining the Nusselt number will naturally depend on the geometry of the flow problem. If, for example, we are considering heat transfer to or from a fluid flowing in a long pipe of internal diameter D, it is natural to select the diameter D as the reference length and in this case the Nusselt number would be defined by $Nu=hD/k$.

We may conclude from the analysis of the requirements for similarity in §9.5 that heat transfer measurements for forced convection, with similar geometric and thermal boundary conditions, may be correlated by expressing either the Nusselt number or the Stanton number as a function of the Reynolds number, the Prandtl number, and the Eckert number. Thus for forced convection

$$Nu = f(Re, Pr, Ec) \tag{9.6.8}$$

or

$$St = f(Re, Pr, Ec). \tag{9.6.9}$$

It is purely a matter of practical convenience whether we choose to work in terms of a Nusselt number or a Stanton number. In many applications with flow at moderate velocities the mechanical energy changes are small compared with the thermal effects and in these circumstances the influence of the Eckert number may be neglected.

We have already noted in §9.5 that the Péclet number provides a measure of the relative magnitude of heat transfer by convection to heat transfer by conduction, and it will be seen from (9.5.9) that

$$\frac{Nu}{St} = Pé = Re \times Pr. \tag{9.6.10}$$

9.7 Heat transfer for steady flow in a pipe or duct

We can now apply the energy equation in the form of (9.2.4) to the important basic case of heat transfer to a fluid flowing through a long tube or duct as indicated in figure 80. We will assume steady-flow conditions in a tube of constant cross-sectional area. Thus the mass velocity ρu_m will remain constant for all sections along the length of the tube. If the fluid enters the tube with bulk temperature T_1 and leaves with temperature T_2 the steady-flow energy equation (9.2.4) may be expressed in the form

$$\rho u_m \pi \frac{D^2}{4} c_p (T_2 - T_1) = Q = \pi D \int_0^L q \, \mathrm{d}x, \tag{9.7.1}$$

where q is the local heat flux at distance x measured from the entry section.

If the temperature of the fluid is T at distance x from the tube entry, and if the tube wall temperature at this point is T_w, the local heat flux q will be related to the local temperature difference by

$$q = h(T_w - T). \tag{9.7.2}$$

Provided the range of the fluid temperature from T_1 at entry to T_2 at exit is not unduly large we can normally assume constant values for the fluid properties μ, c_p, and k. Thus the Reynolds number and the Prandtl number

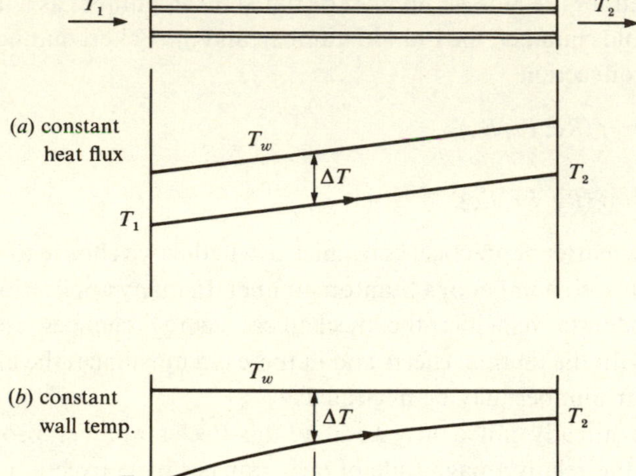

Fig. 80. Heat transfer in a tube, (*a*) with constant heat flux, and (*b*) with constant wall temperature.

will each be constant along the length of the tube. It follows that the Nusselt number, and hence also the factor h, will be constant.

Substituting from (9.7.2) in (9.7.1), the energy equation may be expressed in terms of temperatures in the form

$$\rho u_m c_p (T_2 - T_1) = \frac{4h}{D} \int_0^L (T_w - T)\,\mathrm{d}x \tag{9.7.3}$$

or for the temperature rise $\mathrm{d}T$ of the fluid in a length $\mathrm{d}x$,

$$\rho u_m c_p\,\mathrm{d}T = \frac{4h}{D}(T_w - T)\,\mathrm{d}x. \tag{9.7.4}$$

In figure 80 we indicate two special cases, (a) constant heat flux q from the tube wall to the fluid, and (b) constant wall temperature T_w.

A *constant heat flux* may be achieved if an independently controlled energy input is maintained at the tube wall, for example by the use of electric resistance heating. It follows from (9.7.2) that the temperature difference $\Delta T = T_w - T$ will also be constant in this case. From (9.7.3) we therefore have the result that

$$\rho u_m c_p(T_2-T_1)=4\frac{L}{D}h\,\Delta T. \tag{9.7.5}$$

This may be re-arranged as a dimensionless equation for the temperature change expressed in terms of the Stanton number, i.e.

$$\frac{T_2-T_1}{\Delta T}=4\frac{L}{D}\frac{h}{\rho u_m c_p}=4\frac{L}{D}St. \tag{9.7.6}$$

The special case of a *constant wall temperature* may be achieved if the tube wall is in thermal contact with a large heat reservoir which is maintained at a uniform temperature. In this case it follows from (9.7.4) that

$$\frac{\mathrm{d}T}{T_w-T}=\frac{4}{D}St\,\mathrm{d}x. \tag{9.7.7}$$

Hence, integrating from 0 to x as indicated in figure 80(b),

$$\log_e\left(\frac{T_w-T_1}{T_w-T}\right)=4\frac{x}{D}St \tag{9.7.8}$$

or

$$\frac{T_w-T}{T_w-T_1}=e^{-4\,St(x/D)}, \tag{9.7.9}$$

so that the temperature of the fluid changes in an exponential manner along the length of the tube as shown in the diagram. If we integrate equation (9.7.7) over the complete length from 0 to L we get the result

$$\log_e\left(\frac{T_w-T_1}{T_w-T_2}\right)=4\frac{L}{D}St. \tag{9.7.10}$$

If we now define a *logarithmic mean temperature difference* by

$$\Delta T_{lm}=\frac{T_2-T_1}{\log_e\left[(T_w-T_1)/(T_w-T_2)\right]}, \tag{9.7.11}$$

we can re-write (9.7.10) as

$$\frac{T_2-T_1}{\Delta T_{lm}}=4\frac{L}{D}St. \tag{9.7.12}$$

It will be seen that equation (9.7.12) has exactly the same form as (9.7.6) but with ΔT_{lm} in place of ΔT. The logarithmic mean temperature difference, as defined by (9.7.11), is a useful concept when dealing with the analysis of heat exchangers and will be further developed in chapter 18.

It is of interest to compare equation (9.7.6) or (9.7.12) with the equation for the pressure drop for fully developed flow in a long pipe which was derived in § 8.2. From (8.2.3) we have

$$\frac{\Delta p}{\rho u_m^2} = 2\frac{L}{D}c_f. \tag{9.7.13}$$

If we now eliminate the ratio L/D between equations (9.7.12) and (9.7.13) we find that

$$\frac{T_2 - T_1}{\Delta T_{lm}} = 2\frac{St}{c_f}\frac{\Delta p}{\rho u_m^2}. \tag{9.7.14}$$

There is thus a direct relationship between the temperature change $(T_2 - T_1)$ expressed as a dimensionless ratio, as in (9.7.12), and the pressure-drop coefficient $\Delta p/\rho u_m^2$.

We have seen in § 6.7 that on the basis of the Reynolds analogy for turbulent flow the Stanton number is related to the skin-friction coefficient c_f by $St = \frac{1}{2}c_f$, as in (6.7.21). Under these conditions the relationship (9.7.14) simplifies further to

$$\frac{T_2 - T_1}{\Delta T_{lm}} = \frac{\Delta p}{\rho u_m^2}. \tag{9.7.15}$$

In fact the Reynolds analogy represents an over-simplification of the transfer mechanism and other more accurate models will be reviewed in chapter 17. In each instance, however, the analogy results in an equation relating St to c_f as, for example, in (17.5.6) or (17.6.31).

10

Flow in turbo-machinery

10.1 Classification of turbo-machines

We can define a turbo-machine as a device by means of which mechanical energy in the form of shaft work is transferred to or from a continuously moving fluid stream. We describe the machine as a pump or compressor if there is an energy input *to* the fluid, and as a turbine if energy is extracted *from* the fluid stream. Most turbo-machines may be classified under the two main headings of *radial-flow* machines and *axial-flow* machines depending on the internal geometry and nature of the flow path.

The detailed design of turbo-machinery is a highly specialised branch of mechanical engineering and is outside the scope of an introductory text such as the present book. However, the basic principles of fluid mechanics which we have outlined in chapters 1–9 are directly applicable to the flow of fluids through turbo-machines, and it will therefore be appropriate in this chapter to apply these principles to an elementary analysis of the flow through radial- and axial-flow pumps and compressors. A general understanding of the flow characteristics of turbo-machines is important not only in connection with the design of such machines but also in relation to their application and operation as components within a complete process system. We will place the main emphasis in this chapter on flow through *pumps and compressors* in view of the widespread use of these units in chemical and process engineering applications. The use of turbines as the most important prime movers for electric power generation, including water turbines, steam turbines, and gas turbines, constitutes another major field of application involving the fluid mechanics of turbo-machinery. Limitations of space, however, must preclude any attempt to bring steam and gas turbines within the scope of the present book.

10.2 Basic fluid dynamics of a centrifugal pump

The centrifugal pump is the most widely used type of turbo-machine for handling liquids. The essential feature of a centrifugal pump is the *impeller*, shown diagrammatically in figure 81, which is mounted on a shaft and is surrounded by the pump casing. Fluid enters the pump in an axial direction but is turned into a radial-flow direction at the eye of the impeller. The impeller of a centrifugal pump normally has a number of equally spaced backward-swept blades, as indicated in the diagram, and fluid leaves the impeller at its outer circumference with a relatively large tangential velocity component. The absolute velocity of the fluid leaving the impeller is subsequently reduced in passage through the surrounding *diffuser section* which is sometimes, but not always, equipped with a number of fixed guide vanes. Finally the fluid flows in a spiral manner through the volute passage and leaves the machine at the delivery flange. The direction of flow at exit from the pump will be at right angles to a plane containing the shaft axis.

When analysing the fluid dynamics we have to deal both with absolute and relative fluid velocities. We will follow the normal convention in this chapter of representing the absolute fluid velocity by v and the impeller or blade velocity by u. The velocity of the fluid relative to the impeller will be represented by v_r. Obviously v is the vector sum of u and v_r. The tangential component of the fluid velocity v, usually known as the velocity of whirl, is represented by w. The radial component, which may be described as the velocity of flow, is represented by f. Subscripts 1 and 2 refer to location, e.g. inlet to and outlet from the impeller.

The motion of a liquid flowing through a pump impeller is indicated in figure 81 with the help of velocity diagrams. A slight problem may arise in defining the exact location of the inlet section since the flow is essentially three-dimensional in character in the region of the eye of the impeller. For the purposes of a simple analysis, however, we can make a rather arbitrary choice of the inlet radius r_1 to correspond roughly with the commencement of radial flow in the impeller as indicated in the diagram. This lack of precision in regard to the inlet section is not in fact very serious since we will find that the performance of the pump is determined mainly by the velocities at the outlet section which is clearly defined by the outer radius r_2 of the impeller.

The blade angles at the inner and outer radii r_1 and r_2 are specified by β_1 and β_2 as shown in the diagram. We cannot say, however, that the relative velocity of the fluid v_r will be exactly parallel to the blades. It is found in practice that the angles β'_1 and β'_2 which measure the directions of the relative velocities v_{r1} and v_{r2} will differ from the blade angles β_1 and β_2. This difference is particularly significant in the case of the outlet velocity triangle and will be discussed further in connection with the concept of *slip factor*.

Feed pump work = $\int v\,dp = v\Delta P$ $v \simeq 0.001\ m^3/kg$

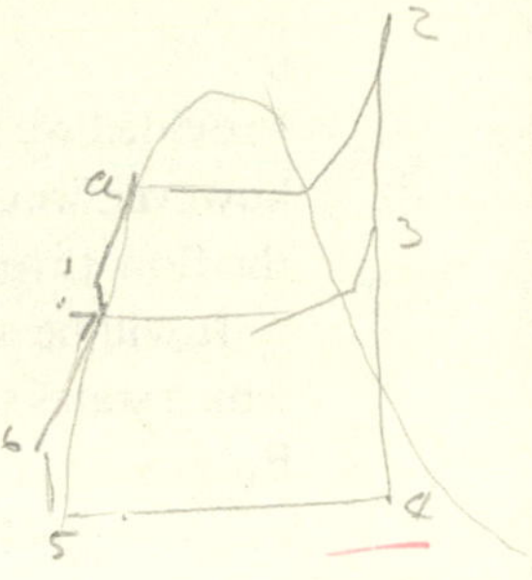

Feed heating (open) $T_B \uparrow$, loss of work $y(h_3 - h_4)$
save external heat input
even after $\frac{W}{\eta}$ consideration.

closed

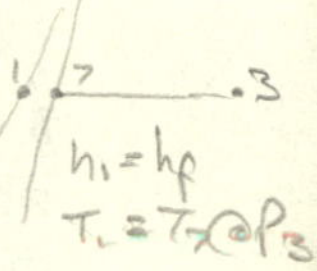

$y_1 h_3 + (1 - y_1) h_{11} = 1 \times h_1$

$\dot{m}\Delta h = m c_p \Delta T$

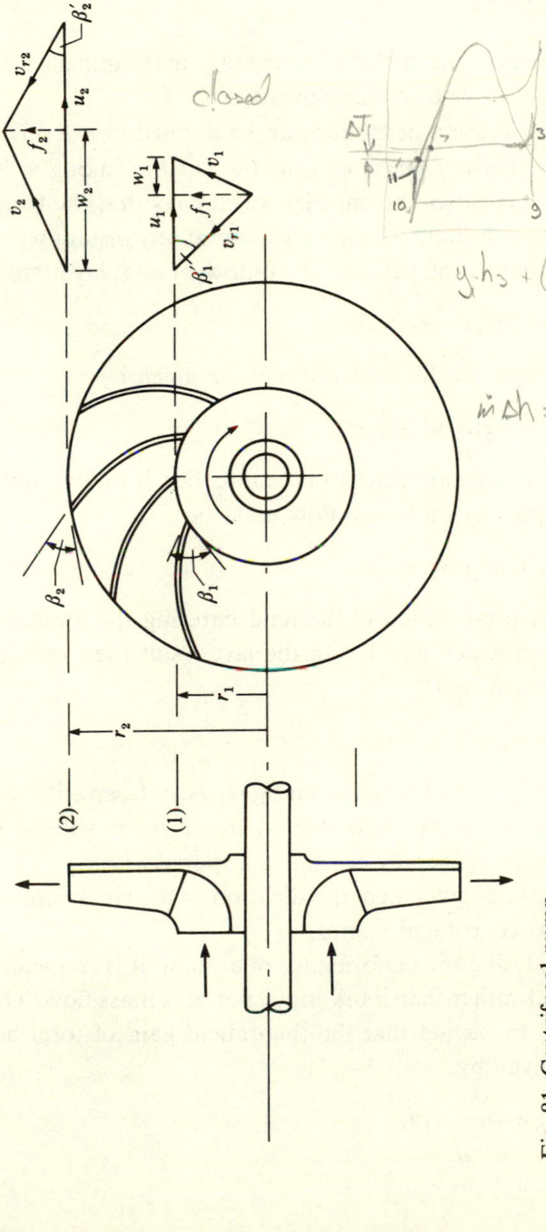

Fig. 81. Centrifugal pump.

Provided we have the actual measured velocities recorded on the diagram, however, we can proceed to apply the basic angular momentum equation to the flow through the impeller between sections (1) and (2).

It will be appropriate to start by taking note of the *continuity equation* which states simply that the volume flow rate Q measured in $m^3\ s^{-1}$ is given by

$$Q = 2\pi r b f, \tag{10.2.1}$$

where b is the axial width of the flow passage in the impeller at radius r and f is the radial-flow velocity component.

The basic equation of motion can be applied most conveniently in the form of the *angular momentum equation* (3.6.6). Taking a control surface which exactly encloses the impeller we can say that the torque exerted by the impeller on the fluid which is enclosed at any instant within the control surface is equal to the net rate of outflow of angular momentum, i.e.

$$T = \rho Q(r_2 w_2 - r_1 w_1). \tag{10.2.2}$$

The power input to the fluid will then be given by

$$T\Omega = \rho Q(w_2 u_2 - w_1 u_1), \tag{10.2.3}$$

where Ω is the angular velocity of the impeller. It follows from (10.2.3) that the work input per unit mass flow must be

$$W_i = w_2 u_2 - w_1 u_1. \tag{10.2.4}$$

If there is no pre-rotation of the fluid entering the impeller we have the special case in which $w_1 = 0$ and the last result then reduces to the very simple statement that

$$W_i = w_2 u_2. \tag{10.2.5}$$

Even if the velocity of whirl at entry w_1 is not zero, its value is usually relatively small. Noting also that $u_1/u_2 = r_1/r_2$ it will be seen that the product $w_1 u_1$ will generally be small compared with $w_2 u_2$. Thus the simple expression (10.2.5) gives a good indication of the work input per unit mass flow for most centrifugal pumps.

When analysing the performance of a pump it is convenient to work in terms of head rather than work input per unit mass flow. Thus we can re-state (10.2.4) by saying that the theoretical gain of total head H_i in the impeller is given by

$$H_i = \frac{w_2 u_2 - w_1 u_1}{g}. \tag{10.2.6}$$

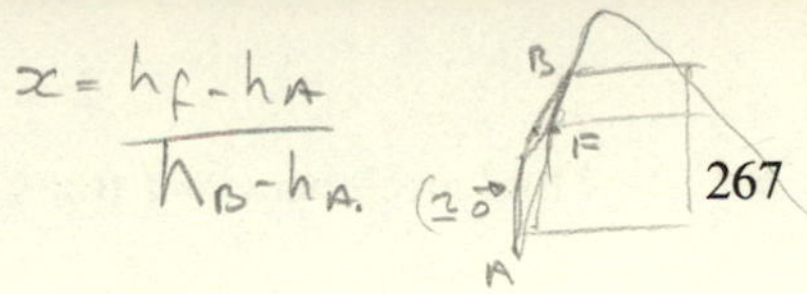

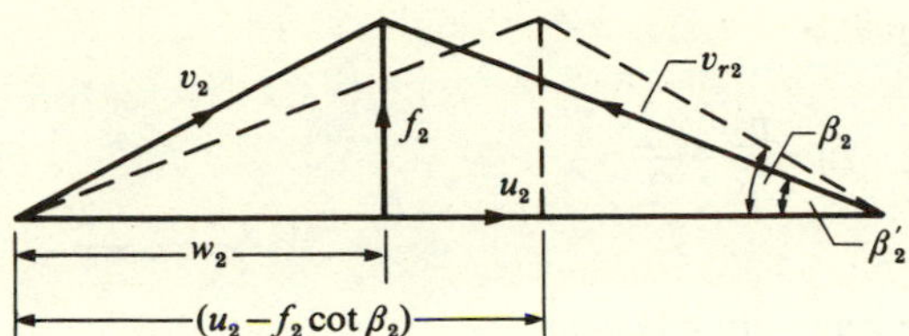

Fig. 82. Velocity triangle at the impeller exit.

For the special case where w_1 is negligible, as in (10.2.5), we have

$$H_i=\frac{w_2 u_2}{g}. \tag{10.2.7}$$

Referring to figure 82, the last result may be written

$$H_i=\frac{u_2(u_2-f_2 \cot \beta'_2)}{g}. \tag{10.2.8}$$

In practice we do not normally know the numerical value of the angle β'_2, although we do know the blade angle β_2. The actual velocity of whirl w_2 will always be less than the ideal value $(u_2-f_2 \cot \beta_2)$ owing to secondary flow effects in the impeller. We can define a slip factor σ such that

$$w_2=\sigma(u_2-f_2 \cot \beta_2). \tag{10.2.9}$$

We can also define a quantity H_v known as the virtual head or *Euler head* by the statement

$$H_v=\frac{u_2(u_2-f_2 \cot \beta_2)}{g}. \tag{10.2.10}$$

It will be evident from the diagram that this is the head that would be gained by the fluid if the relative velocity v_{r2} at exit from the impeller was exactly parallel to the blades of the impeller, i.e. if we had $\beta'_2=\beta_2$. It will also be noted from (10.2.7) and (10.2.9) that $H_i=\sigma H_v$ where σ is the slip factor.

The *mechanical energy equation* can be applied to the overall flow through the pump, from inlet flange to outlet flange, in three steps as follows:

inlet flange to impeller entry

$$\frac{p_0}{\rho}+\frac{v_0^2}{2}-L_e=\frac{p_1}{\rho}+\frac{v_1^2}{2}, \tag{10.2.11}$$

flow through the impeller

$$\frac{p_1}{\rho}+\frac{v_1^2}{2}+W_i-L_i=\frac{p_2}{\rho}+\frac{v_2^2}{2}, \tag{10.2.12}$$

flow through diffuser and volute casing

$$\frac{p_2}{\rho}+\frac{v_2^2}{2}-L_d=\frac{p_3}{\rho}+\frac{v_3^2}{2}, \tag{10.2.13}$$

where L_e, L_i, and L_d are the mechanical energy losses per unit mass flow in the entry section, impeller, and diffuser respectively. Combining the last three equations, we have for the overall flow:

$$\frac{p_3}{\rho}+\frac{v_3^2}{2}=\frac{p_0}{\rho}+\frac{v_0^2}{2}+W_i-(L_e+L_i+L_d) \tag{10.2.14}$$

or, expressed in terms of head, the actual gain of total head H is

$$H=\left(\frac{p_3}{\rho g}+\frac{v_3^2}{2g}\right)-\left(\frac{p_0}{\rho g}+\frac{v_0^2}{2g}\right)=H_i-H_f, \tag{10.2.15}$$

where H_f is the total internal loss of head due to friction and turbulence. Note that $H_f=(L_e+L_i+L_d)/g$.

The energy equation for flow through the impeller (10.2.12) may be investigated further if we substitute for the work input term W_i from (10.2.4). Thus we can say that

$$\frac{p_2}{\rho}+\frac{v_2^2}{2}-w_2u_2=\frac{p_1}{\rho}+\frac{v_1^2}{2}-w_1u_1-L_i. \tag{10.2.16}$$

From the geometry of figure 82, however, we have

$$v_{r2}^2=f_2^2+(u_2-w_2)^2=f_2^2+w_2^2+u_2^2-2w_2u_2$$
$$=v_2^2+u_2^2-2w_2u_2.$$

Hence

$$\frac{v_2^2}{2}-w_2u_2=\frac{v_{r2}^2}{2}-\frac{u_2^2}{2} \tag{10.2.17}$$

and similarly

$$\frac{v_1^2}{2}-w_1u_1=\frac{v_{r1}^2}{2}-\frac{u_1^2}{2}.$$

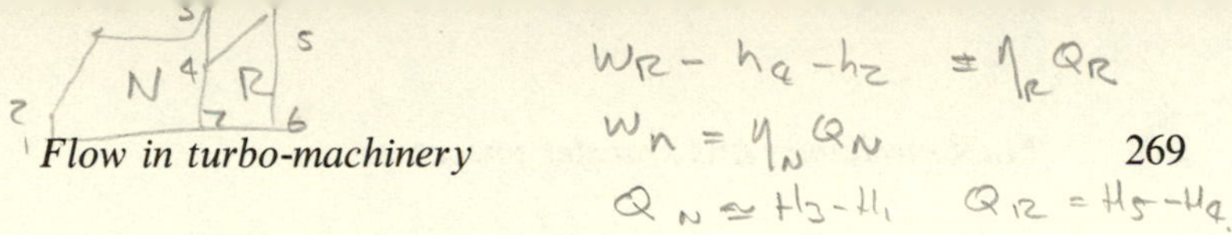

Thus equation (10.2.16) may be re-stated as

$$\frac{p_2}{\rho}+\frac{v_{r2}^2}{2}-\frac{u_2^2}{2}=\frac{p_1}{\rho}+\frac{v_{r1}^2}{2}-\frac{u_1^2}{2}-L_i$$

or

$$\frac{p_2}{\rho}+\frac{v_{r2}^2}{2}-\tfrac{1}{2}r_2^2\Omega^2=\frac{p_1}{\rho}+\frac{v_{r1}^2}{2}-\tfrac{1}{2}r_1^2\Omega^2-L_i. \tag{10.2.18}$$

This may be regarded as the energy equation for a rotating system expressed in terms of fluid velocities measured relative to the impeller. The additional term $\frac{1}{2}r^2\Omega^2$ represents the effect of the radial acceleration or 'centrifugal force' acting on the fluid.

If the pump is running with the delivery valve closed, there will be no flow relative to the impeller and it follows from equation (10.2.18) under these conditions that

$$\frac{p_2-p_1}{\rho}=\tfrac{1}{2}\Omega^2(r_2^2-r_1^2), \tag{10.2.19}$$

which is the expression for the pressure distribution in a *forced vortex* as derived in §3.8.

It is reasonable to assume that, when the pump is running at zero delivery in this manner, the whole of the fluid enclosed between the inlet flange and the outer circumference of the impeller will be rotating with angular velocity Ω about the axis. It follows from (10.2.19) that the head developed by the pump at zero flow, under shut-off conditions, will be given approximately by

$$\frac{p_2-p_0}{\rho g}=\frac{r_2^2\Omega^2}{2g}=\frac{u_2^2}{2g}. \tag{10.2.20}$$

Under normal operating conditions the rate at which energy is transferred from the impeller to the fluid is given by (10.2.3), i.e. by the product of the mass flow rate ρQ and the work input gH_i per unit mass flow. The shaft power required to drive the pump, however, must include the bearing losses and the effects of disc friction between the impeller and the pump casing. We therefore define the *mechanical efficiency* by $\eta_m = \rho Q g H_i/P$. The useful power input into the fluid, however, is given by the product of the mass-flow rate ρQ and the quantity gH where H is the net gain of total head as defined by (10.2.15). We can define the *hydraulic efficiency* as $\eta_h = H/H_i$. The *overall efficiency* η is the ratio of the useful power input to the fluid $\rho Q g H$ divided by the power P required to drive the pump. Thus

$$\eta = \frac{\rho Q g H}{P} = \eta_h \eta_m. \tag{10.2.21}$$

10.3 Operating characteristics of a centrifugal pump

A pump must normally be capable of operating over a range of volume flow rate and it is convenient to plot the performance characteristic in the form of a curve of the actual head H developed by the pump as a function of the volume flow rate Q for a given rotational speed N. A typical head-capacity curve is shown in figure 83. A centrifugal pump is usually started up with the delivery valve closed. When the full rotational speed N has been attained, the delivery valve is opened gradually so that flow can commence. The head developed by the pump at zero flow is known as the *shut-off head* as indicated on figure 83. The value of the shut-off head is given approximately by (10.2.19).

In order to ensure stability of operation at low volume flow rates a falling head-capacity curve, such as that indicated in figure 83, is desirable. This can generally be achieved by designing the impeller with backward-swept blades. It will be seen from equation (10.2.10) that the Euler head H_v will fall with increasing flow rate if the angle β_2 is less than 90°. Conversely, if the blade angle β_2 was greater than 90°, the Euler head H_v would increase with increasing flow rate. Although the relationship between the actual head

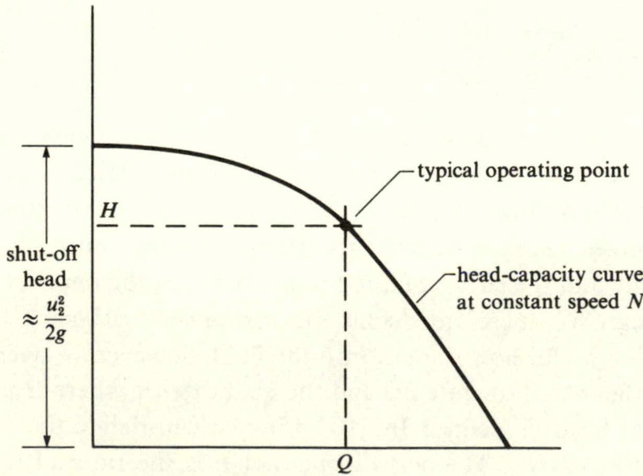

Fig. 83. Typical head-capacity curve.

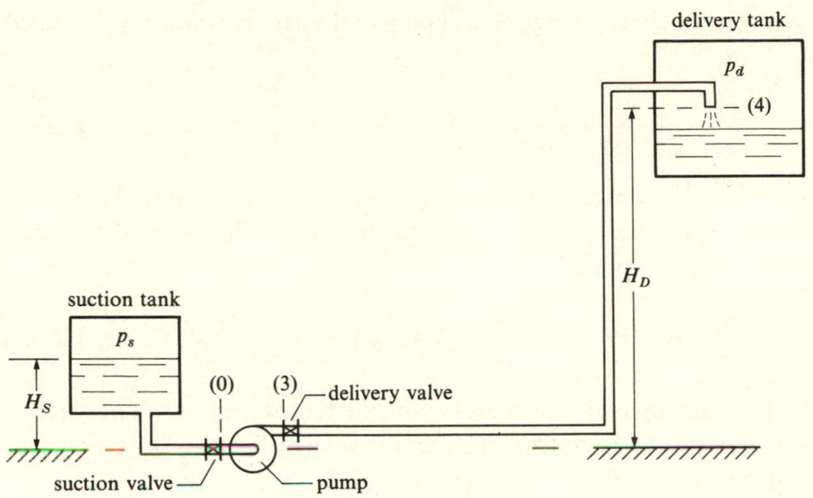

Fig. 84. Simple pumped flow system.

developed H and the hypothetical quantity H_v is not readily calculable, especially at low volume flow rates, a falling head-capacity curve may be expected if backward-swept blades are employed with a sufficiently small value for the outlet angle β_2.

The operation of a pump must be considered in relation to the complete flow system of which it forms a part. A very simple example of such a system is shown diagrammatically in figure 84. The energy equation for flow from the suction tank to the suction flange, i.e. up to section (0), may be stated as

$$\frac{p_s}{\rho g}+H_S-H_{Sf}=\frac{p_0}{\rho g}+\frac{v_0^2}{2g}, \tag{10.3.1}$$

where H_{Sf} is the loss of head due to friction in the suction line. Note that the difference of level between the surface of the liquid in the suction tank and the centre-line of the pump H_S is treated as being positive if the pump is located below the surface level as indicated in the diagram. If, however, the pump is located above the liquid level in the suction tank, i.e. if we have the case of *suction lift*, then H_S will be numerically negative. The energy equation for flow through the pump from section (0) to section (3) is given by equation (10.2.15), i.e.

$$\frac{p_0}{\rho g}+\frac{v_0^2}{2g}+H=\frac{p_3}{\rho g}+\frac{v_3^2}{2g}, \tag{10.3.2}$$

where H is the actual head developed by the pump. The energy equation for flow from the delivery flange of the pump to the delivery tank may be stated as

$$\frac{p_3}{\rho g}+\frac{v_3^2}{2g}=\frac{p_d}{\rho g}+\frac{v_4^2}{2g}+H_3+H_{Df}, \tag{10.3.3}$$

where H_{Df} is the loss of head due to friction in the delivery line. Adding the last three equations together, we get for the overall flow from the suction tank to the delivery tank

$$\frac{p_s}{\rho g}+H_S-H_{Sf}+H=\frac{p_d}{\rho g}+\frac{v_4^2}{2g}+H_D+H_{Df}. \tag{10.3.4}$$

If both the suction tank and delivery tank are vented to atmosphere we have $p_s=p_d=$ atmospheric pressure. If, also, the velocity head $v_4^2/2g$ at discharge into the delivery tank is relatively small, which is usually the case, equation (10.3.4) reduces to

$$H=(H_D-H_S)+(H_{Df}+H_{Sf}), \tag{10.3.5}$$

i.e. the actual head developed by the pump must balance the difference of level between suction and delivery tanks plus the total loss of head due to friction and turbulence in the suction and delivery lines.

The flow conditions prevailing on the suction side of a pump are of special importance and precautions must be taken if the phenomenon known as *cavitation* is to be avoided. If the pressure in the fluid falls locally to a value below the vapour pressure at the prevailing temperature, cavities or bubbles are liable to be formed which will be filled with vapour. These bubbles will generally move with the fluid stream and will subsequently collapse when they reach a region of higher pressure. The collapse of the cavities leads to the formation of pressure waves in the fluid and these are liable to cause mechanical damage in the form of the pitting or corrosion of the impeller. The onset of cavitation in a centrifugal pump can be detected by the sudden increase in noise and vibration and is accompanied by a sudden drop in the head and efficiency curves when the flow rate is increased beyond the point at which cavitation starts.

The *net positive suction head* or *NPSH* is defined as the difference between the total head at the suction flange and the pressure head $p_v/\rho g$ corresponding to the absolute vapour pressure p_v of the liquid at the prevailing temperature in the pump inlet. Referring to figure 84, the total head available at the suction flange is given by (10.3.1) and we can therefore say that:

$$NPSH = \frac{(p_s - p_v)}{\rho g} + H_S - H_{Sf}. \tag{10.3.6}$$

It will also be seen from (10.3.1) that

$$\frac{p_0}{\rho g} + \frac{v_0^2}{2g} = NPSH + \frac{p_v}{\rho g}. \tag{10.3.7}$$

If we now follow the motion of the fluid from the suction flange, where the pressure is p_0 and the velocity v_0, into the eye of the impeller there will be a drop of pressure as the fluid accelerates to locally higher velocities. The flow pattern in this region will inevitably be complex but, if we apply Bernoulli's equation on the assumption that the fluid has not yet taken up any energy from the impeller, we could argue that for an increase of velocity from v_0 to a local critical value v_c the corresponding pressure drop would be given by

$$\frac{p_0 - p_c}{\rho g} = \frac{v_c^2 - v_0^2}{2g}. \tag{10.3.8}$$

It follows from (10.3.7) that in the limiting case, when the local pressure p_c falls to the value of the vapour pressure p_v, we have $NPSH = v_c^2/2g$. If cavitation is to be avoided we want to ensure that $NPSH > v_c^2/2g$. It is impossible to calculate the local value of the velocity v_c with any accuracy but we can argue on dimensional grounds that, for a given geometric design of pump $v_c^2/2g = c_c H$ where c_c is a cavitation coefficient and H is the head developed by the pump. The coefficient c_c can be determined experimentally as a function of the specific speed N_s which is defined in § 10.4. The basic requirement for the avoidance of cavitation, therefore, is that the *NPSH* should be greater than $c_c H$.

10.4 Dimensional analysis applied to pumps and hydraulic turbines

We can use the method of dimensional analysis outlined in § 1.8, and developed more fully in chapter 15, to help in correlating the performance of pumps or hydraulic turbines. We will consider first of all the case of a range of pumps of exactly similar geometric design but of different sizes, characterised by the diameter D of the pump impeller, and driven at different values of the rotational speed N. As we have seen in § 10.3, the performance of a pump can be expressed conveniently in the form of the head-capacity curve measured at constant speed N. Reference to equation (10.2.6) or (10.2.15) will show that the head developed by a pump is inversely proportional to the gravitational constant g although the work input per unit mass flow, as given by (10.2.4), is independent of g. We can avoid introducing the gravitational acceleration as a separate independent variable by working in terms of the product gH rather than the head H. We can therefore start with the basic statement that, for a range of pumps of similar geometric design, the measurable quantity gH will be a function of the fluid properties ρ and μ, the volumetric flow rate Q, the impeller diameter D, and the rotational speed N, i.e.

$$gH = f(\rho, \mu, Q, D, N).$$

This expression involving six physical quantities may be reduced to the following relationship between three dimensionless groups:

$$\frac{gH}{N^2 D^2} = f\left(\frac{Q}{ND^3}, \frac{\rho N D^2}{\mu}\right). \tag{10.4.1}$$

We could equally well start with the basic assumption that the power P required to drive the pump must be a function of ρ, μ, Q, D, and N. This statement can then be reduced by the same procedure to the following relationship between dimensionless groups:

$$\frac{P}{\rho N^3 D^5} = f\left(\frac{Q}{ND^3}, \frac{\rho N D^2}{\mu}\right). \tag{10.4.2}$$

The first group on the right-hand side of (10.4.1) and (10.4.2) may be regarded as a dimensionless capacity or flow coefficient. The second group is in the form of a Reynolds number and provides a measure of the relative magnitude of inertial forces to viscous forces. In the great majority of applications the inertial effects predominate owing to the high degree of turbulence and eddying inside a pump. Except for certain special cases involving the pumping of very viscous fluids, therefore, we can usually ignore the influence of the Reynolds number group and the last two equations can then be re-written in the simpler form:

$$\frac{gH}{N^2 D^2} = f\left(\frac{Q}{ND^3}\right) \tag{10.4.3}$$

and

$$\frac{P}{\rho N^3 D^5} = f\left(\frac{Q}{ND^3}\right). \tag{10.4.4}$$

The overall efficiency of the pump is given by $\eta = \rho QgH/P$ and this is seen to be equal to

$$\left(\frac{Q}{ND^3}\right)\left(\frac{gH}{N^2D^2}\right)\Big/\left(\frac{P}{\rho N^3D^5}\right).$$

It follows that the efficiency η may also be expressed as a function of the dimensionless flow coefficient Q/ND^3. We thus see that the performance of geometrically similar pumps may conveniently be plotted in the form of curves of gH/N^2D^2, $P/\rho N^3D^5$, and η against the flow coefficient Q/ND^3 as shown in figure 85.

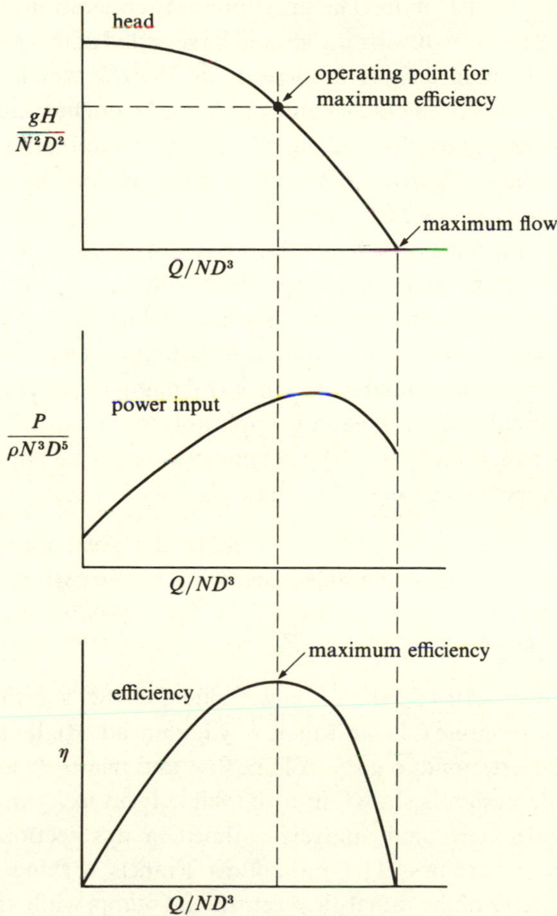

Fig. 85. Dimensionless pump characteristics.

It will be seen from figure 85 that, for any given geometric design, there will be a particular numerical value of the dimensionless ratio gH/N^2D^2, and also of the ratio $P/\rho N^3 D^5$, corresponding to the operation of the pump at the point of maximum efficiency. We can form a new dimensionless ratio from the group gH/N^2D^2 and the flow coefficient Q/ND^3 by eliminating the impeller diameter D and we thus obtain the numerical ratio known as the *specific speed* of the pump N_s measured at the point of maximum efficiency, i.e.

$$N_s = \frac{NQ^{\frac{1}{2}}}{(gH)^{\frac{3}{4}}}. \tag{10.4.5}$$

In this definition of the specific speed N is measured in revolutions per second, Q in $m^3\ s^{-1}$ and H in m. The gravitational acceleration g has the normal value of 9.81 m s^{-2}. A warning should be sounded at this point that the specific speed is often defined incorrectly as $NQ^{\frac{1}{2}}/H^{\frac{3}{4}}$ which is not a dimensionless quantity. Numerical values are also frequently quoted in terms of inconsistent units, for example with the rotational speed in revolutions per minute. Great care must therefore be taken when comparing values for the specific speed.

The specific speed, when correctly defined as in (10.4.5), provides a unique indication of the type of pump which will be appropriate for operation at maximum efficiency with the specified values of H, Q, and N in the particular application to be considered. Although we have developed the treatment in terms of the radial flow centrifugal pump in this chapter the method of dimensional analysis is equally applicable to the case of a mixed-flow or axial-flow pump. Each basic type of pump will have an appropriate range of specific speed which may be summarised as follows:

	N_s
Single-stage radial-flow centrifugal pumps	0.06–0.20
Mixed-flow pumps	0.20–0.60
Axial-flow pumps	0.60–1.00

It will be seen from (10.4.5) that a low value for the specific speed corresponds to low volume flow and relatively high head. High values for the specific speed correspond to large volume flow and relatively low head. The effect on basic design is shown in a simplified form in figure 86.

The method of dimensional analysis outlined in this section is also applicable to water turbines. The radial-flow Francis turbine may be regarded as the inverse of the radial-flow centrifugal pump, while the axial-flow Kaplan turbine is the equivalent of the axial-flow pump. It is in fact

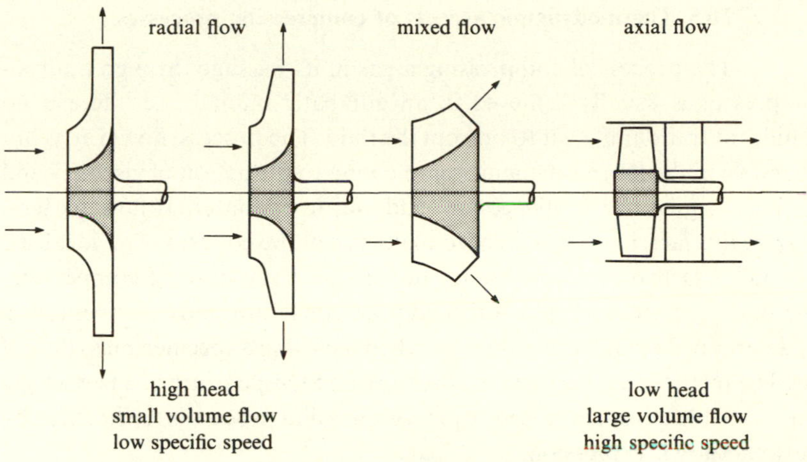

Fig. 86. Variation of pump type with specific speed.

possible to design a turbo-machine which will operate either as a pump or, with reversal of flow direction, as a turbine. Such reversible pump-turbine machines are sometimes employed in pumped-storage hydro-electric installations. Another type of water turbine, for which there is no counterpart in the form of a pump, is the Pelton wheel which is used in very high-head hydro-electric schemes. The Pelton wheel is an impulse turbine where the entire available head is converted into the kinetic energy of the jet or jets which impinge at atmospheric pressure on the buckets of the turbine runner. When analysing the operation of a water turbine we are more interested in the power P developed by the turbine than in the volume flow of water. It is therefore more appropriate to define a specific speed in terms of P by forming a new dimensionless ratio from the groups gH/N^2D^2 and $P/\rho N^3D^5$ with the elimination of the dimension D. In place of (10.4.5) we thus obtain

$$N_s = \frac{NP^{\frac{1}{2}}}{\rho^{\frac{1}{2}}(gH)^{\frac{5}{4}}}. \tag{10.4.6}$$

Typical values for the range of specific speed (as defined by (10.4.6)) for the three main types of water turbine are as follows:

Pelton wheel	up to 0.04,
Francis turbine	0.04–0.40,
Kaplan turbine	0.40–0.80.

10.5 Thermodynamic aspects of compression processes

The process of compressing a gas in its passage through a turbo-compressor is usually achieved in an adiabatic manner, i.e. there is no significant transfer of heat to or from the fluid. The process, however, is not reversible in the thermodynamic sense owing to the action of viscosity and the high degree of turbulence and eddying in the internal flow pattern. Despite this fact, it is convenient to make use of the concept of an idealised frictionless isentropic compression process as a standard of comparison. The basic requirement of a turbo-compressor is normally to maintain a predetermined overall pressure ratio when handling a specified mass flow of gas. The relationship between the pressure and temperature of a perfect gas in a reversible quasi-static isentropic compression process is represented by the well-known expression

$$\frac{p}{p_0}=\left(\frac{\rho}{\rho_0}\right)^{\gamma}=\left(\frac{T}{T_0}\right)^{\gamma/(\gamma-1)}, \tag{10.5.1}$$

where γ is the ratio of the specific heats c_p/c_v. It will be shown in §11.3 that this result is equally applicable to a flow process provided the effects of viscosity and turbulence are negligible.

If we differentiate the isentropic expression (10.5.1) we obtain

$$\frac{\mathrm{d}p}{p}=\frac{\gamma}{\gamma-1}\frac{\mathrm{d}T}{T}. \tag{10.5.2}$$

A real compression process, where the effects of viscosity and turbulence are significant, may be described as a *polytropic* compression process and in this case we must replace (10.5.2) by

$$\frac{\mathrm{d}p}{p}=\eta_P\frac{\gamma}{\gamma-1}\frac{\mathrm{d}T}{T}, \tag{10.5.3}$$

where η_P is the *polytropic efficiency.* It will be seen from (10.5.3) that a polytropic compression process can be represented by the relationship

$$\frac{p}{p_0}=\left(\frac{\rho}{\rho_0}\right)^{n}=\left(\frac{T}{T_0}\right)^{n/(n-1)}, \tag{10.5.4}$$

where the index n is related to η_p by

$$\frac{n}{n-1}=\eta_p\frac{\gamma}{\gamma-1}. \tag{10.5.5}$$

It is convenient, when making thermodynamic calculations involving

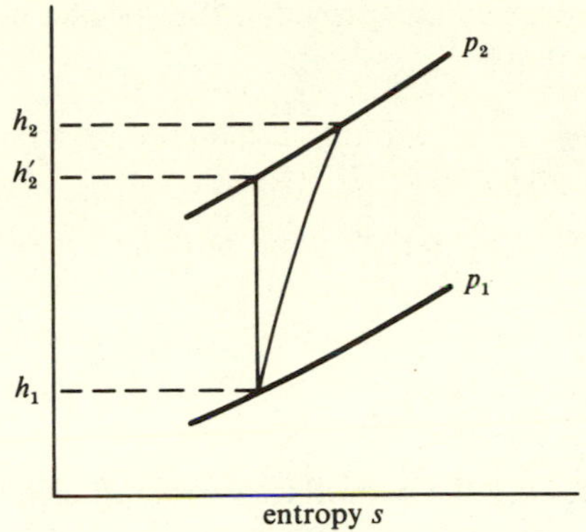

Fig. 87. Mollier diagram for the compression of a gas.

the compression or expansion of a gas, to plot the thermodynamic state of the substance in the form of a Mollier diagram of enthalpy h against entropy s. Lines of constant pressure appear as curves as indicated in figure 87. A reversible isentropic compression process between pressures p_1 and p_2 will be represented by a vertical straight line as indicated in the diagram. On the other hand, an irreversible polytropic compression process over the same pressure ratio involves an increase of entropy and is represented by the slanting line on the diagram.

We have seen in §9.2 that the steady-flow energy equation for 'one-dimensional' flow in a duct with shaft work, but zero heat transfer, is given by (9.2.7), i.e.

$$h_2+\frac{u_2^2}{2}=h_1+\frac{u_1^2}{2}+W_s, \tag{10.5.6}$$

where h is the enthalpy of the gas expressed per unit mass. Thus if the velocity of the gas remains constant, the shaft work per unit mass for compression of the gas from pressure p_1 to p_2 is represented by the difference in the value of the enthalpy h measured on the vertical scale in figure 87. The smallest possible value for the shaft work would be that corresponding to an isentropic frictionless compression process and is represented by the increment $h_2'-h_1$ on the diagram. The actual value for the shaft work will be that corresponding to polytropic compression represented by the increment h_2-h_1. We can therefore define the *isentropic efficiency* by

$$\eta_c=\frac{h_2'-h_1}{h_2-h_1}. \tag{10.5.7}$$

For a perfect gas we have $dh = c_p\,dT$, and provided the specific heat c_p is constant we can therefore say that

$$\eta_c = \frac{T_2' - T_1}{T_2 - T_1}. \tag{10.5.8}$$

Noting from (10.5.1) that $T_2'/T_1 = (p_2/p_1)^{(\gamma-1)/\gamma}$, and making use of (10.5.4) and (10.5.5) we have

$$\eta_c = \frac{(p_2/p_1)^{(\gamma-1)/\gamma} - 1}{(p_2/p_1)^{(\gamma-1)/\eta_p\gamma} - 1}, \tag{10.5.9}$$

which gives the relationship between the overall isentropic efficiency of the process η_c and the small-scale or polytropic efficiency η_p.

In the compression process represented in figure 87 we have assumed that the velocity of the gas remains constant, i.e. $u_1 = u_2$. In reality, however, the velocity of the gas changes very considerably during its passage through a turbo-compressor. For this reason it is convenient to introduce the concept of the *stagnation enthalpy*, h_s, which is defined as the value of the enthalpy that would be measured if the gas stream at any point in the flow was brought to rest in an isentropic compression process. Corresponding values would be measured for the stagnation temperature T_s and the stagnation pressure p_s. At section (1), for example, we would have the value of the stagnation enthalpy defined by

$$h_{s1} = h_1 + \frac{u_1^2}{2}, \tag{10.5.10}$$

while the stagnation temperature and pressure are given by:

$$T_{s1} = T_1 + \frac{u_1^2}{2c_p} \tag{10.5.11}$$

and

$$\frac{p_{s1}}{p_1} = \left(\frac{T_{s1}}{T_1}\right)^{\gamma/(\gamma-1)}. \tag{10.5.12}$$

The compression of a gas stream from pressure p_1 to p_2 and with different values for the gas velocities u_1 and u_2 is shown on the Mollier diagram in figure 88.

10.6 Basic fluid dynamics of a radial-flow compressor stage

The main features of a radial-flow compressor are essentially similar to those of a centrifugal pump. Since higher rotational speeds are

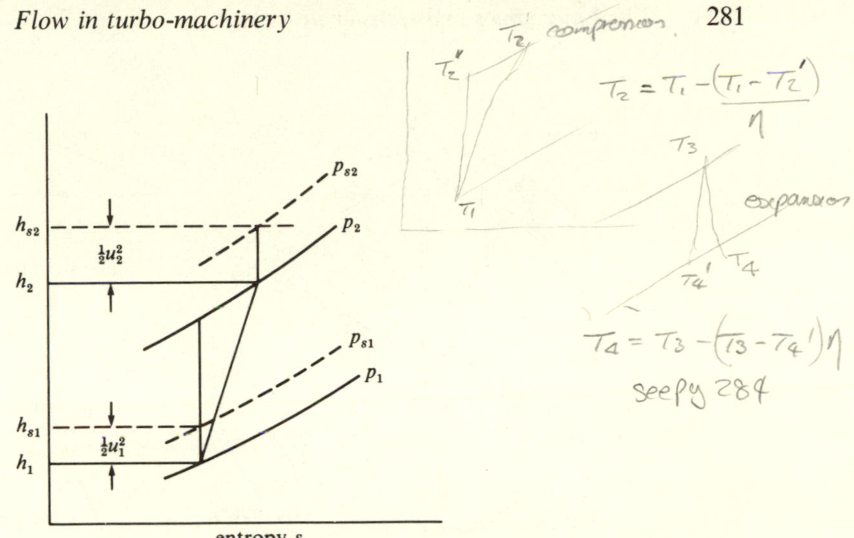

Fig. 88. Mollier diagram for the compression of a flowing gas stream.

normally required, however, the impeller of a compressor may be designed with straight radial blades rather than with curved backward-swept blades because of the high centrifugal stresses which are created. The gas stream enters the eye of the impeller in an axial direction, as with a centrifugal pump, and must be turned through 90° into a radial-flow direction. The flow pattern in this region is necessarily three-dimensional in character and is not amenable to any simple analysis. In high-performance machines the impeller is usually fitted with curved extensions to the blades which act as rotating guide vanes in the region of the eye. Figure 89 shows the general arrangement of a single-stage radial-flow compressor fitted with straight radial blades and velocity triangles are shown both for a nominal inlet section (1) and at outlet from the impeller (2).

If there is no pre-rotation of the gas stream as it enters the eye of the impeller the velocity of whirl w_1 at the inlet section will be zero. In this case, however, the relative velocity v_{r1} would have a tangential component equal to u_1 and the blades of the impeller should be curved at entry. Alternatively, the compressor may be designed with fixed guide vanes in the inlet duct to give pre-rotation so that the relative velocity at entry to the impeller v_{r1} may be nearly radial. With the latter arrangement w_1 will no longer be zero, but it will in any case be small compared with the velocity of whirl w_2 at exit from the impeller.

With straight radial blades, as shown in figure 89, the relative velocity v_{r2} at exit from the impeller should also be nearly radial. Owing to relative circulation and secondary flow within the impeller, as indicated in the diagram, however, the relative velocity v_{r2} will be inclined backwards at

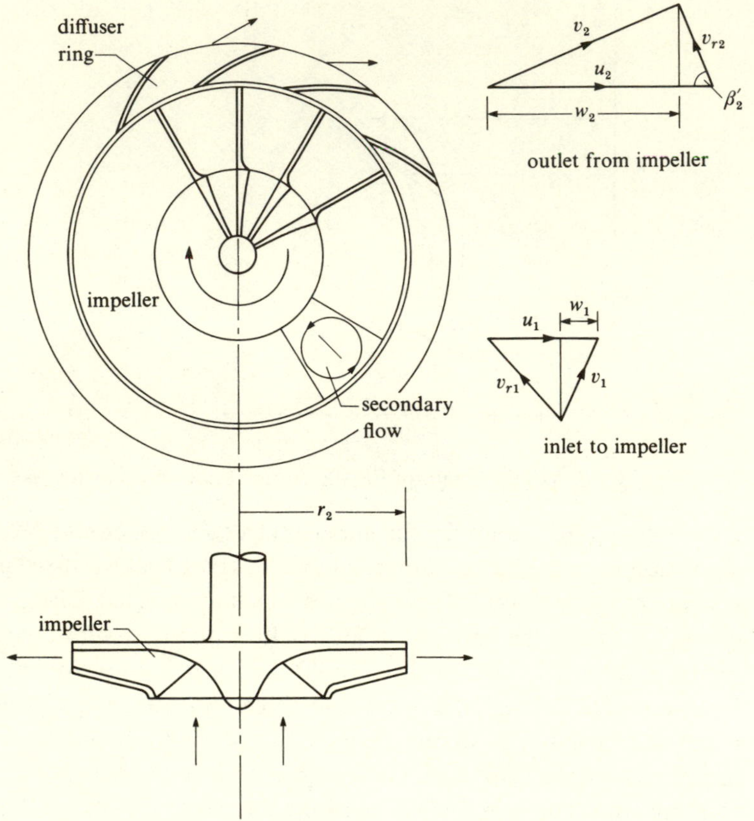

Fig. 89. Radial-flow compressor.

some angle β'_2 which will be less than the blade angle $\beta_2 = 90°$. In other words, the actual velocity of whirl at exit w_2 will be less than the ideal value u_2 and we can say that $w_2 = \sigma u_2$ where σ is a vane efficiency or *slip factor*. This is exactly similar to the slip factor defined in (10.2.9) for the centrifugal pump impeller. The slip factor may be associated with relative circulatory flow of the gas within the spaces between the blades of the impeller as indicated diagrammatically in figure 89. According to a very simple model proposed by Stodola this secondary flow takes the form of circulation with an angular velocity Ω relative to the impeller equal and opposite to the angular velocity of the impeller itself. In other words, the gas in the spaces between the blades is supposed to have zero absolute angular velocity. If the effective radius of this secondary circulatory motion is taken as approximately $\pi r_2/n$, where n is the number of blades, we can say that the secondary flow velocity *relative to the impeller* at the outer edge $= \pi r_2 \Omega / n$

which is equal to $\pi u_2/n$ since $u_2 = r_2\Omega$. Thus the velocity of whirl at exit from the impeller is given by

$$w_2 = \left(1 - \frac{\pi}{n}\right)u_2. \tag{10.6.1}$$

The basic dynamics of the flow through the impeller is clearly similar to that of the centrifugal pump discussed in § 10.2. We can use the angular momentum equation to derive the same expression for the shaft work done on the fluid by the impeller per unit mass, i.e.

$$W_i = w_2 u_2 - w_1 u_1. \tag{10.6.2}$$

In the special case where we have straight radial blades $w_2 = \sigma u_2$. If the term $w_1 u_1$ is sufficiently small so that it may be neglected in comparison with $w_2 u_2$, the last result then simplifies to

$$W_i = \sigma u_2^2. \tag{10.6.3}$$

We can now make use of the steady-flow energy equation (9.2.7) or (10.5.6) for flow through the compressor, and we can represent the process graphically on the Mollier diagram as shown in figure 90. For flow in the

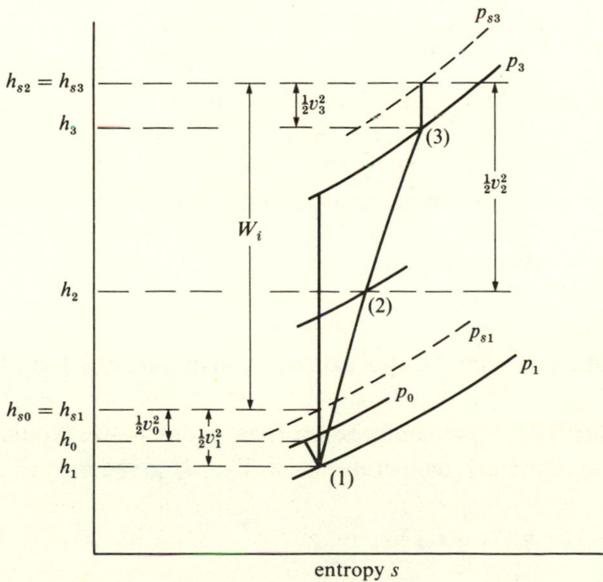

Fig. 90. Mollier diagram for a radial-flow compressor.

entry passage up to section (1) we have

$$h_{s0}=h_0+\frac{v_0^2}{2}=h_1+\frac{v_1^2}{2}=h_{s1}. \tag{10.6.4}$$

For flow through the impeller from (1) to (2) we have

$$h_1+\frac{v_1^2}{2}+W_i=h_2+\frac{v_2^2}{2}=h_{s2}. \tag{10.6.5}$$

For flow through the diffuser ring from (2) to (3) we have

$$h_2+\frac{v_2^2}{2}=h_3+\frac{v_3^2}{2}=h_{s3}. \tag{10.6.6}$$

It will be noted that, unlike the analysis of the centrifugal pump in § 10.2, we are now using the complete energy equation and there are accordingly no terms in (10.6.4), (10.6.5), and (10.6.6) representing energy losses. The effects of viscosity and turbulence are to cause an increase of entropy of the gas in its passage through the compressor as indicated in figure 90. Dissipation of kinetic energy, however, is exactly balanced by a corresponding gain of internal energy. Thus the overall loss is zero.

If we combine the last three equations we have, for the overall energy equation between sections (0) and (3),

$$h_0+\frac{v_0^2}{2}+W_i=h_3+\frac{v_3^2}{2}. \tag{10.6.7}$$

In the special case where $v_0=v_3$ this reduces to

$$h_3-h_0=W_i \tag{10.6.8}$$

or, in terms of the thermodynamic temperature of the gas,

$$T_3-T_0=\frac{W_i}{c_p}=\frac{\sigma u_2^2}{c_p} \tag{10.6.9}$$

assuming that we can use the approximate expression (10.6.3) for the shaft work W_i.

For an ideal isentropic compression from p_0 to p_3 we would have a corresponding isentropic temperature rise T'_3-T_0 given by

$$T'_3-T_0=\eta_c(T_3-T_0)=\eta_c\frac{\sigma u_2^2}{c_p}, \tag{10.6.10}$$

where η_c is the isentropic efficiency as defined in § 10.5. The pressure ratio is therefore given by

$$\frac{p_3}{p_0}=\left(\frac{T'_3}{T_0}\right)^{\gamma/(\gamma-1)}=\left(1+\frac{\eta_c\sigma u_2^2}{c_pT_0}\right)^{\gamma/(\gamma-1)}. \tag{10.6.11}$$

Alternatively, if we prefer to work in terms of the polytropic efficiency η_p, we can say from (10.5.4) that

$$\frac{p_3}{p_0}=\left(\frac{T_3}{T_0}\right)^{n/(n-1)}=\left(1+\frac{\sigma u_2^2}{c_pT_0}\right)^{n/(n-1)}, \tag{10.6.12}$$

where the index n is given by (10.5.5), i.e.

$$n=\frac{1}{1-(\gamma-1)/(\eta_p\gamma)}. \tag{10.6.13}$$

As a simple numerical example we will calculate the pressure ratio that can be developed by a single-stage radial-flow air compressor having an impeller diameter of 1 m driven at 6000 rpm or 100 revolutions per second. We will assume that the impeller has 16 blades giving a slip factor, calculated from (10.6.1), of 0.804. The peripheral speed u_2 of the impeller is $2\pi r_2N=314$ m s^{-1}. Assuming an air inlet temperature $T_0=293$ K, and taking the value of c_p as 1.005×10^3 J kg^{-1} K^{-1}, we find $\sigma u_2^2/c_pT_0=0.269$. Hence, from (10.6.9), $T_3/T_0=1.269$ and $T_3-T_0=79°$. Taking a value of 80% for the polytropic efficiency, and taking $\gamma=1.40$, we have from (10.6.13) $n=1.555$, and $n/(n-1)=2.80$. Hence the pressure ratio is given by (10.6.12), i.e.

$$\frac{p_3}{p_0}=(1.269)^{2.8}=1.95.$$

10.7 Basic fluid dynamics of an axial-flow compressor stage

Flow through an axial-flow compressor is illustrated diagrammatically in figure 91. Most axial-flow compressors are multi-stage machines and comprise alternate rows of fixed and moving blades. Usually the blade height will be relatively small in comparison with the hub or tip radius and it is therefore possible, as a first approximation, to simplify the analysis by treating the problem as a case of *two-dimensional flow*.

In the simple two-dimensional or pitch-line method of analysis we evaluate the gas velocities at a mean radius as indicated in the diagram. We assume that any radial component of velocity is negligible and that there is uniformity of velocity over any circumference. We also ignore the effects of

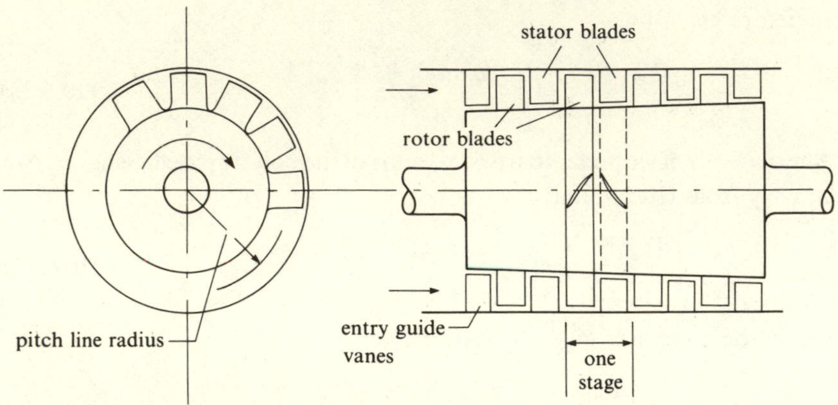

Fig. 91. Axial-flow compressor.

boundary-layer formation on the hub and casing of the compressor. A full three-dimensional analysis would evidently be extremely difficult.

One *stage* of the compressor comprises a row of moving blades and a row of fixed stator blades, and velocity diagrams for a stage are shown in figure 92. We assume that density changes are small and that the axial-flow velocity f remains effectively constant through the stage from section (1) to section (3). Since we are taking a constant value for the mean radius it follows that $u_1 = u_2 = u$.

It should be noted that the blade angle β in figure 92 is measured from a plane drawn through the axis of the machine and that this convention differs from that used in the analysis of radial-flow machines in §§ 10.2 and 10.6. The angle α measures the inclination of the absolute velocity from the axial direction.

The energy equation for flow through the row of moving blades from section (1) to section (2) takes the usual form:

$$h_1 + \frac{v_1^2}{2} + W_i = h_2 + \frac{v_2^2}{2}, \tag{10.7.1}$$

where the work transfer term W_i is given by

$$W_i = (w_2 - w_1)u. \tag{10.7.2}$$

Hence, from (10.7.1) and (10.7.2), we have

$$h_1 + \frac{v_1^2}{2} - w_1 u = h_2 + \frac{v_2^2}{2} - w_2 u. \tag{10.7.3}$$

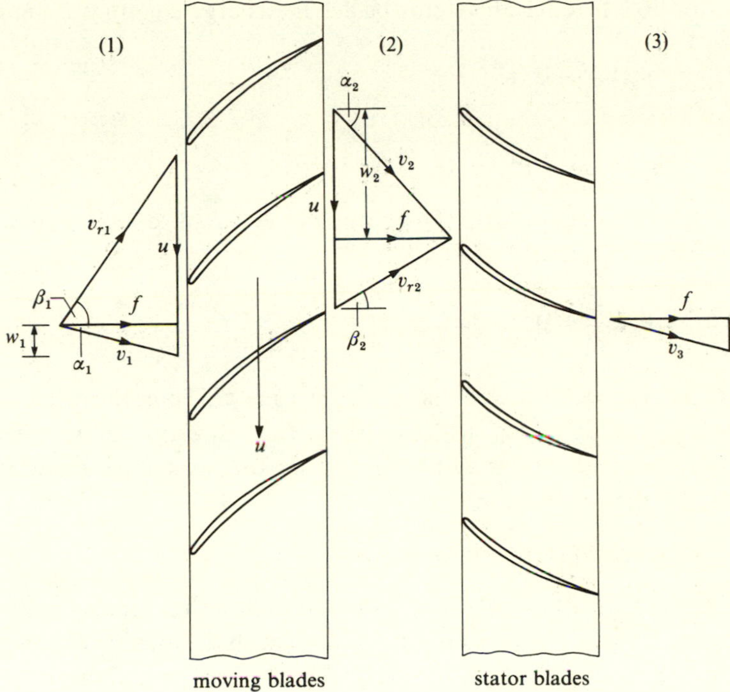

Fig. 92. Velocity triangles for an axial-flow compressor stage.

From the velocity diagrams of figure 92 it will be seen that

$$v_{r1}^2 = (u - w_1)^2 + f^2 = u^2 + v_1^2 - 2w_1 u \tag{10.7.4}$$

and

$$v_{r2}^2 = (u - w_2)^2 + f^2 = u^2 + v_2^2 - 2w_2 u. \tag{10.7.5}$$

Hence

$$\frac{v_1^2}{2} - w_1 u = \frac{v_{r1}^2}{2} - \frac{u^2}{2} \quad \text{and} \quad \frac{v_2^2}{2} - w_2 u = \frac{v_{r2}^2}{2} - \frac{u^2}{2}.$$

If we now substitute in (10.7.3) we obtain the energy equation for flow through the moving blades expressed in terms of the relative velocities, i.e.

$$h_1 + \frac{v_{r1}^2}{2} = h_2 + \frac{v_{r2}^2}{2} \tag{10.7.6}$$

or

$$(h_2 - h_1) = \tfrac{1}{2}(v_{r1}^2 - v_{r2}^2). \tag{10.7.7}$$

For flow through the stator blades the energy equation is simply:

$$h_2+\frac{v_2^2}{2}=h_3+\frac{v_3^2}{2} \tag{10.7.8}$$

or

$$(h_3-h_2)=\tfrac{1}{2}(v_2^2-v_3^2). \tag{10.7.9}$$

The energy equation for flow through the complete stage is obtained by adding (10.7.1) and (10.7.8), i.e.

$$h_1+\frac{v_1^2}{2}+W_i=h_3+\frac{v_3^2}{2}. \tag{10.7.10}$$

If matters are so arranged that $v_3=v_1$ the gas will enter the next stage with the same velocity and with the same flow direction as it entered the preceding stage. This may be regarded as the normal case and equation (10.7.10) then reduces to

$$(h_3-h_1)=W_i=(w_2-w_1)u. \tag{10.7.11}$$

Reference to figure 92 will show that v_{r2} is less than v_{r1} and that v_3 is less than v_2. Thus both the relative motion through the moving blades, and the absolute motion through the stator blades, involve a flow pattern with decreasing velocity and an adverse pressure gradient. The overall pressure rise across the stage is therefore shared between the rotor blades and the stator blades. It is convenient to define the degree of reaction R as the ratio of the enthalpy rise in the rotor blades to the overall enthalpy rise across the stage, i.e.

$$R=\frac{h_2-h_1}{h_3-h_1}. \tag{10.7.12}$$

Substituting from (10.7.7) and (10.7.11) gives

$$R=\frac{\frac{1}{2}(v_{r1}^2-v_{r2}^2)}{u(w_2-w_1)}. \tag{10.7.13}$$

From (10.7.4) and (10.7.5), however, we have

$$v_{r1}^2-v_{r2}^2=2u(w_2-w_1)-(w_2^2-w_1^2)$$

and hence

$$R=1-\frac{1}{2}\frac{(w_2+w_1)}{u}. \tag{10.7.14}$$

It is customary to design an axial-flow compressor with a reaction ratio of approximately 0.5 so that the enthalpy rise is shared equally between the rotor and the stator. It follows from (10.7.14) that, for $R=0.5$, we require $(w_2+w_1)=u$.

We can also express the reaction ratio in terms of the blade angles β_1 and β_2. From (10.7.14) we can say that

$$R=\frac{(u-w_2)+(u-w_1)}{2u}=\frac{f\tan\beta_2+f\tan\beta_1}{2u},$$

i.e.

$$R=\frac{1}{2}\frac{f}{u}(\tan\beta_1+\tan\beta_2)=\frac{f}{u}\tan\beta_m, \tag{10.7.15}$$

where β_m is the mean blade angle defined by $\tan\beta_m=\frac{1}{2}(\tan\beta_1+\tan\beta_2)$. Alternatively we can express the reaction ratio in terms of the angles α_1 and β_2 in the form

$$R=\frac{1}{2}+\frac{f}{u}(\tan\beta_2-\tan\alpha_1). \tag{10.7.16}$$

It follows from (10.7.16) that if $R=\frac{1}{2}$ we must have $\alpha_1=\beta_2$.

The velocity ratio f/u which appears in (10.7.15) is an important design parameter and is known as the flow coefficient. It is found that maximum efficiency is usually achieved when the mean blade angle is approximately 45°, i.e. $\tan\beta_m\approx1.0$. Under these conditions, if the reaction ratio is $\frac{1}{2}$, we require a value of approximately 0.5 for the velocity ratio f/u.

The stage loading factor is defined as the ratio

$$\psi=\frac{W_i}{u^2}=\frac{w_2-w_1}{u}. \tag{10.7.17}$$

From the geometry of the velocity triangles in figure 92 it follows that

$$\psi=\frac{f}{u}(\tan\beta_1-\tan\beta_2) \tag{10.7.18}$$

or alternatively that

$$\psi=1-\frac{f}{u}(\tan\alpha_1+\tan\beta_2). \tag{10.7.19}$$

Enthalpy changes for flow through an axial-flow compressor stage, as determined by equations (10.7.6)–(10.7.11) inclusive, are shown on the Mollier diagram in figure 93.

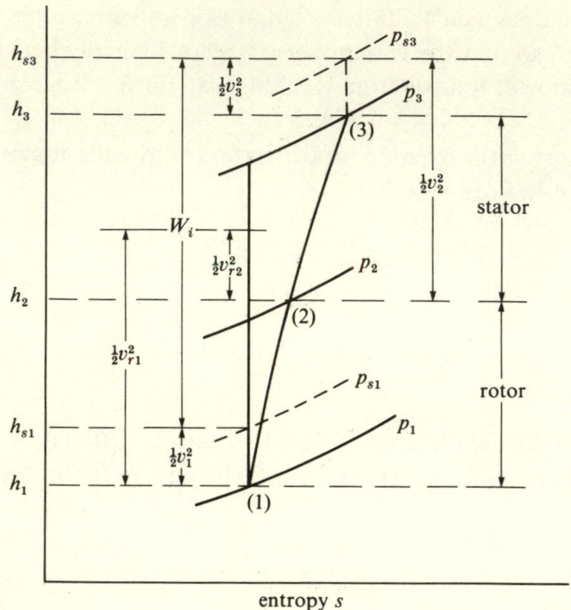

Fig. 93. Mollier diagram for a compressor stage.

The method of analysis based on velocity triangles forms a useful starting point for a study of the flow in an axial-flow compressor stage, but it cannot by itself provide a basis for calculating the required number and dimensions of the blades or for estimating the deflection angles that can actually be realised in practice. Further information may be obtained from aerofoil theory. Figure 94 shows a section of one of the rotor blades with the lift and drag forces L and D acting on the blade element. The figure also indicates the relative velocities v_{r1} and v_{r2} (as defined in figure 92) and the vector mean velocity v_{rm}. It will be seen from the diagram that $v_{rm} = f \sec \beta_m$ and that the angle β_m is defined by $\tan \beta_m = \frac{1}{2}(\tan \beta_1 + \tan \beta_2)$.

The lift force L acting on the rotor blade may be expressed in terms of the lift coefficient C_L by

$$L = C_L \tfrac{1}{2} \rho v_{rm}^2 lc, \tag{10.7.20}$$

where c is the blade chord and l is the blade height. Similarly the drag force D is given by

$$D = C_D \tfrac{1}{2} \rho v_{rm}^2 lc. \tag{10.7.21}$$

The resultant component F_y representing the force acting on the blade in

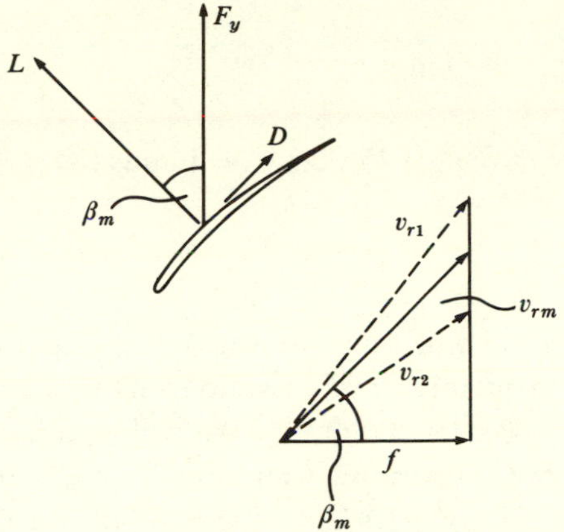

Fig. 94. Lift and drag force acting on a compressor blade.

the plane of rotation is given by

$$F_y = L\cos\beta_m + D\sin\beta_m. \tag{10.7.22}$$

Hence, substituting from (10.7.20) and (10.7.21), we have

$$F_y = \tfrac{1}{2}\rho v_{rm}^2 lc(C_L\cos\beta_m + C_D\sin\beta_m). \tag{10.7.23}$$

The number of blades on a rotor disk is $2\pi r/b$, where r is the pitch line radius and b is the blade pitch measured at the same radius. Hence the torque acting on the rotor disk is given by

$$T = \frac{2\pi r}{b}\times rF_y = \pi r^2\frac{lc}{b}\rho v_{rm}^2(C_L\cos\beta_m + C_D\sin\beta_m).$$

Making the substitution $v_{rm} = f\sec\beta_m$, we have

$$T = \pi r^2\frac{lc}{b}\rho f^2\sec\beta_m(C_L + C_D\tan\beta_m). \tag{10.7.24}$$

Noting that the power, or rate of energy transfer from the rotor to the gas, is $T\Omega$, and that the mass flow m through the compressor is given by $m = 2\pi rl\rho f$, we have

$$T\Omega = \tfrac{1}{2}mr\Omega\frac{c}{b}f\sec\beta_m(C_L + C_D\tan\beta_m). \tag{10.7.25}$$

Hence

$$W_i = \frac{T\Omega}{m} = \frac{1}{2}\frac{c}{b}uf\sec\beta_m(C_L + C_D\tan\beta_m) \tag{10.7.26}$$

or

$$\frac{W_i}{u^2}=\frac{1}{2}\frac{c}{b}\frac{f}{u}\sec\beta_m(C_L+C_D\tan\beta_m). \tag{10.7.27}$$

In the special case where we have $\beta_m=45°$, $\tan\beta_m=1$ and $\sec\beta_m=\sqrt{2}$, i.e.

$$\frac{W_i}{u^2}=\frac{1}{\sqrt{2}}\frac{c}{b}\frac{f}{u}(C_L+C_D). \tag{10.7.28}$$

As a numerical example we will take the case of an axial-flow air compressor stage designed to operate at 50% reaction with flow coefficient $f/u=0.5$ and with a blade speed of 300 m s^{-1}. We will assume a pitch/chord ratio $b/l=1.0$ and a normal operating value for $(C_L+C_D)=0.90$. The stage loading factor, calculated from (10.7.28), is then 0.318. We thus find that $h_3-h_1=c_p(T_3-T_1)=W_i=2.864\times10^4$ J kg^{-1}. Taking $c_p=1.005\times10^3$ J kg^{-1} K^{-1} we have $T_3-T_1=28.5°$ and if the entry temperature is 300 K the temperature ratio $T_3/T_1=1.095$. Assuming a polytropic stage efficiency of 0.87 we have $n=1.489$ from (10.6.13) and hence $p_3/p_1=(T_3/T_1)^{n/(n-1)}=1.318$. The blade angles β_1 and β_2 may be estimated from (10.7.15) and (10.7.18). Putting $R=0.5$ in (10.7.15) and $\psi=0.318$ in (10.7.18) we have $\tan\beta_1+\tan\beta_2=2$ and $\tan\beta_1-\tan\beta_2=0.636$. Hence we find

$$\tan\beta_1=1.318 \quad \text{or} \quad \beta_1=52.8°$$

and

$$\tan\beta_2=0.682 \quad \text{or} \quad \beta_2=34.3°.$$

Note that $\tan\beta_m=1.0$.

It should be noted that experimental measurements of lift and drag coefficients for isolated aerofoil sections are not directly applicable to axial-flow compressor blading. Because of the close pitch of the blades, and interference effects between adjacent blades, it is essential that the basic experimental data is obtained from wind-tunnel tests carried out with *cascades of blades* of the appropriate profile.

10.8 Dimensional analysis applied to turbo-compressors

When dealing with turbo-compressors we must extend the method of dimensional analysis employed in §10.4 by introducing a fourth fundamental dimension $[T]$ for the thermodynamic temperature. A full discussion of the implications of dimensional analysis in problems involving thermal effects will be given in chapter 15, but in the meantime we will simply proceed on the assumption that we require four fundamental dimensions $[m]$, $[l]$, $[t]$, and $[T]$.

We start by assuming that, for a range of geometrically similar machines, the delivery pressure p will be a function of the inlet pressure and temperature, p_0 and T_0, the physical properties of the gas c_p and γ, the rotational speed N, the rotor diameter D, and the mass flow m. We can omit the viscosity of the gas since the flow will be extremely turbulent in character and inertia forces will completely outweigh viscous forces. We do not include the density of the gas since this is not an independent variable but is determined at any point in the flow by the pressure and temperature through the equation of state. The starting point, therefore, is the statement that

$$p = f(p_0, T_0, c_p, \gamma, N, D, m). \tag{10.8.1}$$

There are eight physical quantities and four fundamental dimensions and the statement may therefore be reduced to a functional relationship between four dimensionless groups. Taking p_0, T_0, c_p and D as the primary quantities, we obtain the result that

$$\frac{p}{p_0} = f\left[\frac{m\sqrt{(c_p T_0)}}{D^2 p_0}, \frac{ND}{\sqrt{(c_p T_0)}}, \gamma\right]. \tag{10.8.2}$$

The group $[m\sqrt{(c_p T_0)}]/D^2 p_0$ may be regarded as a dimensionless mass-flow coefficient, and the group $ND/\sqrt{(c_p T_0)}$ as a dimensionless speed coefficient.

Provided we are dealing with gases having the same value for the specific heat ratio γ, we can plot the pressure ratio p/p_0 against the dimensionless mass-flow coefficient for various values of the speed coefficient. This is indicated in figure 95 and represents the operating characteristic for a turbo-compressor in a form equivalent to the dimensionless plot of the head-capacity curve for a pump.

It should be noted, however, that unlike a centrifugal pump a turbo-compressor can only operate over a limited part of the pressure ratio/mass-flow characteristic. For any given value of the speed ratio there is a minimum value for the dimensionless mass-flow ratio below which the flow becomes unstable due to stalling either of the rotor blades, the stator blades, or the diffuser guide vanes. This limit is indicated by the *surge line* shown in figure 95.

It will also be noted from the diagram that the individual characteristic curves for each speed ratio eventually become vertical in shape as the mass-flow rate is increased. This phenomenon is known as *choking* and it occurs

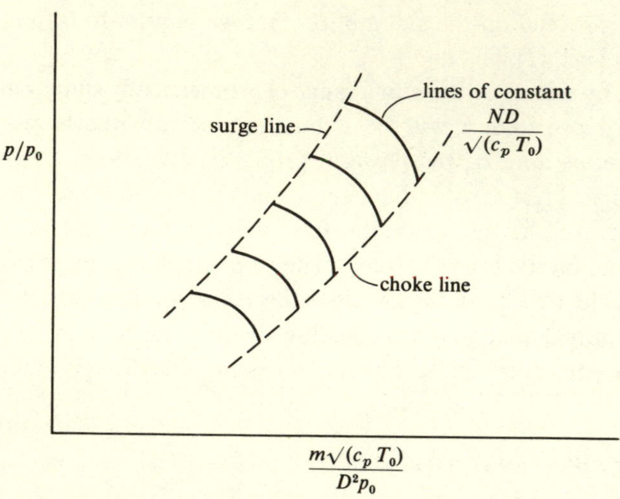

Fig. 95. Pressure ratio as a function of dimensionless mass-flow coefficient.

when the flow velocity reaches the local sonic value at some point in the compressor. The more-important aspects of compressible flow are reviewed in chapter 11 and the conditions under which the flow through a nozzle becomes choked are investigated in §§ 11.8 and 11.9.

11

Gas dynamics

11.1 The effects of compressibility

We have already derived the basic equations which govern the flow of a compressible fluid. It will be appropriate, however, to restate the equations at this point before we proceed to examine some special aspects of high-speed gas flow. The basic laws may be summarised as follows:

Conservation of mass

The equation of continuity for a compressible fluid takes the form of (2.3.3) or (3.3.1), i.e.

$$\frac{\partial \rho}{\partial t}+\operatorname{div}(\rho \mathbf{u})=0. \tag{11.1.1}$$

Equation of motion

The application of Newton's second law to a moving fluid leads to Euler's equation of motion (3.2.1) in the case of an inviscid fluid, i.e.

$$\rho \frac{\mathrm{D}\mathbf{u}}{\mathrm{D}t}=\rho \mathbf{F}-\operatorname{grad} p. \tag{11.1.2}$$

For a fluid with significant viscous stresses, however, the basic Newtonian law of motion takes the more extended form of the Navier–Stokes equation (5.4.2), i.e.

$$\rho \frac{\mathrm{D}\mathbf{u}}{\mathrm{D}t}=\rho \mathbf{F}-\operatorname{grad} p+\mu \nabla^2 \mathbf{u}+\frac{\mu}{3} \operatorname{grad} \operatorname{div} \mathbf{u}. \tag{11.1.3}$$

First law of thermodynamics

When thermal effects are significant in a moving fluid we have to take account of the first law of thermodynamics in the form of equation (9.3.5), i.e.

$$\rho \frac{\mathrm{D}e}{\mathrm{D}t} = k\,\nabla^2 T - p\,\mathrm{div}\,\mathbf{u} + \Phi. \tag{11.1.4}$$

Finally we require the *equation of state* for the fluid in question in order to establish the inter-relationship between the pressure, density, and temperature. When considering the flow of air, and of most other 'permanent' gases, we can use the perfect gas law in the form of equation (1.2.5). It will be convenient to express this as

$$\frac{p}{\rho} = R'T, \tag{11.1.5}$$

where $R' = R/M_r$. Note that R has units J K^{-1} kg^{-1}. It also follows from the properties of a perfect gas that the internal energy e may be expressed as a function of the thermodynamic temperature using the relationship $\mathrm{d}e = c_v\,\mathrm{d}T$.

If we are dealing with the high-speed flow of a gas where there is no external supply or removal of heat and where the effects of viscosity are negligible, we can regard the changes of pressure and density within the gas as taking place in an adiabatic and reversible manner. In these special circumstances we can make use of the expression which relates to *isentropic* changes in a perfect gas:

$$\frac{p}{p_0} = \left(\frac{\rho}{\rho_0}\right)^{\gamma} = \left(\frac{T}{T_0}\right)^{\gamma/(\gamma-1)}, \tag{11.1.6}$$

where γ is the ratio of the specific heats c_p/c_v.

The propagation of a small pulse or pressure wave in a gas may be taken as an example of a nearly isentropic change which will be governed accordingly by equations (11.1.1), (11.1.2), and (11.1.6). We will assume that the gas is initially at rest with uniform temperature and pressure and with a uniform density ρ_0. If a small disturbance spreads through the fluid we can say that the local value of the density at any instant may be expressed as $\rho = \rho_0 + \delta\rho$ where $\delta\rho$ is small. We can also say that the local values of the fluid velocity components u, v, w arising from the propagation of the disturbance will be small. Under these conditions we can simplify equations (11.1.1) and (11.1.2) by ignoring second-order terms involving $\delta\rho$ and the velocity components. Thus the continuity equation (11.1.1) reduces to

$$\frac{\partial\rho}{\partial t} + \rho_0\,\mathrm{div}\,\mathbf{u} = 0. \tag{11.1.7}$$

Similarly, the Euler equation of motion (11.1.2) may be simplified, omitting the gravitational or body force term, to

$$\rho_0 \frac{\partial \mathbf{u}}{\partial t} = -\operatorname{grad} p. \tag{11.1.8}$$

Under isentropic conditions we can say that

$$\operatorname{grad} p = \left(\frac{\partial p}{\partial \rho}\right)_s \operatorname{grad} \rho. \tag{11.1.9}$$

Hence from (11.1.8) and (11.1.9) we have

$$\rho_0 \frac{\partial \mathbf{u}}{\partial t} = -\left(\frac{\partial p}{\partial \rho}\right)_s \operatorname{grad} \rho \tag{11.1.10}$$

or, expressed in terms of the scalar components, we have for the x component

$$\rho_0 \frac{\partial u}{\partial t} = -\left(\frac{\partial p}{\partial \rho}\right)_s \frac{\partial \rho}{\partial x}, \tag{11.1.11}$$

with similar expressions for the y and z components. From (11.1.7) we can say that

$$\frac{\partial^2 \rho}{\partial t^2} = -\rho_0 \left(\frac{\partial^2 u}{\partial x\,\partial t} + \frac{\partial^2 v}{\partial y\,\partial t} + \frac{\partial^2 w}{\partial z\,\partial t}\right) \tag{11.1.12}$$

but, from (11.1.11), we have

$$\rho_0 \frac{\partial^2 u}{\partial x\,\partial t} = -\left(\frac{\partial p}{\partial \rho}\right)_s \frac{\partial^2 \rho}{\partial x^2}, \quad \text{etc.} \tag{11.1.13}$$

Hence, substituting from (11.1.13) in (11.1.12), we conclude that

$$\frac{\partial^2 \rho}{\partial t^2} = \left(\frac{\partial p}{\partial \rho}\right)_s \nabla^2 \rho. \tag{11.1.14}$$

It is immediately apparent that the last result (11.1.14) is in the form of the *wave equation*

$$\frac{\partial^2 \rho}{\partial t^2} = a^2\, \nabla^2 \rho, \tag{11.1.15}$$

where a is the speed of propagation of a wave, or small disturbance, in the density ρ. We thus conclude that the speed of propagation of a small disturbance in a gas, otherwise referred to as the *speed of sound*, is given by the quantity a where

$$a^2 = \left(\frac{\partial p}{\partial \rho}\right)_s. \tag{11.1.16}$$

From (11.1.6), however, we have under isentropic conditions

$$\left(\frac{\partial p}{\partial \rho}\right)_s = \gamma \frac{p_0}{\rho_0}\left(\frac{\rho}{\rho_0}\right)^{\gamma-1} \approx \gamma \frac{p_0}{\rho_0} \quad \text{provided } \frac{\rho}{\rho_0} \approx 1.$$

Thus we can say that, for the propagation of *small* disturbances,

$$a = \sqrt{\left(\gamma \frac{p_0}{\rho_0}\right)} = \sqrt{(\gamma R' T_0)}. \tag{11.1.17}$$

Although we have only considered the simple case in the above derivation of small disturbances in a gas which is initially at rest and with a uniform temperature T_0 and density ρ_0, a similar result applies to the more general case of a non-uniform moving gas. In general we can say that the *local value* of the speed of sound in a perfect gas is given by

$$a = \sqrt{\left(\gamma \frac{p}{\rho}\right)} = \sqrt{(\gamma R' T)}. \tag{11.1.18}$$

The sonic velocity a is an important parameter in high-speed gas flow and may be regarded as a critical velocity having a profound influence on the overall flow pattern. This will be evident if we consider the case of a point-source or disturbance moving at constant velocity U through a uniform static gas as indicated in figure 96.

At a certain instant in time the point source will be located at point P as indicated on the diagram. At a previous instant, t_1 seconds earlier, the source would have been located at point P_1 so that $PP_1 = Ut_1$. Sound waves emitted from the source at point P_1 will have travelled outwards with velocity a reaching the surface of a sphere of radius at_1 in the time interval t_1. Similarly, at an instant t_2 seconds earlier the source would have been located at point P_2 in the diagram, such that $PP_2 = Ut_2$, and sound waves from the source at point P_2 will have travelled outwards to reach the surface of a sphere of radius at_2 in the time interval t_2. Two cases are illustrated in figure 96,

(*a*) with the source moving at subsonic velocity, i.e. with $U < a$,

(*b*) with the source moving at supersonic velocity, i.e. with $U > a$.

It will be clear from figure 96 (*a*) that, provided $U < a$, sound waves or pressure disturbances originating from an earlier position of the moving

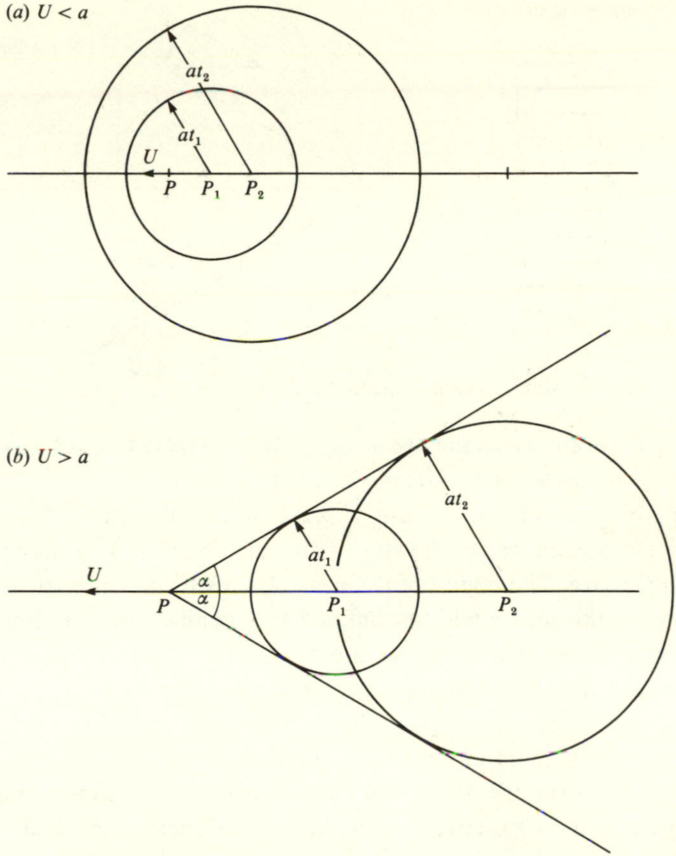

Fig. 96. Effect of a disturbance moving with velocity U, (*a*) $U<a$, (*b*) $U>a$.

source will always have moved ahead of the current position of the source. It will also be seen that the region affected by the sound waves emitted at any instant of time includes the regions affected by the sound waves emitted at later instants. The situation is entirely different, however, if $U>a$, as will be evident from figure 96(*b*). When the source is moving at supersonic velocity the region of space affected by sound waves emitted by the source at earlier instants will be confined to a cone whose apex coincides with the instantaneous position of the source and whose semi-angle α is given by

$$\alpha=\sin^{-1}\frac{a}{U}. \tag{11.1.19}$$

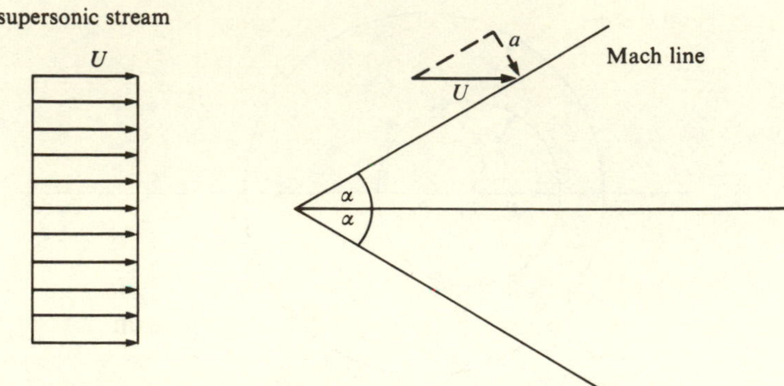

Fig. 97. Mach lines in supersonic flow.

The angle α is known as the *Mach angle*, and the ratio U/a is known as the *Mach number* as we have already noted in (1.8.8).

We obtain a similar picture if we consider the case of a uniform supersonic stream of gas flowing with velocity U past a *fixed* sound-emitting source. The region of the gas stream which is affected by sound waves from the source will be confined to a conical zone of semi-angle α stretching downstream from the source.

The equivalent two-dimensional case of a uniform supersonic gas stream flowing past a line-source, extending in a direction normal to the plane of the flow, is illustrated in figure 97. The line-source of disturbance might represent, for example, the sharp leading edge of an infinitesimally thin plate located in the gas stream with its plane parallel to the stream velocity U. It will be seen from figure 97 that the region of the gas stream which is affected by sound waves or small disturbances originating from the line-source will be confined to a wedge of semi-angle α where α is given by

$$\alpha = \sin^{-1}\frac{a}{U} = \sin^{-1}\frac{1}{Ma}. \tag{11.1.20}$$

The inclined lines in figure 97, which represent the intersection of this wedge with the plane of the picture, are known as Mach waves or *Mach lines*. It will be noted that, if we resolve the stream velocity into components parallel to and normal to a Mach line, the normal component $U \sin \alpha$ must be equal to the sonic velocity a.

11.2 Dynamical similarity in high-speed gas flow

We have seen in chapters 5–8 that, at relatively high flow velocities, the effects of viscosity on the dynamics of the fluid are largely confined to certain restricted regions such as boundary layers. Outside these regions the flow pattern is determined primarily by equation (11.1.2) rather than (11.1.3). If we ignore the effects of gravity, and other possible body forces, Euler's equation of motion (11.1.2) takes the form

$$\rho \frac{\mathrm{D}\mathbf{u}}{\mathrm{D}t} = -\operatorname{grad} p. \tag{11.2.1}$$

If we are dealing with the flow of a perfect gas under isentropic conditions we can make use of the relationship (11.1.9) and equation (11.2.1) may then be re-stated as

$$\rho \frac{\mathrm{D}\mathbf{u}}{\mathrm{D}t} = -\left(\frac{\partial p}{\partial \rho}\right)_s \operatorname{grad} \rho = -a^2 \operatorname{grad} \rho. \tag{11.2.2}$$

We can express this equation in dimensionless form by introducing the following dimensionless ratios, using the same procedure as in §5.7, i.e.

$$\mathbf{u}' = \frac{\mathbf{u}}{U_0}, \quad x' = \frac{x}{L}, \quad \text{etc.,} \quad t' = \frac{U_0 t}{L}, \quad \rho' = \frac{\rho}{\rho_0}, \quad a' = \frac{a}{a_0}.$$

where ρ_0 and a_0 are the values of the density and sonic velocity, respectively, at temperature T_0 in the undisturbed stream, and U_0 is the mainstream velocity.

Substituting in (11.2.2) we obtain the result that

$$\frac{U_0^2}{a_0^2} \rho' \frac{\mathrm{D}\mathbf{u}'}{\mathrm{D}t'} = -a'^2 \operatorname{grad}' \rho', \tag{11.2.3}$$

where grad′ stands for

$$\mathbf{i}\frac{\partial}{\partial x'} + \mathbf{j}\frac{\partial}{\partial y'} + \mathbf{k}\frac{\partial}{\partial z'}.$$

It will be seen that, for given boundary conditions, the solution of equation (11.2.3) depends only on the numerical value of the ratio U_0^2/a_0^2. This dimensionless ratio, which represents the relative magnitude of the inertia forces compared with the net force acting locally on a fluid element as a result of the density gradient, is analogous to the Froude number U_0^2/gL which arises in connection with the effect of gravitational forces acting on the fluid.

We may equally well state that the solution of equation (11.2.3) depends only on the numerical value of the velocity ratio U_0/a_0 which we have already identified as the *Mach number*, *Ma*.

We have seen in §11.1 that the sonic velocity in a perfect gas at temperature T_0 is given by equation (11.1.17), thus the Mach number Ma may be expressed in terms of the gas temperature T_0 in the form

$$Ma = \frac{U_0}{\sqrt{(\gamma R' T_0)}}. \tag{11.2.4}$$

In chapter 9, when analysing energy transfer in a fluid stream, we derived the Eckert number as a measure of the relative magnitude of the kinetic energy compared with the enthalpy of a fluid element. Noting that, for a perfect gas, $R' = c_p - c_v$, we have the relationship that

$$Ec = \frac{U_0^2}{c_p \Delta T_s} = \frac{\gamma R' T_0}{c_p \Delta T_s} Ma^2 = (\gamma - 1) \frac{T_0}{\Delta T_s} Ma^2, \tag{11.2.5}$$

which is the result quoted in (9.5.8).

11.3 One-dimensional flow of a compressible gas in a stream tube

We can derive some useful results relating to the steady flow of a perfect gas with the aid of the simplified concept of 'one-dimensional' flow along a stream tube which we have employed previously in §§ 3.4 and 9.1.

We assume, as in chapters 3 and 9, that the cross-section of the stream tube is sufficiently small to enable us to treat the fluid velocity u as being constant over any cross-section. We will use the letter A to represent the cross-sectional flow area of the stream tube at any point along its length. It should be noted that this involves a change in notation compared with chapters 3 and 9 where we used the lower-case letter a for the flow area. This change is occasioned by the fact that the authors prefer in the present chapter to follow the normal convention of using the lower-case letter a to represent the velocity of sound as defined in equations (11.1.15) and (11.1.16).

The equation of continuity for steady flow in the stream tube takes the same form as (3.3.6), i.e. we can say that

$$\text{mass flow } m = \rho A u = \text{constant}. \tag{11.3.1}$$

If we differentiate this equation with respect to distance measured along the axis of the stream tube we obtain the differential form

$$\frac{\mathrm{d}\rho}{\rho} + \frac{\mathrm{d}A}{A} + \frac{\mathrm{d}u}{u} = 0. \tag{11.3.2}$$

If we can neglect the effects of viscous stresses, we can make use of Euler's equation of motion (11.1.2) for a frictionless fluid. Ignoring the body force **F** and expressing the equation in one-dimensional form, we have

$$\rho u \frac{\mathrm{d}u}{\mathrm{d}x} = -\frac{\mathrm{d}p}{\mathrm{d}x}. \tag{11.3.3}$$

It follows that the differential changes in the velocity u and pressure p, which occur over a short length of the stream tube, are related by

$$u\,\mathrm{d}u + \frac{1}{\rho}\,\mathrm{d}p = 0. \tag{11.3.4}$$

In chapter 9 we derived the general form of the energy equation by combining the first law of thermodynamics (11.1.4) with the mechanical energy equation. For the special case of steady one-dimensional flow along a stream tube, the energy equation may be expressed in the form of (9.1.9) or (9.1.10). Since we are now assuming a frictionless fluid it follows that the shear work done at the edge of the stream tube must be zero. If we also assume that the flow conditions are adiabatic, i.e. that there is no transfer of heat across the wall of the stream tube, equation (9.1.10) reduces to

$$\mathrm{d}h + u\,\mathrm{d}u = 0. \tag{11.3.5}$$

For a perfect gas, noting that $R' = c_p - c_v$ and $c_p/c_v = \gamma$, we have

$$\mathrm{d}h = c_p\,\mathrm{d}T = \frac{\gamma}{(\gamma-1)} R'\,\mathrm{d}T = \frac{\gamma}{(\gamma-1)}\,\mathrm{d}\left(\frac{p}{\rho}\right). \tag{11.3.6}$$

Hence from (11.3.5) and (11.3.6) the steady-flow energy equation may be written in the form

$$\frac{\gamma}{(\gamma-1)}\,\mathrm{d}\left(\frac{p}{\rho}\right) + u\,\mathrm{d}u = 0 \tag{11.3.7}$$

or

$$\frac{\gamma}{(\gamma-1)}\frac{1}{\rho}\,\mathrm{d}p - \frac{\gamma}{(\gamma-1)}\frac{p}{\rho^2}\,\mathrm{d}\rho + u\,\mathrm{d}u = 0. \tag{11.3.8}$$

If we now eliminate the term $u\,\mathrm{d}u$ between the equation of motion (11.3.4) and the energy equation (11.3.8) we obtain the result:

$$\frac{1}{(\gamma-1)}\frac{1}{\rho}\,\mathrm{d}p - \frac{\gamma}{(\gamma-1)}\frac{p}{\rho^2}\,\mathrm{d}\rho = 0$$

or

$$\frac{\mathrm{d}p}{p} = \gamma\frac{\mathrm{d}\rho}{\rho}. \tag{11.3.9}$$

Equation (11.3.9) expresses the relationship which must exist between the pressure p and the density of the fluid ρ in order to satisfy the Euler equation of motion for frictionless flow in the stream tube and the energy equation for steady flow in the stream tube under conditions of zero heat transfer and zero shear work. If we integrate (11.3.9) along the stream tube, from a position where the pressure is p_0 and the density is ρ_0, we obtain the result that

$$\log_e \frac{p}{p_0} = \gamma \log_e \frac{\rho}{\rho_0}$$

and hence

$$\frac{p}{p_0} = \left(\frac{\rho}{\rho_0}\right)^{\gamma}. \tag{11.3.10}$$

This result comes as no surprise since it is identical with the normal condition for *isentropic changes* of state in a perfect gas. In classical thermodynamics, equation (11.3.10) is derived in describing quasi-static reversible changes of state in a perfect gas under adiabatic conditions. We now see that it is equally applicable to the changes of state which occur in a perfect gas which is flowing in a stream tube provided the viscous stresses are negligible and provided there is no transfer of heat within the fluid. We have thus proved the validity of equation (11.1.6) in relation to the ideal frictionless flow of a perfect gas with zero heat transfer.

The Euler equation of motion (11.3.4) for frictionless flow of a perfect gas in a stream tube under isentropic conditions may be expressed in the alternative form:

$$u\,\mathrm{d}u + a^2 \frac{\mathrm{d}\rho}{\rho} = 0, \tag{11.3.11}$$

where

$$a^2 = \frac{\gamma p}{\rho} = \gamma R'T. \tag{11.3.12}$$

It will be seen that equation (11.3.11) follows directly from equations (11.3.4) and (11.3.9). We can equally well obtain equation (11.3.11) from (11.3.4) by making use of the fact that under isentropic conditions $(\mathrm{d}p/\mathrm{d}\rho)_s = a^2$ as in (11.1.16). If we now eliminate the term $\mathrm{d}\rho/\rho$ between the continuity equation (11.3.2) and the equation of motion in the form of (11.3.11), we obtain the result that

$$\left(1 - \frac{u^2}{a^2}\right)\frac{\mathrm{d}u}{u} + \frac{\mathrm{d}A}{A} = 0 \tag{11.3.13}$$

or

$$(1-Ma^2)\frac{\mathrm{d}u}{u}=-\frac{\mathrm{d}A}{A}. \tag{11.3.14}$$

We thus see that if $Ma<1$ the velocity u increases as the flow area A decreases, and vice versa, as in the case of the flow of an incompressible fluid. If, on the other hand, $Ma>1$, the velocity u increases as the flow area A increases. It also follows from (11.3.14) that we can only achieve a local value of $Ma=1$ in the stream tube at a section of minimum area.

The steady flow energy equation, as derived in §9.1, plays an important role in the analysis of compressible flow in a stream tube or duct. For the special case of zero heat transfer, and in the absence of shear work at the edge of the stream tube, the energy equation may be expressed by (9.1.10) or (11.3.5). It is always convenient, when dealing with compressible flow in a stream tube or duct, to picture the stream originating from a *reservoir* where the velocity is zero and where the gas pressure and temperature are p_0 and T_0 respectively. With the help of this concept we can integrate the energy equation (9.1.10) along the stream tube from the reservoir to a section where the pressure, temperature, and gas velocity are p, T, and u respectively. We may thus express the steady-flow energy equation as

$$h_0=h+\tfrac{1}{2}u^2, \tag{11.3.15}$$

where h_0 is the enthalpy of the gas in the reservoir. It will be seen that h_0 is the same as the *stagnation enthalpy*, h_s, as defined in (10.5.10). Provided we are dealing with a perfect gas, we can make use of (11.3.6) and express the energy equation (11.3.15) in the alternative forms:

$$c_pT_0=c_pT+\tfrac{1}{2}u^2 \tag{11.3.16}$$

or

$$\frac{\gamma}{\gamma-1}\frac{p_0}{\rho_0}=\frac{\gamma}{\gamma-1}\frac{p}{\rho}+\tfrac{1}{2}u^2. \tag{11.3.17}$$

We can equally introduce the local velocity of sound as a parameter in place of the gas temperature T noting the relationship that $a^2=\gamma R'T$, as given by (11.1.18) or (11.3.12), and hence:

$$\frac{a_0^2}{\gamma-1}=\frac{a^2}{\gamma-1}+\tfrac{1}{2}u^2. \tag{11.3.18}$$

It will be appreciated that (11.3.16), (11.3.17), and (11.3.18) are simply alternative ways of expressing the energy equation when we are dealing with the flow of a perfect gas.

We can express equation (11.3.18) in dimensionless form by dividing through by a^2, giving the result that

$$\frac{a_0^2}{a^2}=\frac{T_0}{T}=1+\frac{(\gamma-1)}{2}Ma^2, \tag{11.3.19}$$

where Ma is the local value of the Mach number u/a. If we now put $Ma=1$ in equation (11.3.19) we obtain the local value of the gas temperature T^* and the velocity of sound a^* at the point in the flow where sonic velocity is attained, i.e. at the point where the stream velocity u just becomes equal to a. Hence

$$\frac{a_0^2}{a^{*2}}=\frac{T_0}{T^*}=\frac{\gamma+1}{2}$$

or

$$a^*=\sqrt{\left(\frac{2}{\gamma+1}\right)}a_0. \tag{11.3.20}$$

We can regard a^* as a critical velocity for flow in the stream tube. For air, with $\gamma=1.40$, we have the numerical relationships from (11.3.20) that $a^*=0.913\ a_0$, and $T^*=0.833\ T_0$.

Another critical value, or reference velocity u_c, may be identified with the aid of the energy equation. If we imagine a gas stream flowing from a reservoir at pressure p_0 and temperature T_0 into a region of perfect vacuum, i.e. a region where the pressure and thermodynamic temperature are both zero, it follows from (11.3.17) or (11.3.18) that the maximum attainable velocity would be

$$u_c=\sqrt{\left(\frac{2\gamma}{\gamma-1}\frac{p_0}{\rho_0}\right)}=\sqrt{\left(\frac{2}{\gamma-1}\right)}a_0. \tag{11.3.21}$$

Under these conditions all the internal energy associated with the random molecular motion of the gas in the reservoir would be converted into the kinetic energy of the gas stream. From (11.3.20) and (11.3.21) we have:

$$\frac{u_c}{a^*}=\sqrt{\left(\frac{\gamma+1}{\gamma-1}\right)}. \tag{11.3.22}$$

For air, with $\gamma=1.40$, this gives $u_c=2.449\,a^*=2.236\ a_0$. For the flow of air through a nozzle from a reservoir into a vacuum, therefore, the supersonic flow range extends from the critical velocity a^* to a maximum velocity $2.449\ a^*$.

It will be noted that, whereas a^* and u_c are constants for a given flow (i.e. for a given reservoir pressure and temperature) the local value of the velocity of sound a is a variable. It would clearly be possible to choose the ratio u/a^* as a dimensionless velocity ratio in place of the Mach number u/a. It will be seen from (11.3.19) and (11.3.20) that

$$\left(\frac{u}{a^*}\right)^2 = \frac{\gamma+1}{(2/Ma^2)+(\gamma-1)}. \tag{11.3.23}$$

As $$u \rightarrow u_c, \quad \frac{u}{a^*} \rightarrow \sqrt{\left(\frac{\gamma+1}{\gamma-1}\right)} \quad \text{while } \frac{u}{a} \rightarrow \infty.$$

We could equally work in terms of u/u_c as a dimensionless velocity ratio. From (11.3.22) and (11.3.23) we have

$$\left(\frac{u}{u_c}\right)^2 = \frac{\gamma-1}{(2/Ma^2)+(\gamma-1)}. \tag{11.3.24}$$

11.4 Shock waves

One of the distinctive features of supersonic flow is the presence under certain circumstances of discontinuities in the flow pattern where sudden changes are experienced in the velocity, pressure, density, and thermodynamic temperature of the gas. The simplest case is that of the *normal shock wave*, as illustrated in figure 98, where a uniform supersonic gas stream flowing with velocity u_1 changes suddenly, as it passes through the shock wave, into a uniform subsonic stream flowing with velocity u_2.

The basic principles of the conservation of both mass and energy are

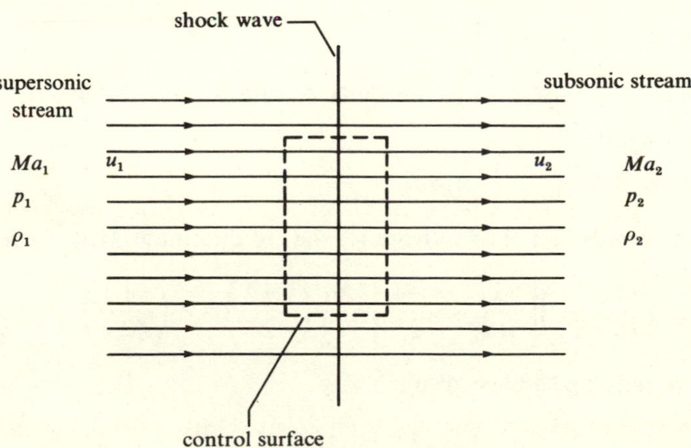

Fig. 98. Flow through a normal shock wave.

applicable to the passage of the gas stream through the shock wave. We can thus make use of the *continuity equation* in the form

$$\rho_1 u_1 = \rho_2 u_2. \tag{11.4.1}$$

We can also use the steady-flow *energy equation* as expressed by (11.3.16) or any of its derivatives. If we choose the version in which the velocity of sound a is used as the parameter, as in (11.3.18), we can say that

$$\frac{a_1^2}{\gamma-1}+\tfrac{1}{2}u_1^2=\frac{a_2^2}{\gamma-1}+\tfrac{1}{2}u_2^2. \tag{11.4.2}$$

Dividing through by u_2^2 and multiplying by $(\gamma-1)$ gives

$$\frac{u_1^2}{u_2^2}\left[\frac{1}{Ma_1^2}+\frac{\gamma-1}{2}\right]=\frac{1}{Ma_2^2}+\frac{\gamma-1}{2}. \tag{11.4.3}$$

Although we cannot apply Euler's equation of motion in the vicinity of the shock wave, we can make use of the integral form of the *momentum equation* as derived in § 3.7. Choosing a control surface as indicated in figure 98, the momentum equation requires that

$$p_1-p_2=\rho_2 u_2^2-\rho_1 u_1^2$$

or

$$p_1+\rho_1 u_1^2=p_2+\rho_2 u_2^2. \tag{11.4.4}$$

Noting from (11.1.18) that for a perfect gas $a^2=\gamma(p/\rho)$, we can re-write (11.4.4) in the form

$$\rho_1(a_1^2+\gamma u_1^2)=\rho_2(a_2^2+\gamma u_2^2)$$

or

$$\rho_1 u_1^2\left(\frac{1}{Ma_1^2}+\gamma\right)=\rho_2 u_2^2\left(\frac{1}{Ma_2^2}+\gamma\right). \tag{11.4.5}$$

Hence, making use of the continuity equation (11.4.1), we have

$$\frac{u_1}{u_2}\left(\frac{1}{Ma_1^2}+\gamma\right)=\left(\frac{1}{Ma_2^2}+\gamma\right). \tag{11.4.6}$$

If we now subtract (11.4.6) from (11.4.3) to eliminate Ma_2^2, we get

$$\left(\frac{u_1^2}{u_2^2}-\frac{u_1}{u_2}\right)\left[\frac{1}{Ma_1^2}+\frac{\gamma-1}{2}\right]-\frac{u_1}{u_2}\left(\frac{\gamma+1}{2}\right)=-\left(\frac{\gamma+1}{2}\right)$$

or, re-arranging the last result,

$$\frac{u_1}{u_2}\left(\frac{u_1}{u_2}-1\right)\left[\frac{1}{Ma_1^2}+\frac{\gamma-1}{2}\right]-\left(\frac{u_1}{u_2}-1\right)\left(\frac{\gamma+1}{2}\right)=0. \tag{11.4.7}$$

There are evidently two possibilities. Either $u_1 = u_2$ and nothing happens, or if we divide through by $(u_1/u_2) - 1$ we get

$$\frac{u_1}{u_2} = \frac{\gamma+1}{(2/Ma_1^2)+(\gamma-1)}, \tag{11.4.8}$$

which represents the velocity ratio across a normal shock wave.

Equation (11.4.8) becomes more intelligible if we make use of the result obtained in (11.3.23). It will be seen that (11.4.8) is equivalent to the statement that

$$\frac{u_1}{u_2} = \left(\frac{u_1}{a^*}\right)^2$$

or

$$u_1 u_2 = a^{*2}. \tag{11.4.9}$$

Thus if u_1 is greater than a^*, representing supersonic flow on the upstream side, u_2 must be less than a^* implying subsonic flow on the downstream side.

Although the laws of conservation of both mass and energy, together with the momentum equation (11.4.4), would be satisfied equally in a hypothetical reverse process involving a sudden change from a subsonic to a supersonic gas stream, such a process is not in fact possible and is not observed in practice. The reason is that flow through a shock wave from a supersonic to a subsonic stream involves an *increase in the entropy* of the gas and is therefore irreversible in the thermodynamic sense. A discontinuity in the flow involving a change from subsonic to supersonic velocity, on the other hand, would require a decrease in the entropy, and a spontaneous change in this direction would contravene the second law of thermodynamics.

The following expressions relating the parameters on either side of a normal shock wave are easily derived from the preceding results:

$$\frac{\rho_2}{\rho_1} = \frac{u_1}{u_2} = \frac{(\gamma+1)Ma_1^2}{2+(\gamma-1)Ma_1^2}, \tag{11.4.10}$$

$$\frac{p_2}{p_1} = \frac{2\gamma Ma_1^2 - (\gamma-1)}{(\gamma+1)}, \tag{11.4.11}$$

$$Ma_2^2 = \frac{(\gamma-1)Ma_1^2 + 2}{2\gamma Ma_1^2 - (\gamma-1)}, \tag{11.4.12}$$

and

$$\frac{T_2}{T_1} = \frac{(\gamma-1)Ma_1^2 + 2}{(\gamma-1)Ma_2^2 + 2}. \tag{11.4.13}$$

The change of entropy of the gas may be calculated from the general expression

$$s_2 - s_1 = c_p \log_e \frac{T_2}{T_1} - R' \log_e \frac{p_2}{p_1}$$

or

$$s_2 - s_1 = c_v \log_e \frac{p_2}{p_1} + c_p \log_e \frac{\rho_1}{\rho_2}. \tag{11.4.14}$$

Hence, substituting from (11.4.10) and (11.4.11),

$$s_2 - s_1 = c_v \log_e \left\{ \frac{2\gamma Ma_1^2 - (\gamma - 1)}{\gamma + 1} \right\} + c_p \log_e \left\{ \frac{(\gamma - 1) Ma_1^2 + 2}{(\gamma + 1) Ma_1^2} \right\}.$$

It is easily verified that the entropy difference $s_2 - s_1$ is positive if Ma_1 is greater than 1.

As a numerical indication of the magnitude of the changes which can occur in a normal shock wave the following figures are readily evaluated for the case of air with $\gamma = 1.40$:

Ma_1	Ma_2	u_2/u_1	p_2/p_1	T_2/T_1
1.1	0.91	0.86	1.25	1.07
1.5	0.70	0.54	2.46	1.32
2.0	0.58	0.38	4.50	1.69
2.5	0.51	0.30	7.12	2.14

It will be seen that the changes in the velocity, pressure, and temperature of a gas in passing through a shock wave can be very large. In order to study the mechanism by which these changes can be brought about it is necessary to abandon the usual concept of a fluid as a continuous medium and to consider the molecular structure and properties of a gas, as in chapter 2, from the standpoint of the kinetic theory. There is evidence that the thickness of a shock wave is typically of the same order of magnitude as the mean free path of a gas molecule.

As an example of the occurrence of a shock wave, we can consider the case of a pitot tube, or any other small blunt-nosed body, located in a supersonic gas stream. A detached shock wave will be formed in the flow at a short distance upstream from the pitot tube. We can calculate the pressure registered by the pitot tube, or the stagnation pressure at the nose of the body, by considering first the flow of the gas through the shock wave, and subsequently the subsonic deceleration of the gas to the point of zero velocity at the mouth of the pitot tube. If the Mach number and pressure in the undisturbed supersonic stream are Ma_1 and p_1, we can determine the values Ma_2 and p_2 on the downstream side of the shock wave from equations (11.4.12) and (11.4.11). For subsonic flow between the downstream side of the shock wave and the mouth of the pitot tube we can use the energy equation in the form of (11.3.19) and say that:

$$\frac{T_3}{T_2}=1+\frac{(\gamma-1)}{2}Ma_2^2, \tag{11.4.15}$$

where T_3 represents the value of the gas temperature at the mouth of the pitot tube where the velocity is zero. With isentropic flow conditions prevailing between the downstream side of the shock wave and the pitot tube, we can say that the pressure ratio p_3/p_2 will be given by

$$\frac{p_3}{p_2}=\left(\frac{T_3}{T_2}\right)^{\gamma/(\gamma-1)}=\left[1+\frac{(\gamma-1)}{2}Ma_2^2\right]^{\gamma/(\gamma-1)}. \tag{11.4.16}$$

Hence, substituting for Ma_2 in terms of Ma_1 in equation (11.4.16), and making use of equation (11.4.11) for the pressure ratio across the shock wave, we find for the *overall pressure ratio*:

$$\frac{p_3}{p_1}=\left[\frac{(\gamma+1)^2Ma_1^2}{4\gamma Ma_1^2-2(\gamma-1)}\right]^{\gamma/(\gamma-1)}\left\{\frac{2\gamma Ma_1^2-(\gamma-1)}{(\gamma+1)}\right\} \tag{11.4.17}$$

or

$$\frac{p_3}{p_1}=\left[\frac{(\gamma+1)}{2}Ma_1^2\right]^{\gamma/(\gamma-1)}\left\{\frac{2\gamma Ma_1^2-(\gamma-1)}{(\gamma+1)}\right\}^{-1/(\gamma-1)}. \tag{11.4.18}$$

11.5 Oblique shock waves

If a uniform supersonic stream is caused to change direction by the geometry of the flow boundary, for example in the case of flow past a wedge or round a concave corner as illustrated in figure 99, an oblique shock wave will be established. The relationship between the flow parameters on either side of the shock wave may be analysed using exactly the same methods as in § 11.4. We can resolve the absolute velocity u_1 on the upstream side of the shock wave into a normal component u_{n1} at right angles to the shock wave, and a tangential component u_{t1} parallel to the line of the shock wave. Similarly the absolute velocity of the gas on the downstream side of the shock wave u_2 (which is parallel to the surface of the wedge) may be resolved into components u_{n2} and u_{t2} as indicated in the diagram.

Application of the continuity, momentum, and energy equations gives

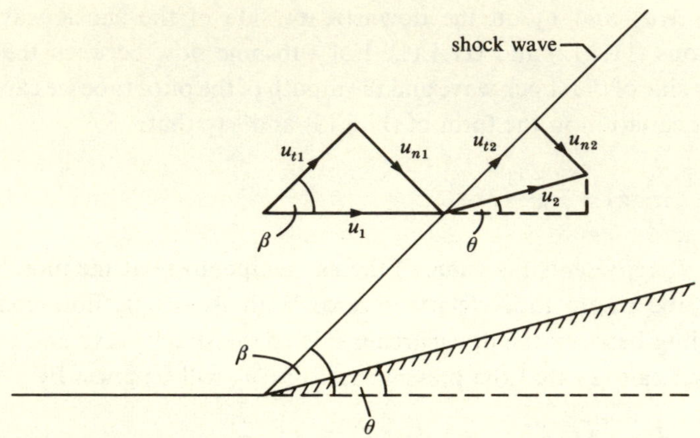

Fig. 99. Flow past a wedge with an oblique shock wave.

the following results:

continuity

$$\rho_1 u_{n1} = \rho_2 u_{n2} \tag{11.5.1}$$

momentum

$$p_1 + \rho_1 u_{n1}^2 = p_2 + \rho_2 u_{n2}^2 \tag{11.5.2}$$

and

$$u_{t1} = u_{t2}$$

energy

$$\frac{a_1^2}{\gamma-1} + \tfrac{1}{2}(u_{n1}^2 + u_{t1}^2) = \frac{a_2^2}{\gamma-1} + \tfrac{1}{2}(u_{n2}^2 + u_{t2}^2) \tag{11.5.3}$$

or

$$\frac{a_1^2}{\gamma-1} + \tfrac{1}{2}u_{n1}^2 = \frac{a_2^2}{\gamma-1} + \tfrac{1}{2}u_{n2}^2.$$

Hence

$$\frac{u_{n1}^2}{u_{n2}^2}\left[\frac{a_1^2}{u_{n1}^2} + \frac{\gamma-1}{2}\right] = \frac{a_2^2}{u_{n2}^2} + \frac{\gamma-1}{2}, \tag{11.5.4}$$

which corresponds to equation (11.4.3) for the normal shock wave.

The momentum equation (11.5.2) may be written in the form

$$\rho_1(a_1^2 + \gamma u_{n1}^2) = \rho_2(a_2^2 + \gamma u_{n2}^2)$$

or

$$\rho_1 u_{n1}^2\left(\frac{a_1^2}{u_{n1}^2} + \gamma\right) = \rho_2 u_{n2}^2\left(\frac{a_2^2}{u_{n2}^2} + \gamma\right). \tag{11.5.5}$$

Hence, making use of the continuity equation (11.5.1), we have

$$\frac{u_{n1}}{u_{n2}}\left(\frac{a_1^2}{u_{n1}^2}+\gamma\right)=\frac{a_2^2}{u_{n2}^2}+\gamma, \tag{11.5.6}$$

which corresponds to equation (11.4.6) for the normal shock wave. If we now subtract (11.5.6) from (11.5.4) in order to eliminate the term a_2^2/u_{n2}^2, and if we disregard the possibility of a non-event with $u_{n1}=u_{n2}$, we obtain an expression relating the normal velocity components on either side of the shock wave, i.e.

$$\frac{u_{n1}}{u_{n2}}=\frac{\gamma+1}{2(a_1^2/u_{n1}^2)+(\gamma-1)}. \tag{11.5.7}$$

It will be seen that this is the equivalent of (11.4.8) for the case of a normal shock wave.

Equation (11.5.7) can be re-arranged in the following form:

$$u_{n1}u_{n2}=\frac{2}{\gamma+1}a_1^2+\frac{(\gamma-1)}{(\gamma+1)}u_{n1}^2.$$

If we now make use of the energy equation, together with the result (11.3.20) for the critical velocity a^*, we can say that

$$a^{*2}=\frac{2}{\gamma+1}a_0^2=\frac{2}{\gamma+1}a_1^2\left[1+\frac{(\gamma-1)}{2}\frac{u_1^2}{a_1^2}\right]$$

or

$$a^{*2}=\frac{2a_1^2}{\gamma+1}+\frac{(\gamma-1)}{(\gamma+1)}u_1^2. \tag{11.5.8}$$

Hence, from (11.5.7) and (11.5.8) we have the relationship

$$u_{n1}u_{n2}=a^{*2}+\frac{(\gamma-1)}{(\gamma+1)}[u_{n1}^2-u_1^2]$$

or

$$u_{n1}u_{n2}=a^{*2}-\frac{(\gamma-1)}{(\gamma+1)}u_t^2. \tag{11.5.9}$$

If the transverse velocity component u_t is zero, we have the case of a normal shock wave and (11.5.9) then reduces to the simple relationship (11.4.9).

The relationship between the angle of the shock wave β and the angle of deflection θ may be obtained by reference to the geometry of figure 99. Noting that $\tan\beta=u_{n1}/u_t$, and that $\tan(\beta-\theta)=u_{n2}/u_t$, we have

$$\tan\beta\tan(\beta-\theta)=\frac{u_{n1}u_{n2}}{u_t^2}. \tag{11.5.10}$$

Hence, substituting from (11.5.9),

$$\tan\beta\tan(\beta-\theta)=\frac{a^{*2}}{u_t^2}-\frac{\gamma-1}{\gamma+1}$$

or, since $u_t=u_1\cos\beta$,

$$\tan(\beta-\theta)=\frac{a^{*2}}{u_1^2}\frac{1}{\sin\beta\cos\beta}-\frac{\gamma-1}{\gamma+1}\cot\beta. \tag{11.5.11}$$

We can express this relationship in terms of the Mach number Ma_1 rather than the velocity ratio u_1/a^* if we make use of (11.3.23) or (11.5.8). From equation (11.5.8) we have

$$\frac{a^{*2}}{u_1^2}=\frac{2}{\gamma+1}\frac{1}{Ma_1^2}+\frac{\gamma-1}{\gamma+1}. \tag{11.5.12}$$

Hence (11.5.11) may be expressed in the form

$$\tan(\beta-\theta)=\frac{2}{(\gamma+1)}\frac{1}{Ma_1^2\sin\beta\cos\beta}+\frac{\gamma-1}{\gamma+1}\tan\beta. \tag{11.5.13}$$

It will be seen from (11.5.13) that in the *limiting case of zero deflection*, putting $\theta=0$, we have $1/Ma_1^2=\sin^2\beta$ or $\beta=\sin^{-1}(1/Ma_1)$ which is the Mach angle α as given by (11.1.19). We can thus regard a Mach wave as the limiting case of a very weak oblique shock wave where the normal velocity component u_{n1} is just equal to the local sonic velocity a.

As a numerical example of the application of equation (11.5.13), the following table gives the calculated values of the angle of deflection θ for a range of values of β from 30° to 50° for the case of a supersonic gas stream with Mach number $Ma_1=2.0$. Note that the least value of β in this case is 30° since this is the value of the Mach angle at $Ma_1=2.0$.

$Ma_1=2.0$						
angle of shock wave	β	30°	35°	40°	45°	50°
angle of deflection	θ	0	5.7°	10.6°	14.8°	18.1°

If, for example, we wish to know the angle of the shock wave formed at the nose of a sharp-edged wedge whose semi-angle is 5° in a supersonic stream with incident Mach number 2.0, we have by interpolation from the above table $\beta=34.4°$.

The trigonometric expression (11.5.13) can be re-arranged in the following form

$$\frac{1}{Ma_1^2} = \sin^2\beta - \frac{(\gamma+1)}{2}\frac{\sin\beta\sin\theta}{\cos(\beta-\theta)}. \tag{11.5.14}$$

The pressure difference across an oblique shock wave may be expressed very easily in terms of the angles β and θ. From the continuity and momentum equations (11.5.1) and (11.5.2) we have

$$p_2 - p_1 = \rho_1 u_{n1}^2 - \rho_2 u_{n2}^2 = \rho_1 u_{n1}(u_{n1} - u_{n2})$$

or

$$\frac{p_2 - p_1}{\rho_1 u_1^2} = \frac{u_{n1}^2}{u_1^2} - \frac{u_{n1}u_{n2}}{u_1^2} = \sin^2\beta - \frac{u_{n1}u_{n2}}{u_1^2}.$$

Making use of (11.5.9) and (11.5.12), this may be expressed as

$$\frac{p_2 - p_1}{\rho_1 u_1^2} = \frac{2}{(\gamma+1)}\sin^2\beta - \frac{2}{(\gamma+1)}\frac{1}{Ma_1^2}. \tag{11.5.15}$$

Hence, substituting from (11.5.14), we have

$$\frac{p_2 - p_1}{\rho_1 u_1^2} = \frac{\sin\beta\sin\theta}{\cos(\beta-\theta)}. \tag{11.5.16}$$

11.6 Supersonic expansion waves

We have seen in the last section that supersonic flow round a concave corner is associated with the presence of an oblique shock wave. The question arises next as to the nature of the flow pattern of a supersonic stream round a *convex corner* as indicated in figure 100. It is clear that the flow involves a zone of *expansion*, as distinct from compression in the previous case. We cannot have a sudden jump involving expansion, for the reasons already given in § 11.4 in connection with the normal shock wave, and the only physical possibility is for the expansion to take place in a gradual and isentropic manner within a zone bounded by the Mach lines appropriate to the steady-flow velocity upstream from the corner and to the steady-flow velocity downstream. This type of flow pattern is known as a *Prandtl–Meyer expansion.*

To investigate the conditions which govern this type of flow we need to consider an infinitesimal increase in the normal velocity component du_n across a Mach line as indicated in figure 100. The corresponding infinitesimal deflection of the stream velocity vector will be $d\theta$ and from the

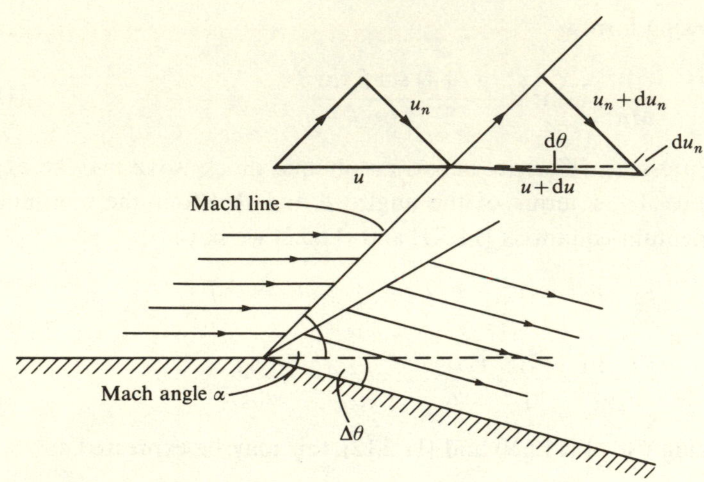

Fig. 100. Supersonic flow past a convex corner.

geometry of the diagram we can say that

$$d\theta = \frac{du_n}{u} \cos \alpha. \tag{11.6.1}$$

Noting also that $du = du_n \sin \alpha$ and that α is the Mach angle, so that $\sin \alpha = 1/Ma$, the expression for $d\theta$ becomes

$$d\theta = \frac{du}{u} \frac{\cos \alpha}{\sin \alpha} = \frac{du}{u} \frac{\sqrt{(1 - \sin^2 \alpha)}}{\sin \alpha} = \frac{du}{u} \sqrt{(Ma^2 - 1)}. \tag{11.6.2}$$

We can now make use of the expression (11.3.23) or (11.5.12), derived from the energy equation, in order to express the Mach number in terms of the velocity ratio u/a^* thus eliminating the variable a. From (11.3.23) we have

$$Ma^2 = \frac{2u^2}{(\gamma + 1)a^{*2} - (\gamma - 1)u^2} \tag{11.6.3}$$

or

$$Ma^2 - 1 = \frac{u^2 - a^{*2}}{a^{*2} - \lambda^2 u^2}, \tag{11.6.4}$$

where

$$\lambda^2 = \frac{\gamma - 1}{\gamma + 1}. \tag{11.6.5}$$

Note from (11.3.22) that $\lambda^2 = a^{*2}/u_c^2$. Hence from (11.6.2) and (11.6.4) we have

$$d\theta = \frac{du}{u} \sqrt{\left(\frac{u^2/a^{*2} - 1}{1 - \lambda^2 (u^2/a^{*2})}\right)}, \tag{11.6.6}$$

which is the basic differential equation governing supersonic flow around a convex corner in a zone of expansion.

In order to integrate (11.6.6) it is convenient to introduce the parameter $r=u^2/a^{*2}$ as the variable in place of the velocity u. After making this substitution, and after some algebraic manipulation, we obtain the result of the integration in the following form:

$$\theta=\frac{1}{2\lambda}\sin^{-1}\left\{\frac{2\lambda^2(u^2/a^{*2})-(1+\lambda^2)}{1-\lambda^2}\right\}$$
$$+\frac{1}{2}\sin^{-1}\left\{\frac{2(a^{*2}/u^2)-(1+\lambda^2)}{1-\lambda^2}\right\}. \tag{11.6.7}$$

If we now revert to the use of the ratio γ rather than the parameter λ, the result takes the form:

$$\theta=\frac{1}{2}\sqrt{\left(\frac{\gamma+1}{\gamma-1}\right)}\sin^{-1}\{(\gamma-1)(u^2/a^{*2})-\gamma\}$$
$$+\frac{1}{2}\sin^{-1}\{(\gamma+1)(a^{*2}/u^2)-\gamma\} \tag{11.6.8}$$

and if we substitute for u/a^* in terms of the Mach number Ma, using equation (11.3.23), we get

$$\theta=\frac{1}{2}\sqrt{\left(\frac{\gamma+1}{\gamma-1}\right)}\sin^{-1}\left\{\frac{(\gamma-1)Ma^2-2\gamma}{(\gamma-1)Ma^2+2}\right\}+\frac{1}{2}\sin^{-1}\left\{\frac{2}{Ma^2}-1\right\}. \tag{11.6.9}$$

The minimum velocity from which a supersonic expansion can commence is represented by $u=a^*$, or $Ma=1$. It is appropriate, therefore, to take a reference or datum angle θ_0 as the value of θ given by the above equations when $u=a^*$ or $Ma=1$. On this basis we can evaluate the difference of angle $(\theta-\theta_0)$ for the case of supersonic expansion from an initial condition where $Ma=1$, and this is plotted in figure 101 for air with $\gamma=1.40$. The *angular deflection* $\Delta\theta$ corresponding to an expansion from any value Ma_1 *to a higher value* Ma_2 can easily be obtained from this diagram.

The maximum angular deflection that is possible would correspond to expansion from $Ma=1$ to $Ma=\infty$ (or to $u=u_c=a^*/\lambda$). It is easily verified from (11.6.7) or (11.6.9) that the value of this maximum angular deflection for a supersonic airstream, expanding from an initial condition of $Ma=1$, is approximately 130°.

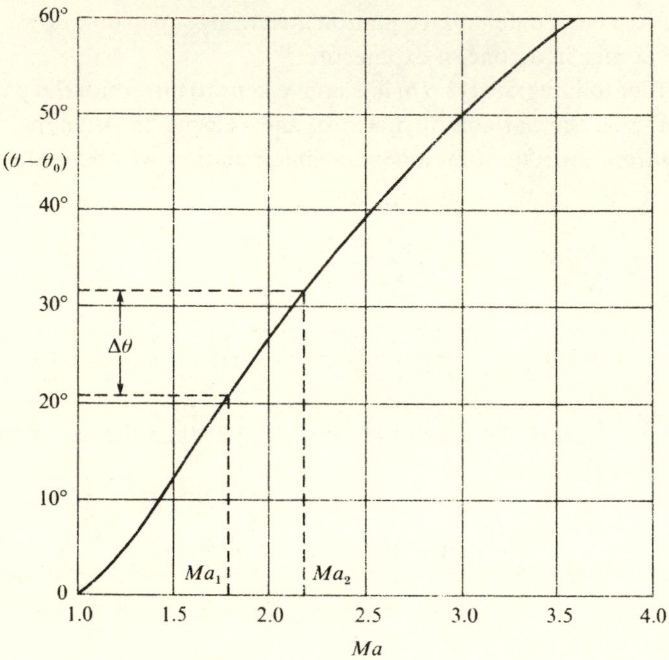

Fig. 101. The angular deflection $\Delta\theta$ as a function of Mach number Ma.

11.7 Supersonic flow patterns

The two basic phenomena of the oblique shock wave, outlined in §11.5, and the Prandtl–Meyer expansion zone outlined in §11.6, play a significant role in the flow patterns which are observed at supersonic velocities. As an example we can consider the case of flow past a thin diamond-shaped aerofoil section at zero angle of incidence as illustrated in figure 102.

Provided the nose angle is sufficiently small, the front shock wave will normally be attached to the nose as indicated in the figure. Flow around the convex corner at the mid-chord position is associated with the existence of a Prandtl–Meyer expansion zone. The return to parallel flow of the airstream behind the trailing edge is associated with the presence of another oblique shock wave as shown in the diagram.

Because of the different relative angles of the oblique shock waves and the Mach waves forming the boundaries of the Prandtl–Meyer expansion wedge, the Mach waves in the expansion zone will eventually intersect the shock waves at some lateral distance from the aerofoil. This leads to further

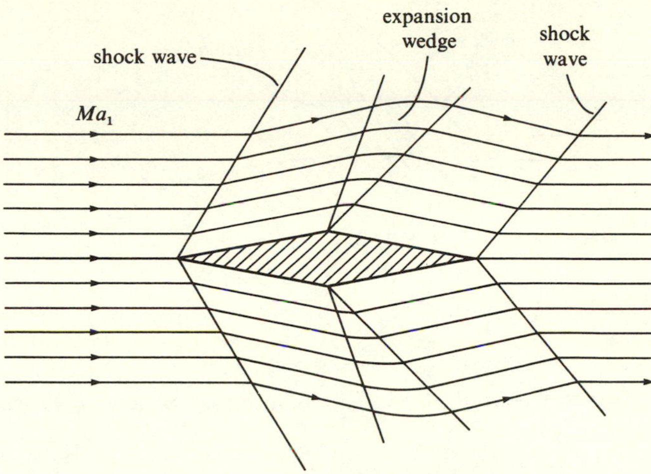

Fig. 102. Supersonic flow past a diamond-shaped aerofoil.

complications in the flow pattern with the shock waves becoming curved, while conditions in the outer regions of the expansion wedge cease to be isentropic.

If instead of the simple diamond shape shown in figure 102 we have a supersonic aerofoil of lenticular section, with curved upper and lower surfaces and a sharp leading edge, the Mach waves in the expansion zone will be spread over the entire curved surface of the aerofoil and will not be confined to a Prandtl–Meyer-type wedge. The front and rear shock waves will generally be curved as a result of their interaction with the Mach waves.

Since there is a limit to the angle of deflection which can be achieved in flow through an oblique shock wave, for any given value of the incident Mach number, as determined by equation (11.5.13), the front shock wave pattern will be different in the case of supersonic flow past a section having a large nose angle or past a blunt-nosed body. In these circumstances a detached shock wave is formed in the flow at a short distance upstream from the nose as indicated in figure 103. Around the nose of the body or aerofoil, immediately behind the detached shock wave, the local velocities must become subsonic, but elsewhere the flow will remain supersonic.

Supersonic flow past a thin flat plate, which is held in position in the air stream at a small angle of incidence ε, is shown diagrammatically in figure 104. We can regard this as a rudimentary example of a supersonic aerofoil and we can make a simple approximate calculation of the lift coefficient C_L on the assumption that the angle ε is sufficiently small.

The pressure p_{2b} acting on the under surface of the plate must be greater

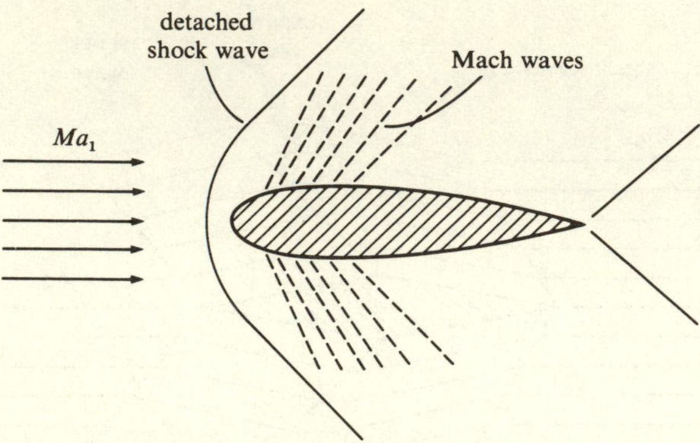

Fig. 103. Supersonic flow past a blunt-nosed body with detached shock wave.

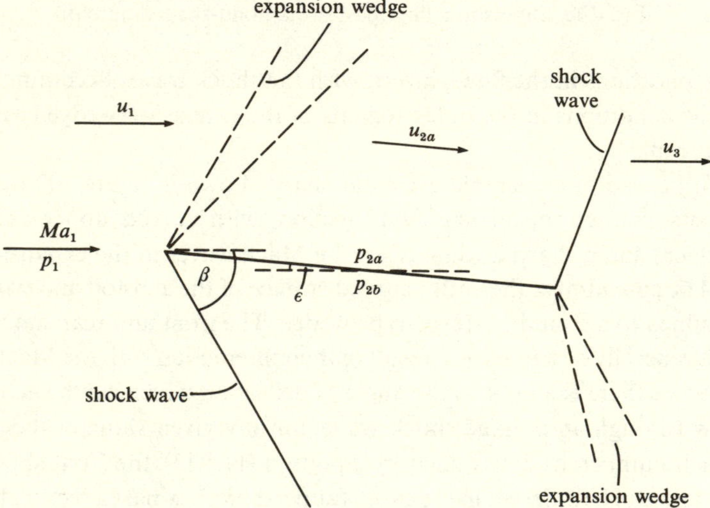

Fig. 104. Supersonic flow past a thin plate at a small angle of incidence.

than the value p_1 in the undisturbed stream as a result of the passage of the air through the oblique shock wave which extends downwards from the leading edge. Similarly the pressure p_{2a} acting on the upper surface must be less than p_1 as a result of passage through the expansion wedge which extends upwards from the leading edge. The total lift force will be equal to $(p_{2b}-p_{2a})c$ per unit span, where c is the chord length.

From (11.5.16), putting $\theta=\varepsilon$, we can say that

$$\frac{p_{2b}-p_1}{\rho_1 u_1^2}=\frac{\sin\beta\sin\varepsilon}{\cos(\beta-\varepsilon)}. \tag{11.7.1}$$

If ε is sufficiently small, this result may be expressed approximately as

$$\frac{p_{2b}-p_1}{\rho_1 u_1^2}=\varepsilon\tan\beta.$$

We can also say that for a very small angle of deflection the angle of inclination β of the shock wave will be only slightly greater than the Mach angle $\alpha=\sin^{-1}(1/Ma)$. Thus, approximately,

$$\frac{p_{2b}-p_1}{\rho_1 u_1^2}=\varepsilon\tan\alpha=\frac{\varepsilon}{\sqrt{(Ma^2-1)}}. \tag{11.7.2}$$

Using the method of analysis of § 11.6 for flow in the expansion zone, it can be shown that the pressure difference p_1-p_{2a} for a small deflection ε through an expansion wedge is given approximately by a similar expression

$$\frac{p_1-p_{2a}}{\rho_1 u_1^2}=\frac{\varepsilon}{\sqrt{(Ma^2-1)}}. \tag{11.7.3}$$

Hence we have, approximately,

$$\frac{p_{2b}-p_{2a}}{\rho_1 u_1^2}=2\varepsilon\tan\alpha=\frac{2\varepsilon}{\sqrt{(Ma^2-1)}} \tag{11.7.4}$$

or

$$C_L=\frac{p_{2b}-p_{2a}}{\frac{1}{2}\rho_1 u_1^2}=4\varepsilon\tan\alpha=\frac{4\varepsilon}{\sqrt{(Ma^2-1)}}. \tag{11.7.5}$$

11.8 Compressible flow through a convergent nozzle

We have already established the basic equations for the 'one-dimensional' flow of a perfect gas along a stream tube in § 11.3. We can make use of this simplified one-dimensional picture to investigate the flow of a gas in a nozzle whose cross-sectional flow area, at any point along the axis of the nozzle, is A. As already noted in § 11.3, it is convenient to picture the flow originating from a reservoir where the pressure and temperature are p_0 and T_0, respectively, and where the gas velocity is zero. The basic flow situation is indicated in figure 105.

The continuity equation takes the usual form for steady one-

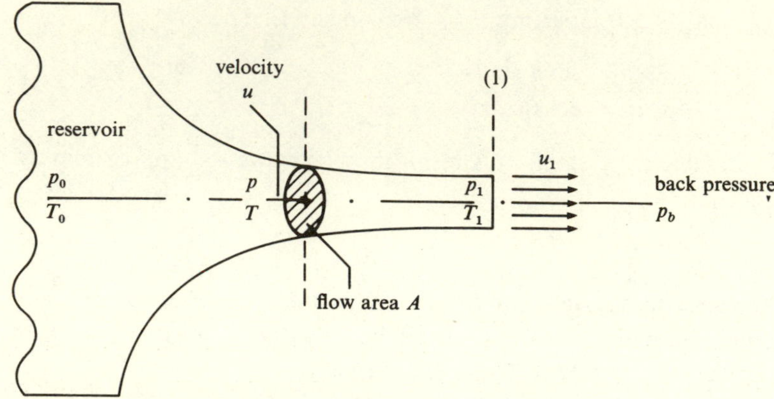

Fig. 105. Compressible flow through a convergent nozzle.

dimensional flow in a tube or duct:

$$\text{mass flow rate } m = \rho A u. \tag{11.8.1}$$

The steady-flow energy equation for a perfect gas under adiabatic (but not necessarily isentropic) conditions is simply

$$c_p T_0 = c_p T + \tfrac{1}{2}u^2, \tag{11.8.2}$$

as in (11.3.16). It may equally well be expressed in terms of the pressure p or the sonic velocity a as parameter as in (11.3.17) or (11.3.18). It is important to appreciate that the energy equation is valid even when viscous stresses or other irreversible effects such as shock waves are present in the flow. We are assuming only that there is no transfer of heat from the wall of the nozzle to the gas and that in this sense the flow is adiabatic.

If we now make the additional assumption that viscous effects can be neglected we can introduce Euler's equation of motion, as in (11.3.4), leading to the isentropic relationship (11.3.10), i.e.

$$\frac{p}{p_0} = \left(\frac{\rho}{\rho_0}\right)^{\gamma} = \left(\frac{T}{T_0}\right)^{\gamma/(\gamma-1)}. \tag{11.8.3}$$

From (11.8.2) and (11.8.3) we can say that for frictionless isentropic flow in the nozzle the velocity u at any point is related to the local value of the temperature T and pressure p by

$$\frac{u^2}{2} = c_p T_0\left(1 - \frac{T}{T_0}\right) = \frac{\gamma}{\gamma-1}\frac{p_0}{\rho_0}\left[1 - \left(\frac{p}{p_0}\right)^{(\gamma-1)/\gamma}\right]. \tag{11.8.4}$$

Substituting for ρ and u in the continuity equation we find that

$$m=\rho_0 a_0 AZ, \tag{11.8.5}$$

where $a_0=\sqrt{[\gamma(p_0/\rho_0)]}$ is the value of the sonic velocity in the reservoir, and Z is a function of the pressure ratio p/p_0 defined by

$$Z=\sqrt{\left(\frac{2}{\gamma-1}\right)}\left[\left(\frac{p}{p_0}\right)^{2/\gamma}-\left(\frac{p}{p_0}\right)^{(\gamma+1)/\gamma}\right]^{\frac{1}{2}}. \tag{11.8.6}$$

This function is plotted against p/p_0 in figure 106 for the case of $\gamma=1.40$.

It is of interest to consider the variation of the mass flow m with a progressive reduction in the back pressure p_b. Starting with a value of p_b equal to the reservoir pressure p_0, the mass flow rate will be zero. If the back pressure is now reduced, the mass flow will increase progressively and will be given by (11.8.5). It is easily verified that the quantity Z, and hence also the mass flow rate m, reaches a maximum at the critical pressure ratio given by

$$\frac{p}{p_0}=\left(\frac{2}{\gamma+1}\right)^{\gamma/(\gamma-1)}. \tag{11.8.7}$$

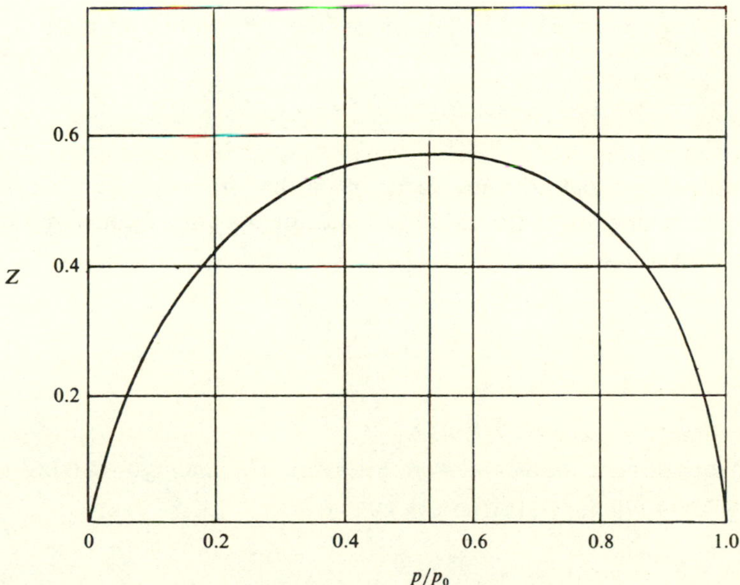

Fig. 106. Plot of function Z against pressure ratio p/p_0.

The corresponding exit velocity can be found from (11.8.4), i.e.

$$u^2 = \frac{2\gamma}{\gamma+1}\frac{p_0}{\rho_0} = \frac{2}{\gamma+1}a_0^2 = a^{*2}. \tag{11.8.8}$$

We thus see that the critical velocity, corresponding to the critical pressure ratio as given by (11.8.7), is the sonic velocity a^*. The same conclusion can be reached if we take the energy equation in the form of (11.3.19) together with the isentropic relationship (11.8.3), i.e.

$$\frac{p_0}{p} = \left[1 + \frac{\gamma-1}{2}Ma^2\right]^{\gamma/(\gamma-1)}. \tag{11.8.9}$$

Putting $Ma = 1$ in (11.8.9) gives

$$\frac{p_0}{p} = \left(\frac{\gamma+1}{2}\right)^{\gamma/(\gamma-1)} \quad \text{or} \quad \frac{p}{p_0} = \left(\frac{2}{\gamma+1}\right)^{\gamma/(\gamma-1)}.$$

Further reduction of the back pressure below the critical value will not result in any increase in the mass-flow rate since a small pressure signal cannot be transmitted upstream when the fluid velocity attains the sonic value. Under these conditions the nozzle is said to be *choked*, the gas velocity remains equal to a^* at section (1) and the stream emerges from the nozzle as a free jet expanding into a region of lower pressure p_b.

If the nozzle is choked, the mass-flow rate will be given by

$$m = \rho_1 A_1 u_1 = \rho_0 A_1 a^* \left(\frac{\rho_1}{\rho_0}\right) = \rho_0 A_1 a^* \left(\frac{p_c}{p_0}\right)^{1/\gamma},$$

where A_1 is the exit area and p_c/p_0 is the critical pressure ratio. Hence, substituting for a^* from (11.3.20) and for the critical pressure ratio from (11.8.7), we have

$$m = \rho_0 A_1 a_0 \sqrt{\left(\frac{2}{\gamma+1}\right)}\left(\frac{2}{\gamma+1}\right)^{1/(\gamma-1)} = \rho_0 A_1 a_0 \left(\frac{2}{\gamma+1}\right)^{(\gamma+1)/[2(\gamma-1)]}. \tag{11.8.10}$$

We can obtain the mass-flow rate in terms of p_0 and ρ_0 by making use of the relationship $a_0 = \sqrt{[\gamma(p_0/\rho_0)]}$. Hence

$$m = \gamma^{\frac{1}{2}}\left(\frac{2}{\gamma+1}\right)^{(\gamma+1)/[2(\gamma-1)]}\sqrt{(p_0\rho_0)}A_1. \tag{11.8.11}$$

For *air*, with $\gamma = 1.40$, equation (11.8.11) becomes

$$m = 0.684\sqrt{(p_0\rho_0)}A_1.$$

For *superheated steam* we can take $\gamma = 1.30$, in which case

$$m = 0.668\sqrt{(p_0\rho_0)}A_1.$$

11.9 Compressible flow through a convergent–divergent nozzle

Compressible flow through a de Laval convergent–divergent nozzle is shown diagrammatically in figure 107. Equations (11.8.4), (11.8.5), and (11.8.6) will be applicable provided the flow is frictionless and isentropic.

We can start by considering the situation when the critical pressure is reached at the throat of the nozzle, denoted by section (1), and the gas

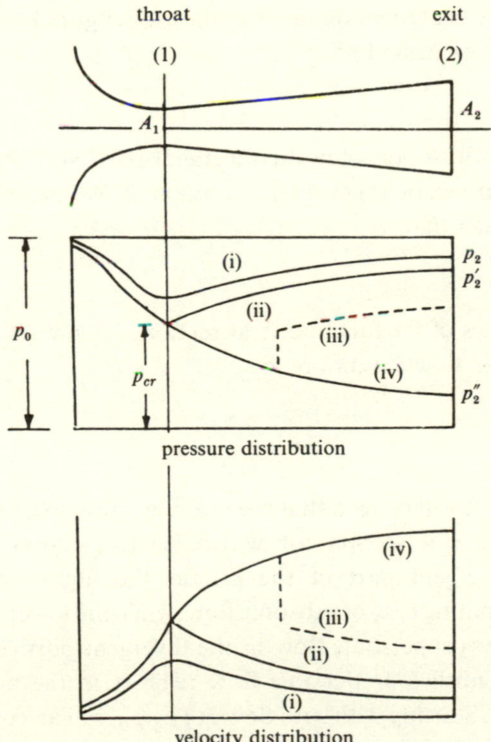

Fig. 107. Pressure and velocity distributions in a convergent–divergent nozzle.

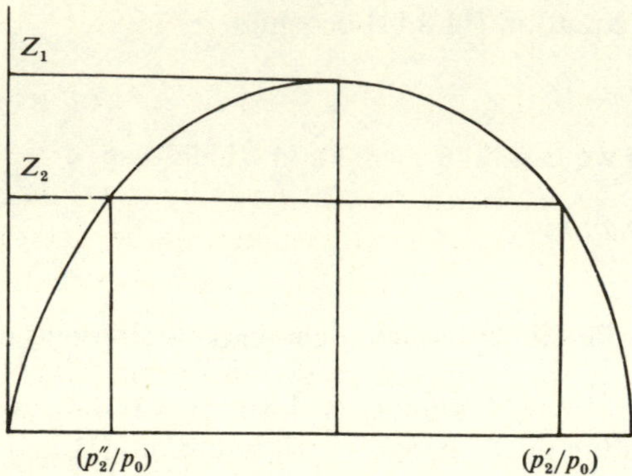

Fig. 108. Continuity requirement for a convergent–divergent nozzle.

velocity $u=a^*$ at this point. The mass-flow rate will be given by (11.8.5) as in the case of the convergent nozzle. With critical conditions prevailing at the throat the value of the function Z will be the maximum value Z_1 corresponding to the critical pressure ratio as shown in figure 108 and the mass-flow rate may be expressed as

$$m=\rho_0 a_0 A_1 Z_1. \tag{11.9.1}$$

If isentropic flow conditions prevail in the divergent portion of the nozzle beyond the throat from section (1) to the exit section (2), we can still apply equation (11.8.5) and say that

$$m=\rho_0 a_0 A_2 Z_2. \tag{11.9.2}$$

It follows that the values of the function Z at sections (1) and (2) must be related under isentropic flow conditions by

$$\frac{Z_1}{Z_2}=\frac{A_2}{A_1}. \tag{11.9.3}$$

Referring to figure 108, it will be seen that there are two possible values for the pressure p_2 at the exit section for which isentropic flow can be maintained in the divergent part of the nozzle, the higher value p_2' corresponding to the limiting case of subsonic flow, while the lower value p_2'' corresponds to shockless supersonic flow in the divergent portion.

We can now distinguish four possible flow regimes in the nozzle as illustrated in figure 107. Starting with zero flow at $p_b=p_0$, we can consider a progressive reduction in the back-pressure as follows:

(i) If $p_b>p_2'$ we have subsonic flow throughout the nozzle and the mass-flow rate is given by (11.8.5) putting $p = p_b$ in (11.8.6) for the function Z.

(ii) If $p_b = p'_2$ the flow will just reach sonic velocity a^* at the throat but will immediately revert to subsonic conditions in the divergent portion. The mass flow will now be given by equation (11.8.10).

(iii) If $p'_2 > p_b > p''_2$ the flow will at first be supersonic beyond the throat, but a shock wave or pattern of shock waves will exist somewhere, either in the divergent portion of the nozzle or at some point beyond the exit. The simplest case of a normal shock wave is indicated in the figure. Subsonic flow will be re-established downstream from the shock wave. The mass-flow rate will be unchanged and will still be given by (11.8.10).

(iv) If $p_b = p''_2$ velocities will be supersonic throughout the divergent portion of the nozzle and the flow will be shockless and isentropic. Further reduction of the back-pressure below the value p''_2 will have no effect on the flow rate.

12

Flow with a free surface

12.1 Surface waves due to gravity

Wave motion on the surface of a liquid is a complex phenomenon but we can investigate some of the more important features of surface gravity waves quite easily if we consider a *two-dimensional* flow picture as indicated in figure 109. We assume that the channel or ocean floor is flat and that the depth of the undisturbed water is h. We assume also that the wave motion is linear and that parallel waves are propagated in the direction of the horizontal x axis. The z co-ordinate is measured positive upwards in the vertical direction, i.e. the undisturbed surface is defined by $z=0$, and the ocean floor by $z=-h$.

We will take a constant value for the density ρ of the fluid and will neglect the effects of viscosity. The unsteady motion of the fluid must be governed by the *equation of continuity* and by *Euler's equation of motion.* The equation of continuity for unsteady incompressible two-dimensional flow (expressed in terms of the co-ordinates x and z) takes the form div $\mathbf{u}=0$, or

$$\frac{\partial u}{\partial x}+\frac{\partial w}{\partial z}=0. \tag{12.1.1}$$

It is reasonable to assume that the flow will be *irrotational* and the appropriate form of Euler's equation, covering unsteady flow conditions, is therefore given by (4.2.11), i.e.

$$\frac{\partial \mathbf{u}}{\partial t}=-\nabla\left(\frac{p}{\rho}+\frac{1}{2}\mathbf{u}^2+gz\right). \tag{12.1.2}$$

With irrotational flow conditions we can introduce a velocity potential as in (4.3.1) such that $\mathbf{u}=-\nabla\phi$. Hence the relevant form of Euler's equation becomes

$$-\nabla\left(\frac{\partial\phi}{\partial t}\right)=-\nabla\left(\frac{p}{\rho}+\frac{1}{2}\mathbf{u}^2+gz\right)$$

or

$$\frac{p}{\rho}+\frac{1}{2}\mathbf{u}^2+gz=+\frac{\partial\phi}{\partial t}. \tag{12.1.3}$$

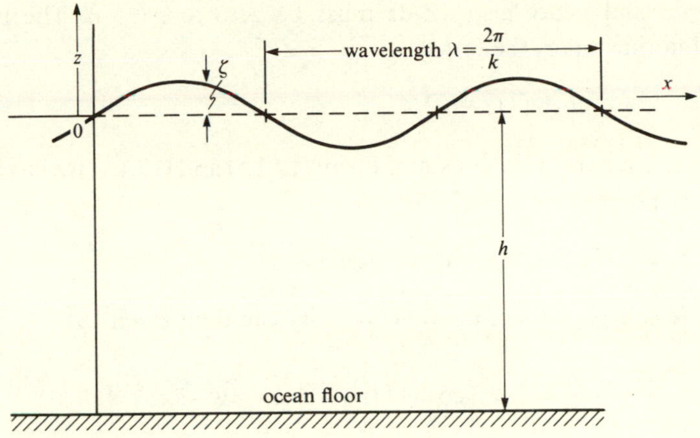

Fig. 109. Two-dimensional surface waves.

It also follows that we can express the equation of continuity in the form of Laplace's equation $\nabla^2\phi = 0$ when we introduce the velocity potential. Thus for two-dimensional flow in the x, z, co-ordinate system we have in place of (12.1.1):

$$\frac{\partial^2 \phi}{\partial x^2} + \frac{\partial^2 \phi}{\partial z^2} = 0. \tag{12.1.4}$$

We can now anticipate a solution to equation (12.1.4), which will describe wave motion in the xz plane, by taking

$$\phi = Z \cos(\omega t - kx), \tag{12.1.5}$$

where Z is a function of z only; ω is the radian frequency of the wave motion; and k is the wave number $= 2\pi/\lambda$, where λ is the wave length. The required form of the function Z is easily found by differentiating (12.1.5) and substituting in (12.1.4), hence we have the differential equation for Z:

$$\frac{\mathrm{d}^2 Z}{\mathrm{d}z^2} - k^2 Z = 0. \tag{12.1.6}$$

The general form of the solution to this equation is

$$Z = A\,\mathrm{e}^{kz} + B\,\mathrm{e}^{-kz}.$$

However, we have to satisfy the boundary condition that the vertical component of the fluid velocity must be zero when $z = -h$. This means that

$-\partial\phi/\partial z$, and hence also $\mathrm{d}Z/\mathrm{d}z$ must be zero at $z=-h$. The required solution thus takes the form:

$$Z=C\cosh\left[k(z+h)\right], \tag{12.1.7}$$

where C is an arbitrary constant. From (12.1.5) and (12.1.7) we have for the velocity potential ϕ:

$$\phi=C\cosh\left[k(z+h)\right]\cos\left(\omega t-kx\right) \tag{12.1.8}$$

and the components of the fluid velocity are then given by

$$u=-\frac{\partial\phi}{\partial x}=-kC\cosh\left[k(z+h)\right]\sin\left(\omega t-kx\right) \tag{12.1.9}$$

and

$$w=-\frac{\partial\phi}{\partial z}=-kC\sinh\left[k(z+h)\right]\cos\left(\omega t-kx\right). \tag{12.1.10}$$

We can now turn attention to Euler's equation (12.1.3) and to the conditions that have to be satisfied at the free surface. As indicated in figure 109, the vertical displacement of the free surface (measured from the plane of the undisturbed surface $z=0$) is represented by the variable ζ. Note that ζ is a function of both x and t. The rate of change of ζ at any position x must be equal to the instantaneous value of the vertical component of the fluid velocity at the surface. Hence, for any value of x, we require that

$$\frac{\partial\zeta}{\partial t}=-\left(\frac{\partial\phi}{\partial z}\right)_{z=\zeta}. \tag{12.1.11}$$

We cannot easily evaluate $\partial\phi/\partial z$ at $z=\zeta$ since ζ is a variable. It is reasonable to assume, however, that the amplitude of the waves will be relatively small when viewed on the overall vertical scale and that we can evaluate $\partial\phi/\partial z$ at $z=0$ with little error. We can therefore replace the exact statement (12.1.11) with the approximate statement:

$$\frac{\partial\zeta}{\partial t}=-\left(\frac{\partial\phi}{\partial z}\right)_{z=0}. \tag{12.1.12}$$

We can obtain an expression for $\partial\zeta/\partial t$ from the Euler equation of motion (12.1.3) by putting $z=\zeta$ and differentiating with respect to t. We note that, at the free surface where $z=\zeta$, the pressure must be constant and equal to the prevailing atmospheric pressure p_0. We may also reasonably assume that the vertical component of the fluid velocity at the surface will be small and that $\mathbf{u}^2$ will consequently be negligible. Euler's equation at the free surface, therefore, may be expressed as:

$$\frac{p_0}{\rho}+g\zeta=\left(\frac{\partial\phi}{\partial t}\right)_{z=\zeta}\approx\left(\frac{\partial\phi}{\partial t}\right)_{z=0}. \tag{12.1.13}$$

Hence, differentiating (12.1.13) with respect to time, we have

$$\frac{\partial\zeta}{\partial t}=\frac{1}{g}\left(\frac{\partial^2\phi}{\partial t^2}\right)_{z=0}. \tag{12.1.14}$$

Thus from (12.1.12) and (12.1.14) the essential free surface condition may be expressed as

$$\frac{\partial^2\phi}{\partial t^2}=-g\frac{\partial\phi}{\partial z}\quad\text{at } z=0. \tag{12.1.15}$$

From (12.1.8) we have at $z=0$

$$\frac{\partial^2\phi}{\partial t^2}=-\omega^2C\cosh\,[kh]\cos\,(\omega t-kx) \tag{12.1.16}$$

and from (12.1.10) at $z=0$

$$\frac{\partial\phi}{\partial z}=kC\sinh\,[kh]\cos\,(\omega t-kx). \tag{12.1.17}$$

Hence, substituting in (12.1.15), we find

$$\omega^2\cosh\,(kh)=gk\sinh\,(kh)$$

or

$$\omega^2=gk\tanh\,(kh) \tag{12.1.18}$$

which expresses the basic relationship between the depth h, the wave number k, and the radian frequency ω for surface gravity waves.

The *wave speed* c is given by

$$c=\frac{\omega}{k}=\left[\frac{g}{k}\tanh\,(kh)\right]^{\frac{1}{2}}. \tag{12.1.19}$$

For large values of kh, $\tanh\,(kh)\rightarrow 1.0$. If, for example, the depth h is greater than a wave length we have $h>2\pi/k$ or $kh>2\pi$ and the value of $\tanh\,(kh)$ is then indistinguishable from 1.0. Thus for waves on *deep water*:

$$c=\left(\frac{g}{k}\right)^{\frac{1}{2}}\quad\text{or}\quad\omega=(gk)^{\frac{1}{2}}. \tag{12.1.20}$$

For *long waves* and *shallow water*, however, where kh is small, we have $\tanh(kh) \to kh$ and hence from (12.1.19):

$$c = (gh)^{\frac{1}{2}} \quad \text{or} \quad \omega = k(gh)^{\frac{1}{2}}. \tag{12.1.21}$$

As a numerical example, if we take $\lambda = 10$ m and $h = 20$ m we have $k = 2\pi/\lambda = 0.628$, $kh = 12.56$, $\tanh(kh) = 1.0$, and hence from (12.1.20) $c = 3.95$ m s^{-1}. If, on the other hand, we take $\lambda = 20$ m and $h = 0.5$ m we have $k = 0.314$, $kh = 0.157$, $\tanh(kh) = 0.156$ and hence from (12.1.19) or (12.1.21) $c = 2.21$ m s^{-1}.

We have seen from (12.1.19) that the wave speed c is in general a function of the wave number k or wavelength λ. In the case of deep water waves we have $c = (g\lambda/2\pi)^{\frac{1}{2}}$, i.e. long waves travel faster than shorter waves. This feature is known as *dispersal*. In the special case of long waves in shallow water, however, it will be noted from (12.1.21) that the wave speed is independent of the wave length and we can describe such a system as being non-dispersive.

It can be shown that the energy of sinusoidal waves is propagated at a velocity, known as the *group velocity*, which differs in the case of a dispersive system from the wave speed c. The group velocity U is given by the relationship:

$$U = \frac{\mathrm{d}\omega}{\mathrm{d}k}. \tag{12.1.22}$$

Hence for waves on *deep water* we have the result from (12.1.20) that $U = \frac{1}{2}c$, i.e. the group velocity is half the wave speed. For long waves in shallow water, however, we see from (12.1.21) that $U = c$, which is the normal state of affairs for a non-dispersive system.

12.2 Flow in open channels

The flow of a liquid in an open channel is indicated in figure 110. The channel floor is assumed to be flat and is inclined at angle α to the horizontal. The co-ordinate z is measured as positive upwards in the vertical direction following our normal convention. The co-ordinate x, however, is measured parallel to the channel floor in the direction of flow and *not* in the usual horizontal direction. The co-ordinate y is measured at right angles to the channel floor. The depth of the stream is shown as h measured at right angles to the channel floor. We will assume steady-flow conditions, i.e. the fluid velocity measured at a fixed point in space will not vary with time. We must distinguish, however, between *uniform flow* where

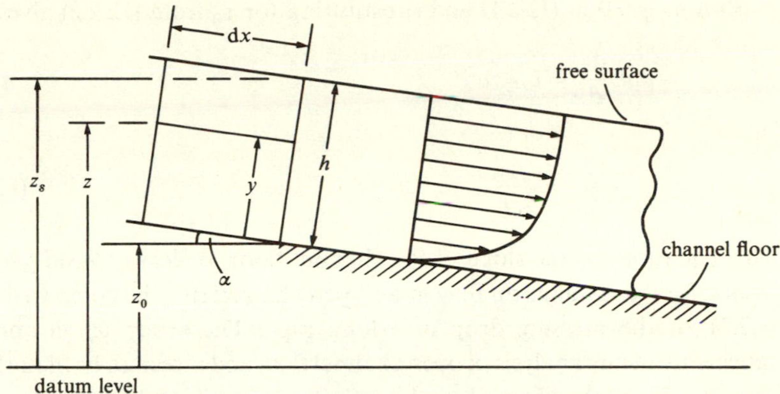

Fig. 110. Uniform flow in an open channel.

the depth h remains constant and the free surface is parallel to the channel floor, and *non-uniform flow* where the depth h varies gradually with the distance x measured in the direction of flow.

We will consider first of all the case of uniform flow in a very wide channel. We will treat the problem as a two-dimensional-flow situation and will neglect the effects of the side walls of the channel on the flow pattern. Referring to figure 110, we can say that for the equilibrium under steady-flow conditions of a layer of fluid having length dx, depth $(h-y)$ measured from the free surface, and unit width,

$$\tau \, \mathrm{d}x = \rho g (h-y) \sin \alpha \, \mathrm{d}x$$

where τ is the shear stress at distance y from the channel floor. Hence

$$\tau = -\rho g (h-y) \frac{\mathrm{d}z_0}{\mathrm{d}x}. \tag{12.2.1}$$

The shear stress will have a maximum value τ_0 at the channel floor where $y=0$ and we can define a skin-friction coefficient by

$$c_f = \frac{\tau_0}{\frac{1}{2}\rho u_m^2}, \tag{12.2.2}$$

where u_m is the mean velocity, i.e. the volume flow rate divided by the cross-sectional flow area.

It may be noted that the definition (12.2.2) is similar to (8.2.1) for flow in a pipe and that the treatment of uniform flow in an open channel is broadly similar to the case of fully developed internal flow in a pipe or duct.

Putting $y=0$ in (12.2.1) and substituting for τ_0 from (12.2.2) gives:

$$-\rho g h \frac{\mathrm{d}z_0}{\mathrm{d}x} = c_f \tfrac{1}{2} \rho u_m^2$$

or

$$-\frac{\mathrm{d}z_0}{\mathrm{d}x} = \frac{c_f u_m^2}{2gh}. \tag{12.2.3}$$

This equation for the slope of the channel floor, or *loss of head per unit length*, for steady uniform flow in an open channel may be compared with (8.2.3) for the pressure drop in a long pipe. The group u_m^2/gh appears repeatedly in any analysis of open channel flow and is seen to be in the form of a Froude number if we adopt the definition given in (1.8.6). Alternatively, if we adopt the definition of the Froude number as a velocity ratio, as in (1.8.7), we could re-write (12.2.3) as

$$-\frac{\mathrm{d}z_0}{\mathrm{d}x} = \tfrac{1}{2} c_f Fr^2 \quad \text{where } Fr = \frac{u_m}{\sqrt{(gh)}}. \tag{12.2.4}$$

If we now turn attention to *non-uniform flow* we can still make use of figure 110 but the depth h will be a function of the distance x measured along the length of the channel. Provided the variation of depth h with distance x is gradual, however, the quantity $(p/\rho g + z)$ will be constant over a cross-section of the stream as in the normal hydrostatic situation. The total head H at level y, where the pressure is p and the velocity is u, will be given by

$$H = \frac{p}{\rho g} + \frac{u^2}{2g} + z = \frac{u^2}{2g} + z_s \tag{12.2.5}$$

if we take the atmospheric pressure to be zero at the free surface, i.e. if we measure p as gauge pressure. From (12.2.5) we can say that

$$H = \frac{u^2}{2g} + z_0 + h \cos \alpha \approx \frac{u^2}{2g} + z_0 + h, \tag{12.2.6}$$

provided α is small and $\cos \alpha \approx 1$. The mean value of the total head H_m is then given by

$$H_m = \frac{\tilde{u}_m^2}{2g} + z_0 + h, \tag{12.2.7}$$

where $\tilde{u}_m^2$ is the mean value of u^2. Referring to equation (12.2.7), the quantity $H_m - z_0$ is known as the specific head. The *specific energy*, E_s, is accordingly defined by

$$E_s = \frac{\tilde{u}_m^2}{2} + gh. \tag{12.2.8}$$

If we differentiate (12.2.7) with respect to distance x, and assuming that the ordinary mean velocity u_m may be used in place of the rms value $\tilde{u}_m$, we have

$$\frac{dH_m}{dx} = \frac{u_m}{g}\frac{du_m}{dx} + \frac{dz_0}{dx} + \frac{dh}{dx}$$

and this must be equal to the loss of head per unit length due to friction. Hence, extending the result which was obtained for uniform flow in (12.2.3), we can say that

$$\frac{u_m}{g}\frac{du_m}{dx} + \frac{dz_0}{dx} + \frac{dh}{dx} = -\frac{c_f u_m^2}{2gh}. \tag{12.2.9}$$

Note that if both the mean velocity u_m and the depth of the stream h remain constant with respect to distance x we revert to the case of uniform flow and equation (12.2.9) then becomes identical with (12.2.3).

The continuity equation requires that the volume flow must be constant under steady-flow conditions, i.e.

$$Q = bhu_m = \text{constant}, \tag{12.2.10}$$

where Q is the total volume flow rate and b is the width of the channel. Hence differentiating (12.2.10) and assuming a constant width of channel, we have the requirement that

$$\frac{dh}{h} + \frac{du_m}{u_m} = 0, \tag{12.2.11}$$

i.e.

$$\frac{u_m}{g}\frac{du_m}{dx} = -\frac{u_m^2}{gh}\frac{dh}{dx}.$$

Substituting in (12.2.9) we get

$$\left(1 - \frac{u_m^2}{gh}\right)\frac{dh}{dx} = -\frac{dz_0}{dx} - \frac{c_f u_m^2}{2gh}. \tag{12.2.12}$$

It will be noted that the numerical value of dh/dx *will become large if the velocity* u_m approaches the value $\sqrt{(gh)}$. As we have already seen in § 12.1, the quantity $\sqrt{(gh)}$ represents the speed of propagation of a surface wave whose wave length is large compared with the depth h of the fluid in the channel. We can regard $\sqrt{(gh)}$ as a critical velocity for flow in an open channel which is analogous in certain respects to the sonic velocity with the compressible flow of a gas in a nozzle. The velocity ratio or Froude number $u_m/\sqrt{(gh)}$ may be compared with the Mach number u/a. If the velocity u_m is less than $\sqrt{(gh)}$ we describe the flow as *tranquil*. If u_m is greater than $\sqrt{(gh)}$ we describe the flow as *rapid*. It must be appreciated, however, that the depth h is itself a variable and it is convenient to define the *critical velocity* u_m^* and the *critical depth* h^* by the condition that

$$u_m^* = \sqrt{(gh^*)}. \tag{12.2.13}$$

From the equation of continuity (12.2.10) we have

$$\frac{Q}{b} = hu_m = h^*u_m^*$$

and we thus find that:

$$u_m^* = \left(\frac{Qg}{b}\right)^{\frac{1}{3}} \tag{12.2.14}$$

and

$$h^* = \left(\frac{Q}{b\sqrt{g}}\right)^{\frac{2}{3}}. \tag{12.2.15}$$

Note that u_m^* and h^* depend only on the volume flow rate per unit width.

In the special case when $\mathrm{d}h/\mathrm{d}x = 0$ equation (12.2.12) becomes identical to (12.2.3) and we have the condition for uniform flow in the channel. We can thus say that, for any given value for the slope of the channel floor $s_0 = -\mathrm{d}z_0/\mathrm{d}x$, and for a fixed volume flow rate per unit width of channel, there must be a *normal depth* h_0 corresponding to uniform flow conditions together with a normal velocity u_{m0}. From (12.2.3) we see that

$$h_0 = \frac{c_f u_{m0}^2}{2gs_0}. \tag{12.2.16}$$

Making use of the equation of continuity $Q/b = h_0 u_{m0}$ we find that:

$$u_{m0} = \left(\frac{2gs_0}{c_f}\frac{Q}{b}\right)^{\frac{1}{3}} \tag{12.2.17}$$

and

$$h_0 = \left(\frac{c_f Q^2}{2gs_0 b^2}\right)^{\frac{1}{3}}. \tag{12.2.18}$$

It will be seen that the normal depth h_0 increases as the slope s_0 decreases. From (12.2.15) and (12.2.18) we have

$$\frac{h_0}{h^*}=\left(\frac{c_f}{2s_0}\right)^{\frac{1}{3}}. \tag{12.2.19}$$

If $h_0>h^*$, i.e. if $s_0<\frac{1}{2}c_f$, we describe the slope as *mild*.

If $h_0<h^*$, i.e. if $s_0>\frac{1}{2}c_f$, we describe the slope as *steep*.

We can now re-write equation (12.2.12), making use of equations (12.2.13) and (12.2.16), to give the result

$$\left(1-\frac{u_m^2}{u_m^{*2}}\frac{h^*}{h}\right)\frac{\mathrm{d}h}{\mathrm{d}x}=s_0\left(1-\frac{u_m^2}{u_{m0}^2}\frac{h_0}{h}\right).$$

Hence, using the equation of continuity (12.2.10), we have

$$\left(1-\frac{h^{*3}}{h^3}\right)\frac{\mathrm{d}h}{\mathrm{d}x}=s_0\left(1-\frac{h_0^3}{h^3}\right), \tag{12.2.20}$$

where s_0 is the slope of the channel floor $=-\mathrm{d}z_0/\mathrm{d}x$. Equation (12.2.20) may be expressed in the alternative form:

$$\frac{\mathrm{d}h}{\mathrm{d}x}=s_0\frac{(h^3-h_0^3)}{(h^3-h^{*3})}. \tag{12.2.21}$$

There are six cases to be considered:

mild slope	$(h_0>h^*)$	$h>h_0$	tranquil flow	$\frac{\mathrm{d}h}{\mathrm{d}x}$ positive
		$h_0>h>h^*$	tranquil flow	$\frac{\mathrm{d}h}{\mathrm{d}x}$ negative
		$h^*>h$	rapid flow	$\frac{\mathrm{d}h}{\mathrm{d}x}$ positive
steep slope	$(h_0<h^*)$	$h>h^*$	tranquil flow	$\frac{\mathrm{d}h}{\mathrm{d}x}$ positive
		$h^*>h>h_0$	rapid flow	$\frac{\mathrm{d}h}{\mathrm{d}x}$ negative
		$h_0>h$	rapid flow	$\frac{\mathrm{d}h}{\mathrm{d}x}$ positive

Some typical surface profiles are indicated in figure 111.

If we write the expression for the specific energy (12.2.8) using the mean velocity u_m in place of the rms value, and if we express u_m in terms of Q using

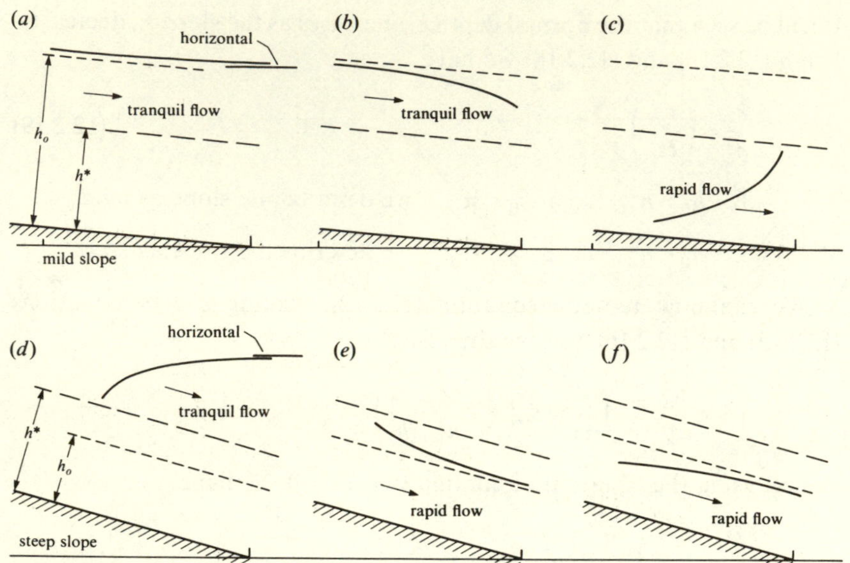

Fig. 111. Surface profiles for non-uniform flow in an open channel.

(12.2.10), we obtain

$$E_s = \frac{Q^2}{2b^2h^2} + gh. \tag{12.2.22}$$

Differentiating this equation with respect to h we find that the minimum specific energy occurs when $h = (Q^2/b^2g)^{\frac{1}{3}}$. This is the critical depth h^* as given by (12.2.15). Making use of (12.2.15), we can re-write (12.2.22) in the form

$$\frac{E_s}{gh^*} = \frac{1}{2}\left(\frac{h^*}{h}\right)^2 + \frac{h}{h^*}. \tag{12.2.23}$$

This relationship is plotted as a specific energy diagram in figure 112. The minimum value of the specific energy at $h = h^*$ is given by

$$E_{s_{min}} = \tfrac{3}{2}gh^*. \tag{12.2.24}$$

12.3 Flow over a weir

Flow of water from a large pond or reservoir over a broad-crested weir is illustrated diagrammatically in figure 113. We will assume that the width of the weir is large and, for simplicity, we will neglect any side-wall effects and will treat the problem as a case of two-dimensional flow.

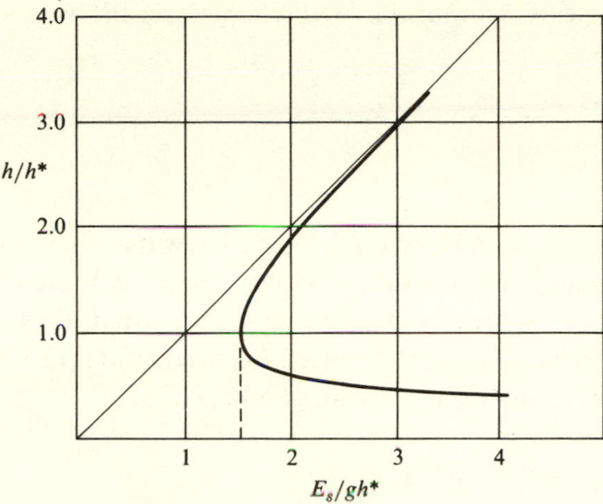

Fig. 112. Specific energy diagram.

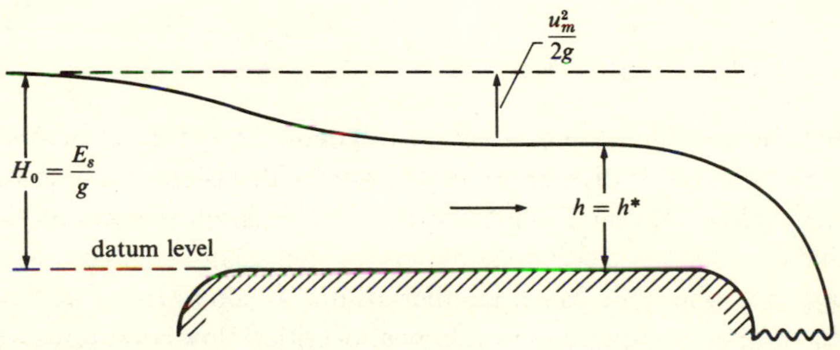

Fig. 113. Flow over a broad-crested weir.

If we neglect frictional effects, the specific energy of the stream will be equal to gH_0 where H_0 is the height of the free surface in the reservoir measured above the crest of the weir which is taken as the datum level. We can use equation (12.2.22) to express the specific energy in terms of the volume flow rate Q and the depth of the stream h, i.e.

$$gH_0 = E_s = \frac{Q^2}{2b^2h^2} + gh \tag{12.3.1}$$

or

$$\frac{Q}{b} = \surd 2h(E_s - gh)^{\frac{1}{2}}. \tag{12.3.2}$$

For maximum discharge at constant specific energy we must have $\mathrm{d}Q/\mathrm{d}h = 0$, i.e.

$$(E_s - gh)^{\frac{1}{2}} - \frac{gh}{2}(E_s - gh)^{-\frac{1}{2}} = 0.$$

Thus the stream depth h for maximum discharge over the weir is given by

$$E_s = \tfrac{3}{2}gh \tag{12.3.3}$$

or

$$h = \frac{2}{3}\frac{E_s}{g} = \frac{2}{3}H_0. \tag{12.3.4}$$

Comparison between (12.3.3) and (12.2.24) will show that the critical depth for maximum discharge at constant specific energy is exactly the same as the critical depth h^* for minimum specific energy at constant flow rate.

Putting the critical value $h = h^*$ from (12.3.4) in equation (12.3.2) gives us an expression for the maximum discharge rate, i.e.

$$\frac{Q_{max}}{b} = \sqrt{g}\, h^{*\frac{3}{2}} = \sqrt{g}\,(\tfrac{2}{3}H_0)^{\frac{3}{2}}. \tag{12.3.5}$$

It will also be noted that

$$u_{m_{max}} = \frac{Q_{max}}{bh^*} = \sqrt{(gh^*)} = u_m^*.$$

We thus conclude that both the critical depth h^* and the critical velocity u_m^* are established at discharge over the weir, with a corresponding value $Fr = u_m/\sqrt{(gh)} = 1$ for the Froude number. The situation is essentially similar to that of the flow of a compressible gas through a choked nozzle, as discussed in §11.8, where the Mach number attains a value $Ma = 1$ at the throat.

Another example of the attainment of critical flow conditions is shown in figure 114 where we have a long channel of mild slope ending in a fall. With tranquil flow conditions in the channel we will have $h_0 > h > h^*$ and accelerating flow with a surface profile corresponding to figure 111(*b*). The

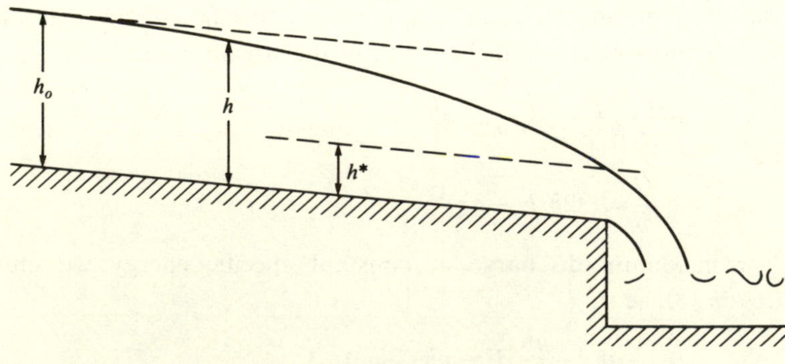

Fig. 114. Flow over a fall at the end of a channel.

flow rate will adjust itself so that the critical depth h^* and critical velocity u_m^* are attained at the discharge end of the channel.

12.4 Hydraulic jumps

In certain circumstances a rapid flowing stream, whose velocity by definition is greater than the critical value u^* and whose depth is less than the critical depth h^*, can change suddenly into a tranquil flowing stream as indicated in figure 115. The sudden change or discontinuity in the surface profile is known as the *hydraulic jump* and the phenomenon is in some ways analogous to the presence of a shock wave in the supersonic flow of a compressible gas.

We can make a simple analysis of two-dimensional flow through a stationary hydraulic jump under steady-flow conditions using the basic equations of continuity and momentum. Referring to figure 115, the following conditions must apply:

Continuity:
$$\frac{Q}{b} = h_1 u_1 = h_2 u_2. \tag{12.4.1}$$

Momentum: taking a control surface extending from section (1) to section (2) and coinciding with the boundaries of the fluid, force on section (1) due to fluid pressure $= [\rho g(h_1/2)]bh_1 = \rho g b(h_1^2/2)$; force on section (2) $= \rho g b(h_2^2/2)$; hence the resultant force acting on the control surface is given by $\frac{1}{2}\rho g b(h_1^2 - h_2^2)$ and this can be equated to the net rate of outflow of momentum which is equal to $(\rho b h_2 u_2^2 - \rho b h_1 u_1^2)$, i.e.

$$\tfrac{1}{2}g(h_1^2 - h_2^2) = h_2 u_2^2 - h_1 u_1^2. \tag{12.4.2}$$

Using the continuity equation (12.4.1), the last result may be expressed as

$$\tfrac{1}{2}g(h_1^2 - h_2^2) = \frac{Q^2}{b^2}\frac{(h_1 - h_2)}{h_1 h_2}. \tag{12.4.3}$$

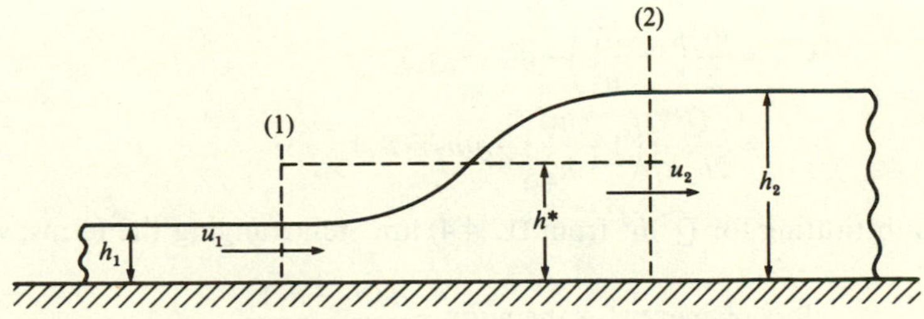

Fig. 115. Hydraulic jump.

There are evidently two possibilities. Either $h_1 = h_2$ and nothing happens, or if we divide through by $(h_1 - h_2)$ we get

$$h_1 h_2 (h_1 + h_2) = \frac{2Q^2}{gb^2}, \tag{12.4.4}$$

which may be re-arranged in the form

$$h_2^2 + h_1 h_2 - \frac{2Q^2}{g h_1 b^2} = 0. \tag{12.4.5}$$

The solution to this quadratic equation for h_2 is

$$h_2 = -\frac{h_1}{2} \pm \frac{1}{2}\sqrt{\left(h_1^2 + \frac{8Q^2}{g h_1 b^2}\right)}. \tag{12.4.6}$$

One positive value of h_2 is possible (apart from $h_1 = h_2$) if we take the plus sign in (12.4.5). This value for h_2 will be greater than h_1 provided

$$\frac{1}{2}\sqrt{\left(h_1^2 + \frac{8Q^2}{g h_1 b^2}\right)} > \frac{3}{2} h_1$$

i.e. if

$$h_1^2 + \frac{8Q^2}{g h_1 b^2} > 9 h_1^2,$$

which simplifies to the condition that $h_1^3 < Q^2/gb^2$, or

$$h_1 < \left(\frac{Q^2}{gb^2}\right)^{\frac{1}{3}}. \tag{12.4.7}$$

Referring to (12.2.15) it will be noted that $(Q^2/gb^2)^{\frac{1}{3}}$ is the *critical depth* h^*. Thus a jump is possible (with $h_2 > h_1$) if h_1 is less than the critical depth h^*.

We can calculate the loss of energy in the jump from the energy equation, i.e.

$$\begin{aligned}
\text{loss} &= \left(\frac{u_1^2}{2} + g h_1\right) - \left(\frac{u_2^2}{2} + g h_2\right) = \tfrac{1}{2}(u_1^2 - u_2^2) + g(h_1 - h_2) \\
&= \frac{u_1^2}{2}\left(1 - \frac{h_1^2}{h_2^2}\right) - g(h_2 - h_1) \\
&= \frac{Q^2}{2b^2 h_1^2}\left(1 - \frac{h_1^2}{h_2^2}\right) - g(h_2 - h_1).
\end{aligned}$$

Substituting for Q^2/b^2 from (12.4.4) and re-arranging the terms, we find:

$$\text{loss of energy for the jump} = \frac{g}{4}\frac{(h_2 - h_1)^3}{h_1 h_2}. \tag{12.4.8}$$

It will be seen that this expression has a positive value, and that an energy loss therefore occurs, when $h_2 > h_1$. Thus a jump is only possible in the direction shown in figure 115, i.e. from a value of h_1 less than h^* to a new depth h_2 which is greater than h_1.

13

Heat conduction and heat transfer

13.1 Introduction

In chapters 3 to 12 we have investigated some of the more important aspects of fluid dynamics. We can now turn attention to the study of transfer processes in moving fluids and to the analysis of heat and mass transfer between a fluid and a solid boundary or surface.

We have already noted in chapter 2 that heat conduction in a static fluid can be represented by the experimental relationship (2.1.2) known as Fourier's law, i.e.

$$\text{heat flux } q = -k\frac{\mathrm{d}T}{\mathrm{d}x}, \tag{13.1.1}$$

where q is the heat flux, or heat flow rate per unit area, in W m^{-2}, and k is the thermal conductivity, measured in W m^{-1} K^{-1}. This experimental law states that the heat flux is proportional to the temperature gradient $\mathrm{d}T/\mathrm{d}x$ and the negative sign in (13.1.1) simply indicates the fact that heat flows down a temperature gradient from a region of higher temperature to a region of lower temperature.

Equation (13.1.1) is the simplest one-dimensional form of Fourier's law but it will be noted that the heat flux $\mathbf{q}$ is in fact a vector quantity, having both magnitude and direction, and the more general three-dimensional statement takes the form

$$\mathbf{q} = -k \operatorname{grad} T. \tag{13.1.2}$$

We have introduced the Fourier conduction law to describe heat transfer in a *static fluid*. The temperature gradients, which are the cause of the heat transfer, however, also have the effect of creating density differences which result in movement or circulation of the fluid. We thus tend to have a situation of *natural convection* rather than one of pure conduction in a static fluid.

Another important class of applications may be described under the heading of *forced convection* where a mainstream fluid velocity is imposed by means of some external agency such as a pump or compressor. In this situation heat transfer by conduction plays a contributory role in the overall process of energy transfer within the moving fluid. We have seen in chapter 9, for example, that the Fourier conduction law appears in the heat transfer term of the first law of thermodynamics applied to a moving fluid (9.3.5) and in the general energy equation (9.4.3). The calculation of heat transfer rates in forced convection, including the effects of conduction, will be pursued further in chapters 16, 17, and 18.

The Fourier heat conduction law (13.1.2) was originally formulated to describe *heat conduction in solids*. For most solid materials the thermal conductivity varies with temperature, but it is usually sufficiently accurate to take an average value corresponding to the mean temperature within the material. We will normally be concerned only with isotropic solids where the thermal conductivity is independent of orientation. It is well known that the thermal conductivity of metals is significantly higher than that of most non-metals. There is some correlation between the thermal and electrical conductivities of metals and the relatively high values are associated with the presence of free electrons in a metal. The theory of thermal conduction through the agency of free electrons in a metal has certain similarities with the theory of energy transport by molecular motion in a gas as outlined in §2.7.

Although the analysis of heat transfer in solids is outside the main subject matter of the present book, we have to take note of the fact that, in most practical examples of heat transfer with fluids, the fluid in question is usually contained within solid boundaries, e.g. the walls of a pipe or duct. In assessing the overall heat transfer rate, therefore, we have to take account of conduction through the solid material of the pipe wall. Similarly, if we are concerned with heat transfer from one fluid to another in a heat exchanger, the transfer of heat by conduction through the solid boundary separating the two fluids must be included in the overall calculation.

Numerical values for the thermal conductivity of different materials range from about 400 $\mathrm{W\,m^{-1}\,K^{-1}}$ for some metals to about 0.02 $\mathrm{W\,m^{-1}\,K^{-1}}$ for gases and a few typical values at 20 °C are given below:

Solids

Copper 386 Mild steel 45 Alumina bricks 4.0 $\mathrm{W\,m^{-1}\,K^{-1}}$,

Liquids

Mercury 8.4 Water 0.60 Toluene 0.15 $\mathrm{W\,m^{-1}\,K^{-1}}$,

Gases

Hydrogen 0.18 Oxygen 0.025 Ammonia 0.024 $\mathrm{W\,m^{-1}\,K^{-1}}$.

The relationship between the thermal conductivity and the coefficient of viscosity for a gas has been noted in §2.7.

13.2 Heat transfer coefficients

The simplest example of heat conduction in a solid is the one-dimensional case of the steady flow of heat through a parallel-sided slab of thickness L as indicated in figure 116. Assuming that the thermal conductivity k is independent of temperature, it follows from the Fourier conduction law (13.1.1) that the heat flux through the slab is given by

$$q = k\,\frac{(T_1 - T_2)}{L} = k\,\frac{\Delta T}{L}. \qquad (13.2.1)$$

Since the heat flux is proportional to the overall temperature difference ΔT, it is convenient to define a heat transfer factor or coefficient h by the relationship

$$h = \frac{q}{\Delta T}. \qquad (13.2.2)$$

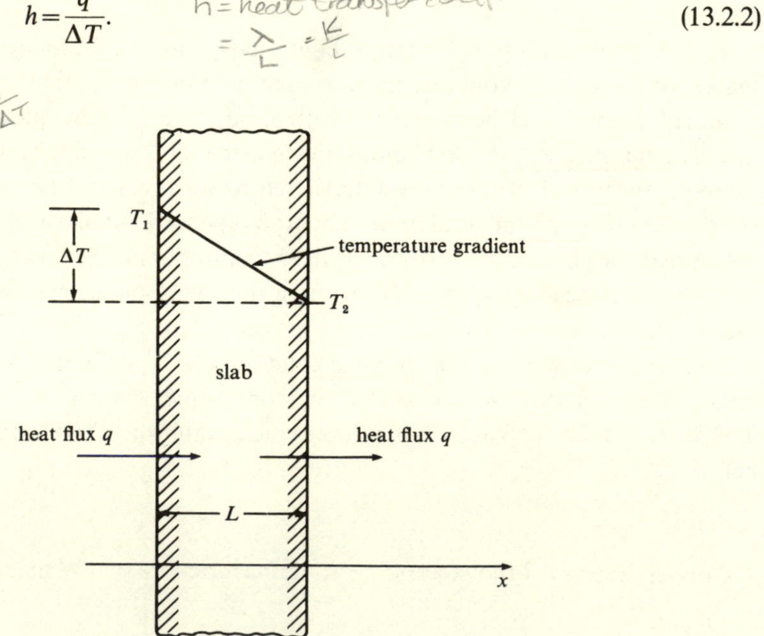

Fig. 116. Heat conduction through a slab.

Note that the dimensions of h are $\mathrm{W\,m^{-2}\,K^{-1}}$. We can obtain a *dimensionless* heat transfer coefficient, however, by taking the ratio hL/k which is known as the *Nusselt number*, Nu. It will be seen from (13.2.1) and (13.2.2) that for the special case of steady conduction through a slab

$$Nu = \frac{hL}{k} = 1. \tag{13.2.3}$$

It is appropriate at this point to compare the concepts of thermal conductivity and the heat transfer coefficient h when applied to a problem of thermal conduction. The thermal conductivity is the ratio of the heat flux to the *temperature gradient* and depends solely on the physical nature of the material. The heat transfer coefficient h is the ratio of the heat flux to the overall *temperature difference* and depends on the geometry of the problem as well as the material. The Nusselt number is a function only of the geometry in any problem of steady heat conduction.

The heat transfer coefficient h defined by (13.2.2) has a wider application to heat transfer between a fluid and a solid boundary. A typical situation is illustrated in figure 117 where a fluid is flowing with main stream temperature T_0 parallel to a solid wall at temperature T_1. We can define a heat transfer coefficient h_f for transfer of heat from the fluid to the solid boundary by the statement

$$h_f = q/(T_0 - T_1) = q/\Delta T_f. \tag{13.2.4}$$

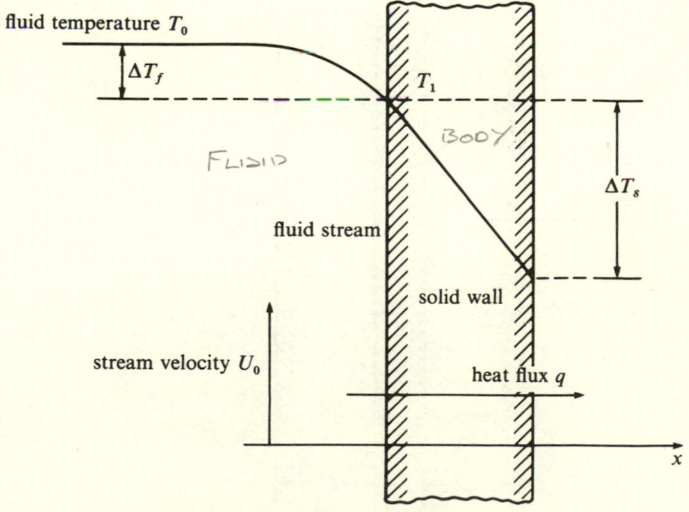

Fig. 117. Heat conduction from a flowing stream to a slab.

The equivalent dimensionless heat transfer coefficient will take the form of a Nusselt number defined in terms of the thermal conductivity of the fluid and in terms of a linear dimension which is characteristic of the geometry of the fluid stream. If, for example, the fluid stream is contained within a circular pipe it would be normal to take the inside diameter d of the pipe to define the Nusselt number, i.e.

$$Nu = \frac{h_f d}{k_f} = \frac{qd}{k_f \, \Delta T_f}. \tag{13.2.5}$$

As will be shown in chapter 15, the Nusselt number for heat transfer by forced convection from a fluid stream to a solid boundary will be a function of the Reynolds number $\rho U_0 d/\mu$, the Prandtl number for the fluid $\mu c_p/k_f$, and the geometry of the boundaries. Having determined the appropriate value of the Nusselt number from the flow parameters and flow geometry, the heat transfer coefficient h_f can be evaluated from (13.2.5).

13.3 Steady one-dimensional flow of heat through several layers

In engineering applications we frequently encounter heat flow by conduction through several layers of different materials as indicated in figure 118. Layer A, for example, might be a refractory lining, layer B might be a structural material, and layer C might be an insulating material. We

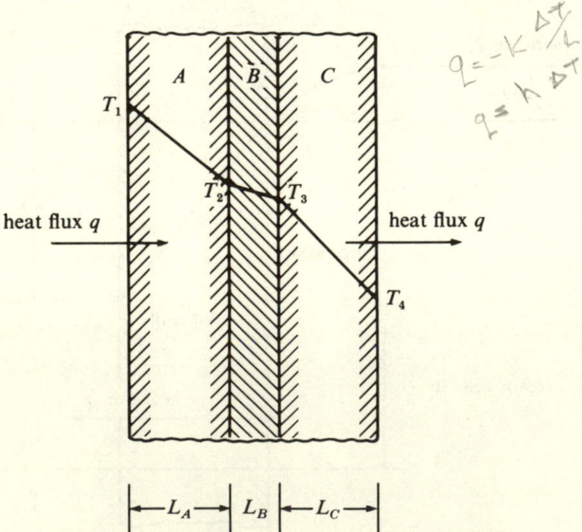

Fig. 118. Heat conduction through slabs in series.

will assume that there are no air gaps between the layers and that there is consequently complete thermal contact at each interface. We can apply equation (13.2.1) and (13.2.2) to each layer in succession and it will be seen that, for continuity of the heat flux q, we must have

$$\begin{aligned} T_1 - T_2 &= q\frac{L_A}{k_A} = \frac{q}{h_A} \\ T_2 - T_3 &= q\frac{L_B}{k_B} = \frac{q}{h_B} \\ T_3 - T_4 &= q\frac{L_C}{k_C} = \frac{q}{h_C} \end{aligned} \quad . \tag{13.3.1}$$

Adding these equations, and thus eliminating the intermediate temperatures T_2 and T_3, we have

$$T_1 - T_4 = q\left(\frac{1}{h_A} + \frac{1}{h_B} + \frac{1}{h_C}\right). \tag{13.3.2}$$

It is often convenient to define an overall heat transfer coefficient h_o as the ratio of the heat flux to the overall temperature difference ΔT which in this case will be $T_1 - T_4$. The overall heat transfer coefficient is more commonly denoted by the symbol U but we prefer h_o to avoid confusion with velocity. Thus

$$q = h_o\,\Delta T = h_o(T_1 - T_4) \tag{13.3.3}$$

and it follows from (13.3.2) that

$$\frac{1}{h_o} = \left(\frac{1}{h_A} + \frac{1}{h_B} + \frac{1}{h_C}\right). \tag{13.3.4}$$

Hence we see that the summation of heat transfer coefficients in series is a matter of adding their reciprocals. Heat transfer coefficients are analogous to electrical conductances and add in the same manner. Another important example of heat flow in series is the case of heat transfer from one fluid to another through a solid wall which separates the two fluids, as indicated in figure 119. We have the following conditions for continuity of the heat flux:

heat transfer from fluid A	$q = h_1(T_A - T_1)$,
heat conduction through wall	$q = \dfrac{k}{L}(T_1 - T_2)$,
heat transfer to fluid B	$q = h_2(T_2 - T_B)$.

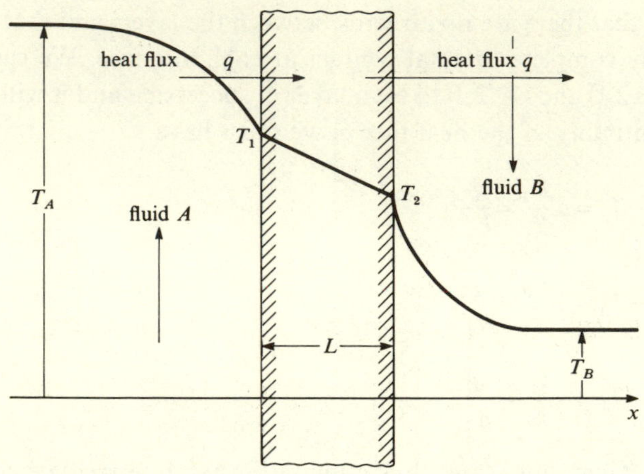

Fig. 119. Heat conduction through a wall between two fluids.

Hence for the overall heat transfer from fluid A at temperature T_A to fluid B at temperature T_B we can say

$$T_A - T_B = q\left(\frac{1}{h_1} + \frac{L}{k} + \frac{1}{h_2}\right)$$

or

$$q = h_o(T_A - T_B), \qquad (13.3.5)$$

where

$$\frac{1}{h_o} = \left(\frac{1}{h_1} + \frac{L}{k} + \frac{1}{h_2}\right).$$

13.4 Steady radial flow of heat in a cylinder

The radial flow of heat by conduction in a cylinder or through the wall of a tube is of special importance in engineering applications. Referring to figure 120 it is important to note that the total heat flow rate Q will be constant in the absence of internal heat generation, but that the heat flux per unit area q will vary inversely with the radius r. For continuity of radial heat flow we have the requirement that

$$Q = 2\pi r L q = \text{constant}, \qquad (13.4.1)$$

where L is the length of the cylinder. Hence, using Fourier's law (13.1.1) we have for the heat flux q

$$q = \frac{Q}{2\pi r L} = -k\frac{\mathrm{d}T}{\mathrm{d}r}. \qquad (13.4.2)$$

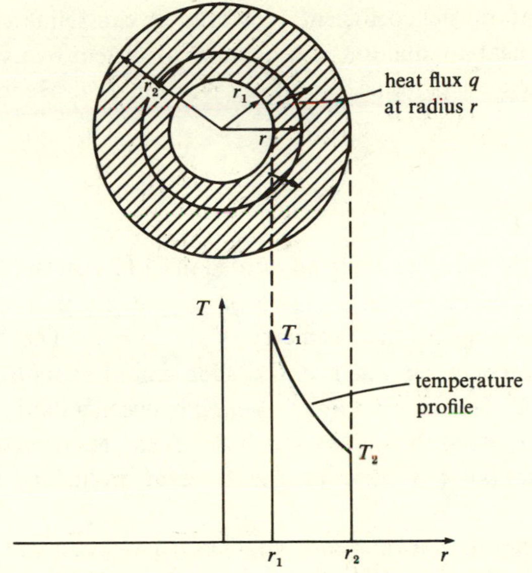

Fig. 120. Heat conduction through a cylindrical shell.

Integrating (13.4.2) from the inside radius r_1, where $T = T_1$, to the outside radius r_2, where $T = T_2$, gives the result

$$\frac{Q}{2\pi L}\log_e\left(\frac{r_2}{r_1}\right) = k(T_1 - T_2). \tag{13.4.3}$$

Thus the total heat flow rate Q is given by

$$Q = \frac{2\pi Lk(T_1 - T_2)}{\log_e r_2/r_1} \tag{13.4.4}$$

and the heat flux q at radius r is given by

$$q = \frac{k(T_1 - T_2)}{r\log_e r_2/r_1}. \tag{13.4.5}$$

It may also be noted from (13.4.3) that the radial temperature profile is represented by

$$\frac{T_1 - T}{T_1 - T_2} = \frac{\log_e r/r_1}{\log_e r_2/r_1}. \tag{13.4.6}$$

With radial flow of heat in a cylindrical system the concept of a heat transfer coefficient defined by $h = q/\Delta T$ is complicated by the fact that q is not constant. We therefore have to specify a reference radius for the evaluation of any heat transfer coefficient. However, we can define a unique Nusselt number for heat conduction in a cylindrical system by using the radius r as the reference length, i.e. by taking $Nu = qr/\Delta Tk$. Hence from (13.4.5) we have

$$Nu = \frac{qr}{(T_1 - T_2)k} = \frac{1}{\log_e r_2/r_1}. \tag{13.4.7}$$

This result confirms the conclusion already noted in § 13.2 that the Nusselt number for *heat transfer by conduction through a solid* is a pure number whose value depends only on the geometry of the system. The Nusselt number for conduction between radii r_1 and r_2 should not be confused with the Nusselt number for *heat transfer by convection* between a fluid and the tube wall at radius r_1 or with the Nusselt number for heat transfer by convection from the outer surface of the tube at radius r_2 to the surroundings.

We can now consider the very common engineering case of heat transfer through a thick-walled tube from a fluid A flowing inside the tube to another fluid B flowing in the space outside the tube as illustrated in figure 121. The following conditions apply

heat transfer from fluid A at r_1 $\quad Q = 2\pi r_1 L h_1 (T_A - T_1)$,

heat conduction through the wall $\quad Q = \dfrac{2\pi Lk(T_1 - T_2)}{\log_e r_2/r_1}$,

heat transfer at outer surface to fluid B $\quad Q = 2\pi r_2 L h_2 (T_2 - T_B)$.

Hence

$$Q = \frac{2\pi L(T_A - T_B)}{[(1/r_1h_1) + (1/k)\log_e r_2/r_1 + (1/r_2h_2)]}. \tag{13.4.8}$$

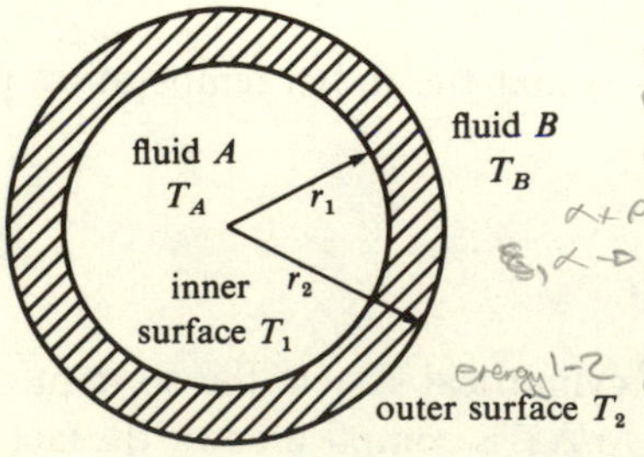

Fig. 121. Heat conduction through a thick-walled tube.

and from equation (13.4.2)

$$qr = \frac{(T_A - T_B)}{[(1/r_1h_1) + (1/k)\log_e r_2/r_1 + (1/r_2k_2)]}. \tag{13.4.9}$$

As discussed above, the heat flux q varies with radius and this causes problems if we attempt to define an overall heat transfer coefficient as in equation (13.3.3). The difficulty can be overcome by arbitrarily selecting a radius, such as the internal radius r_1 and calculating the corresponding heat flux q_1. The overall heat transfer coefficient referred to radius r_1 is then defined by

$$q_1 = h_{o1}(T_A - T_B) \tag{13.4.10}$$

and hence

$$\frac{1}{r_1h_{o1}} = \frac{1}{r_1h_1} + \frac{1}{k}\log_e\left(\frac{r_2}{r_1}\right) + \frac{1}{r_2h_2}. \tag{13.4.11}$$

Very commonly pipes are surrounded by one or more layers of lagging as illustrated in figure 122. An analysis similar to that above yields,

$$\frac{1}{r_1h_{o1}} = \frac{1}{r_1h_1} + \frac{1}{k_1}\log_e\left(\frac{r_2}{r_1}\right) + \frac{1}{k_2}\log_e\left(\frac{r_3}{r_2}\right) + \frac{1}{r_3h_3}. \tag{13.4.12}$$

A numerical example will be appropriate at this point. We will consider the case of a liquid A flowing inside a stainless steel tube with local fluid temperature $T_A = 250\,°\mathrm{C}$ and with a local value for the convective heat transfer coefficient $h_1 = 2000\ \mathrm{W\,m^{-2}\,K^{-1}}$. We will assume that there is another liquid B flowing in the space outside the tube at temperature $T_B = 20\,°\mathrm{C}$ and with a local value for the convective heat transfer coefficient $h_2 = 1200\ \mathrm{W\,m^{-2}\,K^{-1}}$. We will take an inside diameter of 25 mm and an outside

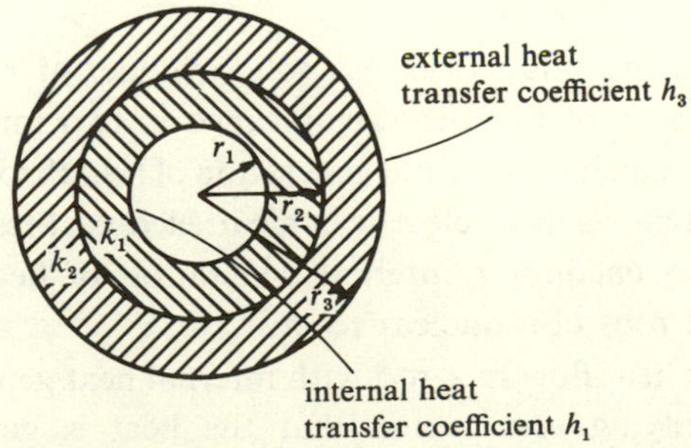

Fig. 122. Heat conduction through a lagged pipe.

diameter of 30 mm for the tube, so that $r_1 = 0.0125$ m and $r_2 = 0.015$ m. For the thermal conductivity of the stainless steel we will take a value $k = 26\ \mathrm{W\,m^{-1}\,K^{-1}}$. Hence we have

$$\frac{r_2}{r_1} = 1.2 \quad \log_e \frac{r_2}{r_1} = 0.1823 \quad \text{and} \quad \frac{1}{k}\log_e \frac{r_2}{r_1} = 7.01 \times 10^{-3},$$

$$r_1 h_1 = 25 \qquad \frac{1}{r_1 h_1} = 40 \times 10^{-3}$$

$$r_2 h_2 = 18 \qquad \frac{1}{r_2 h_2} = 55.55 \times 10^{-3},$$

$$\left(\frac{1}{r_1 h_1} + \frac{1}{k}\log_e \frac{r_2}{r_1} + \frac{1}{r_2 h_2}\right) = 102.56 \times 10^{-3},$$

$$\frac{Q}{2\pi L} = \frac{230}{102.6} \times 10^3 = 2.243 \times 10^3\ \mathrm{W\,m^{-1}},$$

and

$$Q/L = 14.1\ \text{kW per metre length of tube.}$$

$$(T_A - T_1) = \frac{Q}{2\pi L r_1 h_1} = 89.7\ \mathrm{K},$$

$$(T_1 - T_2) = \frac{Q}{2\pi L}\frac{1}{k}\log_e \frac{r_2}{r_1} = 15.7\ \mathrm{K},$$

$$(T_2 - T_B) = \frac{Q}{2\pi L r_2 h_2} = 124.6\ \mathrm{K},$$

hence $T_1 = 160\,°\mathrm{C}$ and $T_2 = 144\,°\mathrm{C}$ approximately. At radius r_1 the heat flux $q_1 = Q/2\pi L r_1 = 179\ \mathrm{kW\,m^{-2}}$, and at radius r_2 the heat flux $q_2 = Q/2\pi L r_2 = 150\ \mathrm{kW\,m^{-2}}$. Hence the overall heat transfer coefficient referred to radius r_1 is given by $q_1/(T_A - T_B) = 778\ \mathrm{W\,m^{-2}\,K^{-1}}$ and $h_{02} = 150/(250-20) = 652\ \mathrm{W\,m^{-2}\,K^{-1}}$.

The radial flow of heat by conduction in a solid cylinder or rod with *internal generation of heat* is of practical significance in two important engineering situations. We can have internal generation of heat through the passage of an electric current, as in an electric cable or electrical resistance heating element. We also encounter internal generation of heat from nuclear fission in the fuel rods of a nuclear reactor.

In order to analyse the heat flow in a rod with internal heat generation we will make the simplifying assumption that the heat is generated uniformly at a constant rate Q_s per unit volume of material. If we now

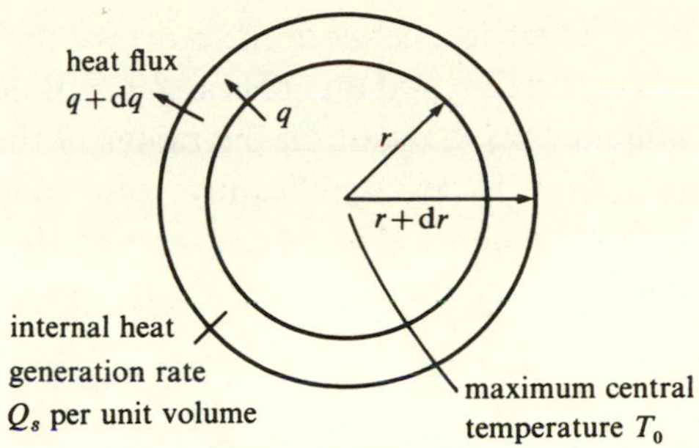

Fig. 123. Heat conduction with internal generation of heat.

consider an annular element of the rod, with inner radius r and outer radius $r+\mathrm{d}r$, as indicated in figure 123, the net rate of outflow of heat per unit length will be $2\pi(q+\mathrm{d}q)(r+\mathrm{d}r)-2\pi qr=2\pi(q\,\mathrm{d}r+r\,\mathrm{d}q)$ if we neglect the second-order term involving the product $\mathrm{d}q\,\mathrm{d}r$. The rate of generation of heat within the volume of the annulus is given by $Q_s 2\pi r\,\mathrm{d}r$ per unit length of rod. Hence for steady-state conduction of heat we must have

$$Q_s r\,\mathrm{d}r=q\,\mathrm{d}r+r\,\mathrm{d}q=\mathrm{d}(rq). \tag{13.4.13}$$

Substituting for q from the Fourier conduction law, the last equation may be written

$$Q_s r=-\frac{\mathrm{d}}{\mathrm{d}r}\left(kr\frac{\mathrm{d}T}{\mathrm{d}r}\right)$$

and hence

$$\int_0^r Q_s r\,\mathrm{d}r=Q_s\frac{r^2}{2}=-kr\frac{\mathrm{d}T}{\mathrm{d}r}. \tag{13.4.14}$$

If the thermal conductivity k is effectively constant, we can integrate (13.4.14) from 0 to r with the result that

$$k(T_0-T)=Q_s\frac{r^2}{4}, \tag{13.4.15}$$

where T_0 is the temperature at the centre of the rod where $r=0$. If the outer surface of the rod is defined by $r=a$ and if the surface temperature is T_s, we have from (13.4.15)

$$T_0-T_s=Q_s\frac{a^2}{4k}=\frac{1}{4\pi k}\frac{Q}{L}, \tag{13.4.16}$$

where Q/L is the rate of heat generation per unit length of rod. We thus see that the overall temperature difference, between the centre of the rod and the outer surface, depends only on the total rate of heat generation per unit length and the thermal conductivity but not on the radius of the rod.

If the thermal conductivity k varies significantly with temperature, however, we have to go back to equation (13.4.14) and express the result of the integration in the form

$$\int_{T}^{T_0} k\,\mathrm{d}T = Q_s\frac{r^2}{4} \tag{13.4.17}$$

and in place of equation (13.4.16) we have the result

$$\int_{T_s}^{T_0} k\,\mathrm{d}T = Q_s\frac{a^2}{4} = \frac{1}{4\pi}\frac{Q}{L}. \tag{13.4.18}$$

The appropriate value of the integral $\int_{T_s}^{T_0} k\,\mathrm{d}T$ has to be obtained from experimental data.

As a numerical example of conduction in a solid cylinder with internal heat generation we can consider the case of a nuclear reactor fuel rod. In the pressurised water reactor (referred to as the PWR) the fuel elements take the form of clusters of small-diameter zircalloy tubes containing rods or pellets of enriched uranium dioxide. Heat is generated through the fission of U235 within the uranium oxide fuel and the rate of energy release is determined by the neutron flux and the degree of enrichment of the fuel. Uranium dioxide is a poor conductor of heat and the internal temperature gradients are consequently very large. A limit is set by the melting point of UO_2 and the maximum internal temperature must be limited to a figure below 2750 °C. The outer surface temperature of the fuel rods, on the other hand, is normally held at a relatively low figure, around 350 °C, since pressurised water flowing at high velocity is used as the coolant in the primary circuit. The numerical value of the integral $\int k\,\mathrm{d}T$ for UO_2 taken between the limits of a typical outer temperature of the fuel pellets and a central temperature equal to the melting point of UO_2 is of the order of 5 kW m^{-1}. Thus from (13.4.18) the maximum permissible value for the linear heat release rate would be about 60–65 kW m^{-1}. In practice a rather lower figure is adopted for design purposes to allow a sufficient safety margin and a typical maximum linear heat rating for a fuel rod would be about 40–45 kW m^{-1}.

Since the neutron flux will have a spatial distribution in both the radial and axial directions within the reactor core, the rate of heat generation in the fuel will also have a spatial distribution with the maximum local rate occurring near the centre of the core. A typical value for the ratio between the peak rate of heat release and the average for the whole reactor core would be about 2.2 to 2.3 for a PWR. Thus if we take a figure of about 42 kW m^{-1} for the maximum linear heat rating, a typical figure for the *average* linear heat rating would be about 18 kW m^{-1}.

In addition to the large radial temperature gradient within the UO_2 fuel pellets, a significant temperature difference of the order of 150°–200° will be established across the small gap between the outer surface of the fuel pellets and the inside surface of the zircalloy tubing. Further temperature differences arise in connection with heat conduction through the zircalloy tube wall and heat transfer by convection from the outer surface of the elements to the pressurised water flowing through the reactor core. A typical figure for the water temperature at entry to the reactor would be about 290–300 °C, and the temperature at exit from the core will usually be around 320–330 °C. Typical dimensions for the fuel rods are as follows:

outer diameter of zircalloy tubing	9.5 mm,
radial thickness of cladding	0.57 mm,
diameter of UO_2 fuel pellets	8.19 mm.

Thus the maximum normal heat flux at the outer surface of the fuel elements, corresponding to a maximum linear heat rating of 42 kW m^{-1}, would be about 1400 kW m^{-2}, while the average surface heat flux with a linear rating of 18 kW m^{-1} would be 600 kW m^{-2}.

The following figures for the temperatures at various points in the fuel element assembly, corresponding to a maximum linear heat rating of 42 kW m^{-1} or a maximum surface heat flux of 1400 kW m^{-2}, can easily be derived from the preceding analysis. We will assume a local water temperature of 315 °C, a figure for the convective heat transfer coefficient $h_f = 4.5 \times 10^4$ W m^{-2} K^{-1} for use in equation (13.2.4), a value of $k = 11$ W m^{-1} K^{-1} for the thermal conductivity of zircalloy, and a very approximate figure of about 2.2 W m^{-1} K^{-1} for the average value of the thermal conductivity of UO_2 for use in equation (13.4.16). On this basis we arrive at the following typical results:

local temperature of the water	315 °C,
outer surface temperature of fuel rods	346 °C,
inner temperature of zircalloy cladding	419 °C,
temperature difference across gap	≈150–200 °C,
outer temperature of UO_2 pellets	≈570–620 °C,
centre temperature of UO_2 pellets	≈2070–2120 °C.

13.5 Steady radial flow of heat in a spherical system

The heat conduction through a spherical shell can be analysed in a manner similar to that used for a cylindrical shell in § 13.4. As can be seen from figure 124, the heat flux q is now inversely proportional to the square of the radius and is related to the total heat flow Q by

$$q=\frac{Q}{4\pi r^2}=-k\frac{\mathrm{d}T}{\mathrm{d}r}. \tag{13.5.1}$$

On integration this gives

$$\frac{Q}{4\pi}\left(\frac{1}{r_1}-\frac{1}{r_2}\right)=k(T_1-T_2), \tag{13.5.2}$$

where T_1 is the temperature at radius r_1 and T_2 is the temperature at $r=r_2$.

A case of particular importance is the heat conduction in an infinite stagnant medium from a spherical surface of radius r_1 at temperature T_1. This result can be obtained by letting $r_2\rightarrow\infty$ so that

$$Q=4\pi r_1 k(T_1-T_2) \tag{13.5.3}$$

and an appropriate heat transfer coefficient can be defined by

$$h_1=\frac{q_1}{(T_1-T_2)}, \tag{13.5.4}$$

where q_1 is the heat flux at radius r_1. Hence

$$h_1=\frac{Q}{4\pi r_1^2(T_1-T_2)}=\frac{k}{r_1}. \tag{13.5.5}$$

For spheres it is customary to use the diameter of the sphere d rather

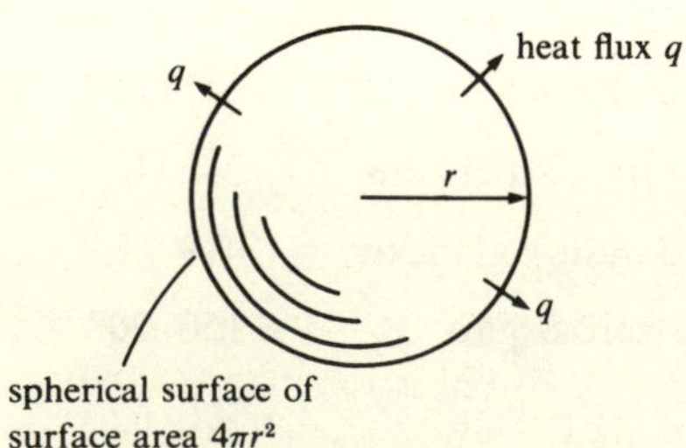

Fig. 124. Heat conduction through a spherical surface.

than the radius when defining a Nusselt number, i.e.

$$Nu=\frac{h_1 d}{k} \tag{13.5.6}$$

and hence for heat conduction from an isolated sphere,

$$Nu=2. \tag{13.5.7}$$

When there is heat generation within a solid sphere a heat balance on a spherical region of radius r gives

$$\tfrac{4}{3}\pi r^3 Q_s=4\pi r^2 q=-4\pi r^2 k\frac{\mathrm{d}T}{\mathrm{d}r} \tag{13.5.8}$$

or

$$-k\frac{\mathrm{d}T}{\mathrm{d}r}=Q_s\frac{r}{3}. \tag{13.5.9}$$

On integration subject to the boundary condition $T=T_s$ on radius $r=a$ this becomes

$$k(T-T_s)=\frac{Q_s}{6}(a^2-r^2). \tag{13.5.10}$$

As in the case of the slab and the solid cylinder, the temperature profile is parabolic and the temperature at the centre, T_0, is given by

$$k(T_0-T_s)=\frac{Q_s a^2}{6}=\frac{Q}{8\pi a}. \tag{13.5.11}$$

It can be seen that this equation has the same algebraic form as equation (13.4.18) but that the constants have different numerical values.

13.6 General heat conduction equations

In the previous sections we considered heat conduction in the three simple geometries of the slab, the cylinder, and the sphere. Because of the symmetry of these geometries we were able to write down the variation of the heat flux with position by inspection and to proceed using only first principles. However, for more complicated systems we have to resort to the use of the heat conduction equation in an appropriate co-ordinate system. We will first derive this equation in general vectorial form and then quote the forms for Cartesian, cylindrical, and spherical co-ordinates.

The outflow of heat per unit volume is given by the divergence of the heat flux vector, div **q**, where **q** is related to the temperature gradient by Fourier's law (equation (13.1.2)),

$$\mathbf{q} = -k \text{ grad } T.$$

A simple heat balance relates the outflow of heat to the rate of accumulation of heat per unit volume, $(\partial/\partial t)(\rho c_v T)$, and to the rate of heat generation per unit volume Q_s, due to the passage of an electric current or chemical or nuclear reaction.

The heat balance gives

$$\frac{\partial}{\partial t}(\rho c_v T) = Q_s - \text{div } \mathbf{q} \tag{13.6.1}$$

or

$$\frac{\partial T}{\partial t} = \frac{Q_s}{\rho c_v} + \frac{1}{\rho c_v} \text{div } (k \text{ grad } T). \tag{13.6.2}$$

If the thermal conductivity k is a constant with respect to position, equation (13.6.2) may be written

$$\frac{\partial T}{\partial t} = \frac{Q_s}{\rho c_v} + \alpha \text{ div grad } T = \frac{Q_s}{\rho c_v} + \alpha \nabla^2 T, \tag{13.6.3}$$

where α is the thermal diffusivity defined by $\alpha = k/\rho c_v$.

Apart from the heat-generation term this is seen to be identical to equation (2.9.9) and equation (9.3.5) also reduces to this form when the velocity is zero.

Equation (13.6.3) can be expressed in various coordinate systems as follows:

Cartesian (x, y, z)

$$\frac{\partial T}{\partial t} = \frac{Q_s}{\rho c_v} + \alpha\left(\frac{\partial^2 T}{\partial x^2} + \frac{\partial^2 T}{\partial y^2} + \frac{\partial^2 T}{\partial z^2}\right). \tag{13.6.4}$$

Cylindrical (r, ϕ, z)

$$\frac{\partial T}{\partial t} = \frac{Q_s}{\rho c_v} + \alpha\left(\frac{\partial^2 T}{\partial r^2} + \frac{1}{r}\frac{\partial T}{\partial r} + \frac{1}{r^2}\frac{\partial^2 T}{\partial \phi^2} + \frac{\partial^2 T}{\partial z^2}\right). \tag{13.6.5}$$

Spherical (r, θ, ϕ)

$$\frac{\partial T}{\partial t} = \frac{Q_s}{\rho c_v} + \alpha\left(\frac{\partial^2 T}{\partial r^2} + \frac{2}{r}\frac{\partial T}{\partial r} + \frac{1}{r^2}\frac{\partial^2 T}{\partial \theta^2} + \frac{\cot\theta}{r^2}\frac{\partial T}{\partial \theta} + \frac{1}{r^2 \sin^2\theta}\frac{\partial^2 T}{\partial \phi^2}\right). \tag{13.6.6}$$

In order to solve these equations we must specify a sufficient number of initial and boundary conditions. The initial condition is normally some known temperature distribution at $t=0$ but the boundary conditions may be expressed either as specified temperatures or as specified temperature gradients on the boundaries. Typical boundary conditions are best explained in terms of the one-dimensional form of equation (13.6.4),

$$\frac{\partial T}{\partial t}=\frac{Q_s}{\rho c_v}+\alpha\frac{\partial^2 T}{\partial x^2}, \tag{13.6.7}$$

and may be classified as follows:

(i) Surface temperature specified, e.g. $T=T_0$ at $x=0$ and $T=T_L$ at $x=L$.

(ii) Specified heat flux at the boundary giving

$$-k\frac{\mathrm{d}T}{\mathrm{d}x}=q. \tag{13.6.8}$$

A common case of this being the perfectly insulated boundary for which

$$\frac{\mathrm{d}T}{\mathrm{d}x}=0. \tag{13.6.9}$$

(iii) Thermal contact with some other conducting medium, requiring continuity of heat flux at the interface,

$$-k_a\frac{\mathrm{d}T_a}{\mathrm{d}x}=-k_b\frac{\mathrm{d}T_b}{\mathrm{d}x}. \tag{13.6.10}$$

(iv) The heat flux at the boundary being determined by the rate of heat transfer by convection or radiation to a surrounding medium at temperature T_0. Here the heat flux is best described in terms of a heat transfer coefficient h and

$$-k\frac{\mathrm{d}T}{\mathrm{d}x}=h(T-T_0). \tag{13.6.11}$$

The solution of the unsteady heat conduction equation, even in its simplest form, equation (13.6.7), is mathematically difficult and two classic results will be presented here without derivation. The heat conduction equation is, however, identical in form to the diffusion equation. This has been studied in great detail by Crank[1], whose book contains solutions for many situations.

(1) We will first consider the case of a large body, initially at a uniform temperature T_0, which is suddenly immersed in a hot liquid at temperature T_1. We can idealise this as a semi-infinite plane body subject to the initial condition

$$t=0, \quad T=T_0 \quad \text{for all } x>0$$

and the boundary conditions

$$t>0, \quad T=T_1 \quad \text{on } x=0,$$

$$t\geqslant 0, \quad T\to T_0 \quad \text{as } x\to\infty.$$

In the absence of heat generation within the body, a similarity solution is appropriate and the resulting temperature distribution is

$$\frac{T_1-T}{T_1-T_0}=\operatorname{erf}\left(\frac{x}{\sqrt{(4\alpha t)}}\right). \tag{13.6.12}$$

The heat flux through the interface is given by

$$q=k(T_1-T_0)/\sqrt{(\pi\alpha t)} \tag{13.6.13}$$

and this is seen to fall off with time as the temperature of the body gradually approaches the external temperature T_1.

(2) As our second example we will consider a finite slab of thickness L. The slab is initially at temperature T_0 and both faces of the slab (at $x=0$ and $x=L$) are kept at this temperature. Up to time $t=0$ there is no heat generation within the slab but for $t>0$ heat is generated at a uniform rate Q_s per unit volume. For this problem a product solution is appropriate giving

$$T=T_0+\frac{Q_s x(L-x)}{2k}-\sum_{n=1}\frac{4Q_sL^2}{n^3\pi^3k}\sin\left(\frac{n\pi x}{L}\right)\mathrm{e}^{-(n^2\pi^2\alpha t/L^2)}. \tag{13.6.14}$$

It can be seen that at large times all the exponential terms in the summation tend to zero and the final temperature distribution is given by

$$T=T_0+\frac{Q_s x(L-x)}{2k}. \tag{13.6.15}$$

This last result could have been obtained more simply by the method outlined in §13.4 and it can be seen that the temperature distribution is parabolic like that in a cylindrical rod, equation (13.4.15).

Frequently the heat conduction equation requires solution by numerical means and details of this can be found in many texts devoted to numerical analysis. For the special case of steady heat conduction without heat generation equation (13.6.3) reduces to Laplace's equation

$$\nabla^2 T = 0 \tag{13.6.16}$$

and all the techniques of potential flow analysis, presented in chapter 4, become available.

Numerical solutions of Laplace's equation are based on the relaxation methods originally derived by Southwell[2] for hand calculations. The advent of high-speed digital computers has, however, enabled these methods to be extended and accelerated considerably.

For steady conduction with heat generation equation (13.6.3) becomes

$$k\,\nabla^2 T = -Q_s, \tag{13.6.17}$$

which is Poisson's equation, the solution of which is hardly more complicated than the solution of Laplace's equation. Furthermore, Poisson's equation can always be reduced to Laplace's equation by the substitution

$$\phi = T + Q_s\frac{x^2}{2k}. \tag{13.6.18}$$

In problems of unsteady heat conduction the known temperature distribution at any particular time can be used to evaluate the spatial derivatives of T and hence $\partial T/\partial t$ can be evaluated from equations (13.6.4) to (13.6.6). Hence we can step forward through a small interval of time δt to give the temperature distribution at time $t+\delta t$. This simple method is, however, unstable unless δt is very small, and reference should be made to standard texts on numerical analysis for improvements to this method.

13.7 Extended surfaces for heat transfer

As we have noted in § 13.1, gases are poor conductors of heat and hence heat transfer coefficients from solid surfaces into gases tend to be low. In order to obtain a reasonable heat transfer rate for a modest temperature difference it becomes necessary to provide a large surface area. This is often achieved by the installation of metal fins onto the surface, the best-known example being the cooling fins clearly visible on most motor-cycle engines. Circular pipes are often fitted with annular fins, but for ease of analysis we will confine ourselves to the case of a plane wall with fins of thickness t and length L attached to the wall at spacing s as shown in figure 125. We will assume that the fins are thin, i.e. that $t \ll L$ and that they are of infinite extent

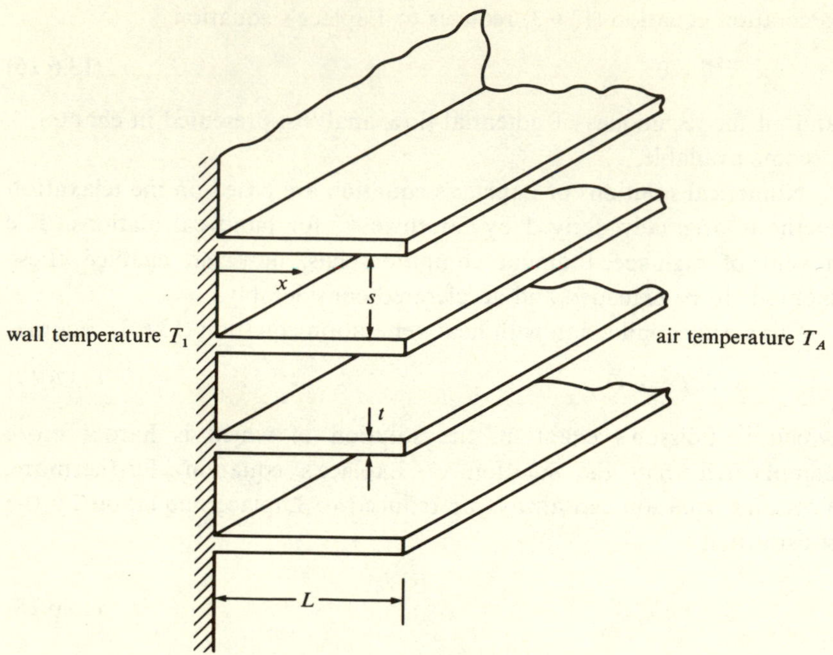

Fig. 125. Plane wall with cooling fins.

normal to the plane of figure 125. The analysis will, however, be conducted for unit width normal to the paper.

We can perform a heat balance on an element of fin as shown in figure 126. The temperature at a distance x from the root of the fin is denoted by T and the surroundings are assumed to be at uniform temperature T_A.

Heat conducted in through section 1
= heat conducted out through section 2
+ heat lost to surroundings.

$$-kt\frac{\mathrm{d}T}{\mathrm{d}x} = -kt\frac{\mathrm{d}}{\mathrm{d}x}\left(T + \frac{\mathrm{d}T}{\mathrm{d}x}\,\delta x\right) + 2h(T - T_A)\,\delta x. \qquad (13.7.1)$$

Here h is the heat transfer coefficient between the surface of the fin and the surrounding gas.

Assuming that h, k, and t are constants we can define the quantity m by

$$m^2 = \frac{2h}{kt} \qquad (13.7.2)$$

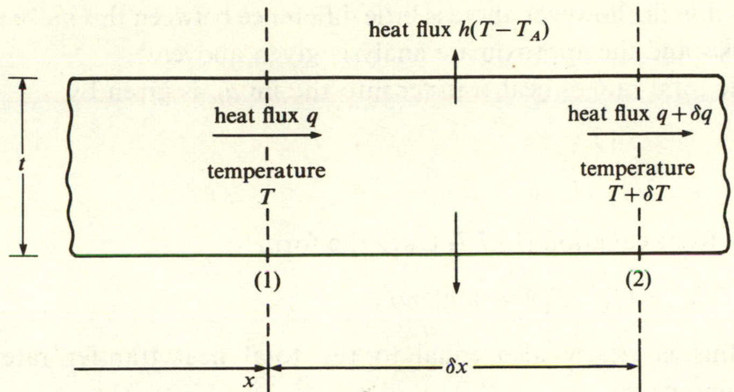

Fig. 126. Heat balance on an element of fin.

and on re-arrangement equation (13.7.1) becomes

$$\frac{d^2T}{dx^2}-m^2T=-m^2T_A. \tag{13.7.3}$$

The solution of equation (13.7.3) can be written as

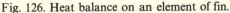

$$T-T_A \cosh mx+B \sinh mx \tag{13.7.4}$$

and the arbitrary constants A and B can be evaluated from the boundary conditions. At $x=0$ the fin temperature will equal that of the wall so that $T=T_1$ and hence

$$A=T_1-T_A. \tag{13.7.5}$$

If the fin is thin the amount of heat transferred through the outer end of the fin will be small compared with the total heat transfer rate, so that to a first approximation we can put $dT/dx=0$ at $x=L$, giving

$$Am \sinh mL+Bm \cosh mL=0$$

or

$$B=-(T_1-T_A)\tanh mL. \tag{13.7.6}$$

The complete solution is therefore

$$\frac{T-T_A}{T_1-T_A}=\cosh mx-\sinh mx \tanh mL. \tag{13.7.7}$$

A more accurate result can be obtained by considering the rate of heat transfer through the far end of the fin. This can be expressed in the form,

$$\text{at } x=L, \quad -k\frac{dT}{dx}=h(T-T_A).$$

For a thin fin, however, there is little difference between this more accurate analysis and the approximate analysis given above.

The total rate of heat transfer into the fin q_{in} is given by

$$-kt\left(\frac{\mathrm{d}T}{\mathrm{d}x}\right)_{x=0}$$

which from equation (13.7.7) takes the form

$$q_{in}=(T_1-T_A)ktm\tanh mL \tag{13.7.8}$$

and this is clearly also equal to the total heat transfer rate to the surrounding gas.

This result is traditionally expressed in two forms. We can define a fin efficiency or effectiveness factor η as the ratio of the heat transfer rate to that which would have occurred if the whole of the fin had been at the wall temperature T_1. Evidently

$$\eta=\frac{q_{in}}{2hL(T_1-T_A)}=\frac{1}{mL}\tanh mL. \tag{13.7.9}$$

It can be seen that this equation has considerable similarity with the expression for the effectiveness factor for a catalyst pore, equation (22.6.8) and the enhancement factor in gas absorption, equation (22.2.15).

Alternatively we can define an enhancement factor E as the ratio of the heat loss from the fins to that lost from the plane wall. With spacing s there are $1/s$ fins per metre and

$$E=\frac{q_{in}}{sh(T_1-T_A)}=\frac{2}{ms}\tanh mL. \tag{13.7.10}$$

Here we have made the assumption that the heat transfer coefficient from the plane wall is the same as that from the side of the fin. This is not necessarily true and allowance can easily be made for any difference when required.

We can evaluate these quantities for copper fins of thickness 1.0 mm, length 50 mm and spacing 10 mm. Assuming a heat transfer coefficient of 20 W m^{-2} K^{-1}, we have

$$m=\left(\frac{2\times 20}{386\times 10^{-3}}\right)^{\frac{1}{2}}=10.2\ \mathrm{m}^{-1},\quad mL=0.509$$

and

$$\eta=\frac{1}{0.509}\tanh 0.509=0.922.$$

This high value of the fin efficiency suggests that the thermal conductivity of the copper is sufficiently high to keep the whole of the fin close to the wall temperature. We can confirm this by calculating the end temperature T_E by putting $x=L$ in equation (13.7.7) giving

$$\frac{T_E - T_A}{T_1 - T_A} = 0.883.$$

The enhancement factor E is found to be 9.20, showing that the presence of such fins gives a worthwhile increase in the heat transfer rate from the wall.

References

[1] Crank, *The Mathematics of Diffusion*, OUP.
[2] Southwell, *Relaxation Methods*, OUP.

14

Diffusion and mass transfer

14.1 Introduction

In chapter 2 we outlined the mechanics of the transport of momentum, energy and matter by the molecular motion in a gas. In § 2.8 we derived the basic diffusion laws in the form of equation (2.8.6) or (2.8.7) for the very simple case of a two-component gas mixture containing marked molecules and ordinary molecules of identical size and mass. When expressed in terms of molar quantities the diffusion law takes the form of a gradient transport equation relating the *molar flux* N to the *concentration gradient* $\mathrm{d}C/\mathrm{d}x$, i.e.

$$N = -\mathscr{D}\frac{\mathrm{d}C}{\mathrm{d}x}, \tag{14.1.1}$$

where N is the molar flux of the marked molecules expressed in $\mathrm{kmol\ m^{-2}\ s^{-1}}$; C is the molar concentration expressed in $\mathrm{kmol\ m^{-3}}$; and $\mathscr{D}$ is the diffusion coefficient expressed in units of $\mathrm{m^2\ s^{-1}}$.

As already noted in chapter 2, this expression is known as *Fick's law of diffusion* and is the counterpart of Fourier's law of heat conduction.

As derived in § 2.8, and as expressed in (14.1.1), the diffusion law appears in its simplest one-dimensional form. However, the molar flux **N** (like the heat flux **q**) is in fact a vector quantity having both magnitude and direction. We can therefore re-state the equation in vector form as

$$\mathbf{N} = -\mathscr{D}\ \mathrm{grad}\ C. \tag{14.1.2}$$

We also considered in § 2.8 the slightly more complex case of the diffusion of molecules of different size in a binary gas mixture under steady-state conditions with zero total molar flux of the gas mixture. We arrived at equations (2.8.15) and (2.8.16) for the molecular flux of components (1) and (2) respectively. It will be seen that these may be expressed in the same form as equation (14.1.1), i.e. for component (1) we have for the molar flux N_1

$$N_1 = -\mathscr{D}_{12}\frac{\mathrm{d}C_1}{\mathrm{d}x}, \tag{14.1.3}$$

where $\mathscr{D}_{12}$ is the coefficient of diffusion of component (1) through component (2) and is given by equation (2.8.17) or (2.8.21), i.e.

$$\mathscr{D}_{12}=\frac{1}{3}\left(\frac{2}{\pi}\right)^{\frac{3}{2}}\frac{(kT)^{\frac{3}{2}}}{P\sigma_{12}^2}\sqrt{\left(\frac{1}{m_1}+\frac{1}{m_2}\right)}, \tag{14.1.4}$$

where T is the absolute temperature of the gas mixture; P is the total pressure of the gas mixture; k is Boltzmann's constant; m_1 and m_2 are the actual molecular masses; and σ_{12} is the mean molecular diameter.

Thus we see that Fick's law (14.1.1) or (14.1.3) applies equally to the more general case of *binary diffusion in a gas mixture* comprising molecules of two different sizes, provided the net molar flux is zero.

If we re-write the expression (14.1.4) for the diffusion coefficient in terms of the relative molecular masses M_{r1} and M_{r2} for the two components, we have as in (2.8.22)

$$\mathscr{D}_{12}=2.1\times 10^{-22}\frac{T^{\frac{3}{2}}}{P\sigma_{12}^2}\sqrt{\left(\frac{1}{M_{r1}}+\frac{1}{M_{r2}}\right)}, \tag{14.1.5}$$

where the pressure P is expressed in N m^{-2} and σ_{12} is measured in metres. Some adjustment to the value of the constant may be necessary, however, in view of the simplifying assumptions which were introduced in the derivation from the kinetic theory of gases.

In many practical applications we are concerned with the diffusion of another gaseous component through air. Air itself is a mixture of gases but may generally be treated as if it was a single component having an average value for the relative molecular mass M_r and an average value for the molecular diameter d. It will be appreciated that the difference between the relative molecular masses of O_2 and N_2 is not very great. Some typical values for the diffusion coefficient through air at a pressure of 1 bar and a temperature of 20 °C are given below for a number of different gaseous substances:

H_2 $\quad 6.5\times 10^{-5}$ m^2 s^{-1},

He $\quad 2.4\times 10^{-5}$ m^2 s^{-1};

CO_2 $\quad 1.6\times 10^{-5}$ m^2 s^{-1},

C_6H_6 $\quad 0.88\times 10^{-5}$ m^2 s^{-1},

C_7H_8 $\quad 0.84\times 10^{-5}$ m^2 s^{-1}.

It will be seen that the diffusion coefficient decreases with increasing molecular size. This follows from the influence of the mean diameter σ_{12}

and the relative molecular mass in equation (14.1.5). It is noteworthy, however, that the effect of the relative molecular mass of the diffusing component is through the group $\sqrt{(1/M_{r1}+1/M_{r2})}$ in (14.1.5). If the relative molecular mass of the diffusing component is much larger than that of the medium (e.g. air) through which it is diffusing, the diffusion coefficient will be primarily a function of the relative molecular mass of the lighter species and will not be significantly affected by that of the diffusing component. Thus, since molecular dimensions increase only slightly with increasing molecular mass, most large molecules have very similar values for the diffusion coefficient through air.

While it is reassuring to have a theoretical basis for the evaluation of the diffusion coefficient in a two-component gas mixture, it has to be noted that equation (14.1.4) or (14.1.5) is of limited use for purposes of direct calculation because of the difficulty of determining the value of the mean molecular diameter. The quantity σ_{12} cannot be measured directly and has to be inferred from other physical measurements. For most commonly encountered gas systems, however, experimental values are available for the diffusion coefficient.

When we turn our attention to diffusion in liquids we have no theoretical basis for an analysis since there is no counterpart to the kinetic theory of gases. As with the other molecular transport processes, however, we find that a simple gradient transport equation in the form of (14.1.1) or (14.1.2) accords well with experimental observation. We may therefore regard Fick's law as an experimentally determined relationship of wide application to both gaseous and liquid systems. Only in the case of some highly non-ideal mixtures does any significant departure occur. In all other cases the minor departures from ideal conditions may be allowed for by treating the diffusivity $\mathscr{D}$ as a mild function of composition.

Diffusivities for liquid systems have to be determined by experimental measurement and are found to be typically an order of magnitude 10^4 smaller than those for gaseous systems. For example the diffusion coefficient for NH_3 through water at 20 °C is found to be approximately $1.7\times10^{-9}\ \mathrm{m^2\ s^{-1}}$.

It should be noted that we are confining our attention to *binary diffusion*. Multi-component diffusion is a much more complicated subject and is outside the scope of this book.

14.2 Analysis of binary diffusion in a gas

We can now consider the more general case of binary diffusion in a gas mixture where the net molar flux is not necessarily zero. For simplicity, however, we will confine the analysis to the one-dimensional case of diffusion and mass flow in the x direction as illustrated diagrammatically in figure 127.

We will assume a constant value for the total pressure of the gas mixture so that the molar concentrations C_1 and C_2 of the two components must be governed by the condition that

$$C_1+C_2=C=\text{constant}. \tag{14.2.1}$$

It follows from (14.2.1) that

$$\frac{dC_1}{dx}=-\frac{dC_2}{dx}. \tag{14.2.2}$$

If we assume that the gas mixture has an overall drift velocity U_0 in the x direction, as indicated in figure 127, we can express the molar flux of each component in the following form, as in (2.8.11) and (2.8.12), i.e.

$$N_1=C_1U_0-\tfrac{1}{3}\bar{\lambda}_1\bar{v}_1\frac{dC_1}{dx} \tag{14.2.3}$$

and

$$N_2=C_2U_0-\tfrac{1}{3}\bar{\lambda}_2\bar{v}_2\frac{dC_2}{dx}. \tag{14.2.4}$$

If we now multiply equation (14.2.3) by C_2 and equation (14.2.4) by C_1 we can eliminate the U_0 term by subtraction and obtain the result

$$N_1C_2-N_2C_1=-\tfrac{1}{3}C_2\bar{\lambda}_1\bar{v}_1\frac{dC_1}{dx}+\tfrac{1}{3}C_1\bar{\lambda}_2\bar{v}_2\frac{dC_2}{dx}.$$

Making use of (14.2.2) we can express this result in the form

$$N_1C_2-N_2C_1=-\mathscr{D}_{12}C\frac{dC_1}{dx}, \tag{14.2.5}$$

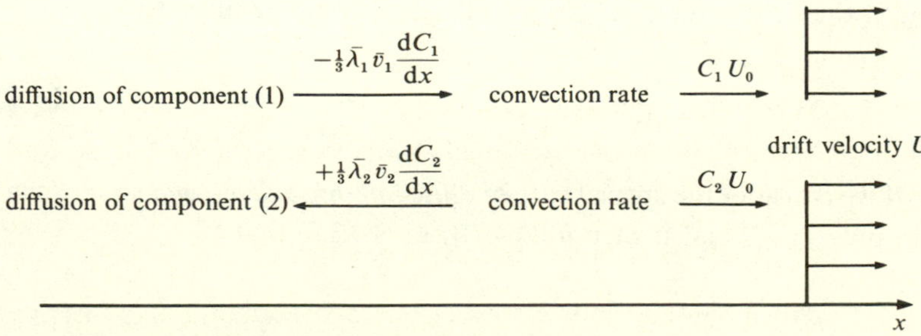

Fig. 127. Diffusion and convection.

where $\mathscr{D}_{12}$ is defined by

$$\mathscr{D}_{12}=\frac{1}{3}\frac{(C_2\bar{\lambda}_1\bar{v}_1+C_1\bar{\lambda}_2\bar{v}_2)}{C_1+C_2}. \tag{14.2.6}$$

It will be seen that this definition of the diffusion coefficient $\mathscr{D}_{12}$ is identical to that of equation (2.8.17).

Equation (14.2.5) may be regarded as the most general form of the diffusion law for a binary gas mixture since we have not introduced any restriction in regard to the net molar flux (N_1+N_2) or to the drift velocity U_0. However, we will now consider three special cases of practical importance.

If we have the condition of *zero net molar flux*, which may be described alternatively as *equal and opposite molar diffusion*, we can say that

$$N_1+N_2=0. \tag{14.2.7}$$

It then follows from (14.2.5) that

$$N_1=-\mathscr{D}_{12}\frac{\mathrm{d}C_1}{\mathrm{d}x}, \tag{14.2.8}$$

which is identical with the result previously quoted as equation (14.1.3).

If we have a *very dilute mixture* in which the molar concentration of the diffusing component C_1 is very small compared with C_2, we can say that $C_2\simeq C$ and it will be evident from equation (14.2.6) that the diffusion coefficient $\mathscr{D}_{12}$ is given approximately by

$$\mathscr{D}_{12}=\tfrac{1}{3}\bar{\lambda}_1\bar{v}_1. \tag{14.2.9}$$

Hence from (14.2.3) we have

$$N_1=C_1U_0-\mathscr{D}_{12}\frac{\mathrm{d}C_1}{\mathrm{d}x} \tag{14.2.10}$$

and if the drift velocity U_0 is zero the last result reduces to

$$N_1=-\mathscr{D}_{12}\frac{\mathrm{d}C_1}{\mathrm{d}x}. \tag{14.2.11}$$

If we consider the special case of *diffusion through a stagnant medium* we can put $N_2=0$ and it then follows from (14.2.5) that

$$N_1=-\frac{C}{C_2}\mathscr{D}_{12}\frac{\mathrm{d}C_1}{\mathrm{d}x} \tag{14.2.12}$$

or

$$N_1 = -\frac{C}{C-C_1}\mathscr{D}_{12}\frac{\mathrm{d}C_1}{\mathrm{d}x}.$$

When dealing with diffusion in gases it is often more convenient to work in terms of partial pressures rather than molar concentrations. Noting that for a perfect gas the partial pressure is related to the molar concentration by

$$p_1 = C_1RT \tag{14.2.13}$$

we can express equation (14.2.5) in the alternative form:

$$N_1p_2 - N_2p_1 = -\mathscr{D}_{12}\frac{P}{RT}\frac{\mathrm{d}p_1}{\mathrm{d}x}, \tag{14.2.14}$$

where P is the total pressure.

As with equation (14.2.5), certain special cases of equation (14.2.14) can be considered.

For diffusion through a stagnant medium, $N_2 = 0$, and equation (14.2.14) becomes

$$N_1 = -\frac{P}{p_2}\frac{\mathscr{D}_{12}}{RT}\frac{\mathrm{d}p_1}{\mathrm{d}x} \tag{14.2.15}$$

and for the case of a very dilute mixture in which $p_1 \ll p_2$ so that $p_2 \simeq P$ we have

$$N_1 = -\frac{\mathscr{D}_{12}}{RT}\frac{\mathrm{d}p_1}{\mathrm{d}x}. \tag{14.2.16}$$

This equation is also valid for the case of zero net molar flux.

A word of caution about units and nomenclature is appropriate at this stage. We can multiply both sides of the special form of the diffusion equation (14.2.8) by the relative molecular mass M_{r1} and obtain a relationship between the mass flux n and the mass concentration c,

$$n_1 = -\mathscr{D}_{12}\frac{\mathrm{d}c_1}{\mathrm{d}x}. \tag{14.2.17}$$

Thus in effect we may use either mass or molar units for this form of the diffusion equation. However, no such manipulation is possible with the general form, equation (14.2.5), for which the use of molar units is essential.

Up to this stage we have denoted the two species with the subscripts 1 and 2 in order to preserve consistency with chapter 2 and the standard texts on the kinetic theory of gases. However, from now on we will reserve the use of numerical subscripts to denote positions and will identify the species by an alphabetical subscript. This is consistent with the convention for naming species in the context of chemical kinetics. Since usually there is little doubt which species are involved in the process the subscripts to $\mathscr{D}$ will frequently be omitted.

Thus the general diffusion equation will normally be written as

$$N_A C_B - N_B C_A = -\mathscr{D} C \frac{\mathrm{d}C_A}{\mathrm{d}x} \tag{14.2.18}$$

or

$$N_A p_B - N_B p_A = -\frac{\mathscr{D} P}{RT} \frac{\mathrm{d}p_A}{\mathrm{d}x}. \tag{14.2.19}$$

14.3 Unidirectional diffusion

As a first example of the use of the diffusion equations let us consider Stefan's classic method of measuring the diffusion coefficient of a vapour through air. In this method a sample of a volatile liquid is placed in the base of a vertical tube as illustrated in figure 128. The liquid evaporates and the vapour diffuses through the effectively stagnant air in the upper part of the tube before escaping to the atmosphere. It is assumed that there is a gentle breeze blowing past the tube so that there is no accumulation of vapour above the top of the tube. Since equilibrium is rapidly attained at

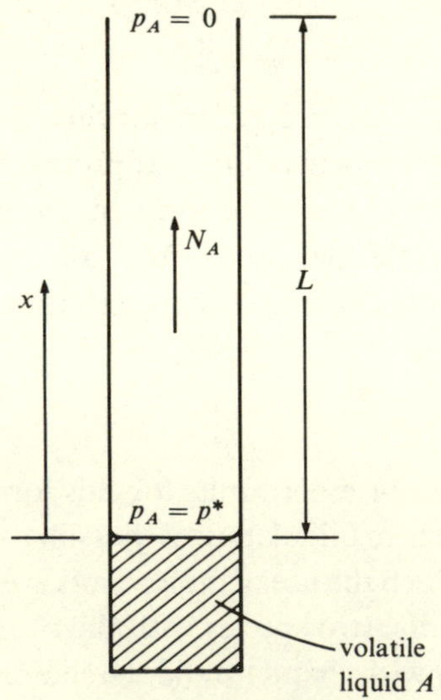

Fig. 128. Stefan's experiment.

the interface, the rate of evaporation is controlled by the diffusion through the air space. It is therefore possible to deduce the diffusion coefficient from the measured rate of weight loss.

Denoting the vapour as species A and the air as B, and noting that the flux of air is effectively zero, we can write the diffusion equation (14.2.19) in the form

$$N_A(P-p_A)=-\frac{\mathscr{D}P}{RT}\frac{\mathrm{d}p_A}{\mathrm{d}x}. \tag{14.3.1}$$

In order to integrate this equation we need to know the partial pressure of A at both ends of the tube. We will assume that equilibrium is maintained at the interface so that at $x=0, p_A=p^*$, the saturated vapour pressure of the liquid. We will also assume that at the top of the tube, $x=L$, the partial pressure falls to zero. We can therefore integrate equation (14.3.1) giving

$$N_A\int_0^L \mathrm{d}x=-\frac{\mathscr{D}P}{RT}\int_{p^*}^0\frac{\mathrm{d}p_A}{P-p_A}$$

or

$$N_AL=\frac{\mathscr{D}P}{RT}\log_e\left(\frac{P}{P-p^*}\right). \tag{14.3.2}$$

The diffusion coefficient can be obtained from this equation since all the other quantities are easily measured. With care this simple experiment can give a reliable measure of the diffusion coefficient.

If the saturated vapour pressure p^* is very much less than the total pressure P, the volatile species will be present only in small amounts and the logarithmic term in equation (14.3.2) can be approximated thus

$$\log_e\left(\frac{P}{P-p^*}\right)\simeq\frac{p^*}{P},$$

so that

$$N_AL=\frac{\mathscr{D}p^*}{RT}$$

or

$$N_A=\frac{\mathscr{D}C^*}{L}, \tag{14.3.3}$$

where

$$C^*=\frac{p^*}{RT}$$

and is the concentration associated with the saturated vapour pressure. This last result could, however, have been obtained more simply by integrating the form of the diffusion equation appropriate to dilute mixtures, equation (14.2.11).

As is expected for dilute mixtures, this result is similar to the corresponding result for heat transfer, equation (13.2.1), and we can define a mass transfer coefficient k_g by analogy with the heat transfer coefficient defined in equation (13.2.4). Thus

$$k_g = \frac{N_A}{\Delta C_A} \tag{14.3.4}$$

and using equation (14.3.3) we have in this case

$$k_g = \frac{\mathscr{D}}{L}. \tag{14.3.5}$$

As with heat transfer coefficients, mass transfer coefficients are often conveniently expressed in dimensionless form and the resulting dimensionless group is called the Sherwood number, *Sh*. This is defined by

$$Sh = \frac{k_g L}{\mathscr{D}} \tag{14.3.6}$$

and it is seen that for the case of Stefan's experiment

$$Sh = 1. \tag{14.3.7}$$

It must, however, be appreciated that the analogy between heat and mass transfer is appropriate for dilute mixtures only. For rich mixtures we cannot define a mass transfer coefficient since the mass flux is not proportional to the concentration difference.

Diffusion in systems with cylindrical and spherical symmetry will be dealt with very briefly as the analysis is barely distinguishable from that for heat transfer already presented in §§13.4 and 13.5.

As can be seen from figure 129, the molar flow rate Q in cylindrical symmetry can be related to the molar flux N_A by an equation analogous to

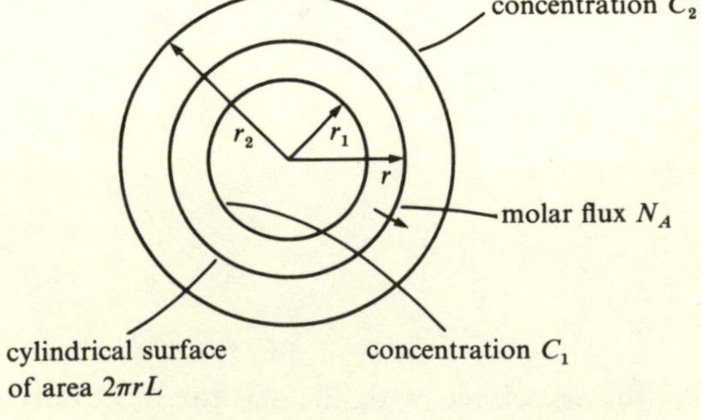

Fig. 129. Diffusion through a cylindrical shell.

equation (13.4.1), i.e.

$$N_A = \frac{Q}{2\pi rL}. \tag{14.3.8}$$

Hence from equation (14.2.12)

$$\frac{Q}{2\pi rL} = -\frac{C\mathscr{D}}{C-C_A}\frac{dC_A}{dr}. \tag{14.3.9}$$

If the concentration of A is C_1 at $r=r_1$ and C_2 at $r=r_2$, this can be integrated to give

$$\frac{Q}{2\pi L}\log_e\left(\frac{r_2}{r_1}\right) = \mathscr{D}C\log_e\left(\frac{C-C_2}{C-C_1}\right). \tag{14.3.10}$$

For the special case when both C_1 and C_2 are very much less than C, this reduces to

$$\frac{Q}{2\pi L}\log_e\left(\frac{r_2}{r_1}\right) = \mathscr{D}(C_1-C_2), \tag{14.3.11}$$

an equation exactly analogous to equation (13.4.3).

A similar analysis is possible for cases of spherical symmetry. As in § 13.5, the only difference from the previous case is that the molar flux is now given by

$$N_A = \frac{Q}{4\pi r^2}, \tag{14.3.12}$$

so that,

$$\frac{Q}{4\pi r^2} = -\frac{\mathscr{D}C}{C-C_A}\frac{dC_A}{dr}. \tag{14.3.13}$$

On integration subject to the boundary conditions $C=C_1$ on $r=r_1$ and $C=C_2$ on $r=r_2$ we have

$$\frac{Q}{4\pi}\left(\frac{1}{r_1}-\frac{1}{r_2}\right) = \mathscr{D}C\log_e\left(\frac{C-C_2}{C-C_1}\right) \tag{14.3.14}$$

or for dilute mixtures where C_1 and $C_2 \ll C$,

$$\frac{Q}{4\pi}\left(\frac{1}{r_1}-\frac{1}{r_2}\right) = \mathscr{D}(C_1-C_2), \tag{14.3.15}$$

which is the mass transfer analogue of equation (13.5.2).

A case of special interest is the diffusion of a volatile sphere into an infinite stagnant medium. We can obtain this result from equation (14.3.14) by letting $r_2 \to \infty$ and interpreting C_2 as the concentration at great distances from the sphere. Very often C_2 will equal zero. Thus

$$Q = 4\pi r_1 \mathscr{D} C \log_e \left(\frac{C - C_2}{C - C_1} \right) \tag{14.3.16}$$

or, if dilute

$$Q = 4\pi r_1 \mathscr{D} (C_1 - C_2). \tag{14.3.17}$$

At the surface of the sphere, $r = r_1$, the molar flux is given by

$$N_A = \frac{\mathscr{D}}{r_1} (C_1 - C_2) \tag{14.3.18}$$

and hence the mass transfer coefficient is given by

$$k_g = \frac{\mathscr{D}}{r_1}. \tag{14.3.19}$$

Expressing this in the form of a Sherwood number based on the diameter of the sphere, we find that

$$Sh = 2, \tag{14.3.20}$$

a result exactly analogous to the result $Nu = 2$ for heat transfer from a sphere.

In general we find that whatever the geometry we get very similar expressions for the Nusselt number for heat conduction and the Sherwood number for dilute diffusion. We will see later that the analogy is equally good for convection. For convective heat transfer it is found that the Nusselt number is a function of the Reynolds number and the Prandtl number. For mass transfer in the same geometry we find that we get the same relationship but with the Sherwood number replacing the Nusselt number and the Schmidt number, $Sc = \nu/\mathscr{D}$, replacing the Prandtl number.

14.4 Counter-diffusion

In the examples considered in the previous section we were dealing with the case when one of the species was stagnant. There exist, however, circumstances of importance in which both species are diffusing. A single example will be sufficient to illustrate the method of solution in this case.

A method has recently been proposed for the measurement of the diffusivity of hydrogen through hydrogen chloride. This involves putting a sample of the liquid metal Gallium in a vessel with a capillary neck of length

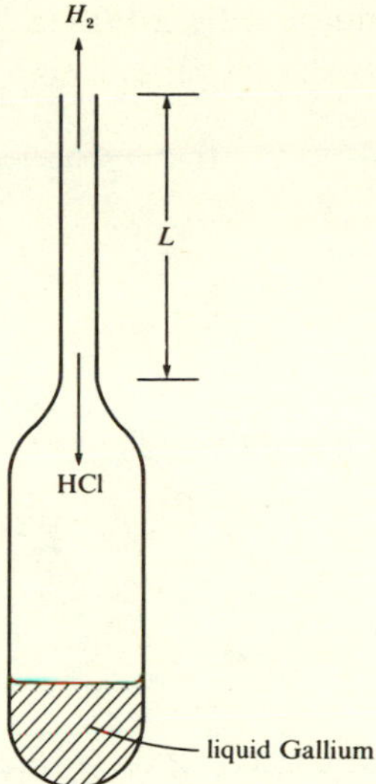

Fig. 130. Apparatus for measuring the diffusivity of HCl through H_2.

L as shown in figure 130. The vessel is suspended in an atmosphere of HCl which diffuses down the capillary and reacts rapidly and irreversibly with the Gallium to liberate hydrogen according to the reaction

$$6\,HCl + 2\,Ga \rightarrow 3\,H_2 + 2\,GaCl_3.$$

The hydrogen diffuses out of the vessel counter-current to the entering hydrogen chloride. Since the reaction is rapid the rate of uptake of chlorine is controlled by the diffusion in the capillary and the diffusion coefficient can be deduced from the measured rate of increase in weight.

This situation can be generalised by considering the reaction in which one mole of the gas A reacts with an involatile substance C to yield n moles of gas B plus an involatile product D,

$$A(g) + C \rightarrow nB(g) + D.$$

For every mole of A entering the vessel, n moles of B leave and hence the molar fluxes of A and B are related by,

$$N_B = -nN_A. \tag{14.4.1}$$

The presence of the minus sign indicating fluxes in opposite directions should be noted.

Substituting into the general diffusion equation (14.2.19) yields

$$N_A(p_B+np_A)=-\frac{\mathscr{D}P}{RT}\frac{\mathrm{d}p_A}{\mathrm{d}x}, \tag{14.4.2}$$

which can be rearranged into the form

$$N_A\int_0^L \mathrm{d}x=-\frac{\mathscr{D}P}{RT}\int_{p_1}^{p_2}\frac{\mathrm{d}p_A}{P+(n-1)p_A}. \tag{14.4.3}$$

Here we have denoted the partial pressure of A in the vessel, $x=0$, as p_1 and the partial pressure of A outside the vessel, $x=L$, as p_2.

On integration this gives

$$N_AL=\frac{\mathscr{D}P}{RT(n-1)}\log_e\left(\frac{P+(n-1)p_1}{P+(n-1)p_2}\right), \tag{14.4.4}$$

a result which can be seen to reduce to the form of equation (14.3.2) for the special case of $n=0$.

14.5 Total evaporation of a volatile sphere

If a sphere of a volatile material of radius a is suspended in an infinite stagnant medium the evaporation rate will be given by equation (14.3.16) which can be written in the form

$$Q=4\pi a\frac{\mathscr{D}P}{RT}\log_e\left(\frac{P}{P-p^*}\right). \tag{14.5.1}$$

Here we have assumed that the partial pressure of the volatile species is negligible at great distances from the sphere and that the partial pressure adjacent to the surface of the sphere is the saturated vapour pressure p^*.

As evaporation takes place the sphere will get smaller and the rate of loss of moles can be related to the evaporation rate by

$$-\frac{\mathrm{d}}{\mathrm{d}t}(\tfrac{4}{3}\pi a^3\rho)=Q, \tag{14.5.2}$$

where ρ is the molar density.

Thus from equation (14.5.1) we have

$$-4\pi a^2\rho\frac{\mathrm{d}a}{\mathrm{d}t}=4\pi a\frac{\mathscr{D}P}{RT}\log_e\left(\frac{P}{P-p^*}\right) \tag{14.5.3}$$

or

$$-\rho a\frac{\mathrm{d}a}{\mathrm{d}t}=\frac{\mathscr{D}P}{RT}\log_e\left(\frac{P}{P-p^*}\right). \tag{14.5.4}$$

We can integrate this equation subject to the initial condition that at time $t=0$ the radius equals the initial radius a_0. This gives

$$\frac{\rho}{2}(a_0^2-a^2)=\frac{\mathscr{D}Pt}{RT}\log_e\left(\frac{P}{P-p^*}\right). \tag{14.5.5}$$

Inspection of equation (14.5.5) shows that t remains finite as the radius tends to zero showing that evaporation is complete at time t_1 given by

$$t_1=\frac{\rho RTa_0^2}{2\mathscr{D}P\log_e[P/(P-p^*)]}. \tag{14.5.6}$$

Thus the mean evaporation rate $\bar{Q}$ is given by

$$\bar{Q}=\frac{4\pi a_0^3\rho}{3t_1}=\frac{8}{3}\pi a_0\frac{\mathscr{D}P}{RT}\log_e\left(\frac{P}{P-p^*}\right). \tag{14.5.7}$$

Comparing this result with equation (14.5.1) shows that the mean evaporation rate is two-thirds of the initial rate.

The apparently similar case of a cooling sphere is in fact quite different for two reasons. In this case the heat transfer driving force ΔT decreases as time passes whereas in the mass transfer case the driving force p^* remains constant. Furthermore the size of the cooling sphere remains constant but the evaporating sphere decreases in size with a corresponding increase in the ratio of surface area to volume.

The rate of heat loss is given by equation (13.5.3) and can be related to the rate of change of temperature by

$$-\frac{\mathrm{d}}{\mathrm{d}t}\left(\frac{4}{3}\pi a^3\rho cT\right)=Q=4\pi ak(T-T_\infty), \tag{14.5.8}$$

which on integration gives

$$\log_e\left(\frac{T_0-T_\infty}{T-T_\infty}\right)=\frac{3kt}{a^2\rho c}. \tag{14.5.9}$$

Unlike the volatile sphere for which evaporation is complete in a finite time, the cooling sphere only approaches its equilibrium temperature exponentially.

We can illustrate some of these results by considering the rate of evaporation of a simple moth-ball. This we will take to be a sphere of naphthalene of initial radius 10 mm. At 20 °C the vapour pressure of naphthalene is about 30 N m^{-2} and the diffusivity through air about 0.8×10^{-5} m^2 s^{-1}. The density of solid naphthalene is 700 kg m^{-3} and the relative molecular mass is 128.

We must first evaluate the quantity $\log_e[P/(P-p^*)]$ which, with $P = 10^5\ \mathrm{N\ m^{-2}}$, is 3.0005×10^{-4}, a number that differs from p^*/P by less than 0.02%, showing that we can use the forms appropriate for dilute mixture without loss of accuracy. Thus the concept of a mass transfer coefficient is meaningful and from equation (14.3.19) this is given by

$$k_g = \frac{\mathscr{D}}{a} = \frac{0.8\times10^{-5}}{10^{-2}} = 0.8\times10^{-3}\ \mathrm{m\ s^{-1}}.$$

The evaporation rate Q is given by equation (14.5.1) which may be simplified, since we are dealing with a dilute mixture, to

$$Q = \frac{4\pi a\mathscr{D}p^*}{RT} = \frac{4\pi\times10^{-2}\times0.8\times10^{-5}\times30}{8314\times293} = 1.238\times10^{-11}\ \mathrm{kmol\ s^{-1}}.$$

The mean evaporation rate is two-thirds of this value,

$$\bar{Q} = 8.25\times10^{-12}\ \mathrm{kmol\ s^{-1}}$$

and the molar content of the original ball is given by

$$\frac{4}{3}\pi a^3\rho = \frac{4\pi}{3}\times10^{-6}\times\frac{700}{128} = 2.29\times10^{-5}\ \mathrm{kmol}.$$

Thus evaporation will be complete after

$$\frac{2.29\times10^{-5}}{8.25\times10^{-12}} = 2.78\times10^6\ \mathrm{s} = 32.1\ \text{days},$$

showing that moth-balls should be renewed monthly.

14.6 Diffusion through stationary media

As was discussed in § 14.2, the simple form of the diffusion equation

$$N = -\mathscr{D}\frac{\mathrm{d}C}{\mathrm{d}x} \tag{14.6.1}$$

is valid either when there is no net flux or when we are concerned with the diffusion of a dilute species through a stagnant second species. Under these circumstances we can express equation (14.6.1) more generally in the vector form

$$\mathbf{N} = -\mathscr{D}\ \mathrm{grad}\ C. \tag{14.6.2}$$

If we now consider a unit volume of space in which the species concerned is being created by chemical reaction at a rate R kmol m^{-3} we can write down a mass balance for that species in the form

rate of creation = rate of accumulation + outflow due to diffusion

$$R = \frac{\partial C}{\partial t} + \operatorname{div} \mathbf{N}.$$

Note that we have no convection terms in this balance since we have already confined our attention to systems with zero, or negligible, net flux.

Substituting from equation (14.6.2) and making the assumption that $\mathscr{D}$ is independent of position gives

$$\frac{\partial C}{\partial t} = R + \mathscr{D} \operatorname{div} \operatorname{grad} C = R + \mathscr{D} \nabla^2 C. \tag{14.6.3}$$

This equation is seen to have an identical mathematical form to the heat conduction equation (13.6.3), the rate of creation of the species R being analogous to the heat production rate Q_s, and the molecular diffusivity $\mathscr{D}$ replacing the thermal diffusivity α.

Like the heat conduction equation, equation (14.6.3) can be expressed in Cartesian co-ordinates

$$\frac{\partial C}{\partial t} = R + \mathscr{D}\left(\frac{\partial^2 C}{\partial x^2} + \frac{\partial^2 C}{\partial y^2} + \frac{\partial^2 C}{\partial z^2}\right) \tag{14.6.4}$$

and the forms for other co-ordinate systems can be written down by analogy with equations (13.6.5) and (13.6.6). Indeed, the analogy is so close that all the arguments presented in §13.6 can be adapted for the mass transfer case simply by replacing α by $\mathscr{D}$ and $Q_s/\rho c_v$ by R.

Very commonly, however, we will be concerned with a species that is being destroyed by an nth-order chemical reaction, so that $R = -kC^n$, giving

$$\frac{\partial C}{\partial t} = -kC^n + \mathscr{D} \nabla^2 C, \tag{14.6.5}$$

this has no counterpart in heat transfer analysis since heat removal per unit volume is rarely if ever proportional to some power of the absolute temperature. Equation (14.6.5) forms the basis of the analysis of gas absorption with chemical reaction which is considered in more detail in chapter 22.

15

Dimensional analysis of transfer processes

15.1 Statement of the basic principle

The use of dimensionless ratios for the presentation of experimental results has been a characteristic feature in the development of fluid mechanics. This practice is closely linked to the concept of dynamical similarity, or the principle of similitude, which was originally formulated by Lord Rayleigh. The basic methods of dimensional analysis, and the procedure for the formulation of dimensionless ratios, were outlined very briefly in § 1.8 and we have already made extensive use of this procedure in chapters 2–12. It will be appropriate at this stage, however, to restate the basic principles in a slightly more formal way before extending the application to a wider range of transfer processes.

The first important point to establish is that, in any equation which expresses a fundamental relationship between a number of different physical quantities, each term in the equation must have the same dimensions. This is the principle of *dimensional homogeneity* or congruence. All physical quantities can be expressed in terms of a limited number of *fundamental dimensions* such as mass, length, time, electric current, absolute temperature, etc. The question of how many fundamental dimensions are necessary in order to describe all known physical quantities raises some interesting points which are beyond the scope of this book. For our present purposes we can simply say that when dealing with mechanical problems in fluid dynamics we require only three fundamental dimensions, mass, length, and time. When dealing with problems of heat transfer, however, we need to add the thermodynamic temperature as a fourth dimension. Furthermore, in the analysis of mass transfer, particularly when we work in terms of molar concentrations and molar fluxes, we require a fifth dimension known as 'amount of substance' which has as its unit, the mole.

The choice of fundamental dimensions is to some extent arbitrary. We could for example select *force* rather than *mass* as a fundamental quantity and proceed to express all other mechanical quantities in terms of the dimensions of force, length, and time. This custom has in fact been widely followed in the past, but currently accepted practice is to choose mass, length, and time as the base quantities for use in all physical problems involving mechanics.

By international convention, as formalised in SI, seven physical quantities are chosen for use as dimensionally *independent base quantities.* All other physical quantities are regarded as being derived from the base quantities. For the purposes of this book we can omit two of the chosen base quantities, namely electric current and luminous intensity, which leaves us with the following five dimensionally independent base quantities:

Physical quantity	Symbol for quantity	SI unit	Symbol for unit
length	l	metre	m
mass	m	kilogram	kg
time	t	second	s
temperature	T	kelvin	K
amount of substance	n	mole	mol

Care must be taken to distinguish between the *symbols for the base quantities* on the one hand and the symbols for the SI units on the other hand.

Although we are not under any obligation to identify our selection of fundamental dimensions with the base quantities listed above, it is obviously convenient to do so and also to represent these dimensions by the same symbols placed in square brackets, e.g. $[l]$, $[m]$, $[t]$, $[T]$, and $[n]$.

Having established the principle of dimensional homogeneity we can state the main proposition of dimensional analysis, which is sometimes referred to as Buckingham's theorem, in the following terms. If a physical quantity Q_1 can be expressed as a function of $(n-1)$ other independent physical quantities or variables Q_2, Q_3, Q_4, ..., Q_n, i.e. if we have a functional relationship between n physical quantities in the form

$$Q_1 = f(Q_2, Q_3, Q_4, \ldots, Q_n) \tag{15.1.1}$$

we can reduce the total number of mathematical variables from n to $(n-p)$, where p is the number of fundamental dimensions required to describe the physical quantities involved, by forming dimensionless ratios according to the following procedure:

(1) pick out p of the $(n-1)$ physical quantities on the right-hand side of (15.1.1) and regard them as *primary quantities*, noting that each of the fundamental dimensions must appear at least once among the primary quantities;

(2) the remaining $(n-p)$ physical quantities which appear in equation (15.1.1) are now expressed as $(n-p)$ dimensionless ratios, which are known as numerics or Π quantities, using the selected primary quantities for the purpose of forming the ratios, i.e. each Π quantity is formed from products or ratios of powers of the primary Qs;

(3) the original equation (15.1.1) may now be replaced by a functional relationship involving only the $(n-p)$ quantities in the form

$$\Pi_1 = F(\Pi_2, \Pi_3, \Pi_4, \ldots, \Pi_{(n-p)}). \tag{15.1.2}$$

15.2 Note on the dimensions for temperature and heat

In the previous section we outlined the principles of dimensional analysis which can be placed on the formal footing of Buckingham's theorem. We also stated, but did not attempt to justify, the proposition that when dealing with problems of heat transfer we need to add the dimension of temperature to the dimensions of mass, length, and time required for the analysis of mechanical systems. This statement is not self-evident and is worthy of further consideration.

According to the kinetic theory of gases the temperature of a gas is measured by the average kinetic energy of its molecules, as expressed by equation (1.2.1) or (2.7.1). If Boltzmann's constant could be regarded as being a dimensionless number the temperature T would have the dimensions of energy and would be measured in joules. The kinetic theory, however, presents only a restricted model and cannot be used to cover all aspects of the physical properties of matter. There are good scientific reasons for regarding the thermodynamic temperature as being a *dimensionally independent base quantity.* When dealing with any application involving thermodynamics or the transfer of heat, therefore, we must introduce the temperature T as a fourth fundamental dimension.

Besides the four independent dimensions of mass, length, time and temperature that are necessary for the description of heat transfer processes, we often find it convenient to introduce a further dimension, namely that of *heat* $[Q]$. According to the first law of thermodynamics, heat and work are interchangeable within certain limits and the dimensions for heat should therefore be $[ml^2t^{-2}]$. In some problems of fluid mechanics and heat transfer there is a measurable interchange of energy between thermal and mechanical quantities, for instance, in the flow of steam through a nozzle. In problems of this sort, therefore, it is appropriate to work in the four dimensions $[m]$, $[l]$, $[t]$, and $[T]$. In other problems, for instance the flow of a liquid through a heat exchanger, there is no measurable interchange between thermal and mechanical quantities. It is true that fluid friction will always produce some thermal effects, but in many cases this is insignificant compared with the heat that is being transferred between the walls of the exchanger and the fluid. In problems of this kind, therefore, it will be appropriate to introduce the fifth dimension $[Q]$ for heat. The introduction of this extra dimension will reduce by one the number of independent dimensionless groups. Typically the missing group will be u^2/c_pT, the ratio of kinetic energy to enthalpy, a group which is clearly irrelevant if there is no interchange between mechanical and thermal energy.

The table below lists the dimensions of the additional derived quantities we will require for analysing heat transfer on the $[m]$, $[l]$, $[t]$, $[T]$ system and on the $[m]$, $[l]$, $[t]$, $[T]$, $[Q]$ system.

	Dimensions in terms of	
Derived quantity	$[m]$, $[l]$, $[t]$, $[T]$	$[m]$, $[l]$, $[t]$, $[T]$, $[Q]$
Heat quantity	$[ml^2t^{-2}]$	$[Q]$
Heat transfer rate Q	$[ml^2t^{-3}]$	$[Qt^{-1}]$
Heat flux q	$[mt^{-3}]$	$[Ql^{-2}t^{-1}]$
Heat transfer coefficient h	$[mt^{-3}T^{-1}]$	$[Ql^{-2}t^{-1}T^{-1}]$
Specific heat c_p	$[l^2t^{-2}T^{-1}]$	$[Qm^{-1}T^{-1}]$
Thermal conductivity k	$[mlt^{-3}T^{-1}]$	$[Ql^{-1}t^{-1}T^{-1}]$
Thermal diffusivity α	$[l^2t^{-1}]$	$[l^2t^{-1}]$

15.3 Forced convection heat transfer

When considering heat transfer to a fluid flowing in a circular pipe, it will be reasonable to assume that the local value of the heat flux q will depend on the temperature difference ΔT between the wall of the pipe and the fluid, on the pipe diameter D, on the mean velocity u_m and mean temperature T_m of the fluid, and on the physical properties k, c_p, ρ, and μ. We can thus say that

$$q = f(\Delta T, k, c_p, \rho, \mu, D, u_m, T_m). \tag{15.3.1}$$

By Buckingham's Π theorem, this connection between nine quantities (one dependent and eight independent variables) can be reduced to a relationship between four dimensionless ratios if we assume five fundamental dimensions $[m]$, $[l]$, $[t]$, $[T]$, $[Q]$.

Using the method of § 15.1 we can take five primary quantities as follows:

Quantity	k	ρ	D	μ	ΔT
Dimensions	$[Ql^{-1}t^{-1}T^{-1}]$	$[ml^{-3}]$	$[l]$	$[ml^{-1}t^{-1}]$	$[T]$

We now require to form dimensionless ratios for q, u_m, c_p, and T_m. For Π_1 we can try

$$\frac{q}{k^a\rho^b D^c \mu^d \,\Delta T^e}.$$

Dimensionally we must have

$$\frac{[Q]}{[l^2t]}=\frac{[Q]^a}{[ltT]^a}\frac{[m]^b}{[l^3]^b}[l]^c\frac{[m]^d}{[lt]^d}[T]^e.$$

Hence for

$$\begin{array}{lll} Q & a=1 & \\ T & 0=-a+e & \therefore\ e=1, \\ t & -1=-a-d & \therefore\ d=0, \\ m & 0=b+d & \therefore\ b=0, \\ l & -2=-a-3b+c-d & \therefore\ c=-1. \end{array}$$

Therefore

$$\Pi_1=\frac{qD}{k\,\Delta T} \quad \text{which is the Nusselt number } Nu.$$

By a similar procedure we arrive at

$$\Pi_2=\frac{\rho u_m D}{\mu} \quad \text{which is the Reynolds number } Re,$$

$$\Pi_3=\frac{\mu c_p}{k} \quad \text{which is the Prandtl number } Pr,$$

and

$$\Pi_4=\frac{T_m}{\Delta T} \quad \text{which is simply a temperature ratio.}$$

We can therefore replace (15.3.1) with the relationship:

$$Nu=f\left(Re, Pr, \frac{T_m}{\Delta T}\right). \tag{15.3.2}$$

It is found experimentally that, provided the temperature difference ΔT is not too large, and provided the variations in the values of the physical properties of the fluid with temperature are not significant, the influence of the dimensionless temperature ratio $T_m/\Delta T$, or its inverse, $\Delta T/T_m$, is small. In these circumstances we can replace (15.3.2) with the simpler relationship:

$$Nu = f(Re, Pr). \tag{15.3.3}$$

We have already noted that the heat transfer coefficient h is defined by $h = q/\Delta T$ and we can therefore write the Nusselt number in the more convenient form, $Nu = hD/k$. Since neither ΔT nor T_m appear in the Reynolds or Prandtl numbers, it will be seen that the use of equation (15.3.3) in place of (15.3.2) implies that in these circumstances the Nusselt number is independent of the thermodynamic temperature of the fluid T_m, and that the heat transfer coefficient h is independent of the temperature difference ΔT. In other words, provided ΔT is reasonably small, the heat flux may be taken as being directly proportional to the temperature difference.

If we had chosen a different set of primary quantities, for instance c_p, ρ, D, u_m, and T, we would have found

$$\Pi_1 = \frac{h}{\rho u_m c_p},$$

which is the Stanton number St as defined in §9.6.

Thus in place of equation (15.3.3) we would have had,

$$St = f(Re, Pr). \tag{15.3.4}$$

The Stanton and Nusselt numbers are alternative dimensionless forms of the heat transfer coefficient for forced convection. It should be noted that

$$St = Nu/(Re\,Pr) \tag{15.3.5}$$

and that equation (15.3.4) gives no information that is not included in equation (15.3.3). The choice between these formulations is one of algebraic convenience only.

A further dimensionless group $c_p \rho u_m D/k$ would have arisen if we had taken k, ρ, D, u_m, and T as our primary quantities. This was defined in §9.5 as the Péclet number $Pé$ and clearly

$$Pé = Re\,Pr. \tag{15.3.6}$$

Like the Stanton number, this introduces no extra information though the formulation

$$Nu = f(Pé, Re) \tag{15.3.7}$$

is sometimes found to be convenient.

Instead of the formal use of Buckingham's theorem, we could have worked by inspection. We are already familiar with the Reynolds number from chapter 5 and the Nusselt number was introduced in chapter 9. Both are clearly relevant to this situation and all that is necessary is to find a dimensionless group containing the missing quantity c_p. Of the many possibilities the Prandtl number is found to be most convenient, as it has a constant value for any given fluid.

Yet another way of deriving these dimensionless groups was presented in §9.5, where it was shown that they arose naturally from the governing equations. However, the techniques of the present section have the advantage that they can be used for situations in which the governing equations are unknown.

Working with the $[m]$, $[l]$, $[t]$, $[T]$ system instead, we require five independent groups of which the temperature ratio $T_m/\Delta T$ and the Reynolds, Nusselt, and Prandtl numbers will suffice for four. As mentioned in § 15.2, the group u_m^2/c_pT_m is one possibility, or alternatively we could use the Eckert number $u_m^2/c_p\,\Delta T$ as defined in §9.5. As was pointed out in that section, both these groups are closely related to the Mach number $u_m/\sqrt{(\gamma RT)}$. It was shown in chapter 11 that the Mach number is only of relevance for high-speed flows, confirming our assumption that, for modest velocities, thermal and mechanical energy can be treated as independent quantities.

Thus in general we can say that

$$Nu=f(Re, Pr, Ec, (T_m/\Delta T)). \tag{15.3.8}$$

However, consideration of the governing equations in §§9.5 and 9.6 lead to equation (9.6.8),

$$Nu=f(Re, Pr, Ec), \tag{15.3.9}$$

showing that the group $T_m/\Delta T$, which is permitted by dimensional considerations, is in fact irrelevant. Thus we see that the expression for the Nusselt number does not contain the absolute temperature explicitly but it must of course be appreciated that the physical properties of the fluid, especially the viscosity, are themselves functions of T_m.

We have further shown that at low speeds the Eckert number is unimportant and hence that the heat transfer coefficient should be independent of the temperature difference as is observed in practice.

Thus we have seen that in most practical applications the Nusselt number is given by equation (15.3.3),

$$Nu = f(Re, Pr). \tag{15.3.10}$$

This equation, however, gives no indication of the mathematical form of the relationship, but for many situations a correlation of the type

$$Nu = C\, Re^a\, Pr^b, \tag{15.3.11}$$

where C, a, and b are constants, is found to be appropriate. For example, it is found experimentally that for the case of fully developed turbulent flow in a long tube, with $Re > 2100$, and for fluids whose Prandtl numbers are in the range from 0.5 upwards, heat transfer coefficients are given within an accuracy of about $\pm 10\%$ by the simple empirical expression

$$Nu = 0.023\, Re^{0.8}\, Pr^{0.4}. \tag{15.3.12}$$

This last result is often referred to as the Dittus–Boelter equation.

The heat transfer coefficient h in the Nusselt number will be a mean value and, as shown in chapter 18, is given by the average heat flux divided by the logarithmic mean temperature difference between the tube wall and the fluid. The physical properties μ, c_p, and k are usually evaluated at the bulk mean temperature of the fluid, i.e. at the arithmetic mean of inlet and outlet temperatures. Minor variations on equation (15.3.12) have been suggested, however, which claim to give better accuracy and involve the evaluation of the physical properties at some hypothetical mean film temperature. Equation (15.3.12) is of surprisingly wide application and covers most ordinary gases and liquids encountered in chemical engineering, but not liquid metals. It can be applied to heat transfer calculations with passages whose cross-sections are other than circular, provided appropriate values are taken for the equivalent diameter. However, the accuracy of the calculations is generally less satisfactory in such cases. For other geometries reference should be made to one of the many tabulations of appropriate correlations such as those of Perry[1], Eckert & Drake[2], or Wong[3].

For fully developed laminar flow in a circular pipe with uniform wall temperature, an analytical solution is possible if we assume constant viscosity and a parabolic velocity distribution. The derivation of this is considered in § 16.4 and the result may be expressed approximately by

$$Nu = \frac{h_{av} D}{k} = 1.62\left(Re\, Pr\, \frac{D}{L}\right)^{\frac{1}{3}} \tag{15.3.13}$$

where L is the pipe length. The appearance of the additional group D/L in this expression reflects the fact that in laminar flow the steady temperature profile is set up much more slowly than in turbulent flow and that distance from the pipe entry is therefore a relevant parameter.

In this expression we take the arithmetic mean temperature difference in evaluating the Nusselt number. In practice the viscosity of a liquid varies appreciably with temperature and the velocity distribution will not be parabolic. The following empirical modification of equation (15.3.13) is found to give better agreement with experimental results,

$$Nu = 1.86\left(\frac{\mu}{\mu_w}\right)^{0.14}\left(Re\,Pr\,\frac{D}{L}\right)^{\frac{1}{3}}, \tag{15.3.14}$$

where μ_w is the viscosity at the wall temperature and μ is the value at the mean bulk temperature of the fluid.

15.4 Forced convection mass transfer

We will deal with the subject of the dimensional analysis of forced convection mass transfer very briefly as the arguments are closely parallel to those for heat transfer presented in § 15.3. Following the method of that section we will assume that the local molar flux N in a circular pipe will depend on the pipe diameter D, the mean velocity u_m, the fluid properties ρ and μ, the diffusivity $\mathscr{D}$, and the concentration difference ΔC. However, we will not include the total concentration C in this list since we found in § 15.3 that the equivalent quantity T_m was unimportant. Thus

$$N = f(\Delta C, D, u_m, \rho, \mu, \mathscr{D}). \tag{15.4.1}$$

On this occasion we have seven quantities and the four dimensions $[m]$, $[l]$, $[t]$, and $[n]$ showing that we have three independent dimensionless groups. However, only two of the parameters, namely N and ΔC, contain the dimension $[n]$ having dimensions $[nl^{-2}t^{-1}]$ and $[nl^{-3}]$ respectively and therefore can only appear as the ratio $N/\Delta C$ which is the mass transfer coefficient k_g (or k_L), as defined in § 14.3.

Thus as an alternative to equation (15.4.1) we can say that

$$k_g = f(D, u_m, \rho, \mu, \mathscr{D}). \tag{15.4.2}$$

This list of six parameters contains only the three dimensions of $[m]$, $[l]$, and $[t]$, leading to the same conclusion that there are three relevant dimensionless groups.

Working by inspection, we can say that the Reynolds number $Re = \rho u_m D/\mu$ as defined in chapter 5 and the Sherwood number $Sh = k_g D/\mathscr{D}$ as defined in chapter 14 are both clearly appropriate. By analogy with the Prandtl number in the heat transfer case of § 15.3 we can define a Schmidt number by $Sc = \mu/\rho\mathscr{D}$. Thus we can say that,

$$Sh = f(Re, Sc), \tag{15.4.3}$$

a result that is entirely equivalent to equation (15.3.3).

Alternatively we could define a modified Stanton number St' by

$$St' = \frac{Sh}{(Re\,Sc)} = \frac{k_g}{u_m} \tag{15.4.4}$$

and a modified Péclet number $P\acute{e}'$ by

$$P\acute{e}' = Re\,Sc = u_m \frac{D}{\mathscr{D}}, \tag{15.4.5}$$

giving rise to the expressions

$$St' = f(Re, Sc) \tag{15.4.6}$$

and

$$Sh = f(P\acute{e}', Re). \tag{15.4.7}$$

These are the mass transfer equivalents to equations (15.3.4) and (15.3.7) and like those equations give no information beyond that already contained in equation (15.4.3).

Mass transfer coefficients are usually correlated by equation (15.4.3) which can often be expressed in the form

$$Sh = C\,Re^a\,Sc^b. \tag{15.4.8}$$

For many situations the constants $C, a,$ and b in equation (15.4.8) are found to be the same as those in the corresponding heat transfer expression, equation (15.3.11). For example, Sherwood *et al.*[4] give the following correlation for mass transfer in a circular pipe.

$$Sh = 0.023\,Re^{0.83}\,Sc^{0.44}. \tag{15.4.9}$$

The difference in the values in this and in equation (15.3.12) is slight and is probably within experimental error.

15.5 Physical significance of the dimensionless groups

It is important to appreciate the physical significance of the dimensionless ratios which were presented in §§ 15.3 and 15.4 and which are commonly employed in the correlation of heat and mass transfer results.

The significance of the Reynolds number as a measure of the ratio of inertial force to viscous forces in the flow has already been described in chapter 5 and will not be repeated here.

The Nusselt number is a dimensionless heat transfer coefficient which gives a measure of the ratio of the actual heat transfer rate to that which would have occurred if the process had been purely conductive. In a body with a typical linear dimension of D we can say that the typical temperature gradient will be $\Delta T/D$ and hence the conductive heat transfer coefficient will be k/D. The Nusselt number is the ratio of the actual heat transfer coefficient to this value and may therefore be regarded as an enhancement factor. It gives a measure of the increase in the heat transfer rate due to the motion of the fluid. As shown in § 13.2, the Nusselt number for conduction through a solid slab is 1.

The Sherwood number can similarly be interpreted as the ratio of the actual mass transfer coefficient to the purely diffusive value of $\mathscr{D}/D$.

The Stanton number is an alternative dimensionless heat transfer coefficient and gives a measure of the ratio of the flux of heat normal to the streamlines, $h\,\Delta T$, to that of convection along the streamlines, $\rho u_m c_p\,\Delta T$. The modified Stanton number can be interpreted in a similar manner and we will see in chapter 17 that these numbers have a close relationship to the friction factor c_f.

The Prandtl number $\mu c_p/k$ can be written as ν/α and the Schmidt number $\mu/\rho\mathscr{D}$ as $\nu/\mathscr{D}$. These numbers are therefore simply the ratios of the kinematic viscosity to the thermal and molecular diffusivities and therefore give a measure of the relative efficiency of the fluid as a conductor of momentum, heat, and solute. As will be seen in chapter 16, these numbers can be derived in a more fundamental way as the ratios of coefficients in the basic transport equations. When these numbers are equal to 1 the velocity, temperature, and concentration profiles are all geometrically similar and the degree of similarity decreases as the values of these numbers depart from 1.

According to the kinetic theory of gases, as presented in chapter 2, the mechanisms giving rise to viscosity, conductivity, and diffusivity are closely related and that the Prandtl and Schmidt numbers for most gases are somewhat less than 1. Equation (2.7.11) predicts that $Pr=0.74$ for a diatomic gas and equation (2.8.10) predicts that $Sc=0.75$ for a mixture of gases of equal molecular masses. These values are in fair agreement with experiment though somewhat larger values of the Schmidt number are found if the minor component of the gas mixture has a greater molecular mass than the major component.

In liquids there is no common mechanism for these effects, viscosity being determined by the degree of entanglement of the molecules and thermal conductivity by the quantity of free electrons. Thus the Prandtl numbers for liquids vary from about 0.05 for liquid metals to several thousand for viscous fluids of poor thermal conductivity. However, for many light fluids the Prandtl number is in the range 1–10, that for water varying between 13.0 at 0 °C and 1.73 at 100 °C.

The low values of diffusivities through liquids result in high Schmidt numbers, with values often lying in the range 500–1000.

Thus the approximate analyses presented in chapter 17, which imply similarity of velocity, temperature, and concentration profiles, work well for gases, are adequate for heat transfer to light liquids but are quite inappropriate for mass transfer to liquids or heat transfer to liquid metals.

15.6 Free convection

It is usual to distinguish between two sorts of convection, forced convection and free convection, though in many cases both processes may be taking place simultaneously. In forced convection the motion of the fluid results from some external agency such as a pump or fan. In free convection, however, the motion is caused by the density differences resulting from the heat or mass transfer process itself. The free convection of heat is a well-known phenomenon but it is not commonly realised that the density differences in the mass transfer case can be just as effective in promoting circulation. Free convection is also known as natural convection, but this phrase is currently out of favour on the argument that there is nothing unnatural about forced convection.

In free convection processes the motion of the fluid is caused by density changes and these can clearly have no effect except in the presence of a gravitational field. We must therefore expect the group $g\,\Delta\rho$ to be of importance. Thus in the heat transfer case we will assume that the heat transfer coefficient h depends on some typical vertical dimension L, the product $g\,\Delta\rho$ and the properties of the fluid ρ, μ, c_p, and k, thus,

$$h = f(L, g\,\Delta\rho, \rho, \mu, c_p, k). \tag{15.6.1}$$

These last four quantities should be evaluated at some typical mean temperature. Using the five dimensions $[m]$, $[l]$, $[t]$, $[Q]$, and $[T]$, the seven quantities of equation (15.6.1) should give rise to only two independent dimensionless groups, but inspection of the dimensions of the relevant parameters shows that the dimensions $[Q]$ and $[T]$ appear only as the ratio

$[Q]/[T]$. Thus we have in effect only four dimensions, $[m]$, $[l]$, $[t]$, and $[Q/T]$ and consequently we get three independent dimensionless groups.

Clearly the Nusselt number $Nu = hL/k$ and the Prandtl number $Pr = \mu c_p/k$ are appropriate and the third group must contain the product $g\,\Delta\rho$. The Grashof number, defined by

$$Gr = \frac{\rho\,\Delta\rho g L^3}{\mu^2}, \tag{15.6.2}$$

serves this purpose, and we can say that

$$Nu = f(Gr, Pr) \tag{15.6.3}$$

and, by analogy, for the mass transfer case,

$$Sh = f(Gr, Sc). \tag{15.6.4}$$

For heat transfer the density changes result from the thermal expansion of the fluid. This is best described in terms of the coefficient of thermal expansion β, defined formally by

$$\beta = -\frac{1}{\rho}\left(\frac{\partial\rho}{\partial T}\right)_P, \tag{15.6.5}$$

so that for small temperature changes

$$\Delta\rho = -\rho\beta\,\Delta T.$$

Ignoring the negative sign, since dimensionless groups are normally taken to be positive, we have the alternative formulation of the Grashof number

$$Gr = \frac{\rho^2\beta\,\Delta T g L^3}{\mu^2}. \tag{15.6.6}$$

For a perfect gas $\rho = P/RT$ and from equation (15.6.5) it follows that $\beta = 1/T$ where T is the absolute temperature.

In the mass transfer case, the density differences result from concentration differences. Assuming ideal gas behaviour, the density of a mixture of A and B is given by

$$\rho = \frac{M_{rA}(P - p_B)}{RT} + \frac{M_{rB}p_B}{RT} = \rho_A\left(1 + \left(\frac{M_{rB} - M_{rA}}{M_{rA}}\right)y\right), \tag{15.6.7}$$

where ρ_A is the density of pure A at the same temperature and total pressure and y is the mole fraction of B. Thus $\Delta\rho = \rho_A\beta'\,\Delta y$ where β' is

$(M_{rB}-M_{rA})/M_{rA}$ and the appropriate formulation of the Grashof number is

$$Gr=\frac{\rho^2\beta'\,\Delta y g L^3}{\mu^2}. \qquad (15.6.8)$$

A theoretical solution, assuming laminar motion in the free convection currents rising from a vertical plate of height L, gives

$$Nu=0.52\,(Gr\,Pr)^{\frac{1}{4}}, \qquad (15.6.9)$$

and the corresponding result for mass transfer is

$$Sh=0.52\,(Gr\,Sc)^{\frac{1}{4}}.$$

It is found experimentally that for values of the product $Gr\,Pr$ in the range 10^4–10^8 the flow is laminar and the average heat transfer coefficient is well correlated by

$$Nu=0.56\,(Gr\,Pr)^{\frac{1}{4}}, \qquad (15.6.10)$$

a result in good agreement with the theoretical prediction. The mass transfer case does not seem to have received much attention experimentally, but one may confidently expect a similar result.

For values of the product $Gr\,Pr$ greater than 10^9, the motion is generally turbulent, and it is found that the Nusselt number is then proportional to $(Gr\,Pr)^{\frac{1}{3}}$. The following correlation is given by Eckert & Drake[2] for turbulent free convection from a vertical plate

$$Nu=0.12\,(Gr\,Pr)^{\frac{1}{3}}. \qquad (15.6.11)$$

The product of Grashof number and Prandtl number which appears in all these expressions is sometimes known as the Rayleigh number, *Ra*.

For heat transfer by free convection from horizontal cylinders, the diameter is used in defining the Nusselt and Grashof numbers and for values of the product $Gr\,Pr<10^8$ the following expression may be used

$$Nu=0.47\,(Gr\,Pr)^{\frac{1}{4}} \qquad (15.6.12)$$

and for $Gr\,Pr>10^8$,

$$Nu=0.10\,(Gr\,Pr)^{\frac{1}{3}}. \qquad (15.6.13)$$

The physical significance of the Grashof number can be interpreted as follows. The pressure difference caused by a density difference $\Delta\rho$ over a height L is $Lg\,\Delta\rho$, and substituting into Bernoulli's equation we can see that this is equivalent to a typical velocity V where V^2 is of order $Lg\,\Delta\rho/\rho$.

Thus

$$Gr = \frac{\rho\, \Delta\rho g L^3}{\mu^2} \sim \left(\frac{\rho\, VL}{\mu}\right)^2$$

and the Grashof number can therefore be interpreted as the square of the Reynolds number of the resulting motion. The transition to turbulence found for Grashof numbers in the range 10^8–10^9 is therefore compatible with the observed transitional Reynolds numbers for external flows in the range 10^4–10^5.

References

[1] Perry, *Chemical Engineers' Handbook*, McGraw-Hill.
[2] Eckert & Drake, *Heat and Mass Transfer*, McGraw-Hill.
[3] Wong, *Heat Transfer for Engineers*, Longmans.
[4] Sherwood, Pigford & Wilke, *Mass Transfer*, McGraw-Hill.

16

Heat and mass transfer in laminar flow

16.1 The basic equations of forced convection

The basic equations which govern the transfer of momentum and energy in a moving fluid have already been established in chapters 2, 5, and 9. In the present chapter we will first of all review these equations with special reference to forced convection problems and we will extend the treatment to include transfer of matter in addition to transfer of energy. As we have noted previously in chapter 5, the basic equations are quite general in their application and would cover the condition of turbulence if we could treat turbulent flow as an extreme example of unsteady motion. In practice, however, we can only make direct use of the equations for purposes of numerical calculation when we are dealing with *laminar flow*.

The general energy equation for a three-dimensional flow system was derived in chapter 9. If we neglect the potential energy term, associated with changes of level, and if we confine attention to *steady flow*, we can take (9.4.9) as the basic form of the energy equation. If we exclude the special case of the flow of a gas through a nozzle or through a turbo-machine where high velocities are encountered, we can say that in most heat transfer processes the kinetic energy $\frac{1}{2}\mathbf{u}^2$ associated with the fluid velocity $\mathbf{u}$ is usually small compared with the changes of the enthalpy h. We can thus simplify equation (9.4.9) by neglecting the $\frac{1}{2}\mathbf{u}^2$ term on the left-hand side. We can also say that shear work only arises as a significant quantity where we have moving boundaries as in the case of flow past moving blades or through turbo-machines. Thus for heat transfer at relatively low velocities through fixed ducts we can express the basic steady-flow energy equation in the form

$$\rho \frac{\mathrm{D}h}{\mathrm{D}t} = k\,\nabla^2 T. \tag{16.1.1}$$

It will be noted that, under steady-flow conditions, the total differential $\mathrm{D}h/\mathrm{D}t$ is identical with the convective differential $\mathbf{u}\cdot\nabla h$, and that the left-hand side of equation (16.1.1) therefore represents the convective rate of change per unit volume of fluid of the enthalpy h.

As we have noted previously in §§2.7 and 9.4, the enthalpy of any substance is defined by $h = e + p/\rho$. In the special case of a perfect gas the enthalpy is a function only of the absolute temperature T and enthalpy changes may be expressed by

$$dh = c_p \, \mathrm{d}T. \tag{16.1.2}$$

This statement is valid for most permanent gases and also applied in the case of liquids at moderate pressures. The specific heat c_p may usually be treated as a constant, at least over a limited range of temperature, and we may therefore re-write equation (16.1.1) as a differential equation for the temperature T as in (9.4.13), i.e.

$$\frac{\mathrm{D}T}{\mathrm{D}t} = \alpha \nabla^2 T, \tag{16.1.3}$$

where α is the thermal diffusivity now defined by $\alpha = k/\rho c_p$.

The derivation of the steady-flow energy equation in §9.4 was based on the application of the first law of thermodynamics to a moving fluid in the absence of any internal generation of heat through chemical or nuclear reactions or through the passage of an electric current in the fluid. If we introduce a *source term* Q_s into the energy equation to represent the rate at which heat is generated internally *per unit volume of fluid*, it will be seen that equation (16.1.3) will be modified to become

$$\frac{\mathrm{D}T}{\mathrm{D}t} = \frac{Q_s}{\rho c_p} + \alpha \nabla^2 T. \tag{16.1.4}$$

It will be seen that (16.1.4) is similar in form to the heat conduction equation for a static medium with internal heat generation derived in chapter 13, but that the total or convective differential $\mathrm{D}T/\mathrm{D}t$ has replaced the partial differential $\partial T/\partial t$ on the left-hand side.

If we now turn attention to the transfer of matter, it will be seen from §§2.9 and 14.6 that the basic diffusion equation for a flowing system, with internal generation of the diffusing component through chemical reaction at a rate R kmol s^{-1} per unit volume, takes an exactly similar form:

$$\frac{\mathrm{D}C}{\mathrm{D}t} = R + \mathscr{D} \nabla^2 C. \tag{16.1.5}$$

It is of interest to compare these two equations (16.1.4) and (16.1.5) with the Navier–Stokes equation derived in chapter 5. The Navier–Stokes equation is essentially a statement regarding the transport of momentum.

For an incompressible fluid it takes the form

$$\frac{\mathrm{D}\mathbf{u}}{\mathrm{D}t} = -\frac{1}{\rho}\,\mathrm{grad}\,p + \nu\,\nabla^2\mathbf{u}. \tag{16.1.6}$$

The similarity between the three equations is obvious and we could interpret the pressure gradient term in (16.1.6) as a source or sink term representing the rate at which momentum is created or destroyed through the action of the pressure field.

Equations (16.1.4), (16.1.5), and (16.1.6), taken in conjunction with the continuity equation (3.3.2), are in principle sufficient to determine the temperature, concentration, and velocity profiles for any specified boundary conditions in the case of steady laminar flow. However, mathematically exact solutions are only possible in the simplest cases and in most practical problems we have to use approximate methods of numerical solution.

The three basic equations relating to energy, matter, and momentum are set out below in tabular form for comparison

	Convection rate	=	*'Source' term*	+	*'Diffusion' term*
Energy	$\frac{\mathrm{D}T}{\mathrm{D}t}$	=	$\frac{Q_s}{\rho c_p}$	+	$\alpha\,\nabla^2 T$
Matter	$\frac{\mathrm{D}C}{\mathrm{D}t}$	=	R	+	$\mathscr{D}\,\nabla^2 C$
Momentum	$\frac{\mathrm{D}\mathbf{u}}{\mathrm{D}t}$	=	$-\frac{1}{\rho}\nabla p$	+	$\nu\,\nabla^2\mathbf{u}$

It will be clear from the table that, *for a given set of similar boundary conditions*, the three equations will have similar solutions provided the source terms are zero (or have the same numerical value at corresponding points), and provided the coefficients α, $\mathscr{D}$, and ν which appear in the diffusion terms are equal.

In the special case of *flow in a boundary layer* past a flat plate with constant external stream velocity the pressure distribution in the fluid is nearly uniform as we have seen in chapter 7. Thus the source term involving grad p in (16.1.6) is effectively zero. We would therefore expect to find similarity between temperature, concentration, and velocity profiles in the boundary layer provided Q_s and R are zero, and provided $\alpha = \mathscr{D} = \nu$. Noting that the Prandtl number is equal to ν/α and that the Schmidt number is equal to $\nu/\mathscr{D}$, we see that the latter condition is equivalent to the requirement that $Pr = Sc = 1$.

16.2 The thermal and concentration boundary layers

Both temperature and concentration equations can be derived for boundary-layer flow on the lines of the momentum equations of §7.6. Considering first the case of flow past a flat surface with heat transfer, we will assume that under certain circumstances the significant changes of temperature in the fluid are confined to a relatively thin region adjacent to the wall. We can regard this as the thermal boundary layer analogous to the velocity or momentum boundary layer of chapter 7. Referring to figure 131, let δ be the thickness of the velocity layer and let δ_T be the thickness of the thermal layer. In figure 131, δ_T is shown as being less than δ but this is not necessarily the case and we will denote by Δ the larger of these two quantities and refer to the region of this thickness as the boundary layer. Unlike the velocity, the temperature does not fall to zero at the wall and it is convenient to work in terms of the temperature difference $T-T_0$ and to base enthalpies on a datum of the wall temperature T_0.

We can define a thermal thickness for this boundary layer by analogy with the momentum thickness discussed in §7.3. Referring to figure 131 we can say that the enthalpy flow within the boundary layer is given by

$$\int_0^\Delta \rho u c_p (T-T_0)\,\mathrm{d}y$$

and we can define a distance δ^{***} so that

$$\int_0^\Delta \rho u c_p (T-T_0)\,\mathrm{d}y = \rho u_1 c_p (T_1 - T_0)(\Delta - \delta^{***}). \tag{16.2.1}$$

Hence

$$\delta^{***} = \int_0^\Delta \left(1 - \frac{u}{u_1}\left(\frac{T-T_0}{T_1-T_0}\right)\right)\mathrm{d}y. \tag{16.2.2}$$

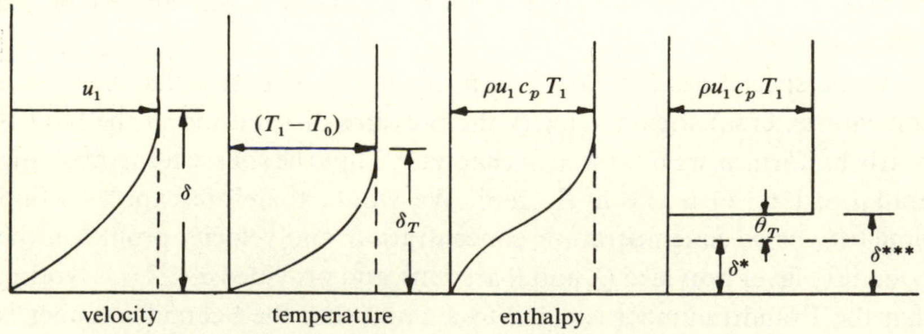

Fig. 131. Velocity, temperature and enthalpy flux profiles in a laminar boundary layer.

As in the case of the momentum thickness, it is usual to define a thermal thickness θ_T by $\theta_T = \delta^{***} - \delta^*$ as shown in figure 131. By combining equations (16.2.2) and (7.3.2) we find that the thermal thickness is given by

$$\theta_T = \int_0^\Delta \frac{u}{u_1}\left(\frac{T_1 - T}{T_1 - T_0}\right) \mathrm{d}y. \tag{16.2.3}$$

In the case when $\delta_T > \delta$, $\Delta = \delta_T$ and in the case when $\delta_T < \delta$, the group $(T_1 - T)/(T_1 - T_0)$ will be zero for $y > \delta_T$. Thus in either case the upper limit can be replaced by δ_T so that

$$\theta_T = \int_0^{\delta_T} \frac{u}{u_1}\left(\frac{T_1 - T}{T_1 - T_0}\right) \mathrm{d}y. \tag{16.2.4}$$

We can similarly define a concentration thickness θ_C by

$$\theta_C = \int_0^{\delta_C} \frac{u}{u_1}\left(\frac{C_1 - C}{C_1 - C_0}\right) \mathrm{d}y, \tag{16.2.5}$$

where δ_C is the thickness of the concentration boundary layer.

In the analysis of the momentum boundary layer presented in §7.6 we carried out a momentum balance on a short length of the boundary layer. We will now carry out an energy balance on a length $\mathrm{d}x$ of the boundary layer as shown in figure 132. The mass-flow rate ρQ and the enthalpy flow H (above a datum at T_0) within the boundary layer at distance x from the leading edge are given from the definitions of δ^* and θ_T by

$$\rho Q = \rho u_1 (\Delta - \delta^*), \tag{16.2.6}$$

$$H = \rho u_1 c_p (T_1 - T_0)(\Delta - \delta^* - \theta_T). \tag{16.2.7}$$

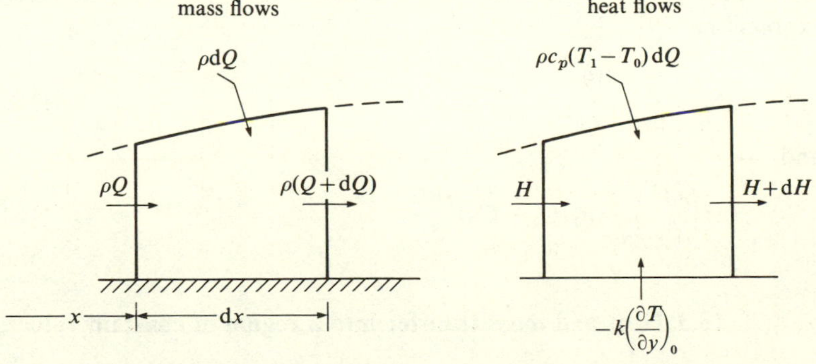

Fig. 132. Mass and enthalpy balances on an elementary length of boundary layer.

The inflow of heat to the element by conduction across the wall is

$$q_0\,\mathrm{d}x = -k\left(\frac{\partial T}{\partial y}\right)_0 \mathrm{d}x \tag{16.2.8}$$

and this can be related to the net outflow of enthalpy. Assuming constant velocity u_1 and temperature T_1 outside the boundary layer we have

$$q_0\,\mathrm{d}x = \mathrm{d}H - \rho c_p(T_1 - T_0)\,\mathrm{d}Q \tag{16.2.9}$$

or

$$\begin{aligned} q_0 = -k\left(\frac{\partial T}{\partial y}\right)_0 &= \frac{\mathrm{d}H}{\mathrm{d}x} - \rho c_p(T_1 - T_0)\frac{\mathrm{d}Q}{\mathrm{d}x} \\ &= \rho c_p u_1(T_1 - T_0)\frac{\mathrm{d}}{\mathrm{d}x}(\Delta - \delta^* - \theta_T) - \rho c_p u_1(T_1 - T_0)\frac{\mathrm{d}}{\mathrm{d}x}(\Delta - \delta^*) \\ &= -\rho c_p u_1(T_1 - T_0)\frac{\mathrm{d}\theta_T}{\mathrm{d}x}. \end{aligned} \tag{16.2.10}$$

Recalling that the heat transfer coefficient h is given by $q_0/(T_0 - T_1)$ and that the Stanton number was defined in § 15.3 by $St = h/\rho u_1 c_p$, this equation can be rewritten as

$$St = \frac{h}{\rho u_1 c_p} = \frac{\mathrm{d}\theta_T}{\mathrm{d}x} \tag{16.2.11}$$

or

$$k\left(\frac{\partial T}{\partial y}\right)_0 = \rho c_p u_1(T_1 - T_0)\frac{\mathrm{d}\theta_T}{\mathrm{d}x}. \tag{16.2.12}$$

These equations are the heat transfer analogue of equation (7.6.7).

By similar arguments we can show that the mass transfer coefficient k_L and the modified Stanton number St' (as defined in § 15.4) are given by the expressions

$$St' = \frac{k_L}{u_1} = \frac{\mathrm{d}\theta_C}{\mathrm{d}x} \tag{16.2.13}$$

and

$$\mathscr{D}\left(\frac{\partial C}{\partial y}\right)_0 = u_1(C_1 - C_0)\frac{\mathrm{d}\theta_C}{\mathrm{d}x}. \tag{16.2.14}$$

16.3 Heat and mass transfer into a region of constant velocity

When the velocity of the fluid is constant, the analysis of the thermal and concentration boundary layers becomes particularly simple and we can work from first principles instead of using the more formal approach of § 16.2. Mass transfer will be considered first because of its great importance in the practical cases of the wetted wall column and the laminar jet apparatus.

We will assume that the liquid supplied to the top of a wetted wall column is free of solute and that it flows down the wall as a smooth film. The film is exposed to a gas containing a soluble component so that the interfacial concentration C_i is determined by the partial pressure of the soluble species. Under these circumstances the solute diffuses inwards from the interface and a concentration boundary layer develops as shown in figure 133.

At low Reynolds numbers, the liquid flows down the wall as a laminar film of thickness δ and the velocity profile in this film can be found using the methods of § 5.8. The driving force for the flow is the specific weight ρg and equation (5.8.6) takes the form,

$$\mu \frac{d^2u}{dy^2} = -\rho g. \tag{16.3.1}$$

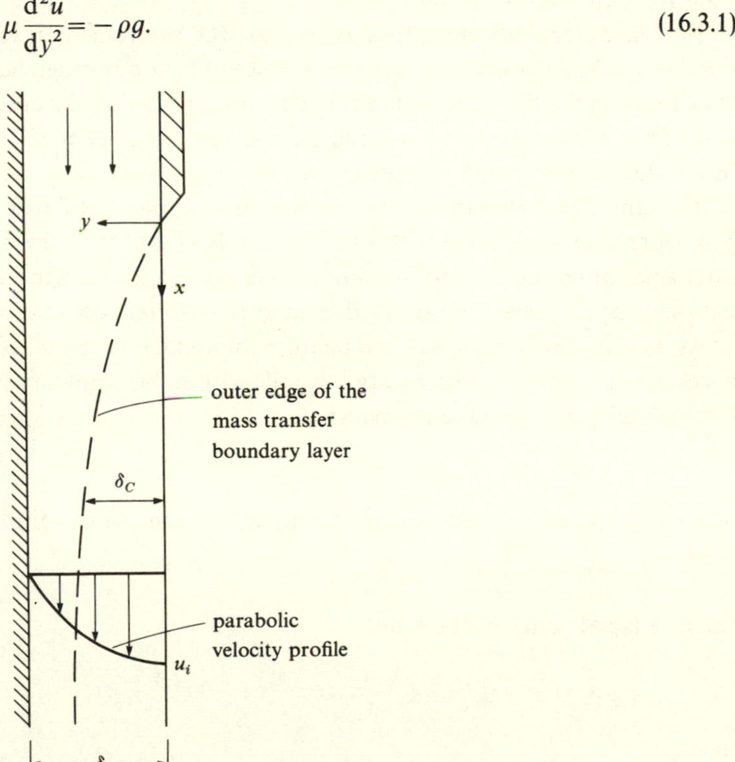

Fig. 133. Mass transfer boundary layer in a falling liquid film.

Measuring y inwards from the interface, this equation can be integrated twice subject to the boundary conditions that at the interface, $y=0$, the shear stress $\mu(\mathrm{d}u/\mathrm{d}y)=0$ and that on the wall, $y=\delta$, the velocity $u=0$. This gives the velocity profile in the film as

$$u=\frac{\rho g}{2\mu}(\delta^2-y^2) \tag{16.3.2}$$

or

$$u=u_i\left(1-\left(\frac{y}{\delta}\right)^2\right), \tag{16.3.3}$$

where u_i is the interfacial velocity $\rho g\delta^2/2\mu$.

Equation (16.3.3) represents the velocity distribution in the film once the disturbance caused by the entry device has died away. Experience shows that at low Reynolds numbers this steady profile is reached very rapidly and we can assume without much loss of accuracy that equation (16.3.3) applies right from the top of the film.

For the outer third of the film, i.e. for $y<\delta/3$, the velocity is effectively constant, falling to only $8u_i/9$ at $y=\delta/3$. Since diffusion through liquids is a slow process, the thickness of the concentration boundary layer δ_C will be small and, provided $\delta_C<\delta/3$, we can treat the problem as one of diffusion into a fluid moving with a constant velocity u_i.

The same result also applies for the laminar jet apparatus shown in figure 134. Here a jet of liquid falls through a soluble gas. Due to the negligible interfacial drag, the velocity is constant across the jet, though variations with distance downstream occur due to gravitational acceleration.

As in the analysis of the momentum boundary layer of §7.7, it is necessary to assume a concentration profile within the boundary layer and for simplicity we will take the parabola,

$$C=a_0+a_1y+a_2y^2. \tag{16.3.4}$$

The constants a_0, a_1, and a_2 can be evaluated from the conditions,

$$y=0, \quad C=C_i$$

the interfacial concentration, and

$$y=\delta_C, \quad C=0, \quad \text{and} \quad \frac{\partial C}{\partial y}=0,$$

giving

$$C=C_i\left\{1-\frac{2y}{\delta_C}+\left(\frac{y}{\delta_C}\right)^2\right\}. \tag{16.3.5}$$

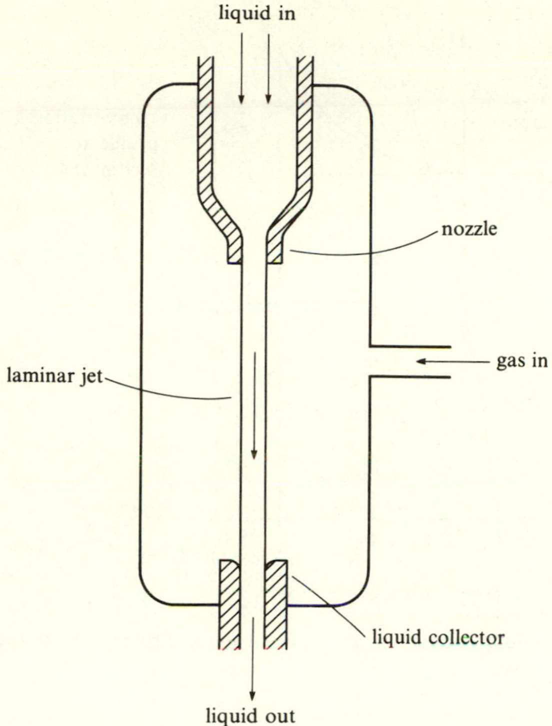

Fig. 134. Laminar jet absorption apparatus.

Other profiles such as a quarter sine wave or a fourth-order polynomial could have been assumed but with negligible improvement in accuracy.

The total flux m of solute across the plane AA of figure 135 is given by

$$m = \int_0^{\delta_C} u_i C \, dy = \frac{u_i C_i \delta_C}{3} \tag{16.3.6}$$

and the flux across the plane a distance dx below AA is given by

$$m + dm = \frac{u_i C_i}{3}(\delta_C + d\delta_C). \tag{16.3.7}$$

The difference dm must equal the quantity transferred from the interface by diffusion, $N_0 \, dx$, where

$$N_0 = -\mathscr{D}\left(\frac{\partial C}{\partial y}\right)_0 = \frac{2\mathscr{D}C_i}{\delta_C}. \tag{16.3.8}$$

Thus

$$\frac{u_i C_i}{3} d\delta_C = \frac{2\mathscr{D}C_i}{\delta_C} dx$$

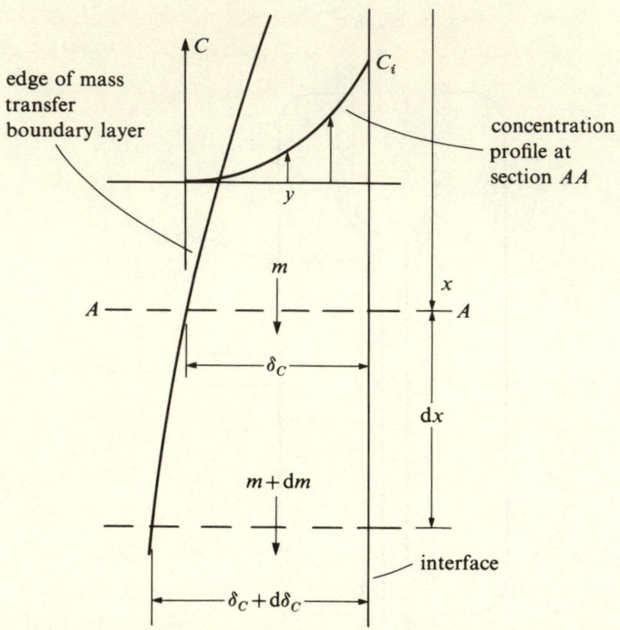

Fig. 135. Solute balance on an elementary length of a mass transfer boundary layer.

or

$$\delta_C \, d\delta_C = \frac{6\mathscr{D}}{u_i} \, dx. \tag{16.3.9}$$

On integration with the assumption that $\delta_C = 0$ at the top of the film, $x = 0$, this becomes

$$\delta_C = \sqrt{\left(\frac{12\mathscr{D}x}{u_i}\right)}. \tag{16.3.10}$$

The thickness of the concentration boundary layer is seen to vary as $\sqrt{x}$ as did the momentum layer considered in § 7.7 and equation (16.3.10) may be thought of as the mass transfer analogue of equation (7.7.7).

We are now in a position to evaluate the absorption rate N_0 by substituting from equation (16.3.10) into equation (16.3.8) giving

$$N_0 = C_i \sqrt{\left(\frac{u_i \mathscr{D}}{3x}\right)} \tag{16.3.11}$$

and the mass transfer coefficient k_L, defined as N_0/C_i, is given by

$$k_L = \sqrt{\left(\frac{u_i \mathscr{D}}{3x}\right)}. \tag{16.3.12}$$

We might note in passing at this stage that an exact solution for steady diffusion into an infinite quantity of fluid moving with constant velocity u_i can be obtained by solving equation (16.1.5) which in this case takes the form

$$u_i \frac{\partial C}{\partial x} = \mathscr{D} \frac{\partial^2 C}{\partial y^2} \tag{16.3.13}$$

and is to be solved subject to the boundary conditions

$$y=0, \quad x>0, \quad C=C_i,$$

$$y \to \infty, \; x \geqslant 0, \quad C \to 0,$$

$$x=0, \quad y>0, \quad C=0.$$

From this it can be shown that

$$k_L = \sqrt{\left(\frac{u_i \mathscr{D}}{\pi x}\right)}. \tag{16.3.14}$$

Thus the difference between our approximate solution and the exact solution is about 2%, which is less than the probable uncertainty in the value of $\mathscr{D}$.

It is seen in equation (16.3.12) that the mass transfer coefficient is a function of x and is therefore a local value, varying with distance down the column. The average mass transfer coefficient $\bar{k}_L$ is of more general utility and this can be evaluated as follows for a column of total height X,

$$\bar{k}_L = \frac{1}{X} \int_0^X k_L \, dx = 2\sqrt{\left(\frac{u_i \mathscr{D}}{3X}\right)}. \tag{16.3.15}$$

This can be put in the dimensionless form

$$Sh_X = \frac{2}{\sqrt{3}} Re_X^{\frac{1}{2}} Sc^{\frac{1}{2}}, \tag{16.3.16}$$

where

$$Sh_X = \frac{\bar{k}_L X}{\mathscr{D}} \quad \text{and} \quad Re_X = \frac{u_i X}{\nu}.$$

A similar analysis can be performed for heat transfer. The local heat transfer coefficient h is found to be

$$h = \sqrt{\left(\frac{u_i k}{3\rho c_p x}\right)} \tag{16.3.17}$$

and the average value $\bar{h}$ is given by

$$\bar{h}=2\sqrt{\left(\frac{u_i k}{3\rho c_p X}\right)}. \qquad (16.3.18)$$

In dimensionless form this becomes

$$Nu_X=\frac{2}{\sqrt{3}}Re_X^{\frac{1}{2}}\,Pr^{\frac{1}{2}}, \qquad (16.3.19)$$

where

$$Nu_X=\frac{\bar{h}X}{k}.$$

The results for both heat and mass transfer can be expressed in terms of the Stanton numbers as follows

$$St\,Pr^{\frac{1}{2}}=St'\,Sc^{\frac{1}{2}}=\frac{2}{\sqrt{3}}Re_X^{\frac{1}{2}}. \qquad (16.3.20)$$

For the laminar jet apparatus the quantity of solute absorbed in length dx is $\pi D k_L C_i\,dx$ where D is the local diameter of the jet.

Substituting from equation (16.3.14) and noting that $\pi D^2 u_i/4$ is the volumetric flow rate Q in the jet, we find that the quantity absorbed dM is given by

$$dM=\pi D C_i\sqrt{\left(\frac{u_i\mathscr{D}}{\pi x}\right)}\,dx=2C_i\sqrt{\left(\frac{Q\mathscr{D}}{x}\right)}\,dx. \qquad (16.3.21)$$

Thus the total quantity absorbed in a jet of length X is

$$M=4C_i\sqrt{(Q\mathscr{D}X)}. \qquad (16.3.22)$$

It is noteworthy that neither the velocity nor the diameter of the jet appear in equation (16.3.22) which relates the diffusivity to easily measured quantities. This equation is well attested experimentally and the laminar jet apparatus is generally taken to be the most reliable method of determining diffusivities through liquids.

16.4 Heat and mass transfer into a linear velocity gradient

We will now consider a thermal or concentration boundary layer growing from a solid wall into a fully developed laminar flow. Adjacent to the wall the velocity will be zero and we will assume that it increases linearly with distance from the wall, i.e.

$$u = by, \tag{16.4.1}$$

where b is a constant equal to the wall shear stress divided by the viscosity, τ_0/μ.

Such a situation might for example occur with laminar flow in a pipe of diameter D. Here the velocity profile is the parabola given by equation (5.8.12), but if the thermal boundary layer is thin the velocity profile within the thermal layer will be almost linear and it can be seen from equations (5.8.11) and (5.8.16) that the shear-strain rate b at the wall is given by

$$b = \frac{8u_m}{D}, \tag{16.4.2}$$

where u_m is the mean velocity.

This problem can be solved from first principles as in § 16.3 but on this occasion we will use the more formal approach of § 16.2.

Assuming a parabolic temperature profile with

$$T = T_0 \quad \text{at} \quad y = 0$$

and

$$T = T_1 \quad \text{and} \quad \frac{\partial T}{\partial y} = 0 \quad \text{at} \quad y = \delta_T$$

we can say that

$$\frac{T - T_0}{T_1 - T_0} = 2\left(\frac{y}{\delta_T}\right) - \left(\frac{y}{\delta_T}\right)^2 \tag{16.4.3}$$

and from the definition of θ_T, equation (16.2.4),

$$\theta_T = \int_0^{\delta_T} \frac{by}{u_1}\left(1 - 2\left(\frac{y}{\delta_T}\right) + \left(\frac{y}{\delta_T}\right)^2\right) \mathrm{d}y = \frac{b\delta_T^2}{12u_1}, \tag{16.4.4}$$

where in this case u_1 should be thought of as an arbitrary reference velocity.

From equation (16.4.3),

$$\left(\frac{\partial T}{\partial y}\right)_0 = \frac{2(T_1 - T_0)}{\delta_T} \tag{16.4.5}$$

and hence substituting into equation (16.2.12) we get

$$\frac{2k(T_1 - T_0)}{\delta_T} = \rho c_p u_1 (T_1 - T_0) \frac{\mathrm{d}}{\mathrm{d}x}\left(\frac{b\delta_T^2}{12u_1}\right)$$

or

$$\frac{k}{\rho c_p b}\,\mathrm{d}x=\frac{\delta_T^2}{12}\,\mathrm{d}\delta_T. \tag{16.4.6}$$

It may be noted that the arbitrary reference velocity u_1 has cancelled by this stage.

On integration, subject to the condition that $\delta_T=0$ at $x=0$, we obtain

$$\delta_T=\left(\frac{36kx}{\rho c_p b}\right)^{\frac{1}{3}}. \tag{16.4.7}$$

Recalling that the heat transfer coefficient h is given by $h=q_0/(T_0-T_1)$ we have, from equation (16.4.5),

$$h=\frac{-k}{(T_0-T_1)}\left(\frac{\partial T}{\partial y}\right)_0=\frac{2k}{\delta_T}=\left(\frac{2\rho c_p bk^2}{9x}\right)^{\frac{1}{3}} \tag{16.4.8}$$

and the average value of h over a length L is given by

$$\bar{h}=\frac{1}{L}\int_0^L h\,\mathrm{d}x=\left(\frac{3\rho c_p bk^2}{4L}\right)^{\frac{1}{3}}. \tag{16.4.9}$$

For the particular case of heat transfer into fully developed laminar flow in a pipe, b is given by equation (16.4.2) and equation (16.4.9) can be re-written as

$$Nu=6^{\frac{1}{3}}\,Re^{\frac{1}{3}}\,Pr^{\frac{1}{3}}\left(\frac{D}{L}\right)^{\frac{1}{3}}, \tag{16.4.10}$$

where $Nu=\bar{h}D/k$, $Re=\rho u_m D/\mu$, and $Pr=\mu c_p/k$. As mentioned in § 15.3, this analysis forms the basis of the semi-empirical correlation for heat transfer to laminar pipe flow known as the Sieder–Tate equation

$$Nu=1.86\,Re^{\frac{1}{3}}\,Pr^{\frac{1}{3}}\left(\frac{D}{L}\right)^{\frac{1}{3}}\left(\frac{\mu}{\mu_0}\right)^{0.14}. \tag{16.4.11}$$

It is seen that there has been a small empirical adjustment to the constant and the insertion of the group $(\mu/\mu_0)^{0.14}$ to allow for changes in the velocity profile caused by the temperature dependence of the viscosity.

For mass transfer equation (16.4.10) takes the form

$$Sh=6^{\frac{1}{3}}\,Re^{\frac{1}{3}}\,Sc^{\frac{1}{3}}\left(\frac{D}{L}\right)^{\frac{1}{3}}, \tag{16.4.12}$$

where $Sh=k_L D/\mathscr{D}$ and $Sc=\nu/\mathscr{D}$.

Both equation (16.4.10) and (16.4.12) can be expressed in terms of the Stanton numbers thus,

$$St\,Pr^{\frac{2}{3}} = St'\,Sc^{\frac{2}{3}} = 6^{\frac{1}{3}}\,Re^{-\frac{2}{3}}\left(\frac{D}{L}\right)^{\frac{1}{3}}, \tag{16.4.13}$$

where $St = \bar{h}/\rho u_m c_p$ and $St' = \bar{k}_L/u_m$.

16.5 Approximate solution for heat transfer in a laminar boundary layer

In the two previous sections we considered thermal and concentration boundary layers growing in a fully developed laminar flow. In this section we will consider a thermal boundary layer developing simultaneously with a momentum boundary layer on a flat plate. The analysis for the momentum layer has already been considered in § 7.7 where it was shown that the most appropriate velocity profile was

$$\frac{u}{u_1} = 2\left(\frac{y}{\delta}\right) - 2\left(\frac{y}{\delta}\right)^3 + \left(\frac{y}{\delta}\right)^4. \tag{16.5.1}$$

From this we showed, equation (7.7.8), that

$$\delta = 5.836\left(\frac{\nu x}{u_1}\right)^{\frac{1}{2}}. \tag{16.5.2}$$

For the thermal layer it is reasonable to assume a similar form for the temperature distribution

$$\frac{T - T_0}{T_1 - T_0} = 2\left(\frac{y}{\delta_T}\right) - 2\left(\frac{y}{\delta_T}\right)^3 + \left(\frac{y}{\delta_T}\right)^4, \tag{16.5.3}$$

where T_0 is the temperature of the wall and T_1 is the temperature of the main stream.

We will consider first the case when the thickness of the thermal layer is less than that of the momentum boundary layer, i.e. $\delta_T < \delta$. Equation (16.5.7) below shows that this occurs when the Prandtl number is greater than 1. This is intuitively reasonable, since a Prandtl number greater than 1 indicates that the kinematic viscosity is greater than the thermal diffusivity and the fluid is therefore a better conductor for momentum than for heat. Hence the effect of the solid wall on the velocity distribution spreads out more rapidly than does the effect on the temperature profile.

Substituting equations (16.5.1) and (16.5.3) into the definition of the thermal thickness, equation (16.2.4), and integrating, we obtain

$$\frac{\theta_T}{\delta_T}=\frac{2}{15}\xi-\frac{3}{140}\xi^3+\frac{1}{180}\xi^4=f(\xi), \tag{16.5.4}$$

where ξ is the ratio δ_T/δ and is less than 1.

Hence from equation (16.2.12) we have

$$\frac{2(T_1-T_0)k}{\delta_T}=\rho c_p u_1(T_1-T_0)f(\xi)\frac{\mathrm{d}\delta_T}{\mathrm{d}x}$$

or

$$\frac{k}{\rho c_p u_1}\mathrm{d}x=f(\xi)\frac{\delta_T}{2}\mathrm{d}\delta_T. \tag{16.5.5}$$

Hence, integrating and assuming that the thermal layer starts from the leading edge, $x=0$, we find that

$$\frac{kx}{\rho c_p u_1}=f(\xi)\frac{\delta_T^2}{4}. \tag{16.5.6}$$

But from equation (16.5.2),

$$\delta^2=34.06\frac{\mu x}{\rho u_1}$$

and therefore equation (16.5.6) can be written,

$$\frac{k}{\mu c_p}=8.515\,\xi^2 f(\xi)$$

or

$$\frac{1}{Pr}=\xi^3(1.135-0.182\,\xi^2+0.047\,\xi^3). \tag{16.5.7}$$

It is seen that when $Pr=1$, $\xi=1.0$ and that when $Pr>1$, $\xi<1$ and hence $\delta_T<\delta$ as was assumed above.

A reasonably good approximate solution to equation (16.5.7) for values of Pr considerably greater than 1 is

$$\xi=Pr^{-\frac{1}{3}}. \tag{16.5.8}$$

Taking this result, the local heat flux can be found from equation (16.2.8) as

$$q_0=-2k(T_1-T_0)\frac{Pr^{\frac{1}{3}}}{\delta}$$

or

$$h=2k\,Pr^{\frac{1}{3}}\,\frac{Re^{\frac{1}{2}}}{5.836\,x}. \tag{16.5.9}$$

Hence the local Nusselt number is given by

$$Nu_x=\frac{hx}{k}=0.343\,Re_x^{\frac{1}{2}}\,Pr^{\frac{1}{3}} \tag{16.5.10}$$

and the average Nusselt number over length L is

$$Nu_L=\frac{\bar{h}L}{k}=0.686\,Re_L^{\frac{1}{2}}\,Pr^{\frac{1}{3}}. \tag{16.5.11}$$

The corresponding result for mass transfer is

$$Sh_L=0.686\,Re_L^{\frac{1}{2}}\,Sc^{\frac{1}{3}}. \tag{16.5.12}$$

Expressed in terms of the Stanton numbers, and recalling from § 7.7 that the average skin-friction coefficient,

$$c_f=1.372\,Re_L^{-\frac{1}{2}}$$

we have

$$St\,Pr^{\frac{2}{3}}=St'\,Sc^{\frac{2}{3}}=\tfrac{1}{2}c_f. \tag{16.5.13}$$

These results were obtained for the case when $\delta_T<\delta$ and this was shown to be so when $Pr>1$. For $Pr<1$, $\delta_T<\delta$ and the integral leading to equation (16.5.4) must be modified. The velocity profile of equation (16.5.1) applies only in the range $y<\delta$ and for $\delta<y<\delta_T$ we must use the relationship $u=u_1$. The function $f(\xi)$ now takes the form

$$f(\xi)=\tfrac{3}{10}-\tfrac{3}{10}\xi^{-1}+\tfrac{2}{15}\xi^{-2}-\tfrac{3}{140}\xi^{-4}+\tfrac{1}{180}\xi^{-5}$$

and

$$\frac{1}{Pr}=\xi^2(2.55-2.55\,\xi^{-1}+1.135\,\xi^{-2}-0.182\,\xi^{-4}+0.047\,\xi^{-5}). \tag{16.5.14}$$

For very small values of Pr we can say that $\xi\simeq 0.626\,Pr^{-\frac{1}{2}}$ from which it follows that

$$Nu_L=1.10\,Re_L^{\frac{1}{2}}\,Pr^{\frac{1}{2}}$$

and

$$0.626\,St\,Pr^{\frac{1}{2}}=\tfrac{1}{2}c_f. \tag{16.5.15}$$

No corresponding result for mass transfer is given since *Sc* is rarely, if ever, very much less than 1.

These results can be compared with the results of the two previous sections. For *Pr* or $Sc > 1$, the thermal and molecular diffusivities are less than the kinematic viscosity. Consequently the thermal and concentration boundary layers are thinner than the momentum boundary layer and therefore grow in a region in which the velocity increases more or less linearly with distance from the wall. It is found that the Stanton number is proportional to $Pr^{-\frac{2}{3}}$ as predicted in § 16.4. For $Pr < 1$ the thermal boundary layer is thicker than the momentum layer and the velocity is constant at u_1 throughout much of the thermal layer. The Stanton number is found to be proportional to $Pr^{-\frac{1}{2}}$ as in § 16.3.

These effects arise from the behaviour of $\xi^2 f(\xi)$ which is plotted in figure 136 together with the two asymptotes $\xi = Pr^{-\frac{1}{3}}$ and $\xi = 0.626\, Pr^{-\frac{1}{2}}$.

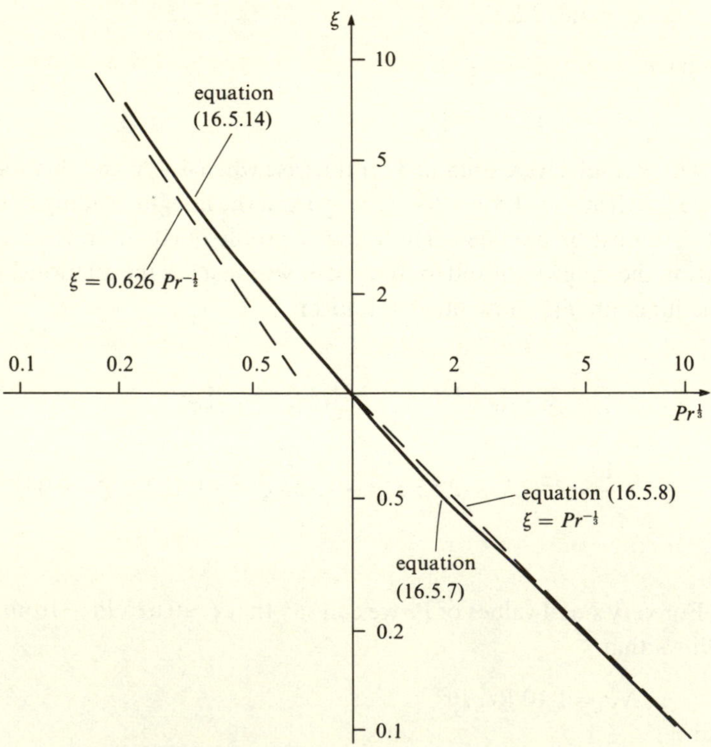

Fig. 136. Dependence of ξ on $Pr^{-\frac{1}{3}}$.

16.6 Exact solution for heat transfer in a laminar boundary layer

Besides the approximate solutions based on assumed temperature and concentration profiles presented in the two previous sections, mathematically exact solutions are possible in several situations. In this section we will consider the exact solutions for heat and mass transfer in a boundary layer on a flat plate and in § 16.7 we will consider heat and mass transfer to laminar flow in a pipe.

The exact solution for the velocity profile in a laminar boundary layer was presented in § 7.4 where it was shown by an order of magnitude analysis that the appropriate forms of the continuity equation and the Navier–Stokes equation were

$$\frac{\partial u}{\partial x}+\frac{\partial v}{\partial y}=0, \tag{16.6.1}$$

$$u\frac{\partial u}{\partial x}+v\frac{\partial u}{\partial y}=\nu\frac{\partial^2 u}{\partial y^2}. \tag{16.6.2}$$

By similar arguments we can reduce the general forms of the equations for heat and mass transfer, equations (16.1.4) and (16.1.5) to the forms

$$u\frac{\partial T}{\partial x}+v\frac{\partial T}{\partial y}=\alpha\frac{\partial^2 T}{\partial y^2}, \tag{16.6.3}$$

$$u\frac{\partial C}{\partial x}+v\frac{\partial C}{\partial y}=\mathscr{D}\frac{\partial^2 C}{\partial y^2}. \tag{16.6.4}$$

In § 7.4 we introduced the substitutions $\eta=\frac{1}{2}(u_1/\nu x)^{\frac{1}{2}}y$ and $f=\psi(u_1\nu x)^{\frac{1}{2}}$ where ψ is the stream function defined in § 4.4. With these substitutions equation (16.6.2) became the ordinary differential equation

$$\frac{\mathrm{d}^3 f}{\mathrm{d}\eta^3}+f\frac{\mathrm{d}^2 f}{\mathrm{d}\eta^2}=0, \tag{16.6.5}$$

which was solved by means of a series solution. This equation may also be written

$$\frac{\mathrm{d}^2 u}{\mathrm{d}\eta^2}+f\frac{\mathrm{d}u}{\mathrm{d}\eta}=0, \tag{16.6.6}$$

since $u/u_1=\frac{1}{2}(\mathrm{d}f/\mathrm{d}\eta)$.

If we make the same substitutions in equations (16.6.3) and (16.6.4) we obtain

$$\frac{d^2T}{d\eta^2}+Pr\,f\,\frac{dT}{d\eta}=0, \tag{16.6.7}$$

$$\frac{d^2C}{d\eta^2}+Sc\,f\,\frac{dC}{d\eta}=0. \tag{16.6.8}$$

For the special case when $Pr=1$ and $Sc=1$ equations (16.6.6), (16.6.7), and (16.6.8) are mathematically identical and with the same boundary conditions the velocity, temperature, and concentration profiles are similar. Thus at a given point (x, y)

$$\frac{u}{u_1}=\frac{T_0-T}{T_0-T_1}=\frac{C_0-C}{C_0-C_1}. \tag{16.6.9}$$

For momentum, heat and mass transfer at the wall we have,

$$\tau_0=\mu\frac{du}{dy},\quad q_0=-k\frac{dT}{dy}\quad\text{and}\quad N_0=-\mathscr{D}\frac{dC}{dy}.$$

Thus from equation (16.6.9)

$$\frac{\tau_0}{\mu u_1}=\frac{q_0}{k(T_0-T_1)}=\frac{N_0}{\mathscr{D}(C_0-C_1)}, \tag{16.6.10}$$

which, with $Pr=Sc=1$, can be rearranged to the form

$$\frac{h}{\rho u_1 c_p}=\frac{k_L}{u_1}=\frac{\tau_0}{\rho u_1^2}. \tag{16.6.11}$$

Hence from equation (7.4.9),

$$St=St'=\tfrac{1}{2}c_f=0.332\,Re_x^{-\frac{1}{2}}. \tag{16.6.12}$$

This result refers to local values of the shear stress and the heat and mass transfer coefficients. As shown in equation (16.3.15), the average values over a length L are twice as large and the corresponding Stanton numbers are given by

$$St=St'=\tfrac{1}{2}c_f=0.664\,Re_L^{-\frac{1}{2}}. \tag{16.6.13}$$

When the Prandtl and Schmidt numbers are not equal to 1, equations (16.6.7) and (16.6.8) must be solved in terms of a series solution in a manner similar to that presented in § 7.4. In this case the coefficients in the series are functions of *Pr* or *Sc*. It is, however, found that for large values of the

Prandtl and Schmidt numbers the results can be expressed in the approximate form

$$St\,Pr^{\frac{2}{3}} = St'\,Sc^{\frac{2}{3}} = \tfrac{1}{2}c_f = 0.664\,Re_L^{-\frac{1}{2}}. \tag{16.6.14}$$

16.7 Exact solution for heat transfer in a circular pipe

We will now consider the case of laminar flow in a circular pipe with heat or mass transfer taking place. Let the temperature of the fluid be T_1 at entry to the heated or cooled section of the pipe. It will be assumed that T_1 is constant across the flow. Let the pipe wall temperature be T_0 over the heated or cooled section and let the temperature at any other point in the flow be T.

Writing $\theta = (T - T_0)/(T_1 - T_0)$ the energy equation without heat generation, equation (16.1.4), becomes

$$\frac{D\theta}{Dt} = \alpha\,\nabla^2\theta \tag{16.7.1}$$

and the boundary conditions which must be satisfied are

$$\theta = 0 \quad \text{at} \quad r = a \quad \text{for all values of } z > 0,$$

$$\theta = 1 \quad \text{at} \quad z = 0.$$

The problem is illustrated diagrammatically in figure 137.

For the corresponding problem in mass transfer, C_1 will be the concentration of the diffusing component at the entry section $z = 0$, C_0 will be the concentration at the wall, and C the value at any other point in the flow. Writing $\psi = (C - C_0)/(C_1 - C_0)$ the mass transfer equation (16.1.5) becomes

$$\frac{D\psi}{Dt} = \mathscr{D}\,\nabla^2\psi \tag{16.7.2}$$

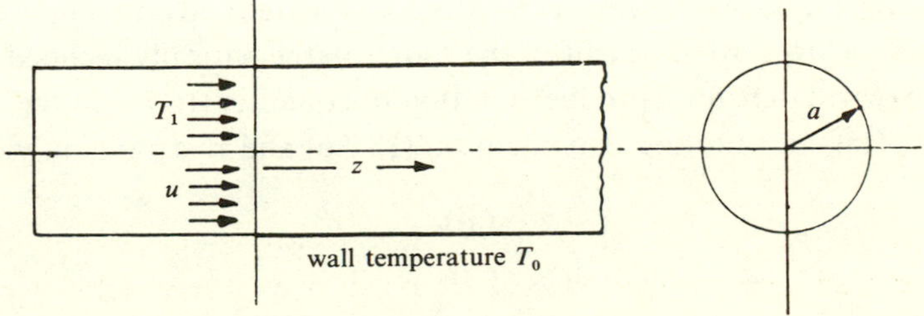

Fig. 137. Heat transfer into laminar flow in a pipe.

and the boundary conditions are the same as in the heat transfer problem. The solution of equation (16.7.1) will therefore apply equally to the mass transfer problem if $\mathscr{D}$ is substituted for α.

In cylindrical polar co-ordinates $\nabla^2\theta = \partial^2\theta/\partial r^2 + 1/r(\partial\theta/\partial r) + (\partial^2\theta/\partial z^2)$. We will consider only the case of steady flow and, using order-of-magnitude arguments similar to those used in § 7.2, we find that the term $\partial^2\theta/\partial z^2$ may be omitted. Equation (16.7.1) therefore becomes

$$u\frac{\partial\theta}{\partial z} = \alpha\left\{\frac{\partial^2\theta}{\partial r^2} + \frac{1}{r}\frac{\partial\theta}{\partial r}\right\}. \tag{16.7.3}$$

There are two special cases to be considered:

(a) Near the entrance to a pipe the velocity will be nearly uniform over the cross-section, i.e. $u = u_m =$ constant. This is referred to as 'rod-like flow' or 'plug flow' and from equation (16.7.3) the equation to be solved for the temperature distribution is

$$u_m\frac{\partial\theta}{\partial z} = \alpha\left\{\frac{\partial^2\theta}{\partial r^2} + \frac{1}{r}\frac{\partial\theta}{\partial r}\right\}. \tag{16.7.4}$$

(b) For fully developed laminar flow at some distance from the pipe entry we may assume that the parabolic Poiseuille velocity distribution applies, i.e. $u = 2u_m(1 - r^2/a^2)$, as in equation (5.8.7), and the equation to be solved for the temperature distribution is

$$2u_m\left(1 - \frac{r^2}{a^2}\right)\frac{\partial\theta}{\partial z} = \alpha\left\{\frac{\partial^2\theta}{\partial r^2} + \frac{1}{r}\frac{\partial\theta}{\partial r}\right\}. \tag{16.7.5}$$

These two equations can be put in the combined form

$$u_m f(r)\frac{\partial\theta}{\partial z} = \alpha\left\{\frac{\partial^2\theta}{\partial r^2} + \frac{1}{r}\frac{\partial\theta}{\partial r}\right\}, \tag{16.7.6}$$

where $f(r) = 1$ for plug flow and $f(r) = 2(1 - r^2/a^2)$ for Poiseuille flow.

The solution of equation (16.7.6) is complicated and only a brief outline is given below for the benefit of the more mathematically inclined reader.

We will look for a product solution to equation (16.7.6) in the form $\theta = R(r)Z(z)$. Substituting into equation (16.7.6) and re-arranging yields,

$$\frac{u_m}{\alpha Z}\frac{\mathrm{d}Z}{\mathrm{d}z} = \frac{1}{Rf(r)}\left\{\frac{\mathrm{d}^2R}{\mathrm{d}r^2} + \frac{1}{r}\frac{\mathrm{d}R}{\mathrm{d}r}\right\} \tag{16.7.7}$$

the left-hand side of this equation is a function of z only and the right-hand side is a function of r only. These can only be equal if they are both equal to a constant which we will denote by $-\beta^2$.

Thus

$$\frac{dZ}{dz}=-\frac{\beta^2 \alpha Z}{u_m} \tag{16.7.8}$$

and

$$\frac{d^2R}{dr^2}+\frac{1}{r}\frac{dR}{dr}+\beta^2 f(r)R=0. \tag{16.7.9}$$

The solution to equation (16.7.8) subject to the condition that $Z=1$ on $z=0$ is

$$Z=e^{-\beta^2 \alpha z/u_m} \tag{16.7.10}$$

and for the plug flow case, $f(r)=1$, the solution to equation (16.7.9) is

$$R=AJ_0(\beta r), \tag{16.7.11}$$

where J_0 is the Bessel function of the first kind of order 0. For the Poiseuille flow case we must look for a series solution to (16.7.9) which we will write as

$$R=AF(\beta r, \beta a). \tag{16.7.12}$$

To satisfy the boundary condition that $\theta=0$ on $r=a$, we must have $J_0(\beta a)=0$ or $F(\beta a)=0$, from which we can evaluate a series of permitted values of β which we will call β_n.

Equation (16.7.9) is of the Sturm–Liouville type and therefore the functions $J_0(\beta_n r)$ are orthogonal as are the functions $F(\beta_n r, \beta_n a)$. The usual techniques of Fourier analysis are therefore applicable and for the plug flow case we have

$$\theta=2\sum_{n=1}^{\infty}\frac{1}{\beta_n a}\frac{J_0(\beta_n r)}{J_1(\beta_n a)}\exp\left\{-\frac{\beta_n^2 \alpha z}{u_m}\right\}. \tag{16.7.13}$$

From this and the corresponding expression for the Poiseuille flow case we can work out the heat flux at the wall, $q_0=k(T_1-T_0)(\partial\theta/\partial r)_{r=a}$ and hence the heat transfer coefficient, Nusselt and Stanton numbers as functions of distance from the pipe entry.

Though mathematically exact these results are based on physical assumptions that may not be achieved in practice. The changing temperature, for example, will affect the viscosity which will disturb the Poiseuille velocity distribution. Also we can assume plug flow near the entry only if the thermal boundary layer is growing much more rapidly than the momentum boundary layer. This will occur when the Prandtl number is very much less than 1. Thus the plug-flow analysis is most likely to be applicable for heat transfer to liquid metals in the entry region of a pipe and the Poiseuille analysis will apply best for mass transfer in fully developed flow.

17

Heat and mass transfer in turbulent flow

17.1 Basic equations of forced convection in turbulent flow

In turbulent flow the temperature, concentration, and velocity will fluctuate. At a point in the flow the instantaneous temperature and concentration may be expressed by

$$T = \bar{T} + T', \tag{17.1.1}$$

$$C = \bar{C} + C', \tag{17.1.2}$$

where $\bar{T}$ and $\bar{C}$ are the mean values and T' and C' are the fluctuating components. The instantaneous velocity components may be similarly expressed as in §6.2.

Taking the energy equation in the form of equation (9.4.11) and neglecting the kinetic energy and the shear work we have

$$\frac{\partial T}{\partial t} + u\frac{\partial T}{\partial x} + v\frac{\partial T}{\partial y} + w\frac{\partial T}{\partial z} = \alpha\,\nabla^2 T + \frac{1}{\rho c_p}\frac{\partial p}{\partial t}. \tag{17.1.3}$$

Substituting from equations (6.2.1) and (17.1.1) and averaging each term over a short interval of time, we have the result that for steady flow,

$$\frac{\mathrm{D}\bar{T}}{\mathrm{D}t} = \alpha\,\nabla^2\bar{T} - \left\{\frac{\partial}{\partial x}(\overline{u'T'}) + \frac{\partial}{\partial y}(\overline{v'T'}) + \frac{\partial}{\partial z}(\overline{w'T'})\right\}. \tag{17.1.4}$$

This is the ordinary energy equation as presented in §16.1 written in terms of mean values but with the addition of the extra terms in brackets on the right-hand side. The additional terms involving the fluctuating components represent the transfer of heat by turbulent convection and are analogous to the Reynolds stress in the equation of motion presented in §6.3. Similar terms also appear in the mass transfer equation which takes the same form as equation (17.1.4) but with C replacing T and $\mathscr{D}$ replacing α.

Taking the simple case of flow parallel to a plane wall with mean velocity $\bar{u}$ which is a function of y and with heat and mass transfer in the y direction only, equation (17.1.4) reduces to

$$\frac{D\bar{T}}{Dt}=\alpha\frac{\partial^2\bar{T}}{\partial y^2}-\frac{\partial}{\partial y}\overline{(v'T')}$$

or

$$\frac{D\bar{T}}{Dt}=\frac{1}{\rho c_p}\frac{\partial}{\partial y}\left\{k\frac{\partial\bar{T}}{\partial y}-\rho c_p\overline{(v'T')}\right\}. \tag{17.1.5}$$

The mass transfer equivalent is

$$\frac{D\bar{C}}{Dt}=\frac{\partial}{\partial y}\left\{\mathscr{D}\frac{\partial\bar{C}}{\partial y}-\overline{(v'C')}\right\}. \tag{17.1.6}$$

The first term on the right-hand side represents the normal transfer of heat or mass by the molecular processes of conduction and diffusion within the fluid and the second term represents the turbulent or eddy transfer. According to the simple mixing length theory, as outlined in §6.7, the eddy heat transfer rate may be expressed by means of a gradient transport equation similar to (6.7.13) i.e.

$$q_T=\rho c_p l'\tilde{v}'\frac{\partial T}{\partial y}, \tag{17.1.7}$$

where l' is the mean value of the mixing length and $\tilde{v}'$ is the root mean square value of the transverse turbulent velocity component v'. As noted in §6.7, the product $l'\tilde{v}'$ may be described as the kinematic eddy viscosity ν_T or ν_τ. The total heat flux q due to the combined effects of molecular conduction and eddy transfer within the fluid at any point is therefore given by

$$q=-\left\{k\frac{\partial\bar{T}}{\partial y}+\rho c_p l'\tilde{v}'\frac{\partial\bar{T}}{\partial y}\right\}$$

or

$$q=-\rho c_p(\alpha+\nu_\tau)\frac{\partial\bar{T}}{\partial y}, \tag{17.1.8}$$

where α is the thermal diffusivity $k/\rho c_p$ and ν_τ is eddy viscosity $l'\tilde{v}'$.

The corresponding equation for mass transfer is

$$N=-(\mathscr{D}+\nu_\tau)\frac{\partial\bar{C}}{\partial y} \tag{17.1.9}$$

and the momentum analogue is given by combining equations (6.5.2) and (6.7.5),

$$\tau=\rho(\nu+\nu_\tau)\frac{\partial\bar{u}}{\partial y}. \tag{17.1.10}$$

Though the mixing length theory does not give an accurate prediction of the value of the eddy viscosity from first principles, this theory does lead to the important conclusion that the same value of ν_τ appears in each of the above equations.

The difference in sign between equation (17.1.10) and the two previous equations is of no consequence and results solely from the convention that q and N represent transfer rates to the fluid but that τ, being a retarding stress, corresponds to momentum transfer from the fluid.

We have seen in chapters 6, 7, and 8 that the nature of turbulent flow in a boundary-layer region or in a pipe is extremely complex. The formulae for the mean velocity distribution derived in §§7.10 and 8.3 are necessarily based on a simplified analysis of the problem. We can, however, extend the methods already used in chapters 7 and 8 to give adequate predictions of transfer rates and temperature and concentration profiles starting from the gradient transport equations (17.1.8) and (17.1.9). Before doing so we will first of all review in §§ 17.2, 17.3, and 17.4 some simpler models which have been widely used in the past in chemical engineering. Not only do these give insight into the assumptions involved but they will also form the basis of the Whitman two-film analysis of § 18.8 and the analysis of mass transfer with chemical reaction presented in chapter 22.

17.2 The film model

We have seen in chapters 6 and 8 that the size of the turbulent eddies becomes large in the central region of fully developed turbulent pipe flow. However, the eddies are suppressed in the vicinity of a solid boundary so that there exists a laminar region adjacent to the wall. In practice there is no sharp boundary between the turbulent and laminar regions as the degree of turbulence increases gradually with distance from the wall. None-the-less we can obtain some useful predictions of the heat and mass transfer coefficients in turbulent flow from the *film model* in which we postulate the existence of a laminar region of thickness δ adjacent to the wall together with a perfectly mixed central core as indicated in figure 138(a). We will show later that in all real situations δ is very much less than the pipe diameter D. The perfect mixing in the turbulent core implies an effectively infinite value for the kinematic eddy viscosity and zero values of the velocity, temperature, and concentration gradients in this central region. By contrast the transfer of momentum, heat, and mass in the laminar layer is assumed to occur by molecular processes only. This model leads to the

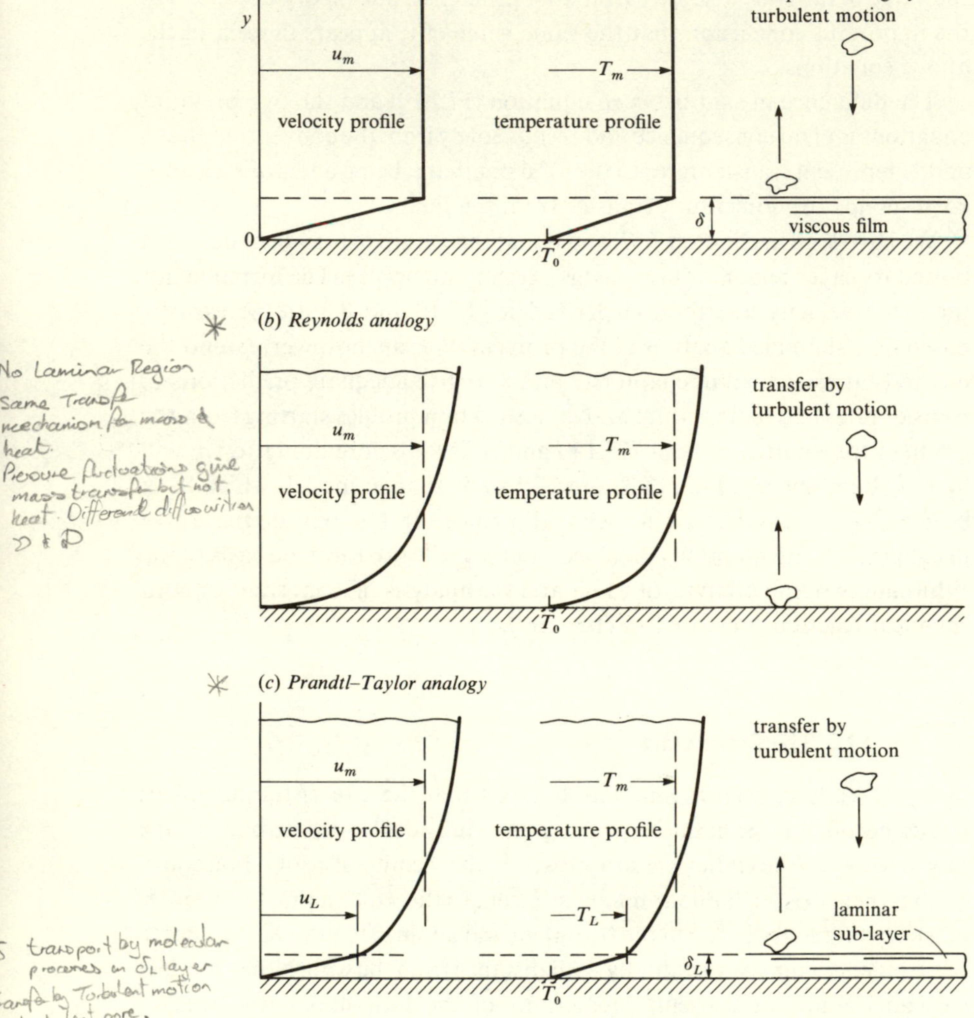

Fig. 138. Models for heat transfer: (a) film model, (*b*) Reynolds analogy, (*c*) Prandtl–Taylor analogy.

idealised velocity, temperature, and concentration profiles shown in figure 138(*a*).

We have assumed that within the well-mixed turbulent core the velocity u, temperature T, and concentration C are constant and, since $\delta \ll D$, these quantities will be close to the mean values u_m, T_m, and C_m for the whole cross-section. We will also assume that in the laminar layer the velocity decreases to zero at the wall. Since the layer is thin we can assume that the velocity profile is linear and hence the velocity gradient is given by u_m/δ. Similarly there will be temperature and concentration gradients $(T_0 - T_m)/\delta$ and $(C_0 - C_m)/\delta$ respectively, where T_0 and C_0 are the temperature and concentration at the wall.

Within the laminar layer transport is presumed to be by molecular processes only and hence the wall shear stress τ_0 is given by

$$\tau_0 = \mu \frac{u_m}{\delta}. \tag{17.2.1}$$

Similarly the rate of heat and mass transfer per unit area into the fluid, q_0 and N_0, are given by

$$q_0 = k \frac{(T_0 - T_m)}{\delta} \tag{17.2.2}$$

and

$$N_0 = \mathscr{D} \frac{(C_0 - C_m)}{\delta}. \tag{17.2.3}$$

Heat and mass transfer coefficients were defined in §§ 13.2 and 14.3 by the equations $h = q_0/(T_0 - T_m)$ and $k_L = N_0/(C_0 - C_m)$. Thus we can see from equations (17.2.2) and (17.2.3) that these are predicted by the film model as

$$h = \frac{k}{\delta} \tag{17.2.4}$$

and

$$k_L = \frac{\mathscr{D}}{\delta}. \tag{17.2.5}$$

Eliminating δ by means of equation (17.2.1) and multiplying through by the pipe diameter D we find that

$$\frac{\tau_0 D}{\mu u_m} = \frac{hD}{k} = \frac{k_L D}{\mathscr{D}}.$$

The second and third parts of this expression are the conventional definitions of the Nusselt and Sherwood numbers as presented in §§ 9.6, 13.2, and 14.3 and the first part can be rearranged as follows

$$\frac{\tau_0 D}{\mu u_m} = \frac{1}{2} \frac{\tau_0}{\frac{1}{2}\rho u_m^2} \cdot \frac{\rho u_m D}{\mu} = \frac{1}{2} c_f \, Re.$$

Thus the result of the film model can be written

$$Nu = Sh = \tfrac{1}{2} c_f \, Re. \tag{17.2.6}$$

Alternatively this result can be presented in terms of the Stanton numbers defined in §§9.6 and 15.3:

$$St\,Pr = St'\,Sc = \tfrac{1}{2}c_f. \qquad (17.2.7)$$

We can now reconsider our assumption that the laminar layer thickness δ was very much less than the pipe diameter D.

From equation (17.2.4) we see that

$$Nu = \frac{hD}{k} = \frac{k}{\delta}\frac{D}{k} = \frac{D}{\delta}$$

and hence

$$\frac{D}{\delta} = \frac{1}{2}c_f\,Re. \qquad (17.2.8)$$

In a pipe the flow is fully turbulent for $Re > 4000$ and within the range $4000 < Re < 10^5$ the friction factor is given by the Blasius expression, equation (8.5.4),

$$c_f = 0.079\,Re^{-\frac{1}{4}}.$$

Thus at $Re = 4000$, $c_f = 0.0099$ and hence $D/\delta \simeq 20$, and at $Re = 10^5$, $c_f = 0.0044$ and $D/\delta \simeq 220$. Thus we see that throughout the turbulent-flow regime the laminar film thickness is very much less than the pipe diameter as was assumed in the derivation of the above results.

As will be seen in §17.4 below, the film model does not in fact give an accurate prediction of the heat and mass transfer coefficients except for gases. We will therefore proceed in the next section to consider the more reliable model known as Reynolds analogy. The importance of the film model is that it forms the starting point for the more sophisticated analyses of heat and mass transfer presented in §§17.5, 17.6, 18.8, and chapter 22.

17.3 Reynolds analogy

As an alternative to the film model, we can use an analysis known as Reynolds analogy. This was presented briefly in §6.7 and will now be considered in greater detail. Reynolds analogy takes a totally different view of the nature of turbulent flow in a pipe and a comparison of the results of this analysis with the predictions of the film model is instructive. In the film model we assumed that there was a perfectly mixed core and that the rates of heat and mass transfer were controlled by conduction and diffusion across a laminar layer. In Reynolds analogy it is assumed that there is no laminar layer and that the transfer rates are determined by the lack of perfection of the mixing in the core. This is equivalent to ignoring the terms α, $\mathscr{D}$, and ν in equations (17.1.8)–(17.1.10).

We assume that the fluid consists of discrete elements which stay in the bulk for sufficient time for their velocity, temperature, and concentration to equilibrate with the mean values in the bulk, u_m, T_m, and C_m. Periodically an element jumps rapidly to the wall, without change in any of its properties, where it stays for long enough to equilibrate with the wall conditions, T_0 and C_0. The element then returns rapidly to the bulk as illustrated in figure 138(*b*).

It is seen that this model is a cruder version of the mixing length analysis of § 6.7. Whereas in that section we considered elements making jumps of random length l', in this case we assume that the elements jump from the bulk right up to the wall. Reynolds analogy also has some similarity with the kinetic theory of gases, discussed in chapter 2. However, in the kinetic theory we considered the motion of individual molecules but in this case we are applying similar ideas to very much larger groups of molecules.

Let m be the rate at which elements arrive at the wall, expressed as a mass-flow rate per unit wall area. The elements bring with them,

momentum (in the downstream direction)	mu_m,
enthalpy	mc_pT_m,
and solute	mC_m/ρ.

On return they take back

momentum	$mu_0=0$,
enthalpy	mc_pT_0,
and solute	mC_0/ρ.

The net rate of momentum transfer to the wall per unit area equals the wall shear stress τ_0. Therefore

$$\tau_0=mu_m \tag{17.3.1}$$

and the rates of heat and mass transfer from the wall are given by

$$q_0=mc_p(T_0-T_m) \tag{17.3.2}$$

and

$$N_0=\frac{m(C_0-C_m)}{\rho}. \tag{17.3.3}$$

Eliminating m between these equations gives

$$\frac{q_0}{c_p(T_0-T_m)}=\frac{N_0\rho}{(C_0-C_m)}=\frac{\tau_0}{u_m}$$

or

$$\frac{h}{\rho u_m c_p}=\frac{k_L}{u_m}=\frac{\tau_0}{\rho u_m^2} \tag{17.3.4}$$

and hence from the definitions of the Stanton numbers,

$$St=St'=\tfrac{1}{2}c_f. \tag{17.3.5}$$

This result can be compared with the prediction of the film model, equation (17.2.7) and it can be seen that the two models are compatible only if the Prandtl and Schmidt numbers are 1. For most gases the Prandtl and Schmidt numbers are close to 1, as was shown in §§2.7 and 2.8 but considerable differences from 1 occur for liquids. Since the models are based on radically different assumptions about the nature of turbulence, the agreement in the case of $Pr=Sc=1$ suggests that they are reliable in this case, and this is confirmed by experiment. For other values of the Prandtl and Schmidt numbers the predictions of the two models differ, often by a considerable amount, but it is usually found that the experimental values lie between the two predictions. This is discussed further in § 17.4 below where a compromise between the two models is presented.

The agreement of the film model and Reynolds analogy for gases is not unexpected. Reynolds analogy considers the random motion of elements of fluid, whereas the kinetic theory of gases considers the random motion of individual molecules. However, the size of the element does not enter the analysis and hence the ratio of the transfer rates predicted by Reynolds analogy for turbulent flow inevitably equals the ratio predicted by the kinetic theory for laminar flow of gases.

Both Reynolds analogy and the film model can be derived directly from the gradient transport equations in the form given in equations (17.1.8)–(17.1.10). We can obtain the equations of the film model by setting v_τ to zero and dividing equation (17.1.10) by (17.1.8) giving

$$\frac{\tau}{q}=-\frac{v}{\alpha c_p}\frac{\mathrm{d}\bar{u}}{\mathrm{d}T}=\frac{\mu}{k}\frac{u_m}{(T_0-T_m)}, \tag{17.3.6}$$

from which equation (17.2.7) follows directly.

If, on the other hand, we assume that $\nu_\tau \gg \nu$ or α, division of equation (17.1.10) by (17.1.8) gives

$$\frac{\tau}{q} = -\frac{\mathrm{d}\bar{u}}{\mathrm{d}T} = \frac{u_m}{T_0 - T_m}, \tag{17.3.7}$$

which is the result of Reynolds analogy, equation (17.3.5). However, this approach does not give us an estimate of the laminar layer thickness δ or the mass interchange rate m.

17.4 The *j*-factor analogy

In the two previous sections we considered the film model and Reynolds analogy. The former gave rise to the result

$$St\ Pr = St'\ Sc = \tfrac{1}{2}c_f \tag{17.4.1}$$

and the latter to the result

$$St = St' = \tfrac{1}{2}c_f. \tag{17.4.2}$$

Clearly these cannot both be correct but since the models take opposite extreme views of the nature of turbulent flow we might suppose that the actual values of the Stanton numbers will lie between the two predictions. For the case of $Pr = Sc = 1$, the two predictions are the same and are found to agree well with experiment.

In § 16.4 we considered the development of thermal and concentration boundary layers into laminar flow near a wall and in § 16.5 we considered the simultaneous development of these layers with a laminar momentum boundary layer. The former analysis gave rise to the result,

$$St\ Pr^{\frac{2}{3}} = St'\ Sc^{\frac{2}{3}} = f(Re) \tag{17.4.3}$$

and the latter to the result

$$St\ Pr^{\frac{2}{3}} = St'\ Sc^{\frac{2}{3}} = \tfrac{1}{2}c_f, \tag{17.4.4}$$

provided the Prandtl and Schmidt numbers were much greater than 1.

Thus in the two cases where rigorous theoretical analysis is possible we see that $St\ Pr^{\frac{2}{3}}$ and $St'\ Sc^{\frac{2}{3}}$ are functions of the Reynolds number. Thus it seems not unreasonable to suppose that the appropriate compromise between equations (17.4.1) and (17.4.2) will be in the same form as equation (17.4.4).

This is the basis of Colburn's *j*-factor analogy in which two dimensionless groups are defined as follows

$$j_H = St\, Pr^{\frac{2}{3}}, \tag{17.4.5}$$

$$j_D = St'\, Sc^{\frac{2}{3}}, \tag{17.4.6}$$

and equation (17.4.4) takes the form

$$j_H = j_D = \tfrac{1}{2}c_f. \tag{17.4.7}$$

Though based more upon intuition than rigorous analysis, this correlation is found to fit the facts in many situations. Care must, however, be exercised when dealing with bluff bodies. In §§ 17.2 and 17.3 we drew an analogy between the shear stress and the rates of heat and mass transfer. There is, however, no analogy between form drag and heat and mass transfer. Thus the quantity c_f appearing in equation (17.4.7) must be a true skin-friction coefficient and not an overall drag coefficient c_D, which also contains a contribution from form drag. In § 5.9 we considered the dependence of the drag coefficient for a sphere on the Reynolds number. Similar results are found for flow round a cylinder and the dependence of both the drag coefficient c_D and the skin-friction component c_f are shown in figure 139 as functions of the Reynolds number. It is seen that $c_D > c_f$ and very much so at high Reynolds numbers.

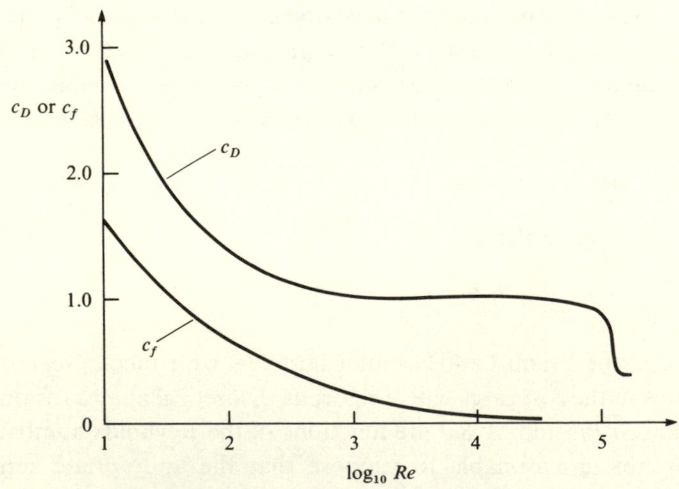

Fig. 139. Dependence of the drag and skin friction coefficients on Reynolds number for a cylinder.

For streamlined bodies and for internal flows no such problems arise and we will compare the predictions of the *j*-factor analogy with the well-established correlation of experimental results for heat transfer to turbulent flow in a pipe, due to Dittus and Boelter,

$$Nu = 0.023\, Re^{0.8}\, Pr^{0.4}. \tag{17.4.8}$$

We can put the *j*-factor analogy, equation (17.4.7), into a similar form by substituting for c_f from equation (6.6.10) and recalling that $St = Nu/Re\,Pr$. This gives

$$Nu = 0.040\, Re^{\frac{3}{4}}\, Pr^{\frac{1}{3}}. \tag{17.4.9}$$

Comparing equation (17.4.8) and (17.4.9) we see that the difference between the indices on the Prandtl number, 0.4 and $\frac{1}{3}$, is small and is probably not significant, especially for modest values of the Prandtl number. Similarly the indices on the Reynolds number, 0.8 and $\frac{3}{4}$, appear to be similar but an explanation is required for the different values of the constant, 0.023 and 0.040. This difference is in fact more apparent than real and results from the small difference in the indices on the Reynolds number. This has an appreciable effect since Re is normally large. The values of the groups $0.023\, Re^{0.8}$ and $0.040\, Re^{\frac{3}{4}}$ are tabulated below and it can be seen that within the normally encountered range of Reynolds numbers the formulations are barely distinguishable and both in fact lie within the scatter of experimental results.

Re	1.0×10^4	5.0×10^4	1.0×10^5
$0.023\, Re^{0.8}$	36.5	132.1	230
$0.040\, Re^{\frac{3}{4}}$	39.5	132.1	222

We have derived the *j*-factor analogy for turbulent flow as a compromise between the film model and Reynolds analogy, drawing on some theoretical analyses for laminar flow to suggest the appropriate dependence of the Stanton numbers on the Prandtl and Schmidt numbers. This is clearly not ideal and the good agreement with experiment is almost certainly fortuitous. We will therefore consider in the next two sections analogies between momentum, heat, and mass transfer based on more realistic views of the nature of turbulent pipe flow.

17.5 The Prandtl–Taylor analogy

The Prandtl–Taylor analogy is an improvement on the Reynolds analogy and is in effect the film model and Reynolds analogy in series. It is assumed that there is a laminar layer of thickness δ_L through which transport is by molecular processes and a turbulent core in which eddy transport dominates. Figure 138(*c*) illustrates the model and shows the corresponding velocity and temperature profiles.

Within the laminar layer equations (17.2.1) and (17.2.2) apply and take the forms

$$\tau_0 = \frac{\mu u_L}{\delta_L}, \tag{17.5.1}$$

$$q_0 = \frac{k(T_0 - T_L)}{\delta_L}, \tag{17.5.2}$$

where u_L and T_L are the velocity and temperature at the edge of the laminar layer. Within the turbulent core we can adapt equations (17.3.1) and (17.3.2) to give

$$\tau_0 = m(u_m - u_L) \tag{17.5.3}$$

and

$$q_0 = mc_p(T_L - T_m). \tag{17.5.4}$$

Eliminating δ_L, m, and T_L between these equations yields

$$T_0 - T_m = \frac{q_0}{\tau_0}\left\{\frac{\mu u_L}{k} + \frac{u_m - u_L}{c_p}\right\}$$

or

$$\frac{\tau_0}{\rho u_m^2} = \frac{h}{\rho u_m c_p}\left\{(Pr-1)\frac{u_L}{u_m} + 1\right\},$$

i.e.

$$\tfrac{1}{2}c_f = St\left\{(Pr-1)\frac{u_L}{u_m} + 1\right\}. \tag{17.5.5}$$

For the special case of $Pr = 1.0$, this reduces to Reynolds analogy, equation (17.3.5).

To proceed further with this method we must obtain an estimate of the ratio u_L/u_m. From the two-part velocity profile discussed in §8.4 we have $u_L = 11.6\ u_\tau$ and from equation (8.3.5) $u_m = u_\tau\sqrt{(2/c_f)}$. Thus $u_L/u_m = 8.2\sqrt{c_f}$ and hence

$$St = \frac{\tfrac{1}{2}c_f}{1 + 8.2\sqrt{c_f}(Pr-1)}. \tag{17.5.6}$$

For flow in a smooth pipe with values of *Re* up to 10^5, equation (6.6.10) gives

$$c_f = 0.079\, Re^{-\frac{1}{4}}$$

and therefore

$$\frac{u_L}{u_m} = 2.3\, Re^{-\frac{1}{8}} \tag{17.5.7}$$

and

$$St = \frac{0.0395\, Re^{-\frac{1}{4}}}{1 + 2.3\, Re^{-\frac{1}{8}}(Pr - 1)}. \tag{17.5.8}$$

However, the two-part profile of §8.4 is an over-simplification and overestimates the thickness of the laminar layer by a factor of about two. Thus $u_L/u_m \simeq 1.0\, Re^{-\frac{1}{8}}$. Complications arise, however, for two reasons. In the first place the sharp division into a laminar layer and an exclusively turbulent core is an over-simplification, though it is an improvement on the simple Reynolds analogy and the film model. Secondly, the velocity and temperature profiles are only similar for the special case of the Prandtl number being equal to 1.0. It is therefore necessary to make an empirical adjustment to equation (17.5.7) and the following expression, due to Hoffmann, may be used

$$\frac{u_L}{u_m} = 1.5\, Re^{-\frac{1}{8}}\, Pr^{-\frac{1}{6}}. \tag{17.5.9}$$

For flow in smooth pipe with value of Re up to 10^5, we can substitute for c_f from equation (6.6.10) and for the velocity ratio u_L/u_m from equation (17.5.9) and the Prandtl–Taylor equation (17.5.5) becomes

$$St = \frac{0.0395\, Re^{-\frac{1}{4}}}{1 + 1.5\, Re^{-\frac{1}{8}}\, Pr^{-\frac{1}{6}}(Pr - 1)}. \tag{17.5.10}$$

For air the Prandtl number is about 0.8, and in the range of Reynolds numbers from 5000 to 50 000, equation (17.5.10) can be represented approximately by the simpler expression

$$St = 0.046\, Re^{-\frac{1}{4}}. \tag{17.5.11}$$

Analogous expressions are found for mass transfer in liquids and for dilute gas mixtures. Equation (17.5.5) for example takes the form

$$\tfrac{1}{2}c_f = St'\left\{(Sc - 1)\frac{u_L}{u_m} + 1\right\}. \tag{17.5.12}$$

For rich mixtures in gases we can analyse the laminar layer using the methods of § 14.3 and we find that

$$\tfrac{1}{2}c_f = St'\left\{\left(\frac{p_{Blm}}{P}Sc-1\right)\frac{u_L}{u_m}+1\right\}, \tag{17.5.13}$$

where P is the total pressure and p_{Blm} is the log mean of the partial pressure of the non-diffusing species B.

The assumed velocity and temperature profiles of the film model, Reynolds analogy, and the Prandtl–Taylor analogy are compared in figure 138. It can be seen that we have made a progression from the highly artificial profile of the film model which assumes a laminar zone and a perfectly mixed turbulent core to the more realistic profile for the turbulent core assumed in Reynolds analogy. Reynolds analogy, however, ignores the effect of the laminar sub-layer and this is introduced in the Prandtl–Taylor analogy. The final stage in this progression, the inclusion of a buffer layer, as proposed by von Karman, is considered in the next section.

17.6 Von Karman's extension of the Prandtl–Taylor analogy

In the absence of a comprehensive theory of turbulence which would permit the direct calculation of the Reynolds stresses and the corresponding heat and mass transfer quantities, recourse must be had to one or other of the analogies which are based on the mixing length theory. The Reynolds and Prandtl–Taylor analogies have been discussed in §§ 17.3 and 17.5. Further refinements are possible such as the following which is due to von Karman.

In § 8.4 it was shown that the turbulent flow in a smooth-walled pipe could be divided into three regions as follows:

(i) a laminar sub-layer extending to $y^+=5$ in which transfer is by molecular processes;

(ii) a buffer layer extending from $y^+=5$ to $y^+=30$ in which both molecular and eddy transfer occurs;

(iii) a turbulent core for $y^+>30$ in which eddy transfer processes predominate.

The quantity y^+ is defined as y/y_τ where y_τ, the friction distance, is $\mu/\sqrt{(\tau_0\rho)}$.

The velocity distribution within these layers is given by the von Karman universal velocity profile

$$u^+=y^+ \qquad y^+<5, \tag{17.6.1}$$

$$u^+ = 5 \log_e y^+ - 3.05 \qquad 5 < y^+ < 30, \tag{17.6.2}$$

$$u^+ = 2.5 \log_e y^+ + 5.5 \quad 30 < y^+, \tag{17.6.3}$$

where $u^+ = u/u_\tau$ and u_τ, the friction velocity, is defined as $\sqrt{(\tau_0/\rho)}$.

As shown in § 17.1, the shear stress and the fluxes of heat and mass are related to the gradients of the mean velocity, temperature, and concentration by equations (17.1.8) to (17.1.10), i.e.

$$\tau_0 = \rho(\nu + \nu_\tau)\frac{\mathrm{d}u}{\mathrm{d}y}, \tag{17.6.4}$$

$$q_0 = -\rho c_p(\alpha + \nu_\tau)\frac{\mathrm{d}T}{\mathrm{d}y}, \tag{17.6.5}$$

$$N_0 = -(\mathscr{D} + \nu_\tau)\frac{\mathrm{d}C}{\mathrm{d}y}. \tag{17.6.6}$$

In these equations we have omitted the bar denoting average value since this is assumed throughout this section.

Equation (17.6.4) can be expressed in terms of the dimensionless velocity and distance defined above and becomes

$$1 = \left(1 + \frac{\nu_\tau}{\nu}\right)\frac{\mathrm{d}u^+}{\mathrm{d}y^+}. \tag{17.6.7}$$

Similar simplifications can be made in the remaining equations by taking dimensionless temperature $T^+ = T/T_\tau$ and concentration $C^+ = C/C_\tau$ where the friction temperature T_τ is defined by

$$T_\tau = \frac{q_0}{\rho c_p u_\tau} \tag{17.6.8}$$

and the friction concentration C_τ is defined by

$$C_\tau = \frac{N_0}{u_\tau}. \tag{17.6.9}$$

The analogy with the friction velocity and distance is made clearer by rewriting the definitions of these quantities as

$$u_\tau = \frac{\tau_0}{\rho u_\tau}, \tag{17.6.10}$$

$$y_\tau = \frac{\mu}{\rho u_\tau}. \tag{17.6.11}$$

Expressed in dimensionless terms equations (17.6.5) and (17.6.6) become

$$1=-\left(\frac{1}{Pr}+\frac{v_\tau}{v}\right)\frac{dT^+}{dy^+}, \tag{17.6.12}$$

$$1=-\left(\frac{1}{Sc}+\frac{v_\tau}{v}\right)\frac{dC^+}{dy^+}. \tag{17.6.13}$$

Accepting von Karman's velocity profile we can evaluate v_τ from equation (17.6.7) and using this in equations (17.6.12) and (17.6.13) we can predict the corresponding temperature and concentration profiles. The same approach can be taken with any other empirical expression for the turbulent velocity profile.

In the laminar sub-layer transfer is by molecular processes only and hence $v_\tau=0$. This result can be seen to be compatible with the velocity profile in the laminar sub-layer by substituting equation (17.6.1) into equation (17.6.7). Similarly, putting $v_\tau=0$ in equation (17.6.12) we obtain

$$\frac{dT^+}{dy^+}=-Pr$$

or

$$T^+=T_0^+-Pr\,y^+. \tag{17.6.14}$$

The concentration profile in the laminar sub-layer is given by,

$$C^+=C_0^+-Sc\,y^+ \tag{17.6.15}$$

and the values T_L^+ and C_L^+ at the edge of the laminar sub-layer are

$$T_L^+=T_0^+-5\,Pr, \tag{17.6.16}$$

$$C_L^+=C_0^+-5\,Sc. \tag{17.6.17}$$

In the buffer layer we can find the value of v_τ by substituting the velocity profile, equation (17.6.2) into equation (17.6.7). This gives

$$\frac{v_\tau}{v}=\frac{y^+}{5}-1.$$

Thus v_τ varies from 0 at $y^+=5$ to $5v$ at $y^+=30$, confirming the assertion that the contributions from eddy and molecular transfer are of comparable magnitude in the buffer layer. Substituting into equation (17.6.12) we obtain

$$1=-\left(\frac{1}{Pr}-1+\frac{y^+}{5}\right)\frac{dT^+}{dy^+}$$

or

$$T^+ - T_L^+ = -\int_5^{y^+} \frac{5\,\mathrm{d}y^+}{y^+ + 5(1/Pr - 1)}. \tag{17.6.18}$$

Hence on integrating and substituting from equation (17.6.16)

$$T^+ = T_0^+ - 5\,Pr - 5\log_e\left\{\frac{y^+\,Pr}{5} + (1 - Pr)\right\} \tag{17.6.19}$$

and the corresponding concentration profile is

$$C^+ = C_0^+ - 5\,Sc - 5\log_e\left\{\frac{y^+\,Sc}{5} + (1 - Sc)\right\}. \tag{17.6.20}$$

The values of the dimensionless temperature and concentration at the outer edge of the buffer layer, i.e. at $y^+ = 30$ are

$$T_B^+ = T_0^+ - 5\,Pr - 5\log_e(1 + 5\,Pr), \tag{17.6.21}$$

$$C_B^+ = C_0^+ - 5\,Sc - 5\log_e(1 + 5\,Sc). \tag{17.6.22}$$

In the turbulent core we assume that turbulent transfer predominates and hence equations (17.6.7) and (17.6.12) become

$$1 = \frac{\nu_\tau}{\nu}\frac{\mathrm{d}u^+}{\mathrm{d}y^+}, \tag{17.6.23}$$

$$1 = -\frac{\nu_\tau}{\nu}\frac{\mathrm{d}T^+}{\mathrm{d}y^+}. \tag{17.6.24}$$

Substituting from equation (17.6.3) into equation (17.6.23) we find that

$$\frac{\nu_\tau}{\nu} = \frac{y^+}{2.5}$$

and hence

$$\frac{\mathrm{d}T^+}{\mathrm{d}y^+} = -\frac{2.5}{y^+}$$

or

$$T^+ - T_B^+ = -2.5\int_{30}^{y^+} \frac{\mathrm{d}y^+}{y^+} = -2.5\log_e\left(\frac{y^+}{30}\right). \tag{17.6.25}$$

From equation (17.6.21) we find that

$$T^+ = T_0^+ - 5\,Pr - 5\log_e(1 + 5\,Pr) - 2.5\log_e\left(\frac{y^+}{30}\right) \tag{17.6.26}$$

and

$$C^+ = C_0^+ - 5\,Sc - 5\log_e(1+5\,Sc) - 2.5\log_e\left(\frac{y^+}{30}\right). \tag{17.6.27}$$

In order to evaluate heat and mass transfer coefficients we must treat the turbulent core somewhat differently. Eliminating ν_τ from equations (17.6.23) and (17.6.24) we obtain

$$\mathrm{d}T^+ = -\mathrm{d}u^+. \tag{17.6.28}$$

If we now make the supposition that at the position where u^+ has the mean value u_m^+, the temperature will also have the mean value T_m^+, equation (17.6.28) becomes

$$T_B^+ - T_m^+ = u_m^+ - u_B^+. \tag{17.6.29}$$

This assumption is not unreasonable since most of the temperature change occurs within the laminar sub-layer, especially when Pr is large, and the temperature is more or less constant within the turbulent core. If equation (17.6.29) is expressed in dimensional form it will be seen to be equivalent to assuming Reynolds analogy for the turbulent core.

Substituting from equation (17.6.21) and recalling that

$$u_B^+ = 5\log_e 30 - 3.05 = 5\log_e 6 + 5$$

we have

$$T_0^+ - T_m^+ = 5\,Pr + 5\log_e(1+5\,Pr) + u_m^+ - 5\log_e 6 + 5$$

or

$$\frac{(T_0 - T_m)\rho c_p u_m}{q_0} = 5(Pr-1) + 5\log_e\left(\frac{1+5\,Pr}{6}\right) + u_m^+. \tag{17.6.30}$$

But by definition $q_0/(T_0 - T_m)$ is the heat transfer coefficient h, $h/\rho c_p u_m$ is the Stanton number St and $u_m^+ = \sqrt{(2/c_f)}$ from equation (8.3.5). Hence we find that

$$St = \frac{\tfrac{1}{2}c_f}{1 + \sqrt{\dfrac{c_f}{2}}\left\{5(Pr-1) + 5\log_e\left(\dfrac{1+5\,Pr}{6}\right)\right\}} \tag{17.6.31}$$

and by analogy

$$St' = \frac{\tfrac{1}{2}c_f}{1 + \sqrt{\dfrac{c_f}{2}}\left\{5(Sc-1) + 5\log_e\left(\dfrac{1+5\,Sc}{6}\right)\right\}}. \tag{17.6.32}$$

It can be seen that for Pr or Sc equal to 1.0 these equations reduce to

Reynolds analogy,

$$St = St' = \tfrac{1}{2}c_f.$$

17.7 Assessment of the analogies between heat, mass, and momentum transfer

It is appropriate at this stage to assess the progress we have made in the prediction of heat and mass transfer coefficients in turbulent flow by analogy with momentum transfer.

The simplest of these predictions is Reynolds analogy (§ 17.3) in which it is assumed that transfer is by turbulent processes only. It follows immediately that Reynolds analogy predicts that the heat and mass transfer coefficients are independent of the conductivity or diffusivity. This is not in accord with experiment and Reynolds analogy is modified to the *j*-factor analogy by incorporating the dependence on the conductivity and diffusivity predicted for *laminar* flow.

Analysis of turbulent flow by the mixing length theory leads to the Prandtl–Taylor analogy or von Karman's extension of it. The prediction of this last model, equation (17.6.31), can be written in the form

$$St\, f(Pr, c_f) = \tfrac{1}{2}c_f$$

and it is instructive to compare $f(Pr, c_f)$ with the corresponding term $Pr^{\frac{2}{3}}$ in the *j*-factor analogy. This can only be done at a particular value of c_f and taking a typical value of 0.0056 (corresponding to a Reynolds number of 4×10^4) the two terms are tabulated below as a function of Pr. It is seen that at least for $Pr < 50$ the correspondence is close and this is perhaps the best theoretical justification for the *j*-factor analogy.

Pr	1	2	5	10	20	50	100	200
$Pr^{\frac{2}{3}}$	1	1.58	2.92	4.6	7.36	13.6	21.5	34.2
$f(Pr, c_f)$	1	1.42	2.44	3.95	7.72	14.9	28.4	55.0

For higher values of Pr the predictions diverge and we must re-examine our assumptions.

In von Karman's model of turbulent flow the eddy kinematic viscosity ν_τ is assumed to vary with dimensionless distance y^+ from the wall as shown in figure 140. This discontinuous distribution is clearly not accurate and there are reasons to believe that close to the wall ν_τ varies roughly as $(y^+)^3$ as shown in the diagram.

Up to some value of y^+ which we will denote by δ^+ (where δ^+ is of order 5.0), $\nu_\tau \ll \nu$ and we can neglect the effects of turbulence in equation (17.6.7).

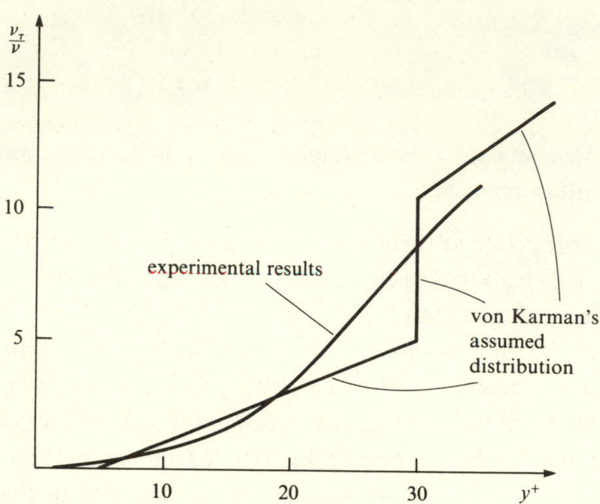

Fig. 140. Variation of eddy kinematic viscosity ν_τ with dimensionless distances from the wall y^+.

From this we deduce that $u^+ \simeq y^+$ as confirmed by experiment. However, in the mass transfer case, equation (17.6.13), we compare ν_τ/ν with $1/Sc$ and, since for liquids Sc is commonly of order 500, the effect of eddy transfer is not negligible as far from the wall as $y^+ = \delta^+$. If we assume that $\nu_\tau \propto (y^+)^3$ we can say that eddy transfer can be neglected in the mass transfer case up to δ_m^+ where $\delta_m^+ \simeq \delta^+ \, Sc^{-\frac{1}{3}}$. Thus we see that the effective laminar layer thickness for mass transfer δ_m is about $\delta \, Sc^{-\frac{1}{3}}$ where δ is the effective laminar layer thickness for momentum transfer. This argument, coupled with the film model of § 17.2, leads directly to the j-factor analogy, but this method still retains the unsatisfactory feature of a clear-cut division between laminar and turbulent regions.

In order to predict the mass transfer rates accurately we need to know how ν_τ varies in the region $\delta_m^+ < y^+ < \delta^+$ and clearly in this region measurements of the velocity profile are insufficiently accurate. Thus there seems to be no alternative but to resort to the use of empirical correlations. Many such correlations for heat transfer are available and are listed in standard reference books such as McAdams[1], Perry[2], Eckert & Drake[3], or Wong[4]. Mass transfer has been less exhaustively studied, but reference may be made to Eckert & Drake[3] and Sherwood[5]. Furthermore, heat transfer coefficients can be adapted for mass transfer use by replacing Nu by Sh, Pr by Sc, etc., but this should be done with caution. Most fluids have Prandtl numbers in the range 1.0–50 and what may be an excellent correlation within this range cannot necessarily be extrapolated to Prandtl and hence Schmidt numbers of around 500.

Finally we can re-examine the film model which is used as the basis of the Whitman two-film model of § 18.8 and the theory of gas absorption with chemical reaction in chapter 22. We have seen that the effective laminar layer thickness for mass and momentum transfer are not the same and this accounts for the failure of the basic film model. However, with the exception of a few light species such as H_2 and He, the diffusivities of most solutes through a given solvent are much the same. Hence we can say that for a particular solvent there will be an effective laminar layer thickness for mass transfer which may be treated as a constant. It must, however, be appreciated that this is not the laminar layer thickness for momentum transfer.

17.8 The penetration model

A different approach to the problem of predicting transfer rates to turbulent flow is represented by the penetration model. This model is most frequently used for predicting gas absorption rates but it is sometimes also applied to heat and mass transfer from solid surfaces. It is assumed that a packet of fluid is brought to the surface by turbulent processes, where it remains stationary for time θ before being replaced by a fresh packet. During the time the packet is stationary at the interface, diffusion takes place and this is controlled by the unsteady laminar diffusion equation (16.1.5) which takes the form

$$\frac{\partial C}{\partial t}=\mathscr{D}\frac{\partial^2 C}{\partial y^2}, \tag{17.8.1}$$

where y is distance from the interface and t is time measured from the arrival of the packet.

Taking the concentration in the newly arrived packet as zero, we can see intuitively that the concentration will change with time as shown in figure 141. Here we have assumed that the interfacial concentration C_i is constant and this implies that there is no gas phase resistance. Assuming that the penetration depth of the solute is small compared with the size of the packet, we have the boundary conditions,

at $t=0$ $C=0$ for all y,

for $t>0$ $C=C_i$ on $y=0$,

for $t\geqslant 0$ $C\to 0$ as $y\to\infty$.

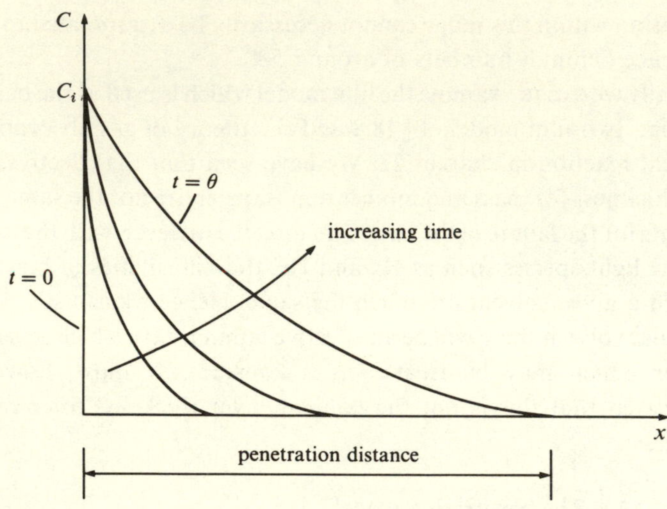

Fig. 141. Concentration profile in the penetration model.

Making the substitution, $\eta = y/\surd(4\mathscr{D}t)$, equation (17.8.1) becomes

$$\frac{d^2C}{d\eta^2} + 2\eta C = 0 \tag{17.8.2}$$

and the boundary conditions are

(i) $\eta = 0, \quad C = C_i,$

(ii) $\eta \to \infty, \quad C \to 0.$

It is seen that C is a function of η only and we have therefore reduced our partial differential equation to an ordinary differential equation.

Putting $dC/d\eta = g$ we have

$$\frac{dg}{g} + 2\eta \, d\eta = 0,$$

which on integration gives

$$\log_e g + \eta^2 = \text{const}$$

or

$$g = \frac{dC}{d\eta} = A \, e^{-\eta^2}, \tag{17.8.3}$$

where A is an arbitrary constant.

Integrating once more using boundary condition (i) we have

$$C = C_i + \frac{A\sqrt{\pi}}{2}\,\mathrm{erf}\,\eta$$

where erf η is the error function. Since erf $\infty = 1.0$, we have from boundary condition (ii), $A = -2C_i/\sqrt{\pi}$. The absorption rate N is given by,

$$N = -\mathscr{D}\left(\frac{\partial C}{\partial y}\right)_0 = -\frac{\mathscr{D}A}{\sqrt{(4\mathscr{D}t)}} = C_i\sqrt{\left(\frac{\mathscr{D}}{\pi t}\right)}.$$

Thus the current mass transfer coefficient is $\sqrt{(\mathscr{D}/\pi t)}$ and the average value up to time θ is given by

$$\bar{k}_L = \frac{1}{\theta}\int_0^\theta \sqrt{\left(\frac{\mathscr{D}}{\pi t}\right)}\,\mathrm{d}t = 2\sqrt{\left(\frac{\mathscr{D}}{\pi\theta}\right)}. \tag{17.8.4}$$

In the earlier version of the penetration model, due to Higbie, it is assumed that all packets stay on the surface for the same length of time θ. This is clearly unrealistic and in a later version by Danckwerts it is assumed that the chance of a packet being taken back into the bulk is independent of the time already spent on the surface. This gives rise to a distribution of resistance times on the surface, given by

$$f(\theta) = s\,\mathrm{e}^{-s\theta},$$

where s is the reciprocal of the mean residence time $\bar{\theta}$ and $f(\theta)\,\mathrm{d}\theta$ is the fraction of packets that stay on the surface for times between θ and $\theta + \mathrm{d}\theta$.

Thus the mean mass transfer coefficient is given by

$$\bar{k}_L = \int_0^\infty s\,\mathrm{e}^{-s\theta}\,2\sqrt{\left(\frac{\mathscr{D}}{\pi\theta}\right)}\,\mathrm{d}\theta = \sqrt{\frac{\mathscr{D}}{\bar{\theta}}}. \tag{17.8.5}$$

Thus in both versions $\bar{k}_L$ is proportional to $\sqrt{\mathscr{D}}$ and the square root of some empirical time.

The dependence on $\sqrt{\mathscr{D}}$ is typical of analyses that make no allowance for the variation of velocity with distance y from the surface as, for example, in §16.3. The penetration model is therefore most successful for gas/liquid interfaces where there is no shear stress and the analogy with momentum transfer is inappropriate. Heat and mass transfer from solid surfaces is usually better described by j-factor or von Karman methods.

Very similar analyses can be performed for the analogous heat transfer situations. From the original version of the penetration model we find that the heat transfer coefficient is given by

$$h = 2\left\{\frac{k\rho c_p}{\pi\theta}\right\}^{\frac{1}{2}} \tag{17.8.6}$$

and from Danckwerts' modification,

$$h = \left\{\frac{k\rho c_p}{\bar{\theta}}\right\}^{\frac{1}{2}}. \tag{17.8.7}$$

References

[1] McAdams, *Heat Transmission*, McGraw-Hill.
[2] Perry, *Chemical Engineers' Handbook*, McGraw-Hill.
[3] Eckert & Drake, *Heat and Mass Transfer*, McGraw-Hill.
[4] Wong, *Heat Transfer for Engineers*, Longmans.
[5] Sherwood, Pigford, & Wilke, *Mass Transfer*, McGraw-Hill.

18

Forced convection transfer processes

18.1 Local and overall transfer rates

In chapters 16 and 17 we have been primarily concerned with the analysis of heat and mass transfer and with the determination of the appropriate local values of the transfer coefficients for use in forced convection problems. The numerical determination of a transfer coefficient, whether from basic analysis or from experimental measurement, is only an intermediate step, however, in the overall procedure of estimating the total heat or mass transfer rate for a complete flow system or for a specific item of engineering process plant.

We have already noted, for example in chapters 13 and 14, that the local heat flux or mass transfer rate per unit area may be expressed as the product of a transfer coefficient and a temperature or concentration difference. Thus for heat transfer from a surface at temperature T_s to a fluid at temperature T we have

$$q = h(T_s - T), \tag{18.1.1}$$

where q is the heat flux and h is the heat transfer coefficient. Similarly for mass transfer in a gas from an interface where the concentration of the diffusing component is C_s, to the bulk or mean flow region of the gas where the concentration is C, the molar flux is given by

$$N_A = k_g(C_s - C). \tag{18.1.2}$$

The temperature difference $(T_s - T)$, or the concentration difference $(C_s - C)$, is often described colloquially as a 'driving force'.

When the temperature or concentration difference is constant throughout the flow system the overall rates of transfer can be obtained very simply by multiplying the transfer coefficient by the driving force. A constant driving force, however, is not normally encountered in most practical engineering situations. Typically the driving force will vary with position within the heat exchanger or mass transfer unit.

The total rate of heat transfer in a heat exchanger, where the local temperature difference is a function of position, may usually be expressed in the form

$$Q = hA\,\Delta T_m, \tag{18.1.3}$$

where h is the heat transfer coefficient which is assumed to be constant; A is the total surface area; and ΔT_m is the appropriate *mean* temperature difference. We have already considered one such example in §9.7 with the case of heat transfer to a fluid flowing in a long tube with the wall temperature maintained at a constant value as in figure 80(*b*). We will see in §18.2 that the total rate of heat transfer in this case is given by

$$Q = hA\,\Delta T_{lm} \tag{18.1.4}$$

where ΔT_{lm} is the logarithmic mean temperature difference as defined in (9.7.11).

Other examples, involving the use of a mean temperature difference or mean concentration difference, will be considered in the later sections of this chapter.

The problem becomes more difficult when we have a situation in which the driving force varies with time. Unsteady transfer problems, however, are generally outside the scope of this book.

18.2 Convective transfer with specified wall properties

Figure 142 illustrates the basic case of heat transfer to a fluid flowing through a long tube with the wall maintained at a constant temperature T_0. These conditions would apply for example to the case of a condenser tube where the steam is condensing on the outside of the tubes, and thus maintains a near-uniform tube wall temperature, and where the cooling water is flowing inside the tubes. As in §9.7 the mean velocity of the fluid flowing inside the tube is represented by u_m, the inside diameter of the tube by D, and the total length by L.

In §9.7 we analysed this situation by starting with the steady-flow energy equation (9.2.2). The same result may be obtained in a more direct manner by performing an energy balance on an element of tube a distance x from the inlet as shown in figure 142. We denote the temperature of the fluid entering the element by T and, if the element is of length δx, the fluid leaving the element will be at temperature $T + \delta T$. Thus the rate of energy gain by the fluid is the product of the mass flowrate $\frac{1}{4}\pi D^2 \rho u_m$ with the enthalpy

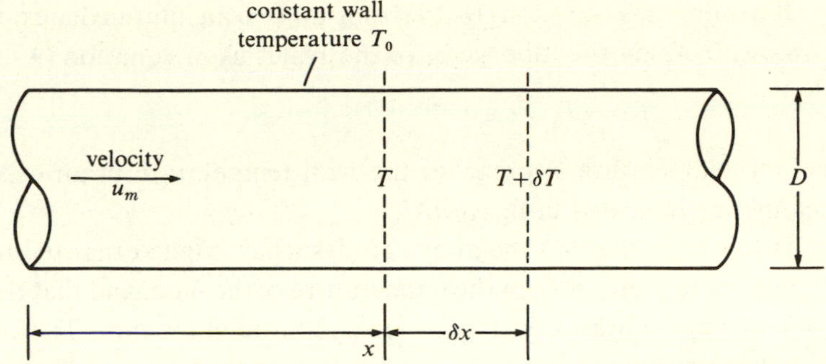

Fig. 142. Enthalpy balance on an elementary length of tube.

change $c_p\,\delta T$. In the absence of heat generation within the fluid this must equal the rate of heat transfer from the wall, $h(T_0-T)\pi D\,\delta x$ so that

$$\frac{\pi}{4}D^2\rho u_m c_p\,\delta T=h(T_0-T)\pi D\,\delta x$$

or

$$\frac{\mathrm{d}T}{T_0-T}=\frac{h}{\rho u_m c_p}\frac{4}{D}\,\mathrm{d}x=\frac{4\,St}{D}\,\mathrm{d}x. \tag{18.2.1}$$

This result is seen to be identical to equation (9.7.4).

Integrating over the entire length of the tube, we have the result as in equation (9.7.10) that

$$\log_e\left(\frac{T_0-T_1}{T_0-T_2}\right)=\frac{4L}{D}\,St$$

or

$$\frac{T_2-T_1}{\Delta T_{lm}}=\frac{4L}{D}\,St, \tag{18.2.2}$$

where ΔT_{lm} is the logarithmic mean temperature difference defined by equation (9.7.11).

The total rate of heat transfer Q is given by

$$Q=\frac{\pi}{4}D^2\rho u_m c_p(T_2-T_1). \tag{18.2.3}$$

Hence from (18.2.2) and (18.2.3) we have

$$Q=\pi DLh\,\Delta T_{lm}. \tag{18.2.4}$$

Thus the total rate of heat transfer is given by the product of the surface area πDL, the heat transfer coefficient h and the logarithmic mean temperature difference ΔT_{lm}.

If we integrate equation (18.2.1) from $x=0$ to an intermediate position at distance x along the tube we have the result as in equation (9.7.9) that

$$(T_0-T)=(T_0-T_1)\,e^{-(4\,St\,x)/D}, \tag{18.2.5}$$

i.e. the temperature approaches the wall temperature in an exponential manner as indicated in figure 143.

It will be noted that the above analysis has implied that the tube wall temperature is higher than the temperature of the fluid, and that the fluid is therefore being heated during its passage through the tube. The analysis is equally applicable, however, to the case where the tube wall temperature is below the temperature of the fluid and where the fluid is consequently cooled during its passage through the tube. Both cases are illustrated in figure 143.

The above analysis of heat transfer from a tube wall at constant temperature T_0 is of industrial importance since it gives an accurate representation of the performance of a condenser. The mass transfer analogue, the mass transfer from a wall where the concentration has the constant value C_0, is of less obvious importance but does represent the case of a liquid flowing through a soluble tube or perhaps more commonly the leaching of the plasticiser from the walls of a plastic tube. The analysis is very similar to the heat transfer case and will therefore be dealt with briefly.

A mass balance on an element such as that shown in figure 142 gives

$$\frac{\pi}{4}D^2u_m\,\delta C=k_L(C_0-C)\pi D\,\delta x, \tag{18.2.6}$$

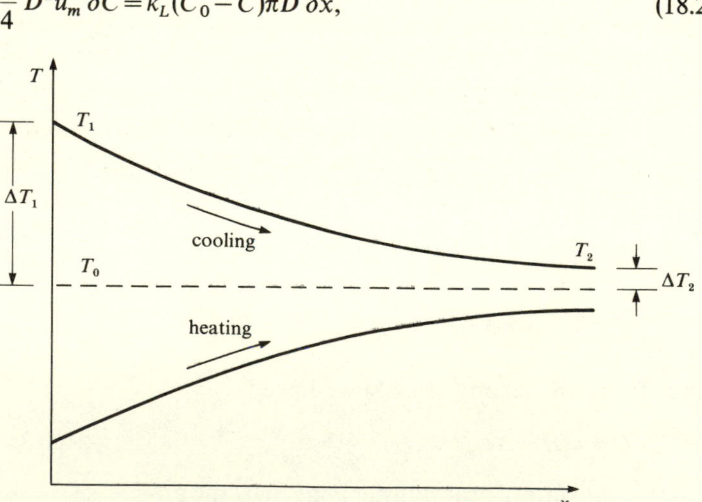

Fig. 143. Temperature profiles in a tube with constant wall temperature T_0.

where k_L is the mass transfer coefficient and C is the mean concentration at a distance x from the inlet. On integration this gives

$$\log_e\left(\frac{C_0-C_1}{C_0-C_2}\right)=\frac{k_L}{u_m}\frac{4L}{D}=\frac{4L}{D}St' \tag{18.2.7}$$

and

$$C_0-C=(C_0-C_1)\,\mathrm{e}^{-(4\,St'\,x)/D}. \tag{18.2.8}$$

These equations are seen to be identical in form to the heat transfer results, equations (18.2.2) and (18.2.5). The total rate of mass transfer is given by

$$Q_m=\pi DLk_L\,\Delta C_{lm}, \tag{18.2.9}$$

where the logarithmic mean concentration difference ΔC_{lm} is defined by analogy with the logarithmic mean temperature difference ΔT_{lm}.

In §9.7 we compared the temperature change of the fluid as given by equation (18.2.2) with the pressure drop predicted by equation (8.2.3) which may be written as

$$\frac{\Delta p}{\rho u_m^2}=\frac{2L}{D}c_f. \tag{18.2.10}$$

From these we deduced that

$$\frac{T_2-T_1}{\Delta T_{lm}}=2\frac{St}{c_f}\frac{\Delta p}{\rho u_m^2}. \tag{18.2.11}$$

We also saw in chapter 17 that the Stanton number is related to the skin-friction coefficient in turbulent flow. According to the Taylor–Prandtl analogy, for example, the relationship is given by equation (17.5.6), while according to the von Karman analogy the relationship takes the form of equation (17.6.31). These expressions, however, are rather awkward to use. As an alternative we can employ the *j*-factor analogy in the form of equation (17.4.4), i.e. we can say approximately that $St\,Pr^{\frac{2}{3}}=\frac{1}{2}c_f$. Hence from equation (18.2.11)

$$\frac{T_2-T_1}{\Delta T_{lm}}=Pr^{-\frac{2}{3}}\frac{\Delta p}{\rho u_m^2}, \tag{18.2.12}$$

which may be compared with equation (9.7.15).

We may equally express the last result in the following form and extend the analogy to cover mass transfer:

$$\frac{\Delta p}{\rho u_m^2}=Pr^{\frac{2}{3}}\log_e\left(\frac{T_0-T_1}{T_0-T_2}\right)=Sc^{\frac{2}{3}}\log_e\left(\frac{C_0-C_1}{C_0-C_2}\right). \tag{18.2.13}$$

It is seen that there is a simple relationship between the temperature change, the concentration change and the pressure drop that is independent of the length, diameter, or roughness of the tube. Any attempt to increase the heat transfer by the provision of a longer tube will automatically result in an increased pressure drop and will have to be paid for in higher pumping costs.

We can illustrate this result with a simple example. Consider water flowing at a velocity of 1.5 m s^{-1} through a tube whose walls are maintained at 100 °C by steam condensing on the outside. The water enters at 20 °C and the tube is of sufficient length for the water to leave at 60 °C. Taking an average value of 5.0 for the Prandtl number we find from equation (18.2.13) that the pressure drop is given by

$$\Delta p = 1000 \times 1.5^2 \times 5.0^{\frac{2}{3}} \log_e \left(\frac{100-20}{100-60}\right) = 4560 \text{ N m}^{-2}.$$

It is seen that we have predicted the necessary pressure drop without any reference to the length or diameter of the tube.

If we were to replace the tube with one of twice the length, it is obvious from equation (18.2.10) that the pressure drop would be doubled to 9120 N m^{-2}, but from equation (18.2.2) we see that it is the logarithm of the temperature difference ratio, $\log_e (T_w - T_1)/(T_w - T_2)$, that is increased by a factor of two, giving an outlet temperature of 80 °C. Clearly there is a relationship between the pressure drop and the outlet temperature and that to obtain an outlet temperature close to the wall temperature would require a large pressure drop.

In the above analysis we have considered the case where the temperature of the wall T_0 was constant. We can now consider the case where the wall temperature is some known function of position x which we will write as $T_0(x)$. Under these circumstances it is convenient to write equation (18.2.1) in the form

$$\frac{D}{4\,St}\frac{\mathrm{d}T}{\mathrm{d}x} + T = T_0(x). \tag{18.2.14}$$

This differential equation will have a complementary function of the form

$$T = A\,\mathrm{e}^{-(4\,St\,x)/D}, \tag{18.2.15}$$

where A is an arbitrary constant and a particular integral the nature of which depends on the form of $T_0(x)$. For the special case of a linear temperature profile, i.e.

$$T_0 = ax + b,$$

where a and b are constants, the particular integral is

$$T = ax + b - \frac{Da}{4\,St}$$

and the complete solution subject to the boundary condition $T = T_1$ at $x = 0$ is

$$T = T_1 + ax + \left(b - \frac{Da}{4\,St}\right)(1 - \mathrm{e}^{-(4\,St\,x)/D}). \tag{18.2.16}$$

18.3 The concentric tube heat exchanger

Whilst heat exchangers consisting of a single tube with a constant wall temperature are encountered in practice, it is much more common to use heat exchangers of more complicated geometry in which heat is transferred from one flowing fluid to another. The constructional details of such exchangers are considered in the next few sections but many of these exchangers can be modelled without much loss of accuracy in terms of the concentric tube exchanger as shown in figure 144.

In the simple heated tube considered in the previous section, the wall temperature was constant or a known function of position. In the present case the wall temperature is neither known nor constant. However, by judicious use of the overall heat transfer coefficient calculated by the methods described in § 13.3, it is possible to avoid having to evaluate the wall temperature at any point, though this can be determined if required.

In this section we will denote the mass-flow rate, specific heat and the temperature of the fluid flowing through the annular gap between the inner and outer tubes by W_a, c_{pa}, and T_a, and we will use the subscript b to denote

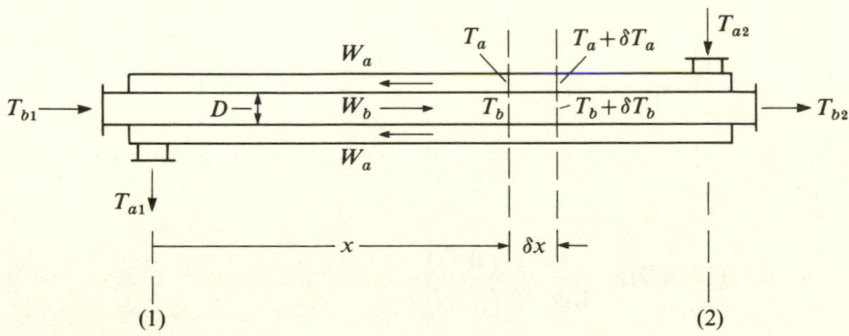

Fig. 144. Enthalpy balance on an elementary length of a concentric tube heat exchanger.

the corresponding quantities for the fluid in the inner tube. An enthalpy balance on the element of the exchanger shown in figure 144 gives

$$\begin{array}{ccccc} \text{heat gained by} & & \text{heat lost by} & & \text{heat transferred} \\ \text{the outer fluid} & = & \text{the inner fluid} & = & \text{through the wall} \\ W_a c_{pa}\,\delta T_a & = & W_b c_{pb}\,\delta T_b & = & \pi D h_0 (T_b - T_a)\,\delta x \end{array}$$

or

$$\frac{\mathrm{d}T_b}{\mathrm{d}x} = \frac{\pi D h_o}{W_b c_{pb}}(T_b - T_a) \tag{18.3.1}$$

and

$$\frac{\mathrm{d}T_a}{\mathrm{d}x} = \frac{\pi D h_o}{W_a c_{pa}}(T_b - T_a). \tag{18.3.2}$$

If we denote the temperature difference between the fluids at any section, $T_b - T_a$, by ΔT and subtract these equations we obtain

$$\frac{\mathrm{d}\,\Delta T}{\mathrm{d}x} = \pi D h_o\,\Delta T\left(\frac{1}{W_b c_{pb}} - \frac{1}{W_a c_{pa}}\right). \tag{18.3.3}$$

Defining the temperature difference at the end $x=0$ by ΔT_1 and that at the end $x=L$ by ΔT_2, we can integrate this equation over the whole length of the exchanger giving

$$\log_e\left(\frac{\Delta T_2}{\Delta T_1}\right) = \pi D L h_o\left(\frac{1}{W_b c_{pb}} - \frac{1}{W_a c_{pa}}\right). \tag{18.3.4}$$

The total heat transfer rate Q is given by

$$Q = W_b c_{pb}(T_{b2} - T_{b1}) = W_a c_{pa}(T_{a2} - T_{a1}), \tag{18.3.5}$$

which can be re-arranged to the form

$$Q\left(\frac{1}{W_b c_{pb}} - \frac{1}{W_a c_{pa}}\right) = (T_{b2} - T_{b1}) - (T_{a2} - T_{a1}) = \Delta T_2 - \Delta T_1. \tag{18.3.6}$$

Eliminating the group $(1/W_b c_{pb} - 1/W_a c_{pa})$ between equations (18.3.4) and (18.3.6) gives

$$Q = \pi D L h_o \frac{(\Delta T_2 - \Delta T_1)}{\log_e(\Delta T_2/\Delta T_1)}. \tag{18.3.7}$$

The logarithmic mean temperature difference ΔT_{lm} was introduced in §9.7 for the special case of constant wall temperature. We will now define a more general form of ΔT_{lm} by the relationship

$$\Delta T_{lm} = \frac{\Delta T_2 - \Delta T_1}{\log_e (\Delta T_2/\Delta T_1)}. \tag{18.3.8}$$

This is seen to reduce to the special form for constant wall temperature presented in §9.7 when $T_{a1} = T_{a2} = T_w$.

Noting that πDL is the surface area A of the tube we can rewrite equation (18.3.7) in the form

$$Q = h_o A\, \Delta T_{lm}. \tag{18.3.9}$$

This equation is commonly referred to as the log mean temperature difference formula and is the basic equation for heat exchanger design. An example of its use is given at the end of this section.

It can be seen from equation (18.3.8) that the value of ΔT_{lm} is unaffected by interchanging the subscripts 1 and 2 and therefore is independent of which direction is considered positive within the exchanger. Similarly, equation (18.3.9) is unaffected by the signs of the flow rates W_a and W_b, and as a result this equation is equally applicable to co- and counter-current exchangers. The temperature profiles in these two cases are, however, quite different as can be seen in figure 145.

Two special cases of equation (18.3.9) are worthy of note

(i) If the temperature of one of the streams is constant due either to a very large flow rate or, more frequently, because of a phase change within the fluid, equation (18.3.9) reduces to that derived in §9.7 for the case of a constant wall temperature.

(ii) If the product of the mass-flow rate and the specific heat is the same for both fluids, i.e. if $W_a c_{pa} = W_b c_{pb}$, we find that $\Delta T_1 = \Delta T_2$ and the value of ΔT_{lm} appears to be indeterminate. However, no problem actually occurs, since under these circumstances $\Delta T_1 = \Delta T_2 = \Delta T_{lm}$. This can be shown by putting $\Delta T_2 = \Delta T_1(1+\varepsilon)$, substituting into equation (18.3.8) and taking the limit as $\varepsilon \to 0$.

In the integration of equation (18.3.3) to give equation (18.3.4) we implicitly made the assumption that D, W_a, W_b, c_{pa}, c_{pb}, and h_o were all constants and this is, in fact, often the case in practice. However, circumstances can occur in which these quantities vary with position. Variable D (tapered tubes) and variable W (leaking tubes) are sufficiently rare to be ignored but the overall heat transfer coefficient h_o often varies along the exchanger. There are two common reasons for this. First, all the physical properties of the fluid are functions of temperature and hence the

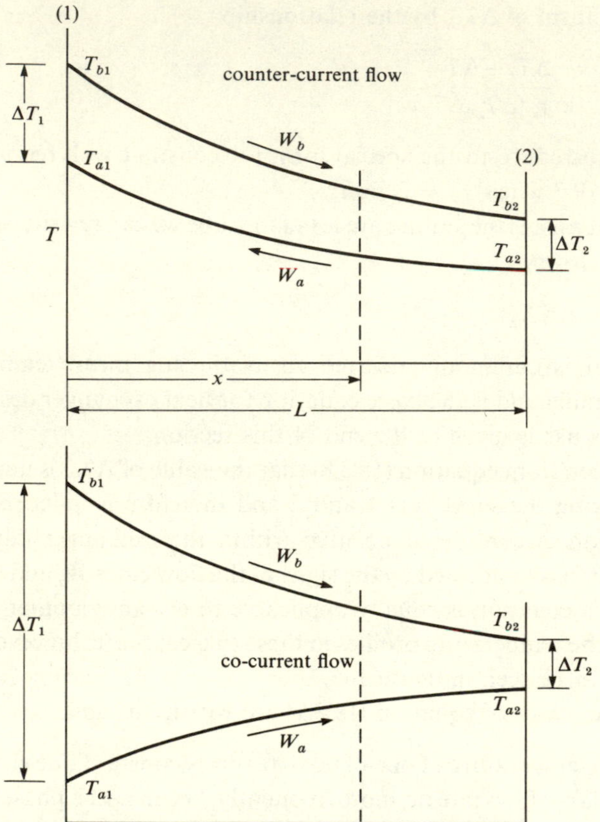

Fig. 145. Temperature profiles in heat exchangers.

overall heat transfer coefficient will also be a function of temperature. Secondly, corrosion or fouling of the tubes may not be uniform, corrosion for example is more likely to occur at the hot end of the exchanger, and in this case h_o may be known as a function of distance along the exchanger. When this is so, equation (18.3.4) takes the form

$$\log_e\left(\frac{\Delta T_2}{\Delta T_1}\right)=\pi DL\left(\frac{1}{W_b c_{pb}}-\frac{1}{W_a c_{pa}}\right)\int_0^L h_o\,\mathrm{d}x$$

$$=\pi DL\bar{h}_o\left(\frac{1}{W_b c_{pb}}-\frac{1}{W_a c_{pa}}\right) \qquad (18.3.10)$$

and the use of an average overall heat transfer coefficient defined by

$$\bar{h}_o=\frac{1}{L}\int_0^L h_o\,\mathrm{d}x$$

is appropriate. In the former case, h_o will be known as a function of temperature and equation (18.3.4) must be written in the form

$$\int_{\Delta T_1}^{\Delta T_2} \frac{d\,\Delta T}{h_o\,\Delta T} = \pi DL\left(\frac{1}{W_b c_{pb}} - \frac{1}{W_a c_{pa}}\right). \tag{18.3.11}$$

In this case no simple average overall heat transfer coefficient can be defined and each example will require separate treatment.

The variation of the specific heats c_p with temperature is normally too small to cause significant error but care must be taken if a phase change, usually boiling, begins part way along the exchanger. During a change of phase, heat is transferred without affecting the temperature of the fluid and consequently the specific heat appears to be infinite. Thus at the point of onset of the phase change we have a discontinuous change in the effective specific heat and as a result the integration of equation (18.3.3) cannot be performed for the exchanger as a whole. Figure 146 shows the temperature profile for an exchanger in which a liquid is heated to its boiling point, evaporated at constant temperature, and then super-heated. Even though the log mean temperature difference formula, equation (18.3.9), may not be used for the exchanger as a whole, the three sections may be considered as independent exchangers and this equation will be applicable for each of the three zones separately. Care must, however, be exercised as the boundaries between the preheating, evaporating, and super-heating sections may not be as clear-cut as implied above.

As an example of the use of equation (18.3.9), let us consider a counter-current concentric tube heat exchanger in which 0.1 kg s^{-1} of toluene is

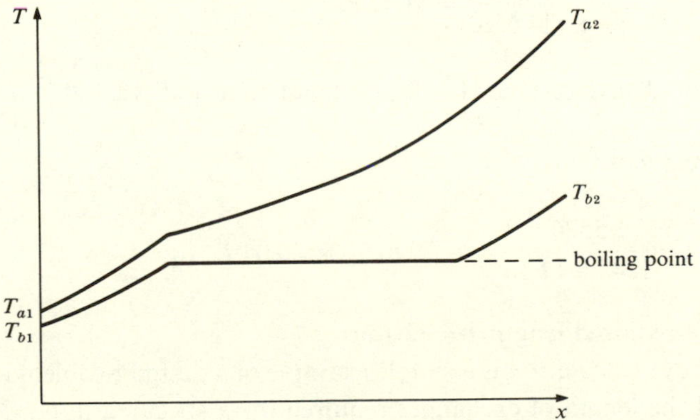

Fig. 146. Temperature profiles in a heat exchanger with change of phase.

cooled from 60 °C to 30 °C by thermal contact with 0.05 kg s^{-1} of water supplied at 20 °C. We will select a tube diameter of 10 mm and will calculate the required length of tube.

Using the following values of the physical properties of toluene, $c_p = 1842$ J kg^{-1} K^{-1}, $\mu = 0.45 \times 10^{-3}$ N m s^{-2}, $k = 0.15$ W m^{-1} K^{-1}, and $\rho = 866$ kg m^{-3}, we can predict the heat transfer coefficient using the Dittus–Boelter correlation, equation (15.3.12), as follows

$$h = \frac{k}{D} \times 0.023 \left(\frac{4W}{\rho \pi D^2} \frac{\rho D}{\mu} \right)^{0.8} \left(\frac{\mu c_p}{k} \right)^{0.4} = 2490 \text{ W m}^{-2} \text{ K}^{-1}.$$

Assuming an equal value on the water-side, the overall heat transfer coefficient h_o in the absence of fouling and on the assumption of a thin-walled tube is given by equation (13.3.4) as $h_o = \frac{1}{2}h = 1245$ W m^{-2} K^{-1}.

The rate of heat transfer Q and the outlet temperature of the water T_1 are given by the enthalpy balances of equation (18.3.5)

$$Q = 0.1 \times 1842(60 - 30) = 0.05 \times 4187(T_1 - 20).$$

Hence $Q = 5.526$ kW and $T_1 = 46.4$ °C.

We can now calculate the temperature differences at the two ends of the exchanger,

$$\Delta T_1 = 60 - 46.4 = 13.6 \text{ K},$$

$$\Delta T_2 = 30 - 20 = 10 \text{ K},$$

and the logarithmic mean is given by

$$\Delta T_{lm} = \frac{13.6 - 10}{\log_e (14.6/10)} = 11.17 \text{ K}.$$

The required surface area A is found from equation (18.3.9),

$$Q = h_o A \, \Delta T_{lm},$$

giving

$$A = \frac{5526}{1245 \times 12.16} = 0.379 \text{ m}^2 = \pi \times 0.01\, L.$$

Hence the required length $L = 12.1$ m.

The above calculation is a simple example of a design problem in which we found the length of exchanger required for a specified duty. Our next example will be a performance calculation in which we predict the outlet temperatures from a specified exchanger.

Let us consider a counter-current concentric tube heat exchanger of diameter 8 mm, length 16 m, and overall heat transfer coefficient 1500 W m^{-2} K^{-1}. Toluene at 70 °C is supplied at a flow rate of 0.15 kg s^{-1} and is cooled by water supplied at 20 °C with a mass-flow rate of 0.1 kg s^{-1}.

The heat balances of equation (18.3.5) give

$$Q = 0.15 \times 1842(70 - T_{b2}) = 0.1 \times 4187(T_{a1} - 20), \tag{18.3.12}$$

where T_{b2} is the outlet temperature of the toluene and T_{a1} is the outlet temperature of the water. Using the log mean temperature difference formula, equation (18.3.9), we have,

$$Q = \pi \times 0.008 \times 16 \times 1500\left(\frac{(70 - T_{a1}) - (T_{b2} - 20)}{\log_e((70 - T_{a1})/(T_{b2} - 20))}\right). \tag{18.3.13}$$

These equations are sufficient for the evaluation of Q, T_{a1}, and T_{b2}. However, the complicated form of equation (18.3.13) seems to suggest that an iterative solution is required and the inexperienced student often loses heart at this stage. No iteration is in fact required, as can be seen from the following manipulations. From (18.3.12),

$$T_{a1} = 20 + 0.66(70 - T_{b2}) \tag{18.3.14}$$

and substituting from (18.3.14) and (18.3.12) into (18.3.13) gives

$$276.3(70 - T_{b2}) = 603.2\left\{\frac{70 - 20 - 0.66(70 - T_{b2}) - (T_{b2} - 20)}{\log_e((70 - T_{a1})/(T_{b2} - 20))}\right\}$$

$$= \frac{603.2(70 - T_{b2})(1 - 0.66)}{\log_e((70 - T_{a1})/(T_{b2} - 20))}.$$

The term in $(70 - T_{b2})$ cancels, giving

$$\log_e\left(\frac{70 - T_{a1}}{T_{b2} - 20}\right) = \frac{603.2 \times 0.34}{276.3} = 0.742$$

or

$$70 - T_{a1} = 2.10(T_{b2} - 20).$$

Thus from equation (18.3.14),

$$50 - 0.66(70 - T_{b2}) = 2.10(T_{b2} - 20)$$

and hence $T_{b2} = 31.8$ °C and $T_{a1} = 45.2$ °C.

It should be noted that in both these examples we have been concerned with modest flow rates of the order of 0.1 kg s^{-1}. The reason for this is that we require a relatively small-diameter tube to achieve good thermal contact between the fluids and that the velocities are limited by the requirement that the pressure drop should not be large. Thus the concentric tube heat exchanger is not capable of handling flow rates of industrial scale and under these circumstances different arrangements of tubes have to be used as described in §18.4.

18.4 The shell-and-tube heat exchanger

We saw at the end of the previous section that the simple concentric tube heat exchanger is unsuitable for most industrial applications. The basic cause of this unsuitability is the limited area available for heat transfer in a single tube of reasonable length. The usual way of overcoming this difficulty is to have many tubes operating in parallel, and these are normally grouped together in a single unit such as that shown schematically in figure 147. Though the diagram shows only four tubes, it is usual for there to be several hundred tubes in a single cylindrical container known as a shell, the whole arrangement being known as a shell-and-tube heat exchanger.

One fluid, commonly called the tube-side fluid, is supplied to the tubes via a common header, and the other fluid, known as the shell-side fluid, is constrained to flow across the tubes by the provision of baffles as shown in the figure. The purpose of the baffles is two-fold. First they act as supports for the tubes, which are often of considerable length, and secondly they control the flow direction of the shell-side fluid. If there are many baffles the velocity on the shell-side will be large, giving rise to much turbulence and

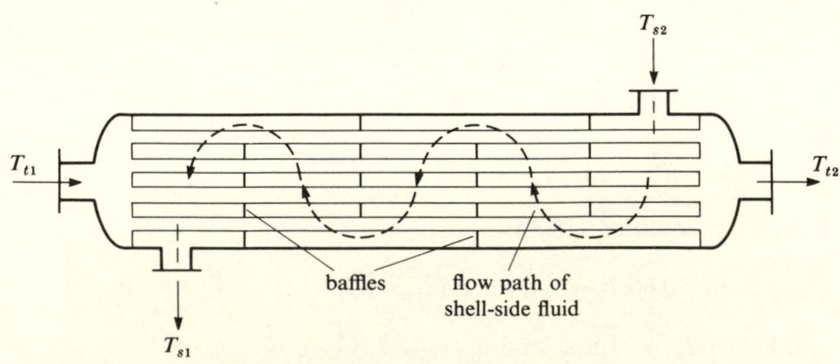

Fig. 147. Shell and tube heat exchanger.

hence to high heat transfer coefficients. Furthermore, with many baffles the flow pattern will approximate to simple counter-current flow and one can treat the temperature of the shell-side fluid as a function of axial distance only. However, increasing the number of baffles increases the shell-side pressure drop and there may be good reasons why this has to be kept small. Thus the number and spacing of the baffles are usually dictated by practical considerations and tend to be fixed more by the designers' experience than by rigorous analysis. Once the baffle spacing has been determined, the cross-wise velocity can be calculated and the shell-side heat transfer coefficient h_s can be found from some appropriate correlation. Perry[1], Eckert & Drake[2], and Wong[3] give useful lists of such correlations.

On the tube side we can calculate the heat transfer coefficient h_t from some relationship such as the Dittus–Boelter correlation, equation (17.4.8),

$$Nu = 0.023\, Re^{0.8}\, Pr^{0.4} \tag{18.4.1}$$

and the two heat transfer coefficients can be combined to give the overall heat transfer coefficient h_0 by the methods of § 13.3. If the tube wall is thick one must also make allowance for the thermal resistance of the wall. Normally, however, the tubes have thin walls of a highly conductive metal such as copper, though other materials may have to be used if the fluids are corrosive and thick-walled tubes are necessary if there is a large pressure difference between the shell- and tube-side fluids. Of more importance is some allowance for fouling due either to corrosion products or the accumulation of dirt or algal growth if impure river water is being used. Some subjective estimate has to be made of the fouling factors that will occur after the exchanger has been in operation for some time. For a thin-walled tube the overall heat transfer coefficient is given by

$$\frac{1}{h_o} = \frac{1}{h_s} + \frac{1}{h_t} + f \tag{18.4.2}$$

and for a thick-walled tube equation (13.4.11) should be used.

Despite the somewhat complex flow pattern in the shell-side, it is usual to treat a shell-and-tube heat exchanger as a pure co- or counter-current exchanger, and hence the total heat transfer rate Q is given by equation (18.3.9),

$$Q = h_o A\, \Delta T_{lm}. \tag{18.4.3}$$

Here A is the area available for heat transfer, given by

$$A = N\pi DL, \tag{18.4.4}$$

where N is the total number of tubes.

We will denote by the subscript t the conditions on the tube-side and by subscript s those on the shell-side. Subscripts 1 and 2 will refer to the sections corresponding to inlet and outlet on the tube-side irrespective of the direction of flow in the shell-side. The temperature differences at the ends of the exchanger are therefore given by

$$\Delta T_1 = T_{t1} - T_{s1}, \tag{18.4.5}$$

$$\Delta T_2 = T_{t2} - T_{s2}, \tag{18.4.6}$$

and the logarithmic mean temperature difference is given by equation (18.3.8),

$$\Delta T_{lm} = \frac{\Delta T_2 - \Delta T_1}{\log_e (\Delta T_2/\Delta T_1)}. \tag{18.4.7}$$

For any specified duty, i.e. if we are given values of the heat transfer rate Q and the external temperatures, we can calculate the required area A provided we have an estimate of the overall heat transfer coefficient h_o. Clearly this area can be provided by many combinations of tube diameter, length, and number. However, not only will certain combinations of tube diameter and number give rise to exchangers of inconvenient length but the selection of D and N also affects the pressure drop on both the tube- and shell-side. The tube-side pressure drop for example can be calculated by the methods of §8.2 as follows.

The mean velocity in the tubes u_m is given by

$$W_t = \frac{N\pi D^2}{4} \rho_t u_m, \tag{18.4.8}$$

where W_t and ρ_t are the mass-flow rate and density of the tube-side fluid and the frictional pressure drop is given by equation (8.2.3) as

$$\Delta p = \frac{2c_f u_m^2 L \rho_t}{D}$$

To this we should add a Bernoulli loss of $\frac{1}{2}\rho_t u_m^2$ giving

$$\Delta p = \left(\frac{4c_f L}{D} + 1\right)\frac{\rho_t u_m^2}{2}. \tag{18.4.9}$$

Thus Δp varies at N^{-2} and D to some power between -4 and -5. Clearly some combinations of D and N will give rise to quite unacceptable pressure drops.

Other constraints must also be borne in mind. Tubes, for example, are available only in a limited number of standard diameters and also for purely geometrical reasons certain numbers of tubes fit more conveniently into a cylindrical container. Many computer algorithms have been developed that search through the available tube diameters and convenient tube numbers to see which combinations give exchangers of reasonable length and with acceptable pressure drops.

One important factor has, however, been glossed over in the above remarks. We have implied that the overall heat transfer coefficient is known and that the design problem is therefore the finding of values of N, D, and L that give a fixed heat transfer area and acceptable values of the length and pressure drop. However, since the heat transfer coefficients depend on the velocities of the fluids, changes in N and D also affect the heat transfer coefficients. No problem arises in a computation based on selected values of N and D since h_o can be calculated for each example, but a fully algebraic analysis becomes complex especially if an attempt is made to obtain optimum values by differentiation.

We will illustrate the arguments of this section by concluding with two examples. In the first we select values of N and D and calculate the required length and the resulting pressure drop and in the second we calculate the necessary number of tubes of specified length and diameter. These calculations are typical of individual stages in the type of computer search described above.

In our first example we will find the length of tube required in a heat exchanger consisting of 50 tubes of internal diameter 12.5 mm designed to cool 8.5 kg s^{-1} of an oil from 80 °C to 40 °C by counter-current contact with water which enters at 20 °C and leaves at 36 °C. We will use the following properties for the oil, $\mu = 2.5 \times 10^{-3}$ N s m^{-2}, $\rho = 890$ kg m^{-3}, $c_p = 2300$ J kg^{-1} K^{-1}, and $k = 0.25$ W m^{-1} K^{-1}.

The velocity of the oil in the tubes is given by

$$u_m = \frac{4W}{N\pi D^2 \rho} = \frac{4 \times 8.5}{50\pi \times 0.0125^2 \times 890} = 1.56 \text{ m s}^{-1}.$$

Hence the Reynolds number $Re = u_m D\rho/\mu = 6942$ and we can predict the friction factor c_f from the Blasius expression, equation (6.6.10),

$$c_f = 0.079\, Re^{-\frac{1}{4}} = 0.008\,65.$$

The Prandtl number for the oil $Pr = \mu c_p/k = 23$ and hence from the j-factor analogy, equation (17.4.7),

$$St=\frac{h}{\rho u_m c_p}=\tfrac{1}{2}c_f\,Pr^{-\frac{2}{3}},$$

giving $St=0.000\,535$ and $h=1708$ W m^{-2} K^{-1}. We will calculate the overall heat transfer coefficient h_o from equation (18.4.2) using a typical value of the shell-side heat transfer coefficient of 4000 W m^{-2} K^{-1} and a typical fouling factor of 0.000 2 m^2 K W^{-1}.

$$\frac{1}{h_o}=\frac{1}{1708}+\frac{1}{4000}+0.0002.$$

Hence $h_o=966$ W m^{-2} K^{-1}.

The temperature differences at the two ends of the exchanger are

$$\Delta T_1=80-36=44\text{ K},$$

$$\Delta T_2=40-20=20\text{ K},$$

and hence, from equation (18.3.8), $\Delta T_{lm}=30.4$ K.

The total heat transfer rate,

$$Q=Wc_p(T_1-T_2)=8.5\times 2300(80-40)=0.782\text{ MW}.$$

Thus from the log mean temperature difference formula, equation (18.3.9),

$$A=\frac{0.782\times 10^6}{966\times 30.4}=26.6\text{ m}^2$$

and the required length $L=26.6/(50\pi\times 0.0125)=13.5$ m.

It should be noted that this length may be inconveniently large and it may be desirable to use either two exchangers in series or a two-pass exchanger as described in § 18.5 below. This example will be continued at the end of that section where we will find the length of a two-pass exchanger designed to perform the same duty.

We can now calculate the frictional pressure drop in the tubes from equation (8.2.3),

$$\Delta p=\frac{2c_f u_m^2 L\rho}{D}=0.405\text{ bar}.$$

This rather large pressure drop occurs because we are passing a viscous oil at fairly high velocity through long narrow tubes. The pressure drop could be reduced by the provision of more tubes since Δp is roughly proportional to u_m^2 and hence to N^{-2}. However, the number of tubes could not be increased by more than a factor of about two as the Reynolds number would then be getting dangerously close to the value at which laminar flow, with its very much poorer heat transfer characteristics, occurs.

We may note that with these specified end temperatures, a co-current heat exchanger is possible. The temperature differences at the two ends are

$$\Delta T_1 = 80 - 20 = 60 \text{ K},$$

$$\Delta T_2 = 40 - 36 = 4 \text{ K},$$

and hence $\Delta T_{lm} = 20.7$ K and the required length $= 13.5 \times (30.4/20.7) = 19.8$ m. A co-current exchanger always requires a greater length than the counter-current exchanger performing the same duty.

As a second example we will evaluate the number of tubes, each of length 6 m and diameter 20 mm, required in a heat exchanger designed to heat 2.5 kg s^{-1} of air from 80 °C to 280 °C by counter-current contact with 2.0 kg s^{-1} of air supplied to the shell-side at 400 °C.

The following average values of the properties of air are appropriate for this range of temperatures, $c_p = 1045$ J kg^{-1} K^{-1}, $\mu = 2.5 \times 10^{-5}$ N s m^{-2}, $k = 0.0346$ W m^{-1} K^{-1}.

The Reynolds number in the tubes is given by

$$Re = \frac{4W}{N\pi D^2 \rho} \frac{D\rho}{\mu} = \frac{6.37 \times 10^6}{N}$$

and the Prandtl number $Pr = \mu c_p / k = 0.755$.

Overall enthalpy balances give

$$Q = 2.5 \times 1045(280 - 80) = 2.0 \times 1045(400 - T_{s0}).$$

Hence $Q = 0.523$ MW and $T_{s0} = 150$ °C,

$$\Delta T_1 = 150 - 80 = 70 \text{ K},$$

$$\Delta T_2 = 400 - 280 = 120 \text{ K},$$

and

$$\Delta T_{lm} = 92.8 \text{ K}.$$

From the Dittus–Boelter correlation, equation (15.3.12),

$$h = 0.023 \frac{k}{D} Re^{0.8} Pr^{0.4} = 9870\, N^{-0.8} \text{ W m}^{-2} \text{ K}^{-1}.$$

Taking a typical shell-side heat transfer coefficient to air of

230 W m^{-2} K^{-1}, the overall heat transfer coefficient is given by

$$\frac{1}{h_0}=\frac{N^{0.8}}{9870}+\frac{1}{230}$$

and from the log mean temperature difference formula, equation (18.3.9),

$$0.523\times10^6\left(\frac{N^{0.8}}{9870}+\frac{1}{230}\right)=A\ \Delta T_{lm}=N\pi\times0.02\times6.0\times92.8$$

or

$$N=1.51\ N^{0.8}+65.0.$$

An iterative solution gives 146 tubes.

18.5 Multi-pass heat exchangers

We have seen in one of the examples of the previous section that the length of tube required for a simple shell-and-tube heat exchanger can be inconveniently large. Under these circumstances one can either operate several smaller exchangers in series or one can double the tubes back on themselves thereby producing a multi-pass heat exchanger. Figure 148 shows the tube arrangement in a two-pass heat exchanger and four- or even six-pass exchangers are commonly encountered in practice. Not only does the multi-pass exchanger give an increased effective tube length but with an even number of passes the connections to the tube-side are all at one end of the exchanger which often has operational and constructional advantages.

When such an exchanger is being used as a condenser or evaporator, the temperature of one of the fluids, normally the shell-side fluid, is constant and the reverse direction of the flow in some of the tubes is irrelevant. Thus

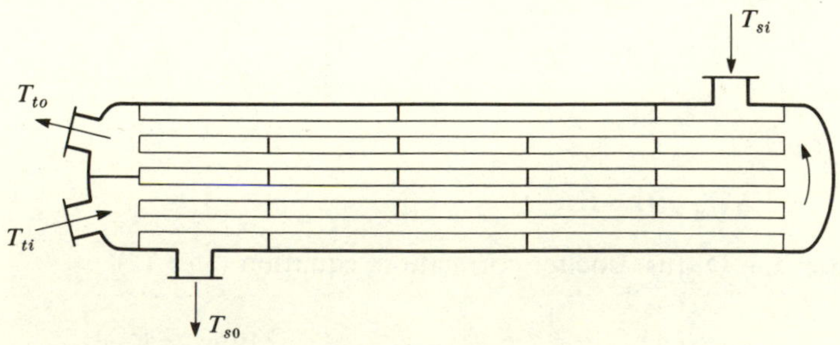

Fig. 148. 2-tube, 1-shell pass heat exchanger.

we can use the log mean temperature difference formula, equation (18.3.9), just as before. However, when both fluids vary in temperature, the simple counter-current flow analysis is inappropriate. None-the-less such exchangers are treated in terms of the equivalent counter-current exchanger and we can define a correction factor Y by

$$Q = h_o A Y \Delta T_{lm}. \tag{18.5.1}$$

The correction factor is found to be a function of the inlet and outlet temperatures as well as the arrangement of the tubes.

As in the previous section, we will denote the temperature of the tube-side fluid by T_t and that of the shell-side fluid by T_s and we will use the subscripts i and o to denote the conditions at inlet and outlet as shown in figure 149. Since we are treating this exchanger in terms of the equivalent counter-current exchanger we define

$$\Delta T_1 = T_{ti} - T_{so}, \tag{18.5.2}$$

$$\Delta T_2 = T_{to} - T_{si}, \tag{18.5.3}$$

and calculate ΔT_{lm} from equation (18.3.8).

It should be noted that in the two-pass arrangement shown in figure 149 the quantity ΔT_1 corresponds to the actual temperature difference at the

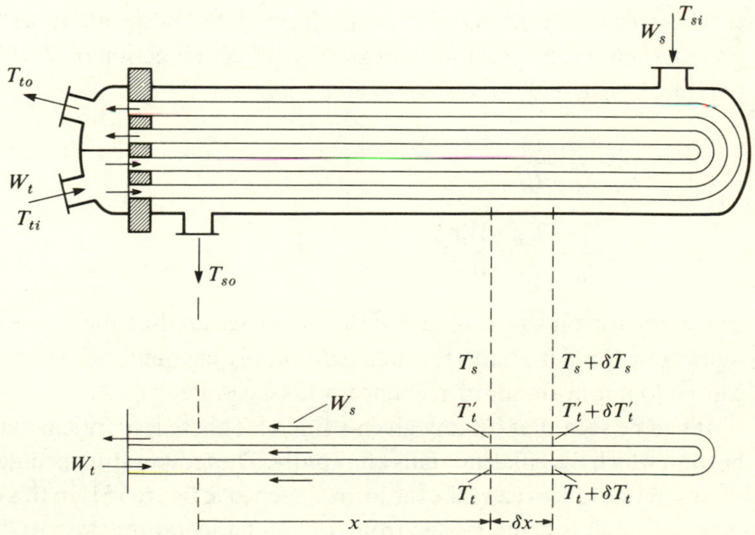

Fig. 149. Enthalpy balance on an elementary length of a 2-tube, 1-shell pass heat exchanger.

tube-side entry but that ΔT_2 is a purely hypothetical temperature difference.

The correction factor Y for a two-pass heat exchanger can be calculated as follows. Considering the exchanger shown in figure 149 we can perform an enthalpy balance between sections at x and $x+\delta x$ giving,

$$N\pi D h_o(T_s - T_t)\,\delta x = W_t c_{pt}\,\delta T_t \tag{18.5.4}$$

and on the reverse pass

$$N\pi D h_o(T_s - T_t')\,\delta x = -W_t c_{pt}\,\delta T_t', \tag{18.5.5}$$

where T_t' is the temperature of the tube-side fluid on the reverse pass and N is the number of tubes in each pass. It will be noted that we have assumed that T_s is a function of x only. This is an approximation but the resulting error will be small if there are many baffles in the shell.

These equations, coupled with an overall enthalpy balance on the region x to $x+\delta x$,

$$W_t c_{pt}(\delta T_t - \delta T_t') = W_s c_{ps}\,\delta T_s, \tag{18.5.6}$$

are sufficient for solution subject to the boundary conditions,

$$\text{at} \quad x=0, \quad T_t = T_{ti}, \quad T_s = T_{so}, \quad \text{and} \quad T_t' = T_{to},$$

$$\text{at} \quad x=L, \quad T_s = T_{si} \quad \text{and} \quad T_t = T_t'.$$

These calculations have been performed by Bowman *et al.*[4], who present their results in the form of graphs of Y as a function of X with Z as a parameter, where X and Z are defined by

$$X = \frac{T_{to} - T_{ti}}{T_{si} - T_{ti}}, \tag{18.5.7}$$

$$Z = \frac{T_{si} - T_{so}}{T_{to} - T_{ti}} = \frac{W_t c_{pt}}{W_s c_{ps}}. \tag{18.5.8}$$

The graph for a heat exchanger with two passes on the tube side is given in figure 150. Similar charts for most commonly encountered arrangements can be found in standard reference books, e.g. Perry[1].

It can be seen that for any given value of Z there is a critical value of X beyond which the efficiency falls off rapidly. The temperature profiles under these circumstances can take the form sketched in figure 151. In this case the tube-side fluid is being heated from T_{ti} to a temperature close to T_{si} in the first pass only to be cooled again in the reverse pass. Such an exchanger is clearly inefficient and a different tube arrangement should be used.

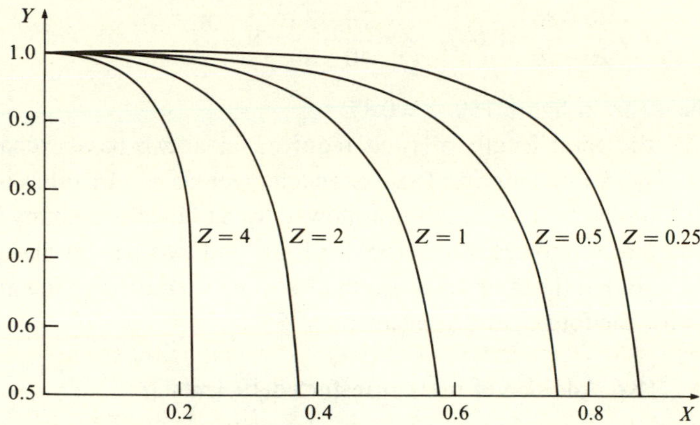

Fig. 150. Correction factor Y for a 2-tube, 1-shell pass heat exchanger.

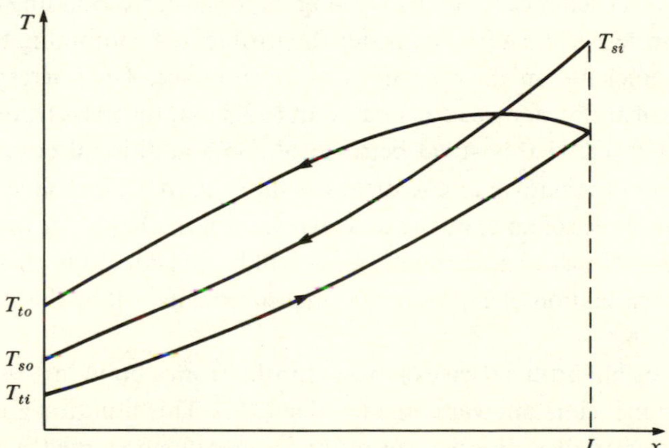

Fig. 151. Temperature profile in a 2-tube, 1-shell pass heat exchanger.

We can illustrate the use of these correction factors by giving further consideration to the heat exchanger we discussed in the first example of the previous section.

In that example we found that the length of tube required in a counter-current exchanger was 13.5 m which is inconveniently large and suggests the use of a two- or even a four-pass exchanger.

With the specified external temperatures,

$$T_{ti}=80\,°\text{C}, \quad T_{to}=40\,°\text{C}, \quad T_{si}=20\,°\text{C}, \quad \text{and} \quad T_{so}=36\,°\text{C},$$

the parameters X and Z are given by

$$X=\frac{40-80}{20-80}=0.667 \quad Z=\frac{20-36}{40-80}=0.4$$

and hence from figure 150, $Y=0.87$.

Thus the total length of tube required in a two-pass exchanger is $13.5/0.87=15.5$ m. Thus in a two-pass exchanger we require tubes of length 7.7 m. A similar procedure can be followed for a four-pass exchanger, but it is found that the correction factors for four- and two-pass exchangers are the same so that the same total length of 15.5 m is required giving an actual length for the four-pass arrangement of 3.9 m.

18.6 Addition of mass transfer coefficients

In the previous sections of this chapter we considered the analysis of heat transfer in forced convection and the performance of some typical designs of heat exchangers. Before passing on to the corresponding analysis of mass transfer we must first consider the problem of combining the mass transfer coefficients on the two sides of an interface. The corresponding problem in heat transfer was considered in § 13.3, but the mass transfer case has been delayed to this stage because of some additional complexities resulting from the nature of chemical equilibrium at an interface and the unfortunate choice of units by the early workers in this field. The analysis of the addition of mass transfer coefficients will be considered in this section and the interpretation of experimental measurements will be the subject of §§ 18.7 and 18.8.

The somewhat artificial case of mass transfer from a boundary where the concentration is constant was considered in § 18.2. This situation is of minor industrial importance and we are more frequently concerned with mass transfer from one solution to another through an interface or a permeable membrane. Mathematically the simplest situation is when transfer of solute occurs through a semi-permeable membrane between different solutions of the same solvent and solute. Under these circumstances mass transfer coefficients add in just the same way as heat transfer coefficients. By analogy with equation (13.3.4) we can deduce that the overall mass transfer coefficient K_L is related to the individual coefficients, commonly known as the film coefficients, in the two solutions, k_{LA} and k_{LB} by

$$\frac{1}{K_L}=\frac{1}{k_{LA}}+\frac{1}{k_{LB}}. \tag{18.6.1}$$

The circumstances in which equation (18.6.1) is valid are commonly encountered in biological systems but are relatively rare in industry. More

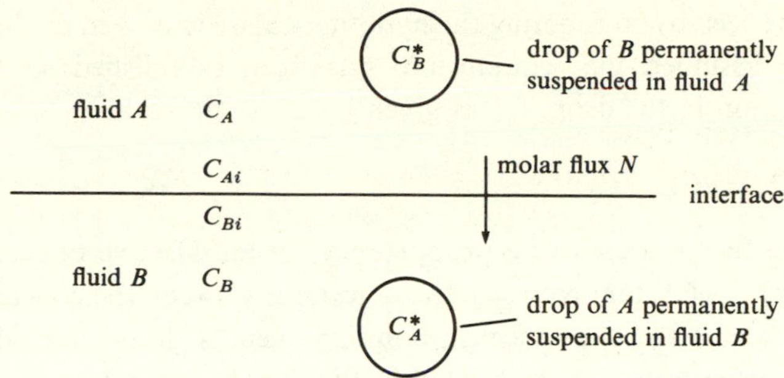

Fig. 152. Mass transfer between two immiscible fluids.

usually the transfer is from one solvent to an immiscible solvent or from a gas to a liquid. In these circumstances the relative solubility in the two phases affects the issue and the analysis of this situation is given below.

Figure 152 illustrates the interface between two immiscible solvents A and B. Also shown in the figure are two hypothetical droplets which are considered to have been suspended in the bulk of the other phase for sufficient time for equilibrium to have been established. Within the two fluids there will be mass transfer coefficients k_{LA} and k_{LB} and, by definition, the molar flux N is given by

$$N = k_{LA}(C_A - C_{Ai}) \tag{18.6.2}$$

and

$$N = k_{LB}(C_{Bi} - C_B), \tag{18.6.3}$$

where C_A and C_B are the molar concentrations in the bulk and C_{Ai} and C_{Bi} are the concentrations at the interface. It is found that the establishment of chemical equilibrium at an interface is very much more rapid than any diffusive process and hence it can be assumed that the interfacial concentrations are related by the equilibrium relationship

$$C_{Ai} = \alpha C_{Bi}, \tag{18.6.4}$$

where α is the partition coefficient.

Eliminating the interfacial concentrations from equation (18.6.2) to (18.6.4) gives

$$N\left(\frac{1}{k_{LA}} + \frac{\alpha}{k_{LB}}\right) = C_A - \alpha C_B \tag{18.6.5}$$

and it is seen that the flux is related not to the overall concentration difference $C_A - C_B$ but to the group $C_A - \alpha C_B$. The physical significance of this can be seen by considering the hypothetical drop of A in the bulk of B. We have assumed that equilibrium has been established so that the concentration in the drop, C_A^*, is given by

$$C_A^* = \alpha C_B. \tag{18.6.6}$$

Hence the flux is seen to be proportional to an effective concentration difference $C_A - C_A^*$, that is to say, the difference between the concentration in the bulk of A and the concentration in solvent A that is in equilibrium with the concentration in the bulk of B. We can therefore define an overall mass transfer coefficient K_{LA} by

$$N = K_{LA}(C_A - C_A^*) \tag{18.6.7}$$

from which it is seen that

$$\frac{1}{K_{LA}} = \frac{1}{k_{LA}} + \frac{\alpha}{k_{LB}}. \tag{18.6.8}$$

As with heat transfer coefficients, the summation of mass transfer coefficients in series is a matter of adding their reciprocals but in this case the reciprocals are weighted by the partition coefficient α.

Dividing equation (18.6.5) by α gives

$$N\left(\frac{1}{\alpha k_{LA}} + \frac{1}{k_{LB}}\right) = \frac{C_A}{\alpha} - C_B = C_B^* - C_B \tag{18.6.9}$$

and an overall mass transfer coefficient referred to solvent B can be defined by

$$N = K_{LB}(C_B^* - C_B) \tag{18.6.10}$$

and hence

$$\frac{1}{K_{LB}} = \frac{1}{\alpha k_{LA}} + \frac{1}{k_{LB}} = \frac{1}{\alpha K_{LA}}. \tag{18.6.11}$$

The greater complexity of this equation compared with equation (18.6.1) or the heat transfer equivalent, equation (13.3.4), results from the nature of the equilibrium relationship. The equilibrium condition for thermal effects is equality of temperature, and this is convenient since temperature is also the driving force for heat transfer. However, the criterion for chemical equilibrium is equality of chemical potential and this is not necessarily the same as equality of concentration which is the quantity of relevance in predicting mass transfer rates.

In principle a very similar analysis can be made for the addition of mass transfer coefficients in gas absorption. The situation is, however, complicated by the unfortunate choice of units by the early workers in this field and it is probably now too late to rationalise the units completely. The problem arises because in a gas it is more convenient to use partial pressure p or mole fraction y as the measure of concentration and hence driving force. This has consequential effects on the value of the mass transfer coefficient. The use of concentration difference, partial pressure difference and mole fraction difference as the driving force gives rise to the three following definitions of the gas-side mass transfer coefficient

$$N = k_g \, \Delta C, \tag{18.6.12}$$

$$N = k_G \, \Delta p, \tag{18.6.13}$$

and

$$N = k_y \, \Delta y, \tag{18.6.14}$$

where the units of k_g are m s^{-1}, the units of k_G are kmol N^{-1} s^{-1}, and the units of k_y are kmol m^{-2} s^{-1}.

The relationship between these coefficients follows directly from the relationship between the measures of concentration. For perfect gases

$$C = \frac{p}{RT} = \frac{Py}{RT},$$

where P is the total pressure, R the universal gas constant, 8314 J kmol^{-1} K^{-1}, and T is the absolute temperature. Hence

$$k_G = \frac{k_g}{RT} = \frac{k_y}{P}. \tag{18.6.15}$$

It must, however, be appreciated that the notation defined above is not universally accepted. Most authors use k_G for all three cases and the reader is expected to deduce from the rest of the equation which of the three is required. Care must be exercised under these circumstances since k_g, having the dimensions of m s^{-1}, is the form that arises naturally from correlations expressed in terms of Sherwood or Stanton numbers but k_G is the form usually required in subsequent calculations. The reader is expected to supply the conversion factor RT even though this does not normally appear in the equations. In this text we will always distinguish between the various forms using the nomenclature defined in equations (18.6.12) to (18.6.14).

Further complications arise from the form of the equilibrium relationship. For many gases the relationship between concentration in the solvent and the equilibrium partial pressure is linear. This result is known as Henry's law and can be expressed algebraically in the form

$$p=HC, \tag{18.6.16}$$

where the units of H are J kmol^{-1}. However, an alternative formulation

$$p=H^{\dagger}x \tag{18.6.17}$$

is often encountered where x is the mole fraction in the liquid. Since $x=C/\rho_L$,

$$H^{\dagger}=H\rho_L \tag{18.6.18}$$

and has the units of N m^{-2}. Here ρ_L is the molar density of the solvent and for water has the value $\frac{1000}{18}=55.6$ kmol m^{-3}.

The values of $H^{\dagger}$ are often numerically large, especially for gases which are only slightly soluble, and values are often quoted in bars rather than N m^{-2}. The table below gives some values of H and $H^{\dagger}$ for solution in water at 20 °C.

	N_2	O_2	CO_2	SO_2	NH_3	
$H^{\dagger}$	81 000	40 000	1420	12.2	0.63	bar
H	146	72.0	2.56	0.022	0.0011	MJ kmol^{-1}

The problem of getting the correct units for mass transfer coefficients is said by some students to be the most difficult aspect of mass transfer. It is recommended that the student works in terms of H and k_G. If a value of $H^{\dagger}$ is obtained from tables it should immediately be converted into H by dividing by ρ_L and values of k_g obtained from the film model or a dimensionless group should be converted to k_G by dividing by RT.

Once the proper units have been established, the calculation of the overall mass transfer coefficient is straightforward. Consider a gas/liquid interface as in figure 153 and again consider a hypothetical drop of liquid suspended in the bulk of the gas and a bubble of gas in the bulk of the liquid. By analogy with equations (18.6.2)–(18.6.6) we can say

$$N=k_G(p-p_i), \tag{18.6.19}$$

$$N=k_L(C_i-C), \tag{18.6.20}$$

$$p_i=HC_i, \tag{18.6.21}$$

$$p^*=HC, \tag{18.6.22}$$

$$p=HC^*. \tag{18.6.23}$$

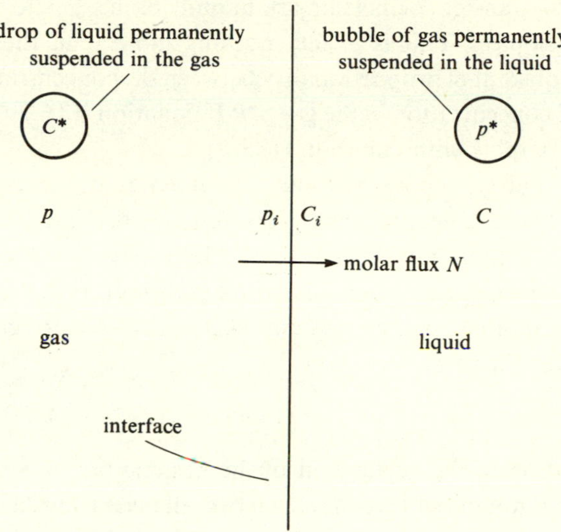

Fig. 153. Mass transfer through a gas/liquid interface.

Defining overall mass transfer coefficients by

$$N = K_G(p - p^*) \tag{18.6.24}$$

and

$$N = K_L(C^* - C) \tag{18.6.25}$$

we have that

$$\frac{1}{K_G} = \frac{1}{k_G} + \frac{H}{k_L} \tag{18.6.26}$$

and

$$\frac{1}{K_L} = \frac{1}{Hk_G} + \frac{1}{k_L} = \frac{1}{HK_G}. \tag{18.6.27}$$

Like the gas-side mass transfer coefficients, the overall mass transfer coefficients can be expressed in other units as follows

$$\frac{K_g}{RT} = K_G = \frac{K_y}{P}. \tag{18.6.28}$$

Substituting from equation (18.6.28) into equation (18.6.26) we find that

$$\frac{1}{K_g} = \frac{1}{k_g} + \frac{H}{RTk_L}. \tag{18.6.29}$$

Here all the mass transfer coefficients are in units of m s^{-1} and the group H/RT is dimensionless. This is in fact nothing more than the partition coefficient, the constant of proportionality between the concentration in the liquid C and the concentration in the gas p/RT. Equation (18.6.29) is seen to be directly comparable with equation (18.6.8).

In the two analyses above we have assumed that the equilibrium relationships are linear as expressed by equations (18.6.4) and (18.6.16) and this is a reasonable approximation for many dilute systems. However, there exist cases in which more complicated equilibrium relationships are required. For example, in the system acetic acid/water/benzene the relationship takes the form

$$C_B = \alpha C_W^2$$

due to dimerisation of the acetic acid in the organic phase. Under these circumstances it is not possible to define an overall mass transfer coefficient and the calculations become very much more difficult. The subject of mass transfer with chemical reaction is treated in chapter 22.

Despite the reactions that occur when NH_3 is dissolved in water and the resulting non-linearity of the equilibrium relationship, it is usual to assume that Henry's law applies in order to make this important topic tractable. It must be borne in mind that any such analysis will only be approximate.

18.7 Evaluation of mass transfer film coefficients

It is rarely possible to get an accurate prediction of the film coefficients k_L and k_G from purely theoretical considerations, though some attempt has been made to do so in chapter 17, and recourse must often be made to correlations of experimental results. However, the experimental determination of these film coefficients is not straightforward since it is rarely possible to measure the interfacial concentration or partial pressure. On the other hand, the measurement of bulk concentrations and partial pressures presents no problems and the experimental determination of overall mass transfer coefficients is easy. The equations relating the overall mass transfer coefficient to the concentrations and partial pressures in the streams entering and leaving an absorption apparatus are derived in § 18.9 below. Decomposition of the overall coefficient into the film coefficients is a necessary preliminary to correlation and this decomposition is facilitated by the wide range of values of the Henry's law constant H that occurs in commonly encountered species. It can be seen, for example, from the table of § 18.6 that the Henry's law constants for N_2 and NH_3 differ by a factor of about 10^5.

For a gas of only modest solubility, the value of H will be large, and from equation (18.6.27)

$$\frac{1}{K_L}=\frac{1}{k_L}+\frac{1}{Hk_G} \tag{18.7.1}$$

it can be seen that

$$K_L \simeq k_L. \tag{18.7.2}$$

The value of H large enough for this approximation to be reasonable depends on the values of k_L and K_G but in all known circumstances the Henry's law constant for oxygen is sufficiently high. Thus the value of k_L can be inferred from the measured value of K_L for oxygen transfer. Traditionally one uses oxygen desorption. A saturated solution is formed by bubbling oxygen through water and the oxygen is then stripped by contact with air in the mass transfer equipment under study.

Similarly it is seen that for a highly soluble gas where H is small, equation (18.6.26) will reduce to

$$K_G \simeq k_G. \tag{18.7.3}$$

However, the Henry's law constant for ammonia into water is only just small enough for this to be a reasonable approximation and it is often wiser to investigate the absorption of ammonia into a dilute acid.

The physical interpretation of equations (18.7.2) and (18.7.3) can readily be visualised. A sparingly soluble gas can only occur in small concentrations in the liquid. Thus the concentration gradients in the liquid are inevitably small and the diffusion rate will therefore be slow. The overall rate of absorption will therefore be controlled by the slow diffusion in the liquid. On the other hand, a very soluble gas is effectively destroyed instantly on contact with the liquid and it is the rate of diffusion through the gas that limits the rate of mass transfer.

The individual film coefficients for these two standard experiments can be deduced as above and, provided it can be assumed that the film coefficients are independent of the nature of the species being transferred, they can be combined to give the overall coefficient K_G^A for species A. From equations (18.6.26), (18.7.2) and (18.7.3),

$$\frac{1}{K_G^A}=\frac{1}{k_G}+\frac{H^A}{k_L}=\frac{1}{K_G^{NH_3}}+\frac{H^A}{K_L^{O_2}}. \tag{18.7.4}$$

where $K_G^{NH_3}$ and $K_L^{O_2}$ are the overall mass transfer coefficients for ammonia absorption and oxygen desorption in the equipment under consideration and H^A is the Henry's law constant for species A. It is of course essential that the overall mass transfer coefficients are measured under conditions of hydrodynamic similarity to the case under investigation.

The assumption that the film coefficients are independent of the nature of the species being absorbed, upon which the above analysis was based, is not ideal and the appropriate corrections are considered in the next section.

18.8 The Whitman two-film theory

The earliest theory to allow for the effect of the properties of the solute on the film coefficients is that due to Whitman. This postulates that on either side of the interface there are laminar layers as shown in figure 154. The film thicknesses will be denoted by δ_L and δ_G and it will be assumed that diffusion across these films is the sole cause of resistance to mass transfer. The films are often referred to as stagnant films, but this is an unnecessary restriction, all that is necessary is that they are laminar.

This model has some similarity with the basic film model that was described in § 17.2 and discussed in § 17.7. In the basic film model it was assumed that the effective laminar layer thicknesses for the transfer of mass and momentum were the same. This is unsatisfactory in a liquid because of the great disparity between the values of the diffusivity and the kinematic viscosity. However, in the Whitman theory we are only making the assumption that the effective film thickness for diffusion is independent of the nature of the species being transferred. This is a not unreasonable

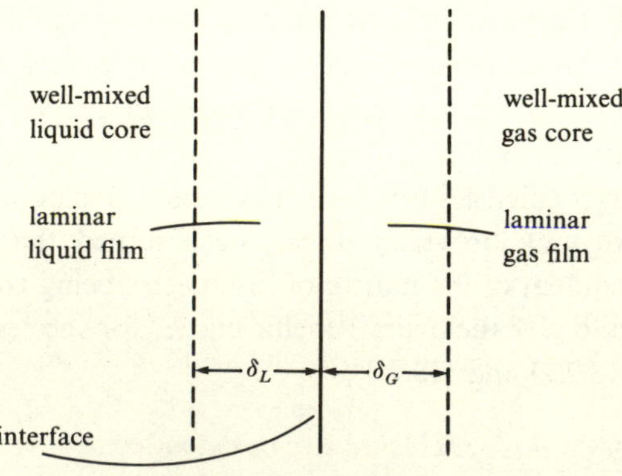

Fig. 154. Whitman's two-film model.

assumption because of the restricted range of diffusivities that occurs in practice.

On the assumption that the sole cause of resistance to mass transfer is diffusion across the laminar films, the film coefficients can be deduced from equation (17.2.5) as

$$k_L = \frac{\mathscr{D}_L}{\delta_L}, \tag{18.8.1}$$

$$k_g = \frac{\mathscr{D}_G}{\delta_G}. \tag{18.8.2}$$

Here the symbol k_g has been used since the units of m s^{-1} are implied by the form of equation (18.8.2). Using equation (18.6.15) this can be put in the more conventional form,

$$k_G = \frac{\mathscr{D}_G}{\delta_G RT}. \tag{18.8.3}$$

It is seen that the mass transfer coefficients are directly proportional to the diffusivity and we can therefore relate the film coefficient k_L^A for the absorption of species A to that for oxygen desorption $k_L^{O_2}$. From the arguments of the previous section this latter quantity is closely equal to the overall mass transfer coefficient for oxygen, $K_L^{O_2}$. Thus

$$k_L^A = k_L^{O_2} \frac{\mathscr{D}_L^A}{\mathscr{D}_L^{O_2}} \simeq K_L^{O_2} \frac{\mathscr{D}_L^A}{\mathscr{D}_L^{O_2}}, \tag{18.8.4}$$

where $\mathscr{D}_L^A$ and $\mathscr{D}_L^{O_2}$ are the diffusivities of A and oxygen through the liquid. Similarly

$$k_G^A = k_G^{NH_3} \frac{\mathscr{D}_G^A}{\mathscr{D}_G^{NH_3}} \simeq K_G^{NH_3} \frac{\mathscr{D}_G^A}{\mathscr{D}_G^{NH_3}}. \tag{18.8.5}$$

Substituting these results into equation (18.6.26) gives

$$\frac{1}{K_G^A} = \frac{1}{K_G^{NH_3}} \frac{\mathscr{D}_G^{NH_3}}{\mathscr{D}_G^A} + \frac{H^A}{K_L^{O_2}} \frac{\mathscr{D}_L^{O_2}}{\mathscr{D}_L^A}. \tag{18.8.6}$$

This equation is an improved version of equation (18.7.4) and should be used in preference to it for predicting K_A^G from measurements on oxygen and ammonia.

The Whitman theory is not perfect and can be extended by incorporating the results of experimental correlations. For example, in some geometries the Sherwood and Gilleland correlation, equation (15.4.9),

$$Sh = 0.023\, Re^{0.83}\, Sc^{0.44} \tag{18.8.7}$$

is appropriate. Under these circumstances the film coefficients are proportional to $\mathscr{D}^{0.56}$ and equation (18.8.6) takes the form

$$\frac{1}{K_G^A} = \frac{1}{K_G^{NH_3}}\left(\frac{\mathscr{D}_G^{NH_3}}{\mathscr{D}_G^A}\right)^{0.56} + \frac{H^A}{K_L^{O_2}}\left(\frac{\mathscr{D}_L^{O_2}}{\mathscr{D}_L^A}\right)^{0.56}. \tag{18.8.8}$$

Fortunately the values of both $\mathscr{D}_L$ and $\mathscr{D}_G$ are confined to narrow ranges and, provided one is not concerned with the transfer of H_2 or He which have unusually high values of $\mathscr{D}$, there is little difference between the predictions of equations (18.8.6) and (18.8.8).

As an example let us consider an absorption column in which we have found the following values of the overall mass transfer coefficients for ammonia and oxygen at 20 °C.

$$K_G^{NH_3} = 7.1 \times 10^{-9} \text{ kmol N}^{-1}\text{ s}^{-1},$$

$$K_L^{O_2} = 4.0 \times 10^{-5} \text{ m s}^{-1}.$$

We will evaluate the laminar film thicknesses and hence predict the overall mass transfer coefficient for SO_2 absorption under the same hydrodynamic conditions. The values of the Henry's law constant for these species are given in the tabulation of § 18.6 and the values of the diffusivities are as follows.

	NH_3	SO_2	O_2	
$\mathscr{D}_L$	1.7×10^{-9}	1.9×10^{-9}	2.5×10^{-9}	$\text{m}^2\text{ s}^{-1}$
$\mathscr{D}_G$	2.34×10^{-5}	1.4×10^{-5}	1.78×10^{-5}	$\text{m}^2\text{ s}^{-1}$

On the assumption that oxygen transfer is liquid-side controlled we can say that $k_L \simeq K_L$ and hence from equation (18.8.1)

$$\delta_L = \frac{\mathscr{D}_L}{K_L} = \frac{2.5 \times 10^{-9}}{4.0 \times 10^{-5}} = 6.25 \times 10^{-5}\text{ m}.$$

Making a similar assumption for ammonia transfer and using equation (18.8.3), we have,

$$\delta_G \simeq \frac{\mathscr{D}_G}{K_G RT} = \frac{2.34 \times 10^{-5}}{7.1 \times 10^{-9} \times 8314 \times 293} = 1.35 \times 10^{-3}\text{ m}.$$

However, it was not in fact necessary to make the above assumptions about which side controls the rate of mass transfer. As an alternative we can use the general relationships given by equations (18.6.26) and (18.6.27) which can be written in the form,

$$\frac{1}{K_G^{NH_3}} = \frac{\delta_G RT}{\mathscr{D}_G^{NH_3}} + \frac{H^{NH_3}\delta_L}{\mathscr{D}_L^{NH_3}},$$

$$\frac{1}{K_L^{O_2}} = \frac{\delta_G RT}{H^{O_2}\mathscr{D}_G^{O_2}} + \frac{\delta_L}{\mathscr{D}_L^{O_2}}.$$

Substituting the appropriate numbers we have,

$$1.41 \times 10^8 = 1.04 \times 10^{11}\ \delta_G + 6.47 \times 10^{11}\ \delta_L$$

and

$$2.5 \times 10^4 = 1.90 \times 10^3\ \delta_G + 4.0 \times 10^8\ \delta_L,$$

from which we find

$$\delta_L = 6.25 \times 10^{-5}\ \text{m},$$

$$\delta_G = 0.97 \times 10^{-3}\ \text{m}.$$

It is seen that our approximate method gave the correct prediction of δ_L, showing that the assumption that oxygen transfer is liquid-side controlled is accurate. However, the two methods give significantly different values for δ_G since in this case we cannot assume that ammonia absorption is controlled only by the resistance of the gas film. It is for this reason that the absorption of ammonia into dilute acid is recommended for the measurement of gas-side mass transfer coefficients.

Once the film thicknesses have been obtained, the mass transfer coefficient for SO_2 absorption can be found from equation (18.6.26)

$$\frac{1}{K_G^{SO_2}} = \frac{0.97 \times 10^{-3} \times 8314 \times 293}{1.4 \times 10^{-5}} + \frac{0.022 \times 10^6 \times 6.25 \times 10^{-5}}{1.9 \times 10^{-9}}$$

$$= 1.687 \times 10^8 + 7.237 \times 10^8 \tag{18.8.9}$$

or

$$K_G = 1.12 \times 10^{-9}\ \text{kmol N}^{-1}\ \text{s}^{-1}.$$

Hence

$$K_L = HK_G = 2.47 \times 10^{-5}\ \text{m s}^{-1}.$$

If we had used the approximate value of 1.35×10^{-3} m for δ_G we would have found that $K_G = 1.04 \times 10^{-9}$ kmol N^{-1} s^{-1}. This result is only marginally in error since the absorption of SO_2 is controlled mainly by the conditions on the liquid side. This can be seen by comparing the magnitudes of the two terms in equation (18.8.9) from which it can be seen that the liquid-side and gas-side resistances are roughly in the ratio of 4:1. Thus the approximate analysis is probably accurate enough for most practical purposes.

18.9 The wetted-wall column

The wetted-wall column may be thought of as the mass transfer analogue of the concentric tube heat exchanger which was considered in §18.3. Like the latter device, it is rarely encountered in industry as the area available for mass transfer is inconveniently small. However, wetted-wall columns are found on the laboratory scale and their analysis forms a sound basis for the study of more complicated situations in the later stages of this chapter.

Figure 155 shows a sketch of such a column which is in effect no more than a vertical tube. Gas is blown up the column with a volumetric flow rate G m^3 s^{-1} and liquid flows down the wall as a film at flow rate L m^3 s^{-1}. In practice there are considerable difficulties in the formation of the liquid film

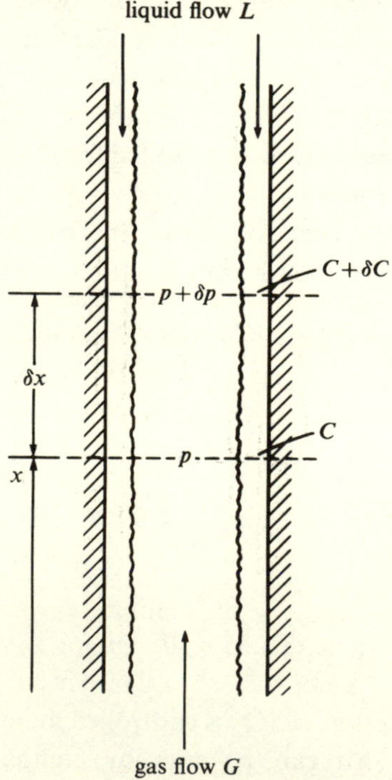

Fig. 155. The wetted-wall column.

and the suppression of waves upon it but these need not concern us here. It will be assumed that the film is uniform and wave-free and that its thickness is very much less than the column diameter D. The analysis will be presented for the absorption of a soluble component from a dilute mixture in the gas. The analysis of desorption from the liquid requires only a change in sign.

Consider a section of the column of height δx a distance x from the base. Within this section the loss of absorbable species from the gas is

$$-\frac{G}{RT}\delta p$$

and this is equal to the gain of that species by the liquid,

$$-L\,\delta C.$$

These quantities are also equal to the amount transferred through the interface

$$\pi D\ \delta x K_G(p-p^*).$$

Noting from Henry's law that $p^*=HC$, we can re-write these as the equations

$$\frac{G}{RT}\frac{\mathrm{d}p}{\mathrm{d}x}=-\pi DK_G(p-p^*), \tag{18.9.1}$$

$$\frac{L}{H}\frac{\mathrm{d}p^*}{\mathrm{d}x}=-\pi DK_G(p-p^*). \tag{18.9.2}$$

These two equations are identical in form to those for the concentric tube heat exchanger, equations (18.3.1) and (18.3.2), and can be manipulated in the same way.

Subtracting equation (18.9.2) from (18.9.1) and denoting the group $p-p^*$ by Δp we have

$$\frac{\mathrm{d}\,\Delta p}{\mathrm{d}x}=-\pi DK_G\,\Delta p\left(\frac{RT}{G}-\frac{H}{L}\right), \tag{18.9.3}$$

which on integration over the whole height of the column becomes

$$\log_e\left(\frac{\Delta p_B}{\Delta p_T}\right)=\pi DZK_G\left(\frac{RT}{G}-\frac{H}{L}\right). \tag{18.9.4}$$

Here the subscripts B and T refer to conditions at the bottom and the top of the column and Z is the total height.

Overall mass balances give the total moles transferred Q_m as

$$Q_m = \frac{G}{RT}(p_B - p_T) = L(C_B - C_T) = \frac{L}{H}(p_B^* - p_T^*), \tag{18.9.5}$$

which on rearranging gives

$$Q_m\left(\frac{RT}{G} - \frac{H}{L}\right) = p_B - p_T - (p_B^* - p_T^*) = \Delta p_B - \Delta p_T. \tag{18.9.6}$$

Eliminating the group $(RT/G - H/L)$ from equations (18.9.4) and (18.9.6) gives,

$$Q_m = \pi DZK_G \frac{\Delta p_B - \Delta p_T}{\log_e(\Delta p_B/\Delta p_T)} = AK_G\,\Delta p_{lm}, \tag{18.9.7}$$

where A is the area available for mass transfer and Δp_{lm} is the logarithmic mean of the values of Δp at the two ends of the column. As is to be expected, this equation is identical in form to the heat transfer equivalent, equation (18.3.9).

Equation (18.9.7) can be re-written in an alternative form. Noting that

$$\Delta p = p - p^* = HC^* - HC = H\,\Delta C,$$

where

$$\Delta C = C^* - C,$$

the equation becomes

$$Q_m = AHK_G \frac{\Delta C_B - \Delta C_T}{\log_e(\Delta C_B/\Delta C_T)}. \tag{18.9.8}$$

However, from equation (18.6.27), $HK_G = K_L$ and hence equation (18.9.8) becomes

$$Q_m = AK_L\,\Delta C_{lm}. \tag{18.9.9}$$

In performing the integration of equation (18.9.3) to give (18.9.4) we implicitly made the assumption that the group $\pi DK_G(RT/G - H/L)$ was independent of position. This is entirely equivalent to the corresponding assumption we made in the heat transfer case and which we discussed in detail at the end of § 18.3. The remarks of that section are equally applicable in the mass transfer case but there exists an additional source of potential error in this case, namely the possibility that Henry's law is not strictly applicable and that H is therefore no longer constant. Special techniques for dealing with the case of a non-linear equilibrium relationship are available and these are discussed in § 18.12.

18.10 Performance of a wetted-wall column

Consideration of the form of equations (18.9.1) to (18.9.3) gives considerable insight into the behaviour of an absorption column. Though these equations were derived specifically for a wetted-wall column, the following remarks are equally valid for other types of absorption column and a very similar analysis is possible for heat exchangers. As in the previous section, we will consider the case of gas absorption. The analysis of desorption requires only certain changes of sign.

The group $(RT/G-H/L)$ appearing in equation (18.9.3) can clearly be positive or negative or coincidently zero depending on the ratio of the volumetric flow rates G and L. Let us consider these three cases separately.

Case 1 $RT/G=H/L$

In this case the flow rates have been adjusted so that

$$\frac{L}{G}=\frac{H}{RT}, \tag{18.10.1}$$

and it is seen from equation (18.9.3) that

$$\frac{\mathrm{d}\,\Delta p}{\mathrm{d}x}=0. \tag{18.10.2}$$

Under these circumstances the partial pressure difference Δp is constant throughout the column. Thus from equations (18.9.1) and (18.9.2) it is seen that both $\mathrm{d}p/\mathrm{d}x$ and $\mathrm{d}p^*/\mathrm{d}x$ are constant and negative. Bearing in mind that $p^*=HC$ and that the liquid is usually supplied to the top of the column free of solute, we can plot the profiles of p and p^* as in figure 156.

It can be seen that the partial pressure profiles are linear and that the absorption rate is uniform throughout the column. Changing the total height Z at constant inlet conditions, i.e. p_B and p_T^* fixed, would result in a different pair of lines and some small change in the partial pressure p_T in the gas leaving the column.

Case 2 $RT/G \gg H/L$

This can alternatively be expressed as

$$\frac{L}{G}\gg\frac{H}{RT} \tag{18.10.3}$$

and corresponds to the case of a large flow of liquid or a small flow of gas. From equation (18.9.3) it is seen that $(\mathrm{d}\,\Delta p)/\mathrm{d}x$ is negative so that Δp

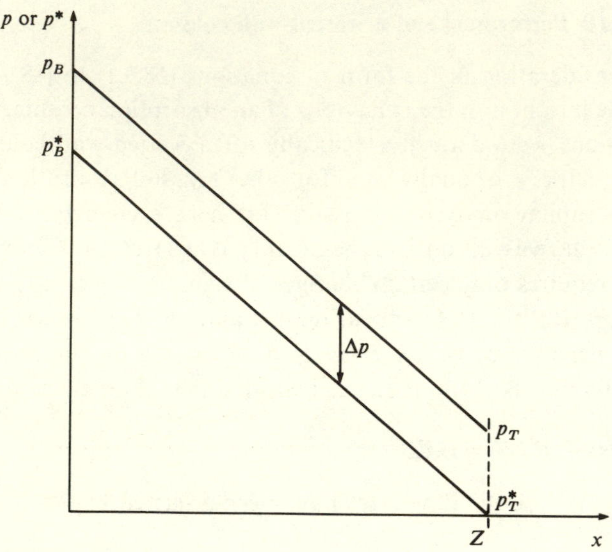

Fig. 156. Partial pressure profiles in a wetted-wall column when $RT/G=H/L$.

decreases as one moves up the column. Thus near the top of the column Δp is small and the two passing streams are therefore close to equilibrium. Thus when solute-free liquid is supplied to the top of the column, the absorption of the soluble component will be almost complete.

Consideration of equation (18.9.1) and (18.9.2) shows that $-\mathrm{d}p/\mathrm{d}x$ and $-\mathrm{d}p^*/\mathrm{d}x$ decrease with increasing height and the partial pressure profiles are as shown in figure 157. Most of the absorption takes place at the bottom of the column where Δp is large and the upper part of the column serves only to remove the last traces of the soluble component.

Case 3 $RT/G \ll H/L$

This case corresponds to

$$\frac{L}{G} \ll \frac{H}{RT}, \tag{18.10.4}$$

that is, to a situation in which the liquid flow rate is small or the gas flow rate large. Case 3 is complementary to case 2 and the partial pressure profiles have the form shown in figure 158. This time the streams are close to equilibrium near the bottom of the column, most of the absorption takes place in the upper part of the column and the absorption is incomplete, i.e.

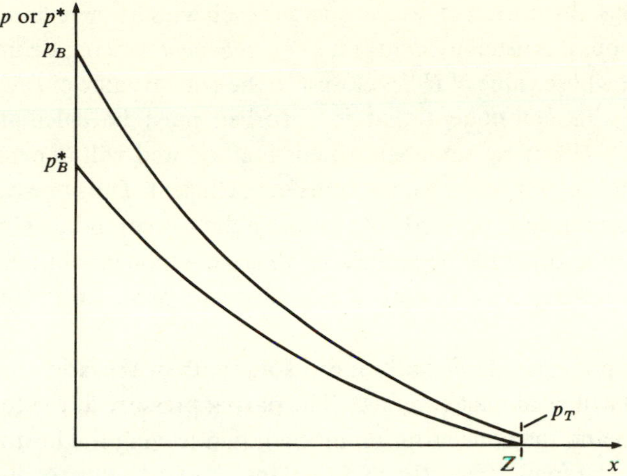

Fig. 157. Partial pressure profiles in a wetted-wall column when $RT/G \gg H/L$.

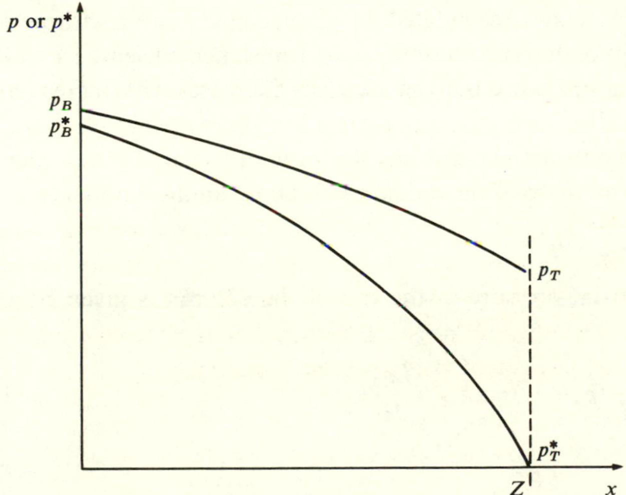

Fig. 158. Partial pressure profiles in a wetted-wall column when $RT/G \ll H/L$.

$p_T \neq 0$. This last result can readily be visualised. With only a small flow of liquid there is insufficient liquid to absorb all the soluble component.

We have seen that for a given ratio of volumetric flow rates, a slightly soluble gas, i.e. one for which $H \gg LRT/G$, will be only partially absorbed but that a more soluble gas for which $H \ll LRT/G$ will be almost completely absorbed. In multi-component absorption we are concerned with the simultaneous absorption of several species, each with its own value of H. In such situations it is usual to identify a *key component* which is defined as the component whose value of H lies closest to the critical value of LRT/G. The amount of this component that is absorbed must be calculated from equation (18.9.7) or by some equivalent method and will depend on the height of the column and the mass transfer coefficient. Thus one can design a column to achieve a specified absorption of the key component. The same equation can in principle be applied to all the components but in practice this is unnecessary and a more straightforward procedure is presented below.

For components significantly more soluble than the key component, absorption will be almost complete. The partial pressure at the top will be almost zero and the concentration in the liquid leaving the bottom of the column can be found from the mass balance,

$$LC_B = \frac{Gp_B}{RT}. \tag{18.10.5}$$

It is seen that we have calculated the outlet concentration without reference to the height of the column or the mass transfer coefficient. The calculation is clearly approximate but the accuracy increases with increasing Z and decreasing H.

For components significantly less soluble than the key, the passing streams at the base of the column will be in equilibrium so that

$$HC_B = p_B \tag{18.10.6}$$

and the partial pressure at the top of the column is given by the mass balance

$$\frac{G}{RT}(p_B - p_T) = LC_B = \frac{Lp_B}{H}, \tag{18.10.7}$$

i.e.

$$\frac{p_T}{p_B} = 1 - \frac{LRT}{GH}. \tag{18.10.8}$$

Thus we can say that the fraction absorbed $(p_B - p_T)/p_B$ is equal to LRT/HG for components considerably lighter than the key and approximately equal to 1 for components much heavier than the key.

In the above analysis we defined the key component as the component for which

$$\frac{H}{RT} = \frac{L}{G}, \tag{18.10.9}$$

where L and G were the volumetric flow rates of liquid and gas. Sometimes it is more convenient to work in terms of the molar flow rates $L^\dagger$ and $G^\dagger$ where

$$L^\dagger = L\rho_L \tag{18.10.10}$$

and

$$G^\dagger = \frac{GP}{RT}. \tag{18.10.11}$$

Recalling from equation (18.6.18) that

$$H^\dagger = H\rho_L \tag{18.10.12}$$

and substituting from these equations into equation (18.10.9) enables us to express the critical value of the Henry's law constant in the alternative form,

$$H^\dagger = \frac{PL^\dagger}{G^\dagger}. \tag{18.10.13}$$

Despite its different form, equation (18.10.13) is entirely equivalent to (18.10.9) and the group HG/RTL can always be replaced by $H^\dagger G^\dagger/PL^\dagger$.

As mentioned at the beginning of this section, an equivalent analysis can be made for the concentric tube heat exchanger. The critical group is the ratio of the products of mass-flow rate and specific heat, $W_a c_{pa}/W_b c_{pb}$. For values of this group significantly different from 1, the outlet temperature of the fluid having the lower value of Wc_p will be close to the inlet temperature of the other fluid provided the heat transfer coefficient and the area of the exchanger are not too small.

18.11 Packed columns

In a wetted-wall column the area available for mass transfer is small and consequently a wetted-wall column designed for an absorption process of industrial scale would be inconveniently large. Attempts are therefore made to increase the interfacial area per unit volume of column and the commonest way of doing this is to use a packed column.

A packed column normally consists of a large cylindrical vessel filled with a packing material over which the liquid trickles as a thin film. The material may be a coarse gravel, coke, or more frequently some patented shape such as the Raschig ring. This consists of a hollow cylindrical shell made of some inert material such as ceramic or stainless steel. These rings commonly exist in the size range 2–20 cm depending on the scale of the column. While the Raschig ring is probably the best-known packing piece, many other, often curiously shaped, pieces are available, each manufacturer claiming special virtues for his own design.

Whilst it is possible to measure, and sometimes even to predict, the interfacial area of the liquid as it flows over the packing, this is of little direct use for the prediction of mass transfer rates. Of more use is the product of the area and the mass transfer coefficient and this can readily be measured for the two standard cases of oxygen desorption and ammonia absorption and adapted for other species by the methods described in §§ 18.7 and 18.8.

Normally the interfacial area is specified in the form of an area per unit volume, a, having the dimensions of m^2/m^3 or m^{-1} and the corresponding products with the mass transfer coefficients have the units,

$K_L a, K_g a$ $\quad s^{-1}$

$K_G a$ $\quad \text{kmol N}^{-1}\,\text{s}^{-1}\,\text{m}^{-1}$

$K_y a$ $\quad \text{kmol m}^{-3}\,\text{s}^{-1}$.

Since a is a constant in any given situation, the gas-side coefficients are related by equation (18.6.28) and hence,

$$\frac{K_g a}{RT} = K_G a = \frac{K_y a}{P}. \tag{18.11.1}$$

Similarly equation (18.8.6) relating the mass transfer coefficient to those measured for oxygen and ammonia takes the form

$$\frac{1}{K_G^A a} = \frac{1}{K_G^{NH_3} a}\frac{\mathscr{D}_G^{NH_3}}{\mathscr{D}_G^A} + \frac{H^A}{K_L^{O_2} a}\frac{\mathscr{D}_L^{O_2}}{\mathscr{D}_L^A}. \tag{18.11.2}$$

The behaviour of a packed column can be analysed by methods similar to those used in § 18.9 for a wetted-wall column. Considering an elemental height δx of a column of diameter D as shown in figure 159, it can be seen that the rate of absorption of the soluble species within the element is given by $-(G/RT)\,\delta p$ and by $-L\,\delta C$ where G and L are the volumetric flow rates of gas and liquid. The rate of absorption must also equal the product of the area available for mass transfer, $\frac{1}{4}\pi D^2 a\,\delta x$, and the mass flux per unit area, $K_G(p-p^*)$. Thus

$$\frac{G}{RT}\frac{dp}{dx} = -\frac{\pi}{4}D^2 K_G a(p-p^*) \tag{18.11.3}$$

and, provided Henry's law holds,

$$\frac{L}{H}\frac{dp^*}{dx} = -\frac{\pi}{4}D^2 K_G a(p-p^*). \tag{18.11.4}$$

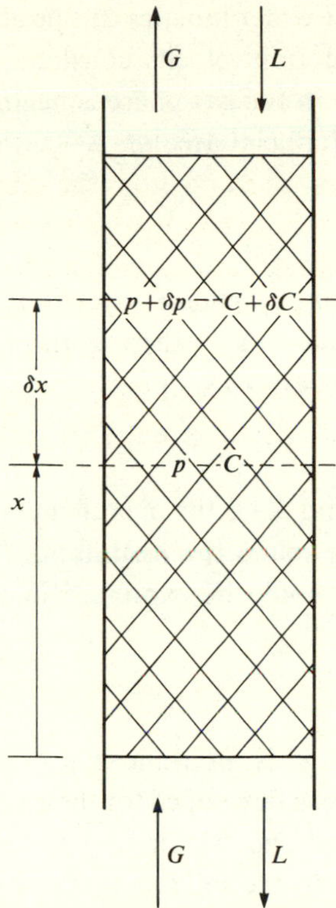

Fig. 159. Mass balance on an elementary length of a packed column.

The close similarity with equations (18.9.1) and (18.9.2) is apparent and these equations can be manipulated in the same way to give the total absorption rate Q_m. The resulting equation is

$$Q_m = VK_Ga\,\Delta p_{lm} = VK_La\,\Delta C_{lm}, \tag{18.11.5}$$

where V is the total volume of the column.

Equation (18.11.5) can be used for the design of a packed column based on experimentally determined values of K_Ga or K_La. We can also use this equation to obtain experimental values of the interfacial area a. Measurement of the flow rates and compositions of the streams entering and leaving the column enables us to calculate the product K_La and if we know the value of K_L the interfacial area can be deduced. In most cases of physical absorption we have to rely on experimentally determined values of K_L and the decomposition of the product K_La is not possible. However, it will be shown in § 22.2 that the mass transfer coefficient K_L for absorption into a liquid with which the gas reacts with first-order kinetics can be accurately predicted to be $\sqrt{(k\mathscr{D}_L)}$ where k is the reaction velocity constant. Thus for reactive systems, provided the resistance in the gas phase is negligible, we can predict K_L and hence obtain the interfacial area a.

18.12 Transfer units

In §18.9 we presented the mass balance equations for an absorption column and manipulated these to give the logarithmic mean partial pressure difference formula, equation (18.9.7),

$$Q_m = K_G A \, \Delta p_{lm}, \tag{18.12.1}$$

where Q_m is the total rate of absorption and K_G is the overall mass transfer coefficient. In deriving this equation we made the assumption that the equilibrium relationship was linear and could be expressed in the form known as Henry's law,

$$p^* = HC. \tag{18.12.2}$$

Cases when the equilibrium relationship is non-linear are relatively common and the following analysis has been developed for these cases. The analysis is, however, general and we will show later that it reduces to equation (18.12.1) when Henry's law applies. We will present the analysis in the form appropriate for a packed column and this can be adapted for a wetted-wall column simply by replacing a by $4/D$.

A mass balance on the gas in an elementary section of a packed bed was performed in §18.11 and gave rise to equation (18.11.3),

$$\frac{G}{RT}\frac{\mathrm{d}p}{\mathrm{d}x} = -\frac{\pi D^2}{4} K_G a (p - p^*) \tag{18.12.3}$$

and this can be re-written in the form

$$Z = \int_0^Z \mathrm{d}x = \frac{4G}{\pi D^2 K_G a R T} \int_T^B \frac{\mathrm{d}p}{p - p^*}, \tag{18.12.4}$$

where the symbols B and T refer to conditions at the bottom and the top of the column. In expressing the mass balance in this form we have made the reasonable assumption that the volumetric flow rate of gas G, the diameter D, the mass transfer coefficient K_G, the interfacial area per unit volume a, and the temperature T are all constant, but we have made no assumption about the nature of the equilibrium relationship. The hypothetical partial pressure p^* is the partial pressure in equilibrium with the concentration in the bulk of the liquid at that section. It can be written formally as a function of the concentration in the form

$$p^* = f_1(C) \tag{18.12.5}$$

and it is only in certain special cases that this general function takes the form of Henry's law, equation (18.12.2).

The group $\int_T^B (\mathrm{d}p/(p-p^*))$ is clearly a dimensionless quantity reflecting the change in the partial pressure of the soluble species and is known as the number of transfer units NTU. The group $4G/\pi D^2 K_G aRT$ has the dimensions of length and is known as the height of a transfer unit HTU. Equation (18.12.4) can therefore be expressed in the form

$$Z = HTU \times NTU. \tag{18.12.6}$$

Since the height of a transfer unit has been defined in terms of the overall gas-side mass transfer coefficient K_G it is more strictly written

$$(HTU)_{OG} = \frac{4G}{\pi D^2 K_G aRT}. \tag{18.12.7}$$

The corresponding number of transfer units is written as $(NTU)_{OG}$ and defined by

$$(NTU)_{OG} = \int_T^B \frac{\mathrm{d}p}{p-p^*}. \tag{18.12.8}$$

It will be recalled from § 18.9 that G is the volumetric flow rate of gas. The gas flow is, however, often expressed in terms of the molar flow rate $G^\dagger$ where $G^\dagger = PG/RT$ or in terms of the mass velocity $w_g = 4\rho G/\pi D^2$ or the mole velocity $w_g^\dagger$ where $w_g^\dagger = 4G^\dagger/\pi D^2$. Thus the height of a transfer unit can be defined in any of the following equivalent ways,

$$(HTU)_{OG} = \frac{4G}{\pi D^2 K_G aRT} = \frac{4G^\dagger}{\pi D^2 K_G aP} = \frac{w_g^\dagger}{K_G aP} = \frac{w_g}{K_G aPM_r} \tag{18.12.9}$$

where M_r is the mean relative molecular mass of the gas mixture. Also, by invoking the relationship between the different forms of the gas-side mass transfer coefficient given by equation (18.6.28), we can express the definition in the following forms,

$$(HTU)_{OG} = \frac{w_g^\dagger}{K_G aP} = \frac{w_g^\dagger}{K_y a} = \frac{w_g^\dagger}{K_g a\rho_M} = \frac{w_g}{K_g a\rho}, \tag{18.12.10}$$

where ρ_M is the molar density and ρ is the mass density of the gas mixture. Clearly many other permutations are also possible.

Since for gas side controlled systems the mass transfer coefficient varies roughly as $G^{0.8}$, the height of a transfer unit is a much less sensitive function of the flow rate than is the mass transfer coefficient. Many authors have therefore preferred to correlate their experiment results in terms of the height of a transfer unit. These can be converted into more conventional form by the use of equations (18.12.9) and (18.12.10).

The number of transfer units can also be expressed in several alternative forms as follows

$$(NTU)_{OG} = \int_T^B \frac{\mathrm{d}p}{p-p^*} = \int_T^B \frac{\mathrm{d}y}{y-y^*} = \int_T^B \frac{\mathrm{d}C^*}{C^*-C}, \tag{18.12.11}$$

where y is the mole fraction in the gas and y^* is the mole fraction in equilibrium with the concentration in the bulk of the liquid.

Alternatively the analysis could have been started with a mass balance on the liquid and this would have given rise to equation (18.11.4) which can be expressed in the form,

$$L\frac{\mathrm{d}C}{\mathrm{d}x} = -\frac{\pi D^2}{4} K_L a(C^* - C) \tag{18.12.12}$$

or

$$Z = \int_0^Z \mathrm{d}x = \frac{4L}{\pi D^2 K_L a} \int_T^B \frac{\mathrm{d}C}{C^*-C} = (HTU)_{OL} \times (NTU)_{OL}, \tag{18.12.13}$$

where $(NTU)_{OL}$ is defined by

$$(NTU)_{OL} = \int_T^B \frac{\mathrm{d}C}{C^*-C} \tag{18.12.14}$$

and $(HTU)_{OL}$ by

$$(HTU)_{OL} = \frac{4L}{\pi D^2 K_L a}. \tag{18.12.15}$$

Comparison of equations (18.12.9) and (18.12.15) gives

$$(HTU)_{OG} = \frac{HG}{LRT}(HTU)_{OL} \tag{18.12.16}$$

and hence it is seen that $(HTU)_{OG} \neq (HTU)_{OL}$ except in the special case when $HG/LRT = 1$. The value of $(NTU)_{OG}$ is similarly only equal to $(NTU)_{OL}$ under the same circumstances. It is therefore essential to use compatible combinations of HTU and NTU. Heights of transfer units have also been defined in terms of the film coefficients k_G and k_L but, since the

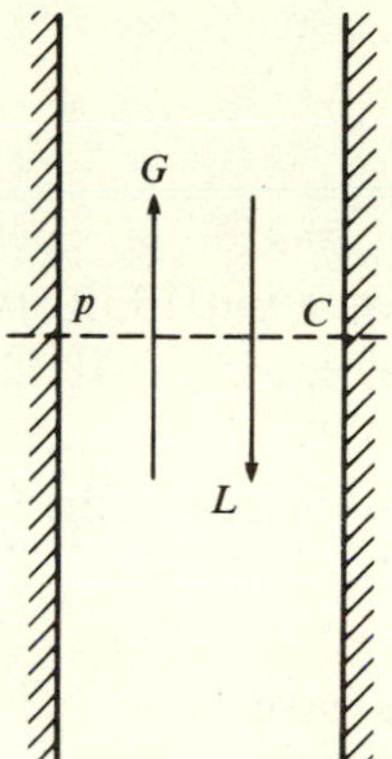

Fig. 160. Passing streams in an absorption column.

corresponding *NTU*s involve the interfacial partial pressure and concentration, these are of minor utility and will not be discussed further here.

In order to proceed with the calculation of the number of transfer units we must first find a relationship between the value of p and p^*. This can be done by relating the net upflow of the absorbable species Q_u to the partial pressure and concentration at an arbitrary section of the column such as that shown in figure 160. From a mass balance,

$$Q_u = G\frac{p}{RT} - LC \tag{18.12.17}$$

and, since Q_u must be independent of position, its value can be found from the specified conditions at either the top or the bottom of the column,

$$Q_u = G\frac{p_B}{RT} - LC_B = G\frac{p_T}{RT} - LC_T \tag{18.12.18}$$

Thus from equation (18.12.17),

$$C = \frac{G}{LRT}p - \frac{Q_u}{L}, \tag{18.12.19}$$

and from equation (18.12.5),

$$p^* = f_1\left(\frac{G}{LRT}p - \frac{Q_u}{L}\right). \tag{18.12.20}$$

The number of transfer units is defined by equation (18.12.8) which now takes the form,

$$(NTU)_{OG}=\int_T^B \frac{\mathrm{d}p}{p-f_1(Gp/LRT-Q_u/L)}. \tag{18.12.21}$$

In many cases the equilibrium relationship of equation (18.12.5) will be of complicated form and numerical integration of equation (18.12.21) will normally be required.

Alternatively, the equilibrium relationship can be put in the form,

$$C=f_2(p), \tag{18.12.22}$$

and from equations (18.12.14) and (18.12.17) we have

$$(NTU)_{OL}=\int_T^B \frac{\mathrm{d}C}{f_2(p-(LRT/G)(C-C_B))-C}. \tag{18.12.23}$$

In this analysis we have, however, glossed over one potential source of error. We have developed these equations for the case of a general equilibrium relationship whilst ignoring the fact that the film coefficients k_L and k_G can only be combined if Henry's law applies, as was discussed in § 18.6. However, for a very soluble species the system will be gas-side controlled and $K_G \simeq k_G$ and is therefore a meaningful quantity. The formulation $Z=(NTU)_{OG}(HTU)_{OG}$ may therefore be used. Similarly, for a sparingly soluble gas we should use $Z=(NTU)_{OL}(HTU)_{OL}$. For a moderately soluble species the gas- and liquid-side resistances are comparable and we cannot in general define an overall mass transfer coefficient. The analysis of such systems is beyond the scope of this book.

If the equilibrium relationship is linear, Henry's law, equation (18.12.2), applies, and we can re-write equations (18.12.17) and (18.12.18) in the form,

$$\varphi p_B - p_B^* = \varphi p - p^* = \varphi p_T - p_T^*, \tag{18.12.24}$$

where $\varphi = GH/LRT$. Hence

$$(NTU)_{OG}=\int_T^B \frac{\mathrm{d}p}{p-p^*}=\int_T^B \frac{\mathrm{d}p}{(1-\varphi)p+p_B^*-\varphi p_B}$$

$$=\frac{1}{1-\varphi}\log_e\left(\frac{p_B-p_B^*}{p_T-p_T^*}\right). \tag{18.12.25}$$

However, from equation (18.12.24)

$$1-\varphi=\frac{\Delta p_B-\Delta p_T}{p_B-p_T}$$

and hence

$$(NTU)_{OG} = \frac{p_B - p_T}{\Delta p_{lm}}. \tag{18.12.26}$$

When this result is substituted into equations (18.12.6) and (18.12.7) the log mean partial pressure difference formula equation (18.11.5) follows directly.

Thus for the case of a linear equilibrium relationship the use of transfer units gives us no extra information. The value of this concept arises from its applicability to cases of non-linear equilibrium. However, the analysis for the linear system enables us to give some physical meaning to the concept of a transfer unit. Let us consider the case when L is large or H small, due to absorption into an excess of liquid or into a liquid with which the species reacts. Under these circumstances φ is small and $p^* \ll p$. Hence the definition of the number of transfer units takes the form

$$(NTU)_{OG} \simeq \int_T^B \frac{\mathrm{d}p}{p} = \log_e \left(\frac{p_B}{p_T}\right). \tag{18.12.27}$$

The height of a transfer unit is then seen to be the height required to reduce the partial pressure in the gas by a factor of e. Though this analysis is not numerically accurate for other cases, the height of a transfer unit can always be thought of as the height required for a specified fractional absorption.

18.13 Transfer units for heat transfer

Transfer units can be defined for heat transfer in a manner analogous to that presented for mass transfer in the previous section. However, their use for heat transfer is less widespread because of the inevitably linear form of the thermal equilibrium relationship which simply takes the form of equality of temperature. As we saw in the previous section, the concept of a transfer unit for a linear system gives rise to an expression which is only a minor modification of the log mean driving force equation. The main advantage of the use of transfer units derives from their applicability to cases of non-linear equilibrium which have no counterpart in heat transfer. None-the-less transfer units are sometimes encountered in the heat transfer literature and are therefore defined below.

In § 18.3 we saw that an enthalpy balance on a section of a heat exchanger gave rise to equation (18.3.2),

$$\frac{\mathrm{d}T_a}{\mathrm{d}x} = -\frac{\pi D h_o}{W_a c_{pa}} (T_a - T_b). \tag{18.13.1}$$

This equation can be put in the following form provided D, h_o, W, and c_p can be taken as constants,

$$L=\int_0^L \mathrm{d}x=\frac{W_a c_{pa}}{\pi D h_o}\int_{T_{a2}}^{T_{a1}}\frac{\mathrm{d}T_a}{T_a-T_b}=(HTU)\times(NTU), \tag{18.13.2}$$

where HTU and NTU are defined by

$$HTU=\frac{W_a c_{pa}}{\pi D h_o}, \tag{18.13.3}$$

$$NTU=\int_{T_{a2}}^{T_{a1}}\frac{\mathrm{d}T_a}{T_a-T_b}. \tag{18.13.4}$$

The close similarity with equations (18.12.7) and (18.12.8) will be noted.

We can relate the values of T_a and T_b by enthalpy balances similar to the mass balances of § 18.12 giving

$$W_a c_{pa} T_{a1} - W_b c_{pb} T_{b1} = W_a c_{pa} T_a - W_b c_{pb} T_b = W_a c_{pa} T_{a2} - W_b c_{pb} T_{b2} \tag{18.13.5}$$

or

$$\varphi T_{a1} - T_{b1} = \varphi T_a - T_b = \varphi T_{a2} - T_{b2}, \tag{18.13.6}$$

where in this case

$$\varphi=\frac{W_a c_{pa}}{W_b c_{pb}}. \tag{18.13.7}$$

Thus

$$\begin{aligned} NTU &= \int_{T_{a2}}^{T_{a1}}\frac{\mathrm{d}T_a}{(1-\varphi)T_a+\varphi T_{a1}-T_{b1}} \\ &= \frac{1}{(1-\varphi)}\log_e\left(\frac{T_{a1}-T_{b1}}{T_{a2}-T_{b2}}\right) \\ &= \frac{T_{a1}-T_{a2}}{\Delta T_{lm}}. \end{aligned} \tag{18.13.8}$$

This result is clearly analogous to equation (18.12.26) and the log mean temperature difference formula, equation (18.3.9), can readily be obtained from it.

References

[1] Perry, *Chemical Engineers' Handbook*, McGraw-Hill.
[2] Eckert & Drake, *Heat and Mass Transfer*, McGraw-Hill.
[3] Wong, *Heat Transfer for Engineers*, Longmans.
[4] Bowman *et al.*, *TASME*, **62**, 283 (1940).

19

Two-phase flow

19.1 Introduction

Figure 161 shows schematically a boiler tube and the changing flow pattern within it as the proportion of vapour to liquid increases. Four basic flow regimes can be recognised and these can be found also in any two-phase mixture whether gas/liquid or liquid/liquid. At low gas flow rates a *bubble flow* regime is found in which the gas is confined to bubbles small in comparison with the tube diameter. The rise velocity of such bubbles is principally a function of their size and is independent of the tube diameter. On increasing the gas flow rate more bubbles are formed and there is a greater tendency for these to collide and coalesce. Thus the average bubble size increases until individual bubbles become comparable in size with the tube diameter. These large bubbles are known as slugs and it can be seen that the transition from the bubble flow regime to the *slug flow* regime is gradual.

The rise velocity of a slug is primarily a function of the tube diameter and independent of the slug volume. Surrounding the slug is a thin film of liquid and when the overall flow rate exceeds a critical value the film suddenly becomes unstable and a chaotic regime known as the *churn flow* regime is found. On further increase of the gas flow the pattern steadies and eventually the liquid forms a film on the tube wall and the gas flows as a continuous core. This pattern is known as the *annular flow* regime. The surface of the annular film is, however, far from smooth. Waves occur and droplets are entrained from the wave crests. The proportion of liquid entrained increases with increasing flow rate but it is improbable that entrainment is complete in an adiabatic system. Thus the *mist flow* regime may never occur except in the presence of heat transfer and is best treated as an extreme form of the annular flow regime.

Figure 162 is a regime map for air and water at *STP* in a 25-mm-diameter pipe. The axes are the superficial velocities j_g and j_f defined by

$$j_g = \frac{4G}{\pi D^2} \qquad (19.1.1)$$

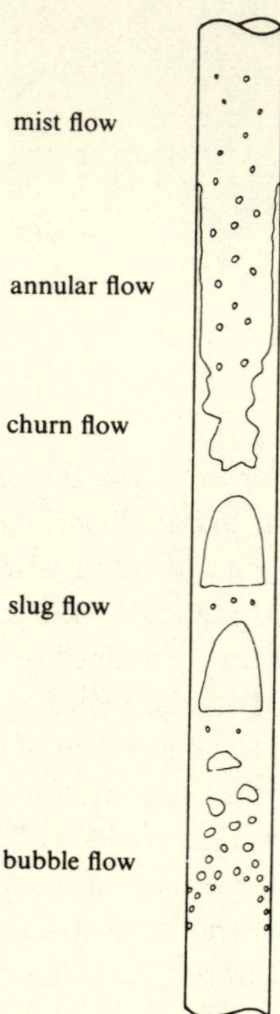

Fig. 161. Flow regimes of two-phase flow.

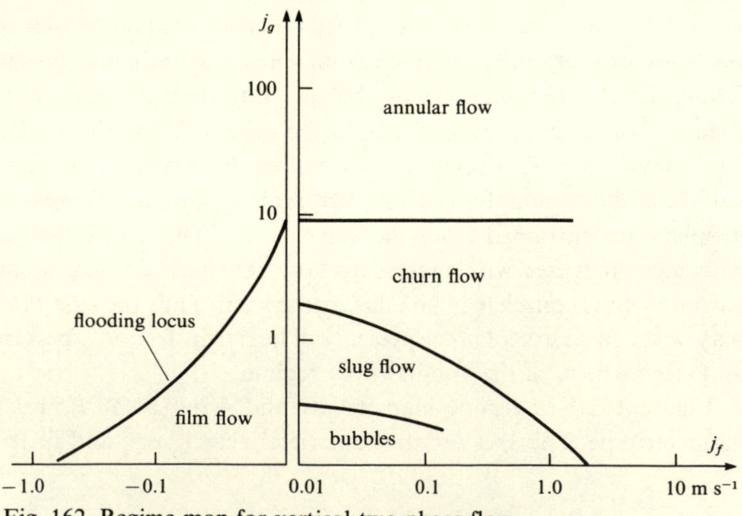

Fig. 162. Regime map for vertical two-phase flow.

and

$$j_f = \frac{4L}{\pi D^2}, \tag{19.1.2}$$

where G and L are the volumetric flow rates of gas and liquid and D is the tube diameter. Also shown in the diagram is the counter-current regime map for the case of gas in upflow and liquid in downflow. Here the principal regime is *falling film flow* where the liquid descends as a film on the tube wall. This film becomes unstable at higher flow rates and the so-called flooding locus is also marked.

It should be borne in mind that many of the regime boundaries are rather subjective and as a result published regime maps show considerable variation. The boundaries are also dependent on the physical properties of the fluids and in particular on the densities and the surface tension. Thus figure 162 should be taken only as a rough guide to the regime boundaries for air and water at STP and should not be used for other materials or at other pressures.

Horizontal two-phase flow is more complicated because of the segregation of the phases due to density differences. For an account of horizontal flow the reader is referred to Wallis[1], whose treatment of vertical flow forms the basis of this chapter. We have also adopted much of Wallis' terminology and in particular the use of j_g and j_f for superficial velocity and j_g^* and j_f^* for dimensionless superficial velocities defined by

$$j_g^* = \frac{j_g}{\sqrt{(gD)}}\sqrt{\left(\frac{\rho_g}{\rho_f - \rho_g}\right)} \simeq j_g\sqrt{\left(\frac{\rho_g}{gD\rho_f}\right)}, \tag{19.1.3}$$

$$j_f^* = \frac{j_f}{\sqrt{(gD)}}\sqrt{\left(\frac{\rho_f}{\rho_f - \rho_g}\right)} \simeq \frac{j_f}{\sqrt{(gD)}}, \tag{19.1.4}$$

where ρ_g and ρ_f are the densities of the gas and liquid.

19.2 Falling film flow

A liquid film flowing down a vertical surface may be in either laminar or turbulent flow. If the flow is laminar the following analysis, due originally to Nusselt, is appropriate.

Consider a film of thickness δ and take a plane at distance y from the wall as shown in figure 163. The shear stress τ on the plane must support the weight of fluid beyond it so that

$$\tau = \rho g(\delta - y). \tag{19.2.1}$$

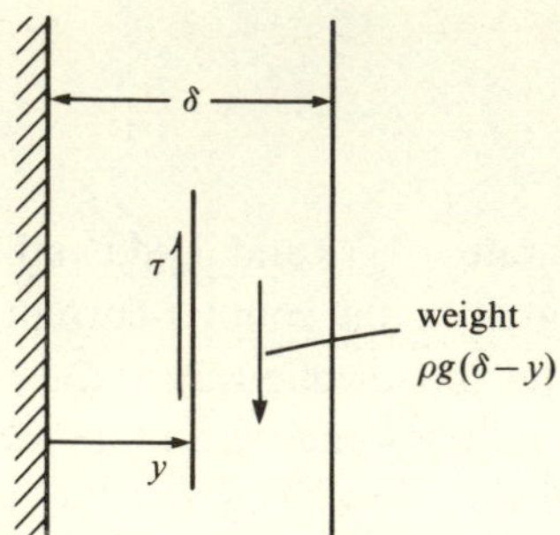

Fig. 163. Force balance in a falling liquid film.

Since we are considering laminar flow

$$\tau=\mu\frac{\mathrm{d}u}{\mathrm{d}y},$$

and on integration subject to the boundary condition $y=0, u=0$, we obtain

$$\mu u=\rho g\left(\delta y-\frac{y^2}{2}\right). \tag{19.2.2}$$

Integration again gives the volumetric flow rate per unit width of surface, Q.

$$Q=\int_0^\delta u\,\mathrm{d}y=\frac{\rho g\delta^3}{3\mu}. \tag{19.2.3}$$

Very commonly the film is formed on the inside surface of a tube of diameter D, so that the total volumetric flow rate is πDQ. Thus the superficial velocity is given by

$$\frac{\pi D^2}{4}j_f=\pi D\frac{\rho g\delta^3}{3\mu}. \tag{19.2.4}$$

A superficial Reynolds number can be defined as $j_f D\rho_f/\mu_f$ which from equation (19.2.4) is seen to be also $4Q/\nu$. This latter definition of the Reynolds number can be used for films on plane surfaces. In terms of the dimensionless variable η, defined below, equation (19.2.4) can be re-written

$$Re=\frac{4}{3}\frac{\rho^2 g\delta^3}{\mu^2}=\frac{4}{3}\eta^2. \tag{19.2.5}$$

The surface of a falling film is normally covered with waves except at very low Reynolds numbers. However, equation (19.2.3) gives a good estimate of the average film thickness up to the point at which turbulence sets in, which is normally at a Reynolds number of about 1000.

Dukler & Bergelin considered turbulent film flow and adapted the universal velocity profile of § 8.4 for this case. From equation (19.2.1) the wall shear stress is given by

$$\tau_w = \rho g \delta \tag{19.2.6}$$

and this is used as the basis of the friction velocity and distance giving,

$$u_\tau = \sqrt{(g\delta)} \tag{19.2.7}$$

and

$$y_\tau = \frac{\nu}{\sqrt{(g\delta)}}. \tag{19.2.8}$$

From equations (17.6.1) to (17.6.3) the dimensionless flow rate is given by

$$Q^+ = \int_0^\eta u^+ \, dy^+ = \int_0^5 y^+ \, dy^+ + \int_5^{30} (5 \log_e y^+ - 3.05) \, dy^+ + \int_{30}^\eta (2.5 \log_e y^+ + 5.5) \, dy^+, \tag{19.2.9}$$

where η is the dimensionless film thickness given by

$$\eta = \frac{\delta}{y_\tau} = \frac{\delta\sqrt{(g\delta)}}{\nu}. \tag{19.2.10}$$

On integrating equation (19.2.9) we obtain

$$\frac{Re}{4} = \frac{Q}{\nu} = 2.5\,\eta \log_e \eta + 3.0\,\eta - 64. \tag{19.2.11}$$

It is noteworthy that when $\eta = 30$ the Reynolds number is 1124, a value that accords well with the observed transition to turbulence.

As an alternative to the use of equation (19.2.11), Wallis recommends the correlation

$$Re = 36.8\,\eta^{1.11} \tag{19.2.12}$$

for the range of Reynolds numbers

$$1000 < Re < 8000.$$

At higher Reynolds numbers one would expect the film thickness to be independent of the viscosity and, noting that both Re and η contain ν^{-1}, we would expect a correlation of the type,

$$Re = A\eta.$$

The best fit to experimental results for $Re > 8000$ is

$$Re = 63.2\,\eta. \tag{19.2.13}$$

This equation can be rearranged to give

$$\frac{\delta}{D} = 0.063\left(\frac{j_f}{\sqrt{(gD)}}\right)^{\frac{2}{3}} = 0.063(j_f^*)^{\frac{2}{3}}. \tag{19.2.14}$$

Bearing in mind that the surface of the film is covered with waves, so that the definition of film thickness is somewhat subjective, the small differences in the predictions of these correlations from the theoretical expression of equation (19.2.11) is probably not significant.

The presence of an upward gas flow normally has little effect as the interfacial shear stresses are small compared with the weight of the film. However, at a critical flow rate the surface of the film becomes highly disturbed and the liquid is carried upwards with the gas. This phenomenon is known as flooding, and the combination of gas and liquid flow rates at the flooding point are given by the correlation

$$(j_g^*)^{\frac{1}{2}} + (j_f^*)^{\frac{1}{2}} = C. \tag{19.2.15}$$

The constant C is normally 1 but lower values can occur if the liquid is supplied in a disturbed state. This form of the correlation is valid for relatively inviscid fluids, i.e. those for which

$$N_f = \frac{g^{\frac{1}{2}} D^{\frac{3}{2}}}{\mu_f} \rho_f > 1000$$

and Wallis gives modifications for more viscous liquids.

19.3 Rise velocity of a single bubble

The rise velocity of a very small bubble can be predicted from Stokes' law which gives the drag force, F, on a bubble of radius a as,

$$F = 4\pi\mu a V, \tag{19.3.1}$$

where V is the rise velocity. This force is two-thirds of that on a solid sphere and, like the result for solid spheres, equation (19.3.1) is only valid for Reynolds numbers less than about 1.

Large bubbles are observed to be of the spherical cap shape as shown in figure 164, and their rise velocity can be predicted by the following potential flow analysis due to Davies & Taylor. The bubble is brought to rest by giving the whole system a velocity V downwards and it is assumed that the

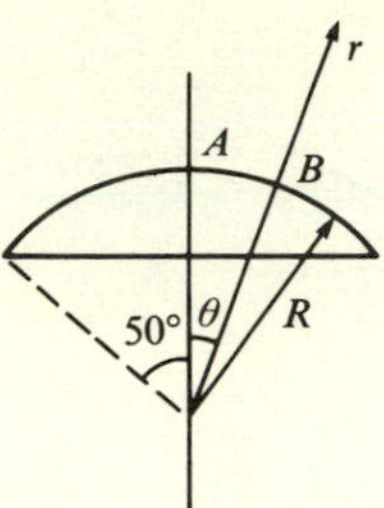

Fig. 164. Spherical cap bubble.

velocity distribution over the front part of the bubble is the same as that in potential flow round a complete sphere of the same radius. The potential function ϕ for flow round a sphere of radius R is given in spherical co-ordinates, as derived in equation (4.3.14), by

$$\phi = V\left(r + \frac{R^3}{2r^2}\right)\cos\theta. \tag{19.3.2}$$

Hence the tangential velocity v_θ on the surface of the bubble is given by

$$v_\theta = -\frac{1}{r}\left(\frac{\partial\phi}{\partial\theta}\right)_{r=R} = \frac{3}{2}V\sin\theta \tag{19.3.3}$$

and applying Bernoulli's equation between points A and B we obtain

$$\frac{1}{2}\left(\frac{3V}{2}\sin\theta\right)^2 = gR(1-\cos\theta). \tag{19.3.4}$$

Thus

$$V^2 = \frac{8gR}{9}\frac{(1-\cos\theta)}{\sin^2\theta} = \frac{8gR}{9(1+\cos\theta)}. \tag{19.3.5}$$

Equation (19.3.5) shows that the assumption that the surface of the bubble is part of a sphere cannot be exact since V, g, and R are constants but θ is a variable. However, since the value of $(1+\cos\theta)^{\frac{1}{2}}$ does not vary much over a considerable range of small θ it is sufficiently accurate to take the limit as $\theta \to 0$ and to interpret R as the radius of curvature at the nose of the bubble. Thus

$$V = \tfrac{2}{3}\sqrt{(gR)}. \tag{19.3.6}$$

There is no theoretical prediction for the angle subtended by a spherical cap bubble but observation indicates that this is usually close to 50°. The volume of such a portion of a sphere is $0.112\,\pi R^3$ and this is equated to $\frac{1}{6}\pi D_e^3$ where D_e is the equivalent diameter defined as the diameter of the sphere of the same volume.

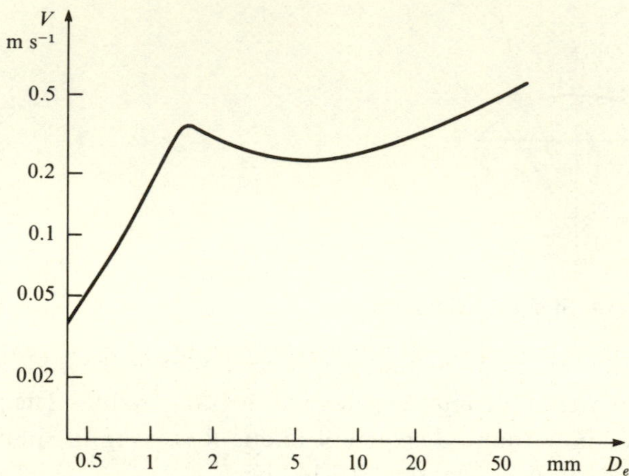

Fig. 165. Rise velocity V of a bubble in water as a function of its equivalent diameter D_e.

Thus $D_e = 0.876\,R$ and substituting into equation (19.3.6) gives

$$V = 0.71\sqrt{(gD_e)}. \tag{19.3.7}$$

Figure 165 shows the form of typical experimental results for the rise velocity of bubbles through water. Excellent agreement with equation (19.3.7) is found for large bubbles and some agreement with equation (19.3.1) for very small bubbles. However, there is a large range of bubble sizes for which neither expression is applicable and in particular there is an extended region from $D_e = 1$ mm to $D_e = 10$ mm in which the rise velocity is almost independent of the bubble size. In this region the bubbles take the form of oscillating ellipsoids and the rise velocity is found to depend on the surface tension, σ.

Wallis[2] has collected experimental results from many sources and has presented the following correlation which is based on fitting five straight lines to the data as follows,

$$v^* = \frac{r^{*2}}{3}, \tag{19.3.8}$$

$$v^* = 0.408\,r^{*1.5}, \tag{19.3.9}$$

$$v^* = \sqrt{2}\,r^{*-\frac{1}{2}}P^{\frac{1}{6}}, \tag{19.3.10}$$

$$v^* = \sqrt{2}\,P^{\frac{1}{12}}, \tag{19.3.11}$$

$$v^* = r^{*\frac{1}{2}}, \tag{19.3.12}$$

where

$$r^* = \frac{D_e}{2}\left(\frac{\rho_f^2 g}{\mu_f^2}\right)^{\frac{1}{3}}, \tag{19.3.13}$$

$$v^* = V\left(\frac{\rho_f}{\mu_f g}\right)^{\frac{1}{3}}, \tag{19.3.14}$$

and

$$P = \frac{\sigma^3 \rho_f}{g \mu_f^4}. \tag{19.3.15}$$

Equations (19.3.8) and (19.3.12) are rearrangements of equations (19.3.1) and (19.3.7) and equations (19.3.9), (19.3.10), and (19.3.11) are empirical fits to the experimental results.

The recommended procedure for the use of this correlation is as follows,

(i) Calculate P, r^*, and B where

$$B = \frac{g D_e^2 \rho_f}{4\sigma}. \tag{19.3.16}$$

If $P > 10^{10}$ or $< 10^2$, special forms of the correlation are required and reference should be made to the original paper.

(ii) If $B > 4$ use equation (19.3.12).

(iii) If $1 < B < 4$ use equation (19.3.11).

(iv) If $B < 1$ and $r^* < 1.5$ use equation (19.3.8).

(v) If $B < 1$ and $r^* > 1.5$ use the lesser of the two values of v^* calculated from equations (19.3.9) and (19.3.10).

The procedure described above is valid for pure fluids. However, the presence of even minute traces of surface-active agents has a profound effect on the rise velocity because such materials accumulate on the surface and immobilise the interface. Consequently such contaminated bubbles behave more like solid spheres. When sufficient surface-active agent (surfactant) is present, and this is usually the case with water, the later stages of the procedure described above should be modified as follows.

(iv) If $B < 1$ and $r^* < 1.5$ use

$$v^* = \tfrac{2}{9} r^{*2}. \tag{19.3.17}$$

(v) If $B < 1$ and $r^* > 1.5$ use the lesser of the two values of v^* calculated from equation (19.3.11) and

$$v^* = 0.307\, r^{*1.21}. \tag{19.3.18}$$

In these equations it has been assumed that the gas density is very much less than the liquid density. Wallis also gives modified forms of these equations for use with drops having density comparable with the surrounding fluid.

19.4 Bubble flow regime

The mean velocities of the gas and liquid, V_g and V_f, are related to the superficial velocities, j_g and j_f, defined in § 19.1, and the void fraction ε by the equations

$$j_g = V_g \varepsilon, \tag{19.4.1}$$

$$j_f = V_f(1-\varepsilon). \tag{19.4.2}$$

Here the void fraction is defined as the fraction of the volume occupied by gas. It is convenient also to define a slip velocity V_s by the relationship

$$V_s = V_g - V_f = \frac{j_g}{\varepsilon} - \frac{j_f}{(1-\varepsilon)}. \tag{19.4.3}$$

Since the gas is permanently trapped within the bubbles, the bubble velocity is V_g and the slip velocity is the bubble velocity as seen by an observer moving with the mean liquid velocity. In many cases it is sufficiently accurate to assume that V_s is equal to the bubble rise velocity V calculated by the methods of the previous section, though there are arguments in favour of the relationship

$$V_s = V(1-\varepsilon)^{n-1}. \tag{19.4.4}$$

This result can be derived from the Richardson and Zaki correlation, presented in § 21.3, and the index n equals 2.39 except for very small bubbles. Since for bubble flow the voidage ε rarely exceeds 0.2, the difference in the predictions of equation (19.4.4) and the simpler relationship

$$V_s = V \tag{19.4.5}$$

is usually small.

The void fraction ε and the bubble rise velocity V_g can be predicted from the superficial velocities by the simultaneous solution of equation (19.4.3) and either (19.4.4) or (19.4.5).

In the latter case ε is given by the quadratic equation

$$j_g(1-\varepsilon) - j_f\varepsilon = \varepsilon(1-\varepsilon)V \tag{19.4.6}$$

and it can easily be shown that there is only one root to this equation for co-current flow.

A graphical solution is to be preferred when equation (19.4.4) is used and this is facilitated by the definition of a hypothetical velocity V_{cd} thus,

$$V_{cd}=\varepsilon(1-\varepsilon)V_s. \tag{19.4.7}$$

Substituting from equation (19.4.4) gives

$$V_{cd}=\varepsilon(1-\varepsilon)^n V \tag{19.4.8}$$

and V_{cd} can be plotted as a function of ε as in figure 166. Equation (19.4.3) can be re-arranged to give

$$V_{cd}=j_g(1-\varepsilon)-j_f\varepsilon \tag{19.4.9}$$

which is a straight line on figure 166 extending from the point $\varepsilon=0$, $V_{cd}=j_g$, to the point $\varepsilon=1$, $V_{cd}=-j_f$. The value of ε is found from the intersection of these lines.

It must, however, be appreciated that in the analysis above we have assumed that the velocities and the void fraction are uniform across any cross-section. This is normally found to be sufficiently accurate for liquid downflow and for modest values of the void fraction in upflow. However, at voidages greater than about 0.15 the bubbles tend to accumulate in the fast-

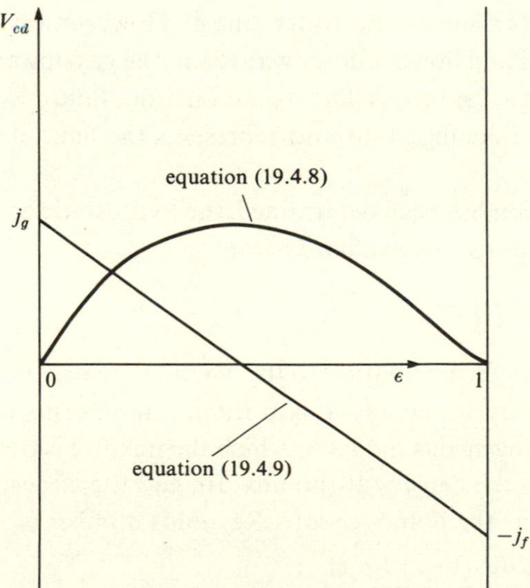

Fig. 166. The Wallis plot for predicting void fractions.

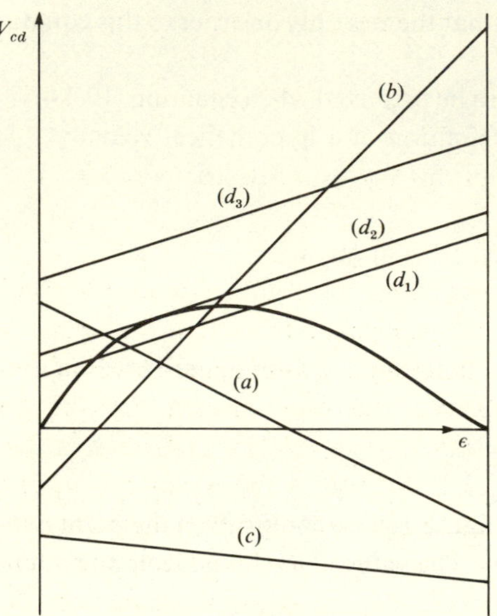

Fig. 167. The Wallis plot illustrating the existence of solutions in co- and counter-current flow.

moving fluid near the centre of the pipe and the gas velocity is greater than that predicted by this analysis.

It is clear from figure 167 that there is only one root for co-current upflow (line *a* of figure 167) and similarly for co-current downflow (line *b*) there is also one root. For counter-current flow with the gas flowing downwards and the liquid upwards there are no roots (line *c*). However, for counter-current flow with the liquid flowing downwards and the gas upwards there may be two roots (line d_1), no roots (line d_3), or one root (line d_2). This last case corresponds to a flooding point and represents the limit of counter-current operation.

Once the void fraction has been determined, the hydrostatic component of the pressure gradient can be evaluated from

$$\left(\frac{\mathrm{d}p}{\mathrm{d}l}\right)_H = \rho_f g(1-\varepsilon) + \rho_g g\varepsilon. \tag{19.4.10}$$

This is normally the major contribution to the overall pressure gradient but sometimes it is necessary to include a skin-friction term. This is usually calculated from a homogeneous model in which the mixture is treated as a single-phase fluid with the density of the mixture and the viscosity of the continuous phase. Thus the homogeneous Reynolds number is

$$Re_h = \frac{(j_g + j_f)D(\rho_f(1-\varepsilon) + \rho_g\varepsilon)}{\mu_f} \tag{19.4.11}$$

and the frictional pressure gradient is given by

$$\left(\frac{\mathrm{d}p}{\mathrm{d}l}\right)_f = \frac{4c_f(j_g+j_f)^2(\rho_f(1-\varepsilon)+\rho_g\varepsilon)}{2D}, \tag{19.4.12}$$

where c_f is calculated from the Blasius relationship of §6.6,

$$c_f = 0.079\ Re^{-\frac{1}{4}}. \tag{19.4.13}$$

Since the frictional pressure gradient is usually only a small fraction of the overall pressure gradient, this approximate analysis is normally adequate.

The size of the individual bubbles is a function primarily of the nature of the gas distributor. However, the only property of the bubbles of importance for this analysis is the rise velocity V. As seen in the previous section, this is effectively constant over a thousand-fold range of bubble volumes and a detailed knowledge of the bubble volume is not required. Unless the distributor produces very small or very large bubbles, the assumption that the bubble rise velocity is given by equation (19.3.11) will not give significant error.

With the bubble rise velocity remaining constant, an increase in the gas flow rate can only result in an increase in the void fraction. When ε exceeds about 0.2, the chances of two bubbles colliding and coalescing greatly increase. Thus large bubbles are formed and the bubble flow regime merges gradually into the slug flow regime. No objective criterion based on gas flow rate is possible since the rate of coalescence depends markedly on the presence of surface-active contaminants. Furthermore, if the gas flow is pulsed, slug flow can be achieved at gas flows that would otherwise result in bubble flow.

19.5 Slug flow

The characteristic feature of this regime is the presence of large gas bubbles known as slugs which occupy almost the whole cross-section of the tube as shown in figure 168. The rise velocity of such bubbles through stagnant liquid is found experimentally to be given by

$$V = 0.35\sqrt{(gD)}, \tag{19.5.1}$$

a result in almost perfect agreement with the theoretical prediction of Dumitrescu[3]. It is noteworthy that the rise velocity is a function solely of the tube diameter, D, and is independent of the bubble volume or the fluid properties. Equation (19.5.1) is, however, only valid if $\rho D^2 g/\mu > 70$ and

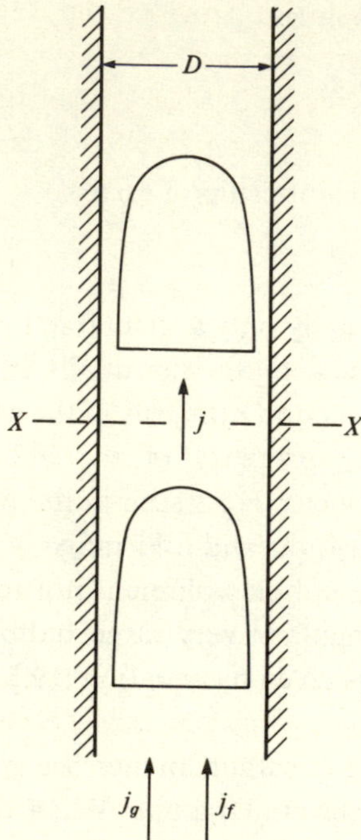

Fig. 168. Slug flow in a vertical pipe.

$N_f = g^{\frac{1}{2}} D^{\frac{3}{2}} \rho / \mu > 550$. For lower values of these dimensionless groups reference should be made to the correlation of White & Beardmore[4].

Equation (19.5.1) represents the rise velocity relative to the fluid ahead of the bubble, so that in a flowing system an additional term must be introduced. A balance up to a section passing through clear fluid between the bubbles, such as section XX of figure 168, shows that the mean velocity j is given by

$$j = j_g + j_f. \tag{19.5.2}$$

Nicklin argues that the important velocity at this section is the centre line velocity since the Dumitrescu analysis is almost entirely concerned with conditions near the nose of the slug. Normally the liquid will be in turbulent flow and from the $\frac{1}{7}$th power law of §§ 6.6 and 8.4 we find that the centre line velocity is 1.2 times the mean velocity. Thus in steady flow we expect to find that the slug rise velocity, V_b, is given by

$$V_b = 1.2(j_g + j_f) + 0.35\sqrt{(gD)}. \tag{19.5.3}$$

This result is in excellent accord with experiment provided the Reynolds number for the flow between the slugs is greater than 2500.

Since the gas is permanently trapped within the bubbles the gas velocity V_g is equal to the bubble velocity V_b and the overall void fraction is therefore given by

$$\varepsilon=\frac{j_g}{V_g}=\frac{j_g}{1.2(j_g+j_f)+0.35\sqrt{(gD)}}. \tag{19.5.4}$$

From this we can calculate the hydrostatic component of the pressure gradient,

$$\left(\frac{\mathrm{d}p}{\mathrm{d}l}\right)_H=\varepsilon\rho_g g+(1-\varepsilon)\rho_f g. \tag{19.5.5}$$

This is the major component of the overall pressure gradient, since not only are the wall shear stresses small but those between the slugs are in the opposite direction to those in the film surrounding the slugs as illustrated in figure 169. Thus the small contributions from wall shear cancel each other out to some extent.

Unless the slug is very short, the film surrounding it will tend to some equilibrium thickness that can be predicted by the analyses presented in § 19.2. Let us denote this equilibrium thickness by δ' and the superficial flow rates in the parallel-sided section of the slug by j'_g and j'_f. As illustrated in figure 170, j'_f is normally measured positive downwards.

An overall balance up to section YY gives

$$j'_g-j'_f=j \tag{19.5.6}$$

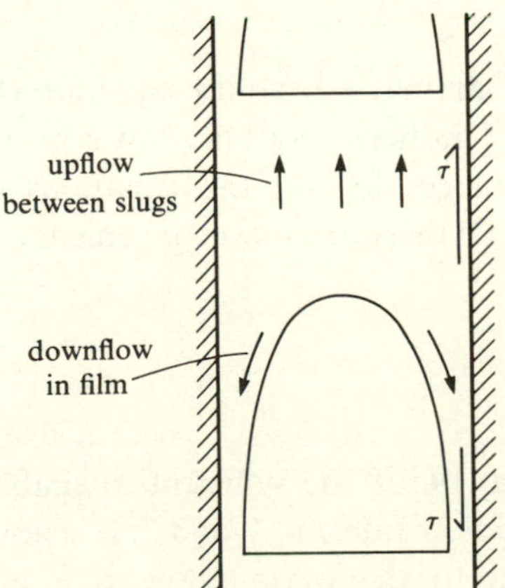

Fig. 169. Shear stress directions in slug flow.

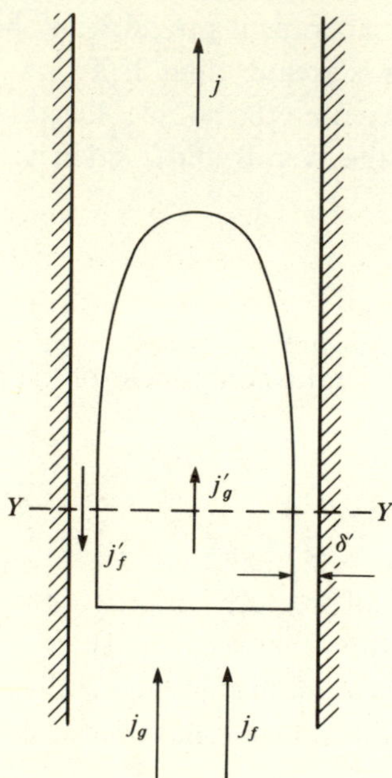

Fig. 170. Mass balance on the cylindrical portion of a slug.

and the superficial gas velocity is related to the slug rise velocity by

$$\frac{\pi D^2}{4} j_g' = \frac{\pi (D-2\delta')^2}{4} V_b. \tag{19.5.7}$$

Substituting from equation (19.5.3) gives

$$j_g' = \left(1 - \frac{2\delta'}{D}\right)^2 (1.2\, j + 0.35\sqrt{(gD)}). \tag{19.5.8}$$

If the flow in the liquid film is laminar we can use equation (19.2.4) to relate δ' to j_f' but unless the liquid is rather viscous the flow is more usually turbulent. Under these circumstances any of the equations (19.2.11), (19.2.12), or (19.2.14) may be used. Of these, the most convenient is (19.2.14) which can be re-written as

$$\frac{\delta'}{D} = 0.063 \left(\frac{j_f'}{\sqrt{(gD)}}\right)^{\frac{2}{3}}. \tag{19.5.9}$$

Equations (19.5.6), (19.5.8), and (19.5.9) are sufficient to enable one to predict the local film thickness and flow rates, δ', j_f', and j_g'. It is noteworthy that these quantities are related to the overall flow rate j and are independent of the separate gas and liquid flow rate j_g and j_f.

On increasing the total flow rate j, both j'_g and j'_f increase and eventually a combination is reached at which flooding occurs. The critical combination can be predicted from some flooding correlation such as equation (19.2.15). Thus it is seen that there exists a critical value of j at which the film surrounding the slug becomes unstable and this gives a sharp upper limit to the slug flow regime. The critical value of j is a function of the fluid properties and the tube diameter only. For air and water in a 30-mm-diameter tube at *STP* the critical value is about 1.8 m s^{-1} and increases roughly as $\sqrt{D}$.

19.6 Churn flow

At velocities somewhat above the upper limit of the slug flow regime a highly chaotic regime is found and this is often referred to as the churn flow regime. Because of the random nature of the flow, little theoretical prediction is possible and recourse must be made to empirical correlations. Of these, the best known is that due to Lockhart & Martinelli[5]. They correlate the pressure gradient in horizontal two-phase flow, $(\mathrm{d}p/\mathrm{d}l)_{TP}$, with the pressure gradients $(\mathrm{d}p/\mathrm{d}l)_G$ and $(\mathrm{d}p/\mathrm{d}l)_L$ that would have occurred if the two phases had flowed separately.

Two parameters are defined as follows,

$$\varphi_g^2 = \left(\frac{\mathrm{d}p}{\mathrm{d}l}\right)_{TP} \Bigg/ \left(\frac{\mathrm{d}p}{\mathrm{d}l}\right)_G, \tag{19.6.1}$$

$$\chi^2 = \left(\frac{\mathrm{d}p}{\mathrm{d}l}\right)_L \Bigg/ \left(\frac{\mathrm{d}p}{\mathrm{d}l}\right)_G. \tag{19.6.2}$$

The two-phase multiplier, φ_g, and the void fraction, ε, are correlated against χ as shown in the table below.

χ	0.04	0.07	0.10	0.20	0.40	0.70	1.0	2.0	4.0	7.0	10.0	20.0
ε	—	0.96	0.95	0.91	0.86	0.81	0.77	0.69	0.60	0.52	0.47	0.34
φ_g	1.54	1.71	1.85	2.23	2.83	3.53	4.20	6.20	9.56	13.7	17.5	29.5

This presentation refers to the most probable case of both fluids being in turbulent flow. If either or both of the fluids are in laminar flow a different form of the correlation applies, details of which are given in the original paper. The Martinelli correlation has an accuracy of about $\pm 30\%$ and modifications to this method due to Chisholm[6] and Baroczy[7] are said to be more reliable.

Though this correlation was developed for horizontal flow it can be used for vertical flow if the hydrostatic contribution to the pressure gradient is included

$$\left(\frac{\mathrm{d}p}{\mathrm{d}l}\right)_{TP}=\varphi_g^2\left(\frac{\mathrm{d}p}{\mathrm{d}l}\right)_G+\rho_f g(1-\varepsilon)+\rho_g g\varepsilon. \tag{19.6.3}$$

In this form its use is straightforward. The single-phase pressure gradients are predicted by the methods outlined in §8.2. From these, χ can be calculated and hence φ_g and ε are obtained from the correlation.

The Martinelli correlation, or the Chisholm or Baroczy modifications are often recommended for use in all the flow regimes of horizontal flow. However, for the bubble and slug flow regimes of vertical flow the hydrostatic terms dominate and these are found more accurately by the analyses of §§19.4 and 19.5. For vertical flow the use of the Martinelli correlation is normally confined to the churn and annular flow regimes.

19.7 Annular flow

On increasing the gas flow rate the random movements of the churn flow regime gradually subside and eventually an ordered regime is obtained in which the liquid flows as an annular film on the tube wall and the gas flows as a central core. The transition is gradual but seems to be complete at about

$$j_g^*=1. \tag{19.7.1}$$

The surface of the film is covered by random waves and it is therefore not possible to make a theoretical prediction of the dependence of the interfacial shear stress on the flow rates. However, the interfacial shear stress can be related to the pressure gradient as follows

$$\pi(D-2\delta)\tau_i=-\frac{\pi}{4}(D-2\delta)^2\left(\frac{\mathrm{d}p}{\mathrm{d}l}+\rho_g g\right) \tag{19.7.2}$$

or for a thin film,

$$\tau_i=-\frac{D}{4}\left(\frac{\mathrm{d}p}{\mathrm{d}l}+\rho_g g\right). \tag{19.7.3}$$

In these expressions ρ_g is the density of the gas core including any entrained liquid. At higher flow rates significant proportions of the liquid can be entrained and ρ_g may be very much greater than the density of the gas itself.

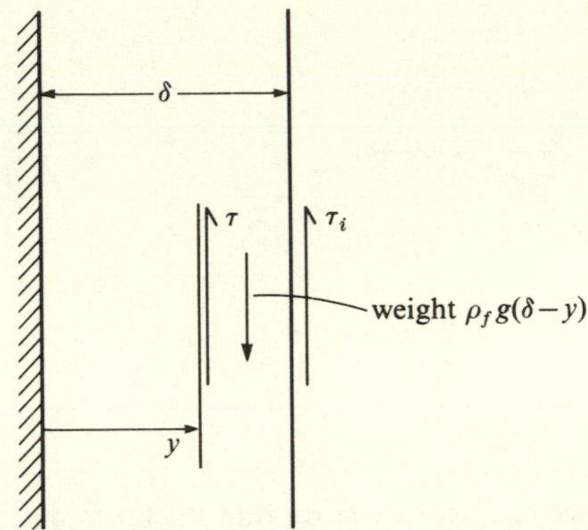

Fig. 171. Force balance on a climbing liquid film.

A set of analyses known as the triangular relationships can be used to relate the pressure gradient to the liquid flow rate and the film thickness. The simplest version is as follows.

From figure 171 it can be seen that the shear stress τ a distance y from the wall is given by

$$\tau = \tau_i - \rho_f g(\delta - y). \tag{19.7.4}$$

For laminar flow $\tau = \mu(\mathrm{d}u/\mathrm{d}y)$ and substituting from equation (19.7.3) we obtain

$$\mu\frac{\mathrm{d}u}{\mathrm{d}y} = -\frac{D}{4}\left(\frac{\mathrm{d}p}{\mathrm{d}l} + \rho_g g\right) - \rho_f g(\delta - y). \tag{19.7.5}$$

Integrating twice gives the volumetric flow rate of liquid per unit run of wall, Q,

$$Q = -\frac{D\delta^2}{8\mu}\left(\frac{\mathrm{d}p}{\mathrm{d}l} + \rho_g g\right) - \frac{\rho_f g \delta^3}{3\mu}. \tag{19.7.6}$$

Here, Q is measured positive upwards and it must be remembered that $\mathrm{d}p/\mathrm{d}l$ is a negative quantity. From this relationship the mean film thickness δ can be predicted as a function of the flow rate and pressure gradient. Figure 172 shows the relationship between Q and δ and it can be seen that there are usually two positive roots for δ. Of these, the smaller value is normally found in practice. Figure 172 also shows that there is an upper limit to the liquid flow rate, Q_{max}, for any specified pressure gradient. It must, however, be remembered that this analysis only provides a relationship between δ, $\mathrm{d}p/\mathrm{d}l$, and Q. It is insufficient to predict the pressure gradient or the film thickness independently.

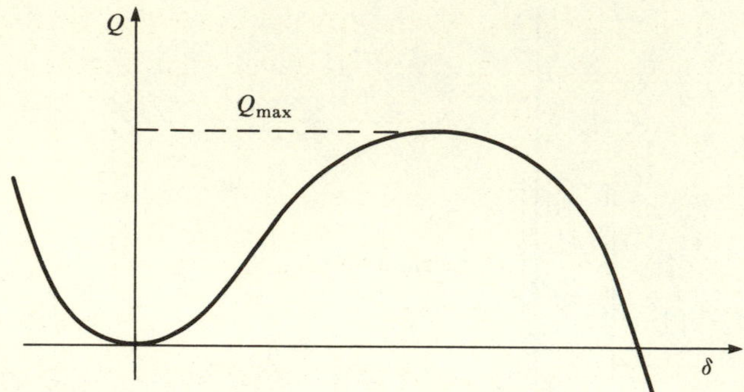

Fig. 172. The dependence of flow rate Q on film thickness δ.

Alternative versions of this analysis have been presented in which allowance has been made for the curvature of the wall and also for the effects of turbulence. Dukler presents an analysis for the latter case in which the shear stress is given by

$$\tau=(\mu+\rho v_\tau)\frac{\mathrm{d}u}{\mathrm{d}y}, \tag{19.7.7}$$

as in § 17.6, and some assumption is made about the dependence of the eddy viscosity v_τ on distance from the wall.

The pressure gradient may be adequately predicted by the Martinelli or related correlations as described in the previous section. However, an alternative correlation is presented by Wallis which seems to be based more soundly on physical principles.

An interfacial friction factor c_{fi} can be defined by

$$c_{fi}=\frac{\tau_i}{\frac{1}{2}\rho_g V_g^2}, \tag{19.7.8}$$

where V_g is the mean gas velocity in the core and can be related to the superficial gas velocity j_g and the voidage ε by

$$j_g=\varepsilon V_g. \tag{19.7.9}$$

Substituting from equation (19.7.2) and noting that

$$D-2\delta=D\sqrt{\varepsilon}$$

we have

$$c_{fi}=-\frac{D\varepsilon^{\frac{5}{2}}}{2\rho_g j_g^2}\left(\frac{\mathrm{d}p}{\mathrm{d}l}+\rho_g g\right). \tag{19.7.10}$$

The interfacial friction factor will depend primarily on the amplitude and frequency of the waves on the surface of the film and these will in turn be functions of the film thickness. Wallis finds that the following correlation applies over a considerable portion of the annular flow regime.

$$c_{fi}=0.005\left(1+\frac{300\,\delta}{D}\right)\simeq 0.005(1+75(1-\varepsilon)). \tag{19.7.11}$$

Thus we find that

$$-\left(\frac{\mathrm{d}p}{\mathrm{d}l}+\rho_g g\right)=0.01\frac{j_g^2\rho_g}{D}\frac{(1+75(1-\varepsilon))}{\varepsilon^{\frac{5}{2}}}. \tag{19.7.12}$$

Defining a dimensionless pressure gradient ΔP^* by

$$\Delta P^*=\frac{-(\mathrm{d}p/\mathrm{d}l+\rho_g g)}{(\rho_f-\rho_g)g} \tag{19.7.13}$$

and using j_g^* defined by equation (19.1.3), equation (19.7.12) can be rewritten as

$$\Delta P^*=0.01\,j_g^{*2}\frac{(1+75(1-\varepsilon))}{\varepsilon^{\frac{5}{2}}}. \tag{19.7.14}$$

A force balance on the tube as a whole gives

$$\pi D\tau_w=-\frac{\pi D^2}{4}\left(\frac{\mathrm{d}p}{\mathrm{d}l}+\varepsilon\rho_g g+(1-\varepsilon)\rho_f g\right) \tag{19.7.15}$$

and Wallis quotes experimental evidence that the wall-friction factor defined by

$$c_{fw}=\frac{\tau_w}{\frac{1}{2}\rho_f V_f^2}=\frac{\tau_w(1-\varepsilon)^2}{\frac{1}{2}\rho_f j_f^2} \tag{19.7.16}$$

is always close to 0.005 for turbulent flow.

Combining these equations and the definition of j_f^* (equation (19.1.4)) we obtain

$$\Delta P^*=(1-\varepsilon)+\frac{0.01\,j_f^{*2}}{(1-\varepsilon)^2}. \tag{19.7.17}$$

Equations (19.7.14) and (19.7.17) represent a simultaneous pair from which the pressure gradient and the void fraction can be predicted for any given combination of j_g and j_f.

19.8 Entrainment

The prediction of the quantity of liquid entrained as droplets in the gas core is exceedingly complex and is beyond the scope of this book.

Butterworth & Hewitt[8] present experimental evidence that up to half the liquid may be entrained and correlations are presented by Wallis[1].

One feels intuitively that for given gas and liquid flow rates there will be an equilibrium entrainment at which the rate of deposition, due to random impact of the droplets on the film, just balances the rate of entrainment from the waves. However, experiments show that the rate of approach to this equilibrium is very slow with characteristic lengths (the distance over which the departure from equilibrium is reduced by a factor of e) of the order of 5 m. Since pressure gradients in annular flow are large, there will be a considerable pressure drop over this distance and a corresponding change in the volumetric flow rate of gas. Thus there may be no steady state and entrainment will always lag, often by a considerable distance, behind the equilibrium value appropriate to the local flow rates. This has far-reaching effects on the behaviour of an evaporator, as will be discussed in §20.5.

References

[1] Wallis, *One-Dimensional Two-Phase Flow*, McGraw-Hill.
[2] Wallis, *Int. J. Multiphase Flow*, **1**, 491 (1974).
[3] Dumitrescu, *Z. Angew. Math. Mech.*, **23**, 139 (1943).
[4] White & Beardmore, *Chem. Eng. Sci.*, **17**, 351 (1962).
[5] Lockhart & Martinelli, *Chem. Eng. Progr.*, **45**, 39 (1949).
[6] Chisholm, *Int. J. Heat and Mass Transfer*, **16**, 347 (1973).
[7] Baroczy, *Chem. Eng. Prog. Symp. Ser.*, **62**, 232 (1966).
[8] Butterworth & Hewitt, *Two-phase Flow and Heat Transfer*, OUP.

20

Condensation and evaporation

20.1 Film condensation

Condensation of a vapour in a surface condenser can take place in two different ways. The normal mechanism is that of film condensation, i.e. the condensed liquid forms a continuous film on the cooling surface and the vapour actually condenses on the liquid film. The other mechanism is that of dropwise condensation when individual drops are formed which detach themselves from the surface. Dropwise condensation can result in very high heat transfer coefficients with values of h of 50 kW m^{-2} K^{-1} and upwards. However, it can only occur if the liquid does not wet the surface of the condenser and in practice it is very rarely achieved.

The mechanism for film condensation can be understood from the following analysis for condensation on a vertical surface, due to Nusselt.

It was shown in § 19.2 that the volumetric flow rate Q per unit width for laminar flow is given by

$$Q=\frac{\rho g\delta^3}{3\mu}, \tag{20.1.1}$$

where δ is the film thickness. The mass flow per unit width is therefore

$$M=\frac{\rho^2 g\delta^3}{3\mu}. \tag{20.1.2}$$

With condensation taking place, the liquid film must increase both in mass flow and thickness as it flows down the surface. If $\mathrm{d}M$ is the condensation rate in a length $\mathrm{d}x$ of surface, as shown in figure 173, a heat balance will give

$$\lambda\,\mathrm{d}M=q\,\mathrm{d}x, \tag{20.1.3}$$

where λ is the latent heat and q is the local heat flux. Assuming heat flow by conduction across the liquid film, with a linear temperature profile, we can say

$$q=\frac{k\,\Delta T}{\delta}, \tag{20.1.4}$$

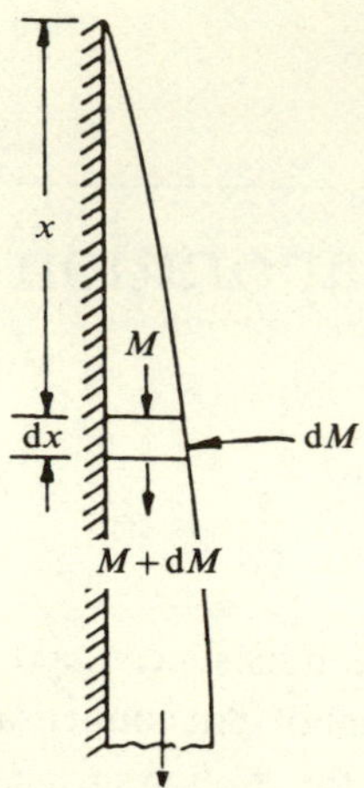

Fig. 173. Film condensation.

where ΔT is the local transverse temperature difference and k is the thermal conductivity of the liquid. Hence from equation (20.1.2)

$$q=k\,\Delta T\left(\frac{\rho^2 g}{3\mu M}\right)^{\frac{1}{3}} \tag{20.1.5}$$

and substituting into equation (20.1.3),

$$M^{\frac{1}{3}}\,\mathrm{d}M=\frac{k\,\Delta T}{\lambda}\left(\frac{\rho^2 g}{3\mu}\right)^{\frac{1}{3}}\mathrm{d}x. \tag{20.1.6}$$

Integrating from 0 to x, assuming ΔT is constant,

$$M^{\frac{4}{3}}=\frac{4k\,\Delta Tx}{3\lambda}\left(\frac{\rho^2 g}{3\mu}\right)^{\frac{1}{3}}. \tag{20.1.7}$$

The total rate of heat flow is λM and therefore the average heat transfer coefficient $\bar{h}$ is given by

$$\bar{h}=\frac{M\lambda}{x\,\Delta T} \tag{20.1.8}$$

and the average value of the Nusselt number for height x of the surface is

$$Nu=\frac{\bar{h}x}{k}=\frac{M\lambda}{k\,\Delta T}. \tag{20.1.9}$$

Substituting from equation (20.1.7) we have

$$Nu=\left(\frac{4x}{3}\right)^{\frac{3}{4}}\left(\frac{\lambda}{k\,\Delta T}\right)^{\frac{1}{4}}\left(\frac{\rho^2 g}{3\mu}\right)^{\frac{1}{4}} \tag{20.1.10}$$

or

$$Nu = 0.943\left(\frac{\lambda \rho g}{k\,\Delta T v}\right)^{\frac{1}{4}} x^{\frac{3}{4}}. \tag{20.1.11}$$

This result may be expressed in an alternative form in terms of the mass-flow rate M at distance x from the top by eliminating x from equations (20.1.7) and (20.1.8) giving

$$\bar{h} = \frac{M\lambda}{\Delta T} \cdot \frac{4k\,\Delta T}{3\lambda M^{\frac{4}{3}}} \left(\frac{\rho^2 g}{3\mu}\right)^{\frac{1}{3}} \tag{20.1.12}$$

or

$$\frac{\bar{h}}{k}\left(\frac{v^2}{g}\right)^{\frac{1}{3}} = 1.47\left(\frac{4M}{\mu}\right)^{-\frac{1}{3}}. \tag{20.1.13}$$

Note that $(v^2/g)^{\frac{1}{3}}$ has the dimensions of length and the group on the left-hand side of equation (20.1.13) therefore has the form of a Nusselt number. Furthermore, the group $4M/\mu$ is the conventional definition of a film Reynolds number, as given in § 19.2, and we can therefore write

$$\frac{\bar{h}}{k}\left(\frac{v^2}{g}\right)^{\frac{1}{3}} = 1.47\,Re^{-\frac{1}{3}}. \tag{20.1.14}$$

It is found experimentally that flow in the film is usually turbulent for $Re > 1000$ and in these circumstances the following correlation due to Colburn may be used,

$$\frac{\bar{h}}{k}\left(\frac{v^2}{g}\right)^{\frac{1}{3}} = 0.0077\,Re^{0.4}. \tag{20.1.15}$$

Alternatively the heat transfer coefficient across turbulent films can be calculated by the methods of § 20.2 below.

We can modify the above analysis for condensation on the outside of a cylindrical tube of diameter D as shown in figure 174. At angular position θ the relationship between mass-flow rate and film thickness is given by a modified form of equation (20.1.2), i.e.

$$M = \frac{\rho^2 g \sin\theta\delta^3}{3\mu} \tag{20.1.16}$$

and equation (20.1.3) takes the form

$$\lambda\,\mathrm{d}M = q\frac{D}{2}\,\mathrm{d}\theta. \tag{20.1.17}$$

The rest of the analysis follows similar lines and gives the following result

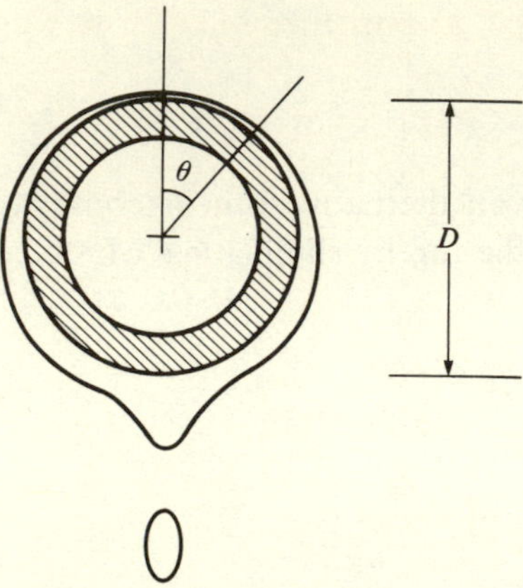

Fig. 174. Condensation on a horizontal cylinder.

$$Nu = \frac{\bar{h}D}{k} = 0.725\left(\frac{\lambda\rho^2 g}{k\,\Delta T\mu}\right)^{\frac{1}{4}} D^{\frac{3}{4}}. \qquad (20.1.18)$$

Commonly condenser tubes are arranged in a vertical tier and for a tier of N tubes

$$Nu = 0.725\left(\frac{\lambda\rho^2 g}{Nk\,\Delta T\mu}\right)^{\frac{1}{4}} D^{\frac{3}{4}}. \qquad (20.1.19)$$

20.2 Heat transfer across a turbulent film

In the previous section we considered a laminar film across which the heat transfer coefficient could be expressed simply as k/δ. In a turbulent film we must allow for the effects of eddy transfer and we can extend the Dukler–Bergelin analysis of § 19.2 as follows.

We assume that the velocities are given by the universal velocity profile of § 8.4,

$$\begin{aligned} u^+ &= y^+ && y^+ < 5 \\ u^+ &= 5\log_e y^+ - 3.05 && 5 < y^+ < 30 \\ u^+ &= 2.5\log_e y^+ + 5.5 && 30 < y^+ < \eta \end{aligned}, \qquad (20.2.1)$$

where $u^+ = u\sqrt{(\rho/\tau_0)}$, $y^+ = y\sqrt{(\tau_0\rho)}/\mu$, and η is the dimensionless film thickness given by

$$\eta = \frac{\delta\sqrt{(\tau_0\rho)}}{\mu}.$$

Following von Karman's method, as presented in § 17.6 we find from equation (17.6.26)

$$T_0^+ - T_i^+ = 5\,Pr + 5\log_e(1+5\,Pr) + 2.5\log_e\left(\frac{\eta}{30}\right). \tag{20.2.2}$$

Recalling from equation (20.2.1) that the dimensionless interfacial velocity u_i^+ is given by

$$u_i^+ = 2.5\log_e \eta + 5.5 = 5\log_e 6 - 5 + 2.5\log_e\left(\frac{\eta}{30}\right),$$

equation (20.2.2) can be re-arranged to give

$$T_0^+ - T_i^+ = u_i^+ + 5(Pr-1) + 5\log_e\left(\frac{1+5\,Pr}{6}\right). \tag{20.2.3}$$

Noting that $T_0^+ - T_i^+ = (T_0 - T_i)(\rho c_p u^*/q) = \rho c_p u^*/h$, we can re-write equation (20.2.3) in the form

$$\frac{1}{St} = u_i^+\left(u_i^+ + 5(Pr-1) + 5\log_e\left(\frac{1+5\,Pr}{6}\right)\right), \tag{20.2.4}$$

where St is the Stanton number based on the interfacial velocity,

$$St = \frac{h}{\rho c_p u_i}.$$

Though the value of the Prandtl number for water is about 7.0 at room temperature this falls rapidly with increasing temperature to 1.73 at 100 °C and 0.94 at 300 °C. Thus for the condensation of steam, except at reduced pressure, we can assume that $Pr \simeq 1$ and equation (20.2.4) becomes

$$\frac{h}{\rho c_p u_i} = \frac{u^{*2}}{u_i^2} = \frac{\tau_0}{\rho u_i^2}$$

or

$$h = \frac{\tau_0 c_p}{u_i}. \tag{20.2.5}$$

Alternatively for large values of Pr we can assume that conduction across a laminar sub-layer of thickness δ_T dominates so that

$$h = \frac{k}{\delta_T},$$

where from the analysis of § 17.7 we can say that

$$\delta_T \simeq 5 \frac{\mu}{\sqrt{(\tau_0 \rho)}} Pr^{-\frac{1}{3}}.$$

These analyses coupled with the Dukler–Bergelin relationship, equation (19.2.11),

$$\frac{Q}{\nu} + 64 = 2.5\,\eta \log_e \eta + 3.0\,\eta$$

can be used to predict the average heat transfer coefficient in turbulent film condensation, following the method of §20.1. However, the form of the equations is unsuited to algebraic analysis and recourse must be made to numerical methods.

Dukler[1] has considered the more general case when there exists also an interfacial shear stress τ_i and presents his results graphically, plotting $h(\mu^2/\rho^2 k^3 g)^{\frac{1}{3}}$ against Re with $\beta = \tau_i/(\rho\mu^2 g^2)^{\frac{1}{3}}$ as a parameter. These results are sketched in figure 175.

20.3 Effect of non-condensable gases

With the high heat transfer coefficients found in condensers there is a rapid flow of the condensable species towards the interface and this carries with it any non-condensable gas that happens to be present. The non-

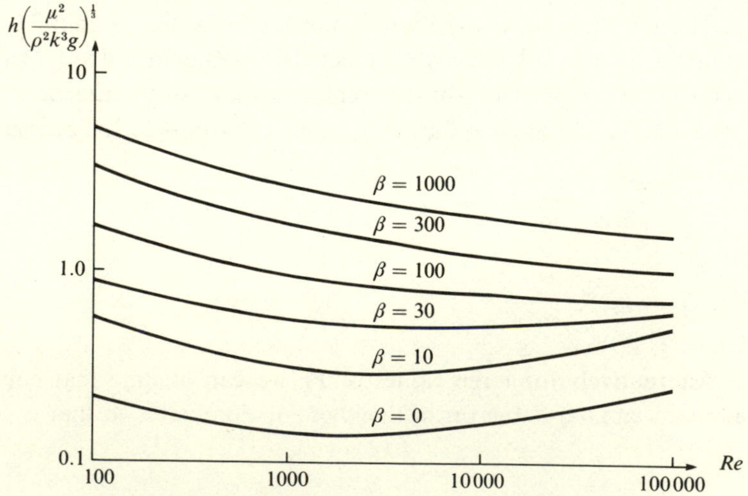

Fig. 175. Heat transfer coefficient h through a climbing film as a function of Reynolds number Re.

condensable gas accumulates at the interface until the concentration gradient becomes large enough to give an equal and opposite diffusive flux. The resulting partial pressure profile can be considered in terms of the film model as follows. If the molar condensation rate is N_A, the partial pressure profile of the non-condensable species B will be given by equation (14.2.19), and since N_B must be zero,

$$N_A p_B = \frac{P\mathscr{D}}{RT}\frac{\mathrm{d}p_B}{\mathrm{d}z}, \tag{20.3.1}$$

where P is the total pressure and $\mathscr{D}$ is the diffusion coefficient. Hence on integrating across a laminar film of thickness δ we have

$$p_{Bi} = p_{Bb}\,\mathrm{e}^{RTN_A\delta/P\mathscr{D}}, \tag{20.3.2}$$

where p_{Bi} and p_{Bb} are the partial pressures of the non-condensable species at the interface and in the bulk. For typical values of the condensation rate and film thickness it is found that p_{Bi}/p_{Bb} is large so that, even with small overall concentrations, the partial pressure of the non-condensable species at the interface may be large. This will lower the condensation temperature and adversely affect the performance of the condenser. Consequently the heat transfer rates achieved in practice are rarely as favourable as those predicted above and it is advisable to ensure that non-condensable gases are eliminated as completely as possible from systems with condensers.

20.4 Pool boiling

Heat transfer from a flat metal surface to a boiling liquid is often called pool boiling and there are three possible stages that must be considered.

(i) Sometimes heat is transferred to the liquid at a temperature below the local boiling point. This will occur, for instance, over the lower portion of a submerged heating surface where boiling is suppressed locally owing to the hydrostatic head of liquid above. For free convection to a liquid from horizontal or vertical heating surfaces in this manner the usual expressions for heat transfer given in §15.6 may be used.

(ii) The term nucleate boiling is used to describe the process whereby small bubbles of vapour form on the heating surface and subsequently detach themselves and rise through the liquid. With nucleate boiling the heat transfer coefficient h varies with the temperature difference ΔT

approximately in the following manner

$$h \propto \Delta T^{2.5},$$

so that the heat flux q is proportional to $\Delta T^{3.5}$.

However, nucleate boiling does not occur immediately the liquid temperature exceeds the boiling point T_B. Due to surface tension σ the pressure inside a spherical bubble of radius r exceeds that in the surrounding liquid by $2\sigma/r$. Hence from the Clausius–Clapeyron equation

$$\frac{\mathrm{d}}{\mathrm{d}T}\log_e p = \frac{\lambda}{RT^2},$$

where λ is the latent heat, we have

$$\frac{2\sigma}{rP} \approx \frac{\lambda\theta}{RT_B}$$

or

$$\theta \approx \frac{2\sigma T_B}{\lambda r \rho_v}, \tag{20.4.1}$$

where θ is the elevation of the boiling point and ρ_v is the density of the vapour in the bubble.

It is seen that the elevation of the boiling point is inversely proportional to the radius. Thus it is difficult to nucleate a small bubble within the bulk of the liquid and bubbles are normally formed on suspended particles or at irregularities on the surface. Once formed, a bubble can grow rapidly and this accounts for the bumping that is commonly observed in the boiling of pure liquids adjacent to a smooth heating surface. The heat transfer coefficient depends on the number of nucleation sites on the surface and therefore can rarely be predicted from first principles.

(iii) If the temperature of the heating surface is sufficiently high, a continuous film of vapour may be formed on the surface and the heat transfer coefficient will fall sharply from the value that it would have under conditions of nucleate boiling. In film boiling the heat must be conducted across a film of poorly conducting gas, though at very high temperature differences radiative heat transfer becomes appreciable.

The three zones of single-phase heating, nucleate boiling, and film boiling are shown in figure 176 which gives the general form of the curve of heat flux q plotted against the temperature difference ΔT between the heating surface and the liquid. There will generally be a definite maximum value for the heat flux q_{max} occurring at a certain critical temperature difference. This critical temperature difference depends on the physical

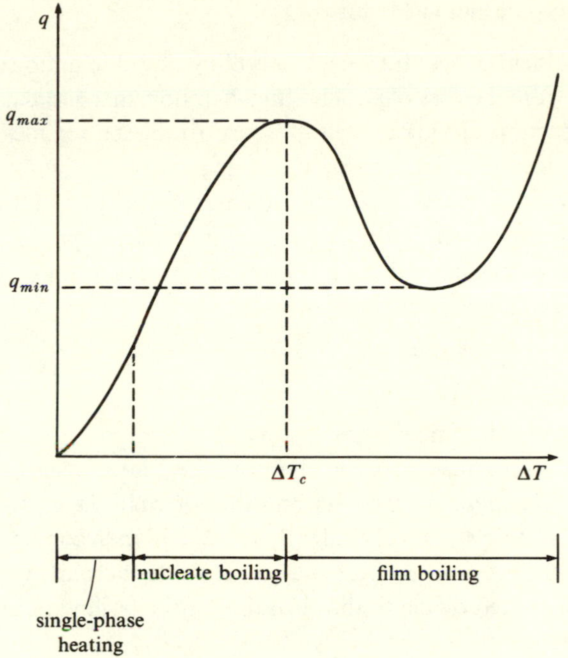

Fig. 176. Heat flux q as a function of temperature difference ΔT.

properties of both the liquid and the heating surface and factors such as contact angle, surface roughness and the presence of surface-active agents. Theoretical prediction of the critical temperature difference is therefore impossible but much empirical information is available, for example in Hsu & Graham[2].

One must distinguish between the performance of evaporators in which the temperature difference is controlled, e.g. when the heating medium is some condensing vapour, and those that operate with a specified heat flux as will occur in a chemical or nuclear reactor. In the former case we can operate at any point on the curve of figure 176, though it is advisable from the point of view of minimising the heat transfer area to work in the vicinity of ΔT_c. For a specified heat flux q we must work with q considerably less than q_{max} since should q increase accidentally beyond q_{max} a very large increase in ΔT would occur. This may result in over-heating of the surface and burnout may occur, causing irreparable damage to the surface. Even if the surface is not damaged it may be necessary to reduce q below q_{min} to restore nucleate boiling. Hence the whole region between q_{min} and q_{max} is potentially unstable and caution must be used when operating under these conditions.

20.5 Evaporation inside tubes

The problem of heat transfer to a boiling liquid in a tube differs in certain features from the case of pool boiling. For flow in a vertical tube the increasing proportion of vapour will give rise to all the regimes of two-phase flow in succession as illustrated schematically by figure 161 and described in § 19.1. The liquid is assumed to enter the bottom of the tube at some temperature below the boiling point and in the lower part of the tube the heat transfer coefficient will be given by the ordinary expressions for single-phase forced convection such as

$$Nu = 0.023\, Re^{0.8}\, Pr^{0.4} \tag{20.5.1}$$

for turbulent flow.

When the tube wall temperature is somewhat above the local boiling point, nucleate boiling may occur in the boundary layer region near the wall. Bubbles of vapour formed within the thermal layer will move outwards into the main stream of colder fluid and will then condense again. As explained in the previous section, the liquid in the thermal layer will be somewhat super-heated owing to the elevation of the boiling point caused by surface tension.

The evaporation of even a small fraction of the liquid causes a large increase in the mean velocity and a corresponding thinning of the thermal boundary layer and increase in the heat transfer coefficient. When the thermal boundary layer is sufficiently thin, nucleation at the wall is suppressed. The reason for this can be seen from the following simplified version of the analysis of Hsu & Graham.

It is assumed that bubbles are formed over cavities in the wall as shown in figure 177. We will assume that the bubble is always hemispherical with its centre at the centre of the mouth of the cavity. As evaporation proceeds the bubble grows until it is large enough to be carried away by its own buoyancy or by the motion of the liquid. We will also assume that when the bubble departs is leaves a small hemispherical bubble of radius equal to that of the mouth of the cavity r_c.

The model postulates that a bubble will only grow if the liquid temperature at all points on the surface of the bubble exceeds the effective boiling point T_E which from equation (20.4.1) is given by

$$T_E = T_B\left(1 + \frac{2\sigma}{\lambda r \rho_v}\right). \tag{20.5.2}$$

The temperature in the thermal boundary layer will be a function of the distance y from the wall but, since the temperature of importance is that at

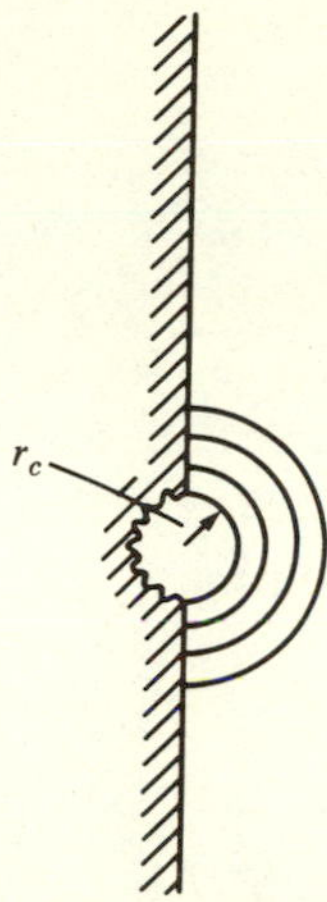

Fig. 177. Growth of a hemispherical bubble at a cavity of radius r_c.

the farthest extremity of the hemispherical bubble, y and r are synonymous and we can express the temperature profile in the liquid in the form

$$T = f(r). \tag{20.5.3}$$

Plotting equations (20.5.2) and (20.5.3) as in figure 178, we see that the curves can cut at two points r_1 and r_2 and that between these $T > T_E$. Thus bubbles in cavities of radius $r_c > r_1$ will grow up to size r_2. This is usually large enough for the bubble to detach. Hence cavities of radius less than r_1 are ineffective as nucleation sites.

As the thermal boundary layer gets thinner on increase of the volumetric flow rate, the temperature gradient becomes steeper. As a result, r_1 increases and nucleation from some of the smaller cavities is suppressed. Eventually the temperature gradient can become so steep that nucleation is totally suppressed as shown by the dotted line of figure 178. Thereafter evaporation takes place on the surface of existing large bubbles within the bulk. It must be appreciated that the above analysis is only qualitatively correct. The bubbles are not hemispherical and in his full analysis Hsu makes allowance for this and for the effect of contact angle.

As the bubbles grow by evaporation and coalescence they develop into slugs, i.e. bubbles comparable in size with the tube diameter. These later break down into the churn flow regime as described in § 19.6. Within the slug flow regime a particular section of the tube wall is alternately in contact with the liquid between the slugs and with the film surrounding the slugs. The former situation may be treated as a single-phase heat transfer problem

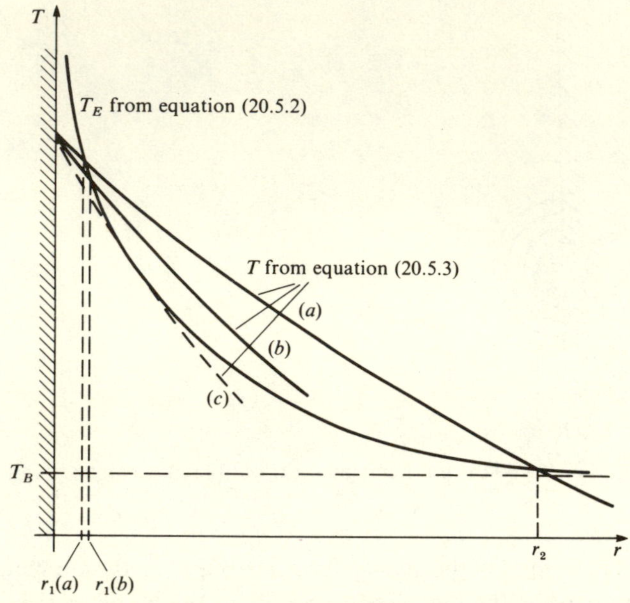

Fig. 178. Hsu and Graham's analysis for the suppression of nucleate boiling.

and for the latter the analysis of § 20.2 is appropriate. Within the churn flow regime only empirical results are available.

Further evaporation produces stabilisation of the flow pattern into annular flow for which Dukler's analysis mentioned in § 20.2 is appropriate. The heat transfer coefficients in this regime and also in the slug and churn flow regimes are high, as the heat has only to be conducted across a thin liquid film between the wall and the evaporating interface. Values as high as 200 kW m^{-2} K^{-1} have been reported and the heat transfer rate is therefore normally controlled by external resistances.

As explained in § 19.1, the liquid film in annular flow is usually disturbed by large waves from which droplets are entrained into the gas phase. Under isothermal conditions an equilibrium may be set up between entrainment and deposition but these processes are slow and the actual entrainment will always lag behind the equilibrium value. With sufficient heat flux the evaporation of the film can be more rapid than the droplet deposition rate and the film can dry out while evaporation is still incomplete.

Beyond the *dry-out point* the tube wall is unwetted and the heat transfer coefficient falls to the low value associated with solid-gas heat transfer, typically 10–50 W m^{-2} K^{-1}. Evaporation beyond the dry-out point is therefore a very slow process.

If the source of heat is a medium at constant temperature, evaporation beyond the dry-out point is merely protracted but if a constant heat flux is maintained, as would be the case if a reactor, either chemical or nuclear, were being cooled by a boiling liquid, very high wall temperatures would occur and the dry-out point might be identified as a *burn-out point*. Thus evaporation beyond dry-out should be avoided and only partial evaporation is normally attempted, the resulting two-phase mixture being separated and the liquid re-cycled.

References

[1] Dukler, *Modern Chemical Engineering*, ed. Acrivos, Reinhold.
[2] Hsu & Graham, *Transport Processes in Boiling and Two Phase Systems*, McGraw-Hill.

21

Solid particles and fluidisation

21.1 Shape factors and mean diameters

Consider a particle of irregular shape having a volume V and surface area A. It is useful to take as a comparison the equivalent-volume sphere, i.e. a sphere whose diameter D_s is given by

$$\frac{\pi}{6} D_s^3 = V$$

or

$$D_s = \left(\frac{6V}{\pi}\right)^{\frac{1}{3}}. \qquad (21.1.1)$$

D_s is sometimes referred to as the nominal or equivalent diameter of the particle. The surface area of the equivalent-volume sphere A_s is πD_s^2 and the shape factor λ is then defined by

$$\lambda = \frac{A}{A_s}, \qquad (21.1.2)$$

i.e. λ is the ratio of the surface area to that of the sphere of the same volume. For the special case of a spherical particle $\lambda = 1$, but for all other shapes λ will be greater than 1. The quantity $1/\lambda$ is sometimes referred to as the sphericity.

If the particle diameter is specified in some other way (e.g. by the maximum linear dimension of the particle) and denoted by D_p, the volume V and surface area A may be expressed by

$$V = \varphi_v \frac{\pi}{6} D_p^3 \quad \text{where } \varphi_v \text{ is a volume factor}$$

and

$$A = \varphi_a \pi D_p^2 \quad \text{where } \varphi_a \text{ is an area factor.}$$

Note that

$$\varphi_v = \left(\frac{D_s}{D_p}\right)^2 \quad \text{and} \quad \varphi_a = \lambda\left(\frac{D_s}{D_p}\right)^2.$$

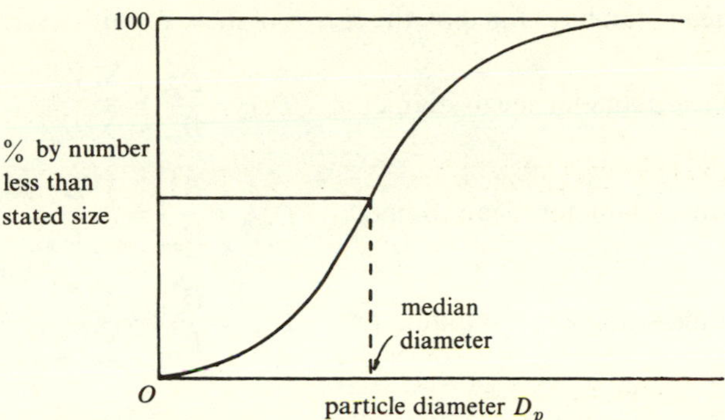

Fig. 179. Cumulative distribution of particle diameters D_p.

In practice one usually encounters mixtures of particles of different sizes. Consider first of all the case of a mixture of N particles of similar shape but of different diameters specified by D_p. If we could plot the percentage (by number) of particles less than a stated size against stated size, the result would be a cumulative size distribution curve such as that shown in figure 179. The median diameter is defined as the particle diameter for which 50% of the particles in the mixture are less than the stated size.

Other mean diameters for a mixture of particles may be defined as follows:

Arithmetic mean diameter $$D_{am} = \frac{\sum D_p}{N} = \overline{D_p},$$

Geometric mean diameter $$D_{gm} = \sqrt[N]{(D_{p1}, D_{p2}, \ldots, D_{pN})},$$

or $$\log_e D_{gm} = \frac{\sum \log_e D_p}{N},$$

Surface mean diameter $$D_{sm}^2 = \frac{\sum D_p^2}{N} = \overline{D_p^2},$$

Volume mean diameter $$D_{vm}^3 = \frac{\sum D_p^3}{N} = \overline{D_p^3}.$$

The surface mean diameter D_{sm} for instance is the diameter of a particle having surface area equal to the average for all particles in the mixture. The volume mean diameter D_{vm} is the diameter of the particle having the average volume.

Further means can be defined by the ratios of these quantities as follows:

Surface-diameter mean diameter $D_{sdm}=\frac{D_{sm}^2}{D_{am}}=\frac{\sum D_p^2}{\sum D_p}$,

Volume-diameter mean diameter $D_{vdm}^2=\frac{D_{vm}^3}{D_{am}}=\frac{\sum D_p^3}{\sum D_p}$,

Volume-surface mean diameter $D_{vsm}=\frac{D_{vm}^3}{D_{sm}^2}=\frac{\sum D_p^3}{\sum D_p^2}$.

21.2 Particle size distributions

If the particle size distribution for a mixture followed the normal Gaussian or error distribution, we would define the deviation from the mean diameter by $x=D_p-D_{am}$, where D_{am} is the arithmetic mean diameter which, for a Gaussian distribution, is also the most probable diameter. We would then have for the number of particles dN with size deviation between x and $x+\mathrm{d}x$

$$\mathrm{d}N=\frac{N}{\sigma\sqrt{2\pi}}\,\mathrm{e}^{-x^2/2\sigma^2}\,\mathrm{d}x, \tag{21.2.1}$$

where N is the total number of particles and σ is the standard deviation defined by

$$\sigma^2=\frac{\sum x^2}{N}. \tag{21.2.2}$$

With the symmetrical Gaussian distribution the number of particles bigger than a stated size D_p, i.e. with values of the deviation x greater than $x=D_p-D_{am}$, is obtained by integrating equation (21.2.1). Thus the number of particles larger than size D_p

$$=\frac{N}{\sigma\sqrt{(2\pi)}}\int_x^\infty \mathrm{e}^{-x^2/2\sigma^2}\,\mathrm{d}x=\frac{N}{2}\left[1-\mathrm{erf}\left(\frac{\sqrt{2x}}{\sigma}\right)\right]$$

and the number of particles smaller than size D_p is

$$\frac{N}{2}\left[1+\mathrm{erf}\left(\frac{\sqrt{2x}}{\sigma}\right)\right].$$

The symmetrical or Gaussian distribution for the particle size in a mixture, however, would imply negative values for some particle diameters, which is obviously impossible. A more plausible distribution would be the

Gaussian distribution for $\log_e D_p$, and this is found to fit the facts for many cases of particle mixtures.

For this distribution we define the deviation x_l by

$$x_l = \log_e D_p - \log_e D_{gm} \qquad (21.2.3)$$

and x_l is presumed to be normally distributed with standard deviation σ_l. Note that when $D_p \to 0$, $x_l \to -\infty$.

The normal or Gaussian distribution of x_l gives the 'logarithmic' or 'skew probability' distribution of particle size. However, this distribution is most frequently referred to as the 'log-normal' distribution. Whilst there is rarely, if ever, any physical reason why the particles in a sample should satisfy the log-normal distribution, it is often found that this distribution is an acceptable fit to the actual distribution. It is convenient to evaluate the parameters of the best fit log-normal distribution as the various mean diameters defined in § 21.1 can be calculated simply from these parameters.

By analogy with equation (21.2.1), the number of particles dN with deviations between x_l and $x_l + \mathrm{d}x_l$ is given by

$$\mathrm{d}N = \frac{N}{\sigma_l\sqrt{(2\pi)}}\,\mathrm{e}^{-x_l^2/2\sigma_l^2}\,\mathrm{d}x_l, \qquad (21.2.4)$$

where

$$\sigma_l^2 = \frac{\sum x_l^2}{N}$$

and the number of particles n with size less than x_1 is

$$n = \frac{N}{2}\left[1 + \mathrm{erf}\left(\frac{x_l}{\sqrt{2}\sigma_l}\right)\right]. \qquad (21.2.5)$$

The log-normal distribution can conveniently be fitted to experimental data by plotting the particle size D_p against the cumulative probability p_n defined by $p_n = n/N$. If we plot $\log_e D_p$, which is the deviation x_l plus some as yet unknown constant $\log_e D_{gm}$, against y where

$$\mathrm{erf}\, y = 2p_n - 1 \qquad (21.2.6)$$

we should get a straight line. This procedure is facilitated by the use of commercially available log-probability graph paper. This paper has a logarithmic scale for the ordinate on which D_p can be plotted directly and the abscissa is linear in y but normally labelled with the corresponding value of p_n as shown in figure 180. If the size distribution approximates to the log-normal distribution the data points will lie close to a straight line.

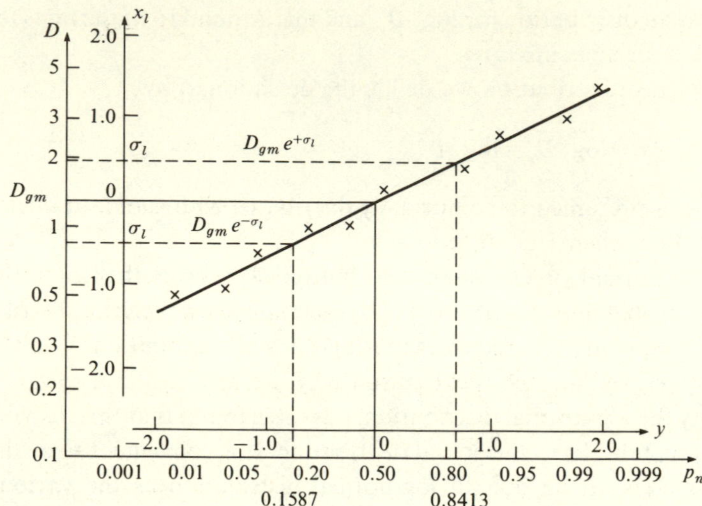

Fig. 180. Log probability distribution of particle diameters.

The best-fit straight line can usually be drawn in by eye with sufficient accuracy.

By definition $D_p = D_{gm}$ when $p_n = 0.5$ and hence the geometric mean can be read directly off the graph as shown in figure 180. Furthermore, in the normal distribution the probability of being more than one standard deviation from the mean is 0.1587, so that the points on the line at $p_n =$ 0.1587 and 0.8413 correspond to $x_l = -\sigma_l$ and $x_l = +\sigma_l$ respectively and, from equation (21.2.3),

$$D_{p,0.1587} = D_{gm}\, e^{-\sigma_l} \tag{21.2.7}$$

and

$$D_{p,0.8413} = D_{gm}\, e^{\sigma_l}. \tag{21.2.8}$$

The value of σ_l can be found from either of these points but it is usual to use both to provide a check.

The usefulness of the log-normal distribution lies in the ease with which it is possible to calculate the mean diameters defined in the previous section.

By definition the mean of D_p^m is given by

$$\overline{D_p^m} = \int_{-\infty}^{+\infty} D_p^m \frac{e^{-x_l^2/2\sigma_l^2}}{\sqrt{(2\pi)}\sigma_l}\, dx_l, \tag{21.2.9}$$

which from equation (21.2.3) can be written

$$\overline{D_p^m} = D_{gm}^m \int_{-\infty}^{+\infty} \frac{e^{mx_l}\, e^{-x_l^2/2\sigma_l^2}}{\sqrt{(2\pi)}\sigma_l}\, dx_l. \tag{21.2.10}$$

Making the substitution $\alpha = x_l - m\sigma_l^2$ equation (21.2.10) becomes

$$\overline{D_p^m} = D_{gm}^m \, e^{\sigma_l^2 m^2/2} \int_{-\infty}^{+\infty} \frac{e^{-\alpha^2/2\sigma_l^2}}{\sqrt{(2\pi)}\sigma_l} \, d\alpha \tag{21.2.11}$$

and since

$$\int_{-\infty}^{+\infty} \frac{e^{-\alpha^2/2\sigma_l^2}}{\sqrt{(2\pi)}\sigma_l} \, d\alpha = 1,$$

we have

$$\overline{D_p^m} = D_{gm}^m \, e^{\sigma_l^2 m^2/2}. \tag{21.2.12}$$

Hence we can evaluate $\overline{D_p^m}$ for any value of m from the values of D_{gm} and σ_l obtained from the log-probability plot.

Nowadays there exist computerised particle size analysers using television cameras that can count the number of particles below a specified size. Some of these are also programmed to fit a log-normal distribution and display the value of any selected mean. However, it is still common practice to use the established technique of sieve analysis. By sieving the sample through a series of sieves of progressively finer mesh, the quantity within any size range can be measured. It is, however, much more convenient to measure the mass of particles in a size range rather than the number of particles.

Clearly the mass dM of particles in the size range x_l to $x_l + dx_l$ is given by

$$dM = \frac{\pi}{6} D_p^3 \varphi_v \rho \, dN = \frac{\pi}{6} D_p^3 \varphi_v \rho \frac{N}{\sqrt{(2\pi)}\sigma_l} \, e^{-x_l^2/2\sigma_l^2} \, dx_l. \tag{21.2.13}$$

Noting that $D_p^3 = e^{3x_l} D_{gm}^3$ from equation (21.2.3) and making the substitution $\alpha = x_l - 3\sigma_l^2$ we obtain

$$dM = \frac{\pi}{6} D_{gm}^3 \varphi_v \rho \frac{N}{\sqrt{(2\pi)}\sigma_l} \, e^{9\sigma_l^2/2} \, e^{-\alpha^2/2\sigma_l^2} \, dx. \tag{21.2.14}$$

Thus the distribution by mass is normal about $\alpha = 0$ with the same standard deviation σ_l. The criterion $\alpha = 0$ gives $x_l = 3\sigma_l^2$, i.e.

$$\log_e D_p = \log_e D_{gm} + 3\sigma_l^2$$

and hence the median particle size for the log-normal distribution by mass D_{mm} is given by

$$\log_e D_{mm} = \log_e D_{gm} + 3\sigma_l^2$$

or

$$D_{mm} = D_{gm} \, e^{3\sigma_l^2}. \tag{21.2.15}$$

Thus if, instead of using the log-probability paper with abscissa p_n, we use abscissa p_m defined by

$$p_m = \frac{m}{M}, \tag{21.2.16}$$

where M is the total mass in the sample and m is the mass with size less than D_p, we should still obtain a straight line. The value of D_p at $p_m = 0.5$ is D_{mm} and the values at $p_m = 0.1587$ and 0.8413 are $D_{mm}\,\mathrm{e}^{-\sigma_l}$ and $D_{mm}\,\mathrm{e}^{\sigma_l}$ respectively.

The mean values defined in § 21.1 can therefore be obtained as follows. From equation (21.2.12) putting $m = 1$ we have

$$D_{am} = D_{gm}\,\mathrm{e}^{\sigma_l^2/2}$$

and from equation (21.2.15)

$$D_{am} = D_{gm}\,\mathrm{e}^{\sigma_l^2/2} = D_{mm}\,\mathrm{e}^{-5\sigma_l^2/2}. \tag{21.2.17}$$

Similarly,

$$D_{sm} = D_{gm}\,\mathrm{e}^{\sigma_l^2} = D_{mm}\,\mathrm{e}^{-2\sigma_l^2}, \tag{21.2.18}$$

$$D_{vm} = D_{gm}\,\mathrm{e}^{3\sigma_l^2/2} = D_{mm}\,\mathrm{e}^{-3\sigma_l^2/2}. \tag{21.2.19}$$

From these we can also obtain

$$D_{sdm} = D_{gm} = D_{mm}\,\mathrm{e}^{-3\sigma_l^2}, \tag{21.2.20}$$

$$D_{vdm} = D_{gm}\,\mathrm{e}^{2\sigma_l^2} = D_{mm}\,\mathrm{e}^{-\sigma_l^2}, \tag{21.2.21}$$

$$D_{vsm} = D_{gm}\,\mathrm{e}^{5\sigma_l^2/2} = D_{mm}\,\mathrm{e}^{-\sigma_l^2/2}. \tag{21.2.22}$$

21.3 Solid particles in fluid streams

The behaviour of small spherical particles in a fluid stream can be predicted with reasonable accuracy. The motion of small solid particles of other shapes, however, is more difficult to analyse.

The force or drag F on a sphere is usually expressed in terms of a drag coefficient c_D defined by

$$c_D = \frac{F}{\frac{1}{2}\rho u^2(\pi/4)D_p^2}, \tag{21.3.1}$$

where u is the velocity of the fluid relative to the sphere and D_p is the diameter. The drag coefficient is a function of the Reynolds number $Re = uD_p\rho/\mu$ as indicated for example in figure 139.

In the viscous range, with values of Re less than 2, Stokes' law, equation (5.9.4), applies, i.e.

$$c_D = \frac{24}{Re}. \tag{21.3.2}$$

For large values of Re the experimental curve for the drag coefficient diverges from Stokes' law and some empirical relationship must be used. Clift *et al.*[12] quote 12 such relationships of which

$$c_D = \frac{24}{Re}(1 + 0.15\, Re^{0.687}) \tag{21.3.3}$$

is probably the most accurate for $Re < 800$.

Alternatively the simpler expression

$$c_D = 18.0\, Re^{-0.6} \tag{21.3.4}$$

may be used in the range $2 < Re < 200$.

The curve for c_D subsequently flattens out to a value of $c_D = 0.44$ at $Re = 2000$ and varies little from this value in the range $2000 < Re < 200\,000$. At a value of Re somewhere between 10^5 and 10^6 there is a sharp drop in the value of c_D due to transition to turbulence in the boundary layer. The critical value depends on the roughness of the surface and the degree of turbulence in the approaching fluid. It is, however, well outside the usual range of Reynolds numbers appropriate to solid particles in fluid streams.

The terminal settling velocity u_t of a spherical particle under gravity is given by

$$\frac{\pi}{6} D_p^3(\rho_s - \rho)g = c_D \frac{1}{2} \rho u_t^2 \frac{\pi}{4} D_p^2, \tag{21.3.5}$$

where ρ is the density of the fluid, ρ_s the density of the solid particle, and D_p the diameter of the particle. It is sometimes convenient to use the density ratio s defined by

$$s = \frac{\rho_s}{\rho} \tag{21.3.6}$$

and hence

$$u_t = \left[\frac{4}{3} D_p \frac{(s-1)}{c_D} g\right]^{\frac{1}{2}}. \tag{21.3.7}$$

This expression is not convenient for predicting u_t since c_D is itself a function of the Reynolds number and hence u_t. It is more convenient to eliminate u_t from the Reynolds number and the drag coefficient thereby defining a dimensionless group N_D thus,

$$N_D = Re^2 c_D = \frac{4}{3} D_p^3 \frac{(s-1)g\rho^2}{\mu^2}. \tag{21.3.8}$$

Correlations between N_D and both Re and c_D are available in the literature. Since N_D can readily be calculated, this method gives Re and hence u_t directly. The similarity between N_D and r^* of § 19.3 may be noted but N_D is probably best interpreted as a Grashof number as defined in § 15.6.

In the Stokes' law range, for $Re < 2$, there is no need to resort to empirical correlations and substitution of equation (21.3.2) into equation (21.3.7) yields

$$u_t = \frac{D_p^2}{18}(s-1)\frac{\rho g}{\mu}. \tag{21.3.9}$$

The maximum particle diameter for Stokes law to apply in the case of gravity settling can be found as follows

$$Re = D_p \frac{u_t}{\mu}\rho = \frac{D_p^3}{18}(s-1)g\frac{\rho^2}{\mu^2}$$

and putting $Re = 2$ for the critical case

$$D_{p\,max} = \left[\frac{36\mu^2}{\rho^2(s-1)g}\right]^{\frac{1}{3}}. \tag{21.3.10}$$

These results may be applied to non-spherical particles using D_s, the diameter of the equivalent-volume sphere, in place of D_p with a correction factor applied to the drag coefficient. The greater the departure from the spherical shape, however, the less will be the value of this method of analysis. Detailed analyses of this situation are given by Heywood[2] and Clift[1].

The results above refer to the case of a single particle in a large quantity of liquid. The settling velocity u of the particles in a concentrated suspension may be found from a correlation due to Richardson & Zaki[3]

$$u = u_t \varepsilon^n, \tag{21.3.11}$$

where ε is the volume fraction of fluid and n is a function of the Reynolds number based on the terminal velocity u_t

$$Re_t = \frac{u_t D_p \rho}{\mu}.$$

The dependence of n on Re_t for wide vessels is as follows

$$Re_t < 0.2 \quad n = 4.65$$

$$0.2 < Re_t < 1.0 \quad n = 4.35\, Re_t^{-0.03}$$

$$1.0 < Re_t < 500 \quad n = 4.45\, Re_t^{-0.1}$$

$$500 < Re_t \quad n = 2.39$$

The velocity u in equation (21.3.11) is the settling velocity in a vessel through which there is no net flux. In the case of steady flow the techniques described in the section on bubble swarms, § 19.4, may be used.

For heat transfer by conduction between a spherical particle and an infinite surrounding fluid the following expression was derived in § 13.5:

$$Nu = \frac{hD_p}{k} = 2.0, \tag{21.3.12}$$

where h is the heat transfer coefficient and k is the thermal conductivity. This result will also apply with flow round the sphere at very low values of the Reynolds number, $uD_p\rho/\mu$. In terms of a Stanton number, $h/\rho uc_p$, equation (21.3.12) takes the alternative form

$$St = \frac{2}{Re\, Pr}. \tag{21.3.13}$$

At high Reynolds numbers empirical expressions must be used for the heat transfer coefficient and the following form may be used for Reynolds number above about 20:

$$Nu = C\, Re^a\, Pr^b, \tag{21.3.14}$$

or

$$St = C\, Re^{1-a}\, Pr^{1-b}.$$

The values usually assigned to these constants are $C = 0.37$, $a = 0.6$, $b = 0.3$ and this last value is seen to be close to the value of $\frac{1}{3}$ predicted by the j-factor analogy of § 17.4.

For $Re < 20$, the simple combination of equations (21.3.12) and (21.3.14),

$$Nu = 2 + C\, Re^a\, Pr^b \tag{21.3.15}$$

is usually adequate but if greater accuracy is required reference should be made to the more extensive tabulations given for example by Clift[1] or Wong[4].

For mass transfer the analogous expressions

$$Sh = \frac{k_l D_p}{\mathscr{D}} = 2.0 \tag{21.3.16}$$

and

$$Sh = 2 + C\,Re^a\,Sc^b \tag{21.3.17}$$

may be used.

21.4 Viscous flow through a packed bed

Consider a packed bed of solid particles. The porosity or void fraction is defined by

$$\varepsilon = \frac{\text{volume of voids}}{\text{total volume}} = \frac{A_e}{A},$$

where A is the total cross-sectional area of the bed and A_e is the effective cross-sectional flow area of the voids.

If a fluid is flowing through the bed at a total volumetric flow rate Q, the effective mean axial velocity component u_e will be

$$u_e = \frac{Q}{A_e} = \frac{Q}{\varepsilon A} = \frac{u_m}{\varepsilon},$$

where u_m is the mean approach or superficial velocity of the fluid defined as Q/A. The velocity u_e is also referred to as the interstitial velocity. Owing to the tortuous nature of the paths through the bed, the effective length l_e of a channel will be greater than the thickness of the bed l. The effective mean absolute velocity of the fluid v_e passing through the spaces between the particles will be greater than u_e as shown in figure 181 and is given by

$$v_e = u_e \frac{l_e}{l} = \frac{u_m l_e}{\varepsilon l}.$$

Referring back to the Hagen–Poiseuille law for viscous flow in pipes, equation (5.8.17), we might expect the pressure drop for flow through a packed bed to be given by an expression of the form

$$\Delta p = \frac{k_0 \mu v_e l_e}{m^2}, \tag{21.4.1}$$

where m is the hydraulic radius for the void passages, v_e is the mean velocity in the void passages, and k_0 is a constant. Comparing equation (21.4.1) with (5.8.17) and noting that $4m$ takes the place of the pipe diameter D, the

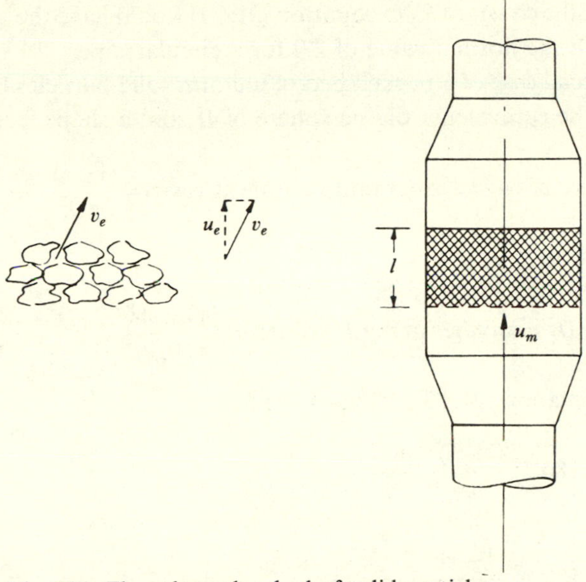

Fig. 181. Flow through a bed of solid particles.

constant k_0 should have the value 2. This value, however, applies strictly only to passages of circular cross-section.

For the packed bed the hydraulic radius of the void passages is given by

$$m = \frac{\text{flow area}}{\text{wetted perimeter}} = \frac{\text{volume of fluid in the bed}}{\text{wetted surface}},$$

i.e.

$$m = \frac{\varepsilon A l}{\text{wetted surface}} = \frac{\varepsilon}{S},$$

where S is the wetted surface of the particles per unit volume of the vessel.

Substituting for v_e and m in equation (21.4.1) we have the result

$$\Delta p = k_0 \frac{S^2}{\varepsilon^3} \left(\frac{l_e}{l}\right)^2 \mu u_m l, \tag{21.4.2}$$

which is known as the Carman–Kozeny equation. This may be written

$$\Delta p = k \frac{S^2}{\varepsilon^3} \mu u_m l, \tag{21.4.3}$$

where $k = k_0(l_e/l)^2$. It is found experimentally that the value of the constant k is about 5 in many practical cases. This would imply that, if the average angle of the fluid passage through the bed is inclined at 45° to the axis so that $l_e/l=\sqrt{2}$, the constant k_0 in equation (21.4.1) would have the value 2.5 compared with the normal value of 2.0 for a circular pipe.

For the special case of a packed bed of uniform solid particles having a diameter for the equivalent-volume sphere of D_s and a shape factor λ,

$$\text{number of particles per unit volume of vessel}=\frac{(1-\varepsilon)}{\frac{1}{6}\pi D_s^3},$$

therefore,

$$S=\lambda\pi D_s^2\times\text{number per unit volume}=\frac{6(1-\varepsilon)\lambda}{D_s}. \tag{21.4.4}$$

Hence from equation (21.4.3), taking $k=5.0$,

$$\Delta p=180\,\lambda^2\,\frac{(1-\varepsilon)^2}{\varepsilon^3}\,\frac{\mu u_m l}{D_s^2} \tag{21.4.5}$$

or

$$\frac{\Delta p}{\rho u_m^2}=180\,\lambda^2\,\frac{(1-\varepsilon)^2}{\varepsilon^3}\,\frac{l}{D_s}\left(\frac{\nu}{u_m D_s}\right). \tag{21.4.6}$$

For spherical particles equation (21.4.5) takes the form

$$\Delta p=180\,\frac{(1-\varepsilon)^2}{\varepsilon^3}\,\frac{\mu u_m l}{D_p^2}. \tag{21.4.7}$$

It is sometimes more convenient to work in terms of a pressure gradient in which case equation (21.4.7) can be re-expressed, paying due attention to signs as

$$-\frac{\mathrm{d}p}{\mathrm{d}l}=180\,\frac{(1-\varepsilon)^2}{\varepsilon^3}\,\frac{\mu u_m}{D_p^2} \tag{21.4.8}$$

and this result is often called the Carman–Kozeny equation though this title more properly belongs to equation (21.4.2).

The result that the pressure gradient is proportional to the flow rate was found by Darcy in 1846 and is often referred to as Darcy's law. Darcy's work is often thought of as the starting point of modern fluid mechanics.

Darcy's law can conveniently be written

$$\frac{\mathrm{d}p}{\mathrm{d}l}=-Ku_m$$

and the result can also be expressed vectorially as

$$\text{grad}\,p=-K\mathbf{u}_m. \tag{21.4.9}$$

Combining this with the continuity equation,

$$\operatorname{div} \mathbf{u}_m = 0,$$

which was derived in §3.3, we find that

$$\operatorname{div} \operatorname{grad} p = 0$$

or

$$\nabla^2 p = 0. \qquad (21.4.10)$$

This is Laplace's equation, which was first derived in §4.3 but with the pressure replacing the potential function ϕ. Thus all the techniques of potential flow analysis described in chapter 4 can be used to predict the pressure and velocity distributions within a packed bed.

For a bed of particles of mixed size but of similar shape,

$$\text{total particle surface area} = \lambda\pi \sum D_s^2 = \lambda\pi N D_{sm}^2$$

$$\text{total volume of solid matter} = \frac{\pi}{6} \sum D_s^3 = \frac{\pi N}{6} D_{vm}^3.$$

Hence the particle surface per unit volume of bed, S, is given by

$$S = \frac{6\lambda\pi N}{\pi N D_{vm}^3} D_{sm}^2 (1-\varepsilon) = \frac{6\lambda(1-\varepsilon)}{D_{vsm}}.$$

Comparison with equation (21.4.4) shows that under this circumstance the appropriate diameter is the volume-surface mean diameter D_{vsm} and this quantity replaces D_s in equations (21.4.5) and (21.4.6).

Whilst processes involving the percolation of a fluid through a porous medium are of common occurrence throughout industry, the most important commercial example must surely be the extraction of oil from an oil-bearing stratum. If as an example we consider the flow of oil towards a cylindrical bore hole of radius r_1 in a level stratum of thickness H, we can relate the superficial velocity u_m to the radial pressure gradient $\mathrm{d}p/\mathrm{d}r$ by equation (21.4.3),

$$\frac{\mathrm{d}p}{\mathrm{d}r} = \frac{kS^2}{\varepsilon^3} \mu u_m. \qquad (21.4.11)$$

The group ε^3/kS^2 is characteristic of the solid medium and is defined as the permeability K so that equation (21.4.11) can be written

$$u_m = \frac{K}{\mu} \frac{\mathrm{d}p}{\mathrm{d}r}. \qquad (21.4.12)$$

In oil-field practice the unit of permeability is called the darcy when μ is measured in centipoise, u_m in cm s^{-1}, and dp/dr in atm cm^{-1}. The darcy differs from its SI equivalent by a factor of 0.99×10^{-12} so that a typical permeability of 100 millidarcies is equivalent to about 10^{-13} m^2.

The velocity at radius r is given by $Q/2\pi rH$ where Q is the volumetric flow rate and hence on integrating equation (21.4.12) we have

$$Q = \frac{2\pi HK}{\mu} \frac{(p_2 - p_1)}{\log_e (r_2/r_1)}. \qquad (21.4.13)$$

Here p_1 is the pressure in the well and p_2 is the pressure at radius r_2. Typical values of p_1 and r_1 are 250 bar and 0.18 m with the pressure 100 m from the well being about 350 bar. Thus, taking K as 10^{-13} m^2, μ as 10^{-3} N s m^{-2}, and H as 15 m, we have

$$Q = \frac{2\pi \times 15 \times 10^{-13}}{10^{-3}} \times \frac{100 \times 10^5}{\log_e (100/0.18)} = 0.015 \text{ m}^3 \text{ s}^{-1},$$

or 8100 barrels per day.

It will be noted that the flow rate is insensitive to the value of r_2 since this appears only in the logarithmic term. In practice r_2 is usually taken to be the radius of the circle of area equal to the area being drained by the well.

If we now consider a well in a porous level stratum containing both water and oil, it is of interest to know at what rate the oil can be extracted without also extracting water. Water will be drawn up in the form shown in figure 182 due to the pressure gradient in the flowing oil and this process is known as coning. If there is no extraction of water, the pressure gradient in the water will be zero and the pressure gradient in the oil must therefore be balanced by hydrostatic pressure changes. Thus the maximum allowable pressure difference is $(\rho_w - \rho_0)gH$ which in our example is about $(1000 - 800) \times 9.81 \times 15 = 0.3$ bar, a value very much less than the value of 100 bar assumed. Thus the presence of water in a horizontal stratum would greatly reduce the maximum production rate. However, the stratum is unlikely to be level and will normally be inclined at some angle α to the horizontal. Under these circumstances the water will not be drawn along the stratum if the pressure gradient in the oil is less than $g(\rho_w - \rho_0) \tan \alpha$. Hence from equation (21.4.12), Q must be less than

$$\frac{2\pi rHK}{\mu} (\rho_w - \rho_0) g \tan \alpha$$

and, substituting the values of this example with $\alpha = 45°$, we see that the

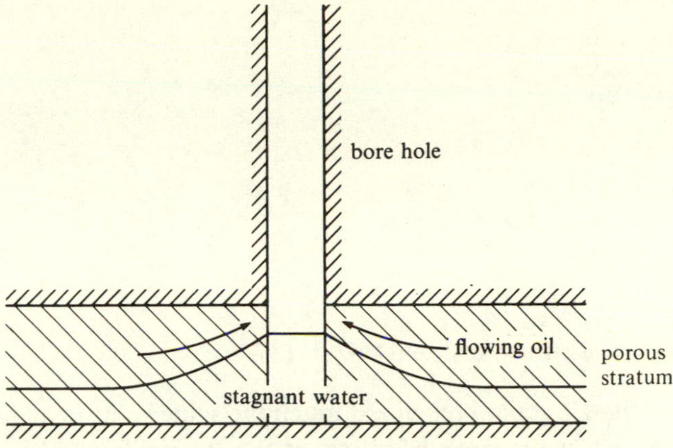

Fig. 182. Flow from an oil-bearing stratum into a bore hole.

original flow rate of 0.015 $m^3\ s^{-1}$ can be maintained provided the oil/water interface is at a radius greater than 800 m from the well.

21.5 General analysis of flow through packed beds

We can proceed as before by analogy with flow through a pipe. Using the skin-friction coefficient c_f, the pressure drop for flow in a pipe is given by equation (8.2.3), i.e.

$$\frac{\Delta p}{\rho u_m^2} = 2c_f \frac{l}{D}, \tag{21.5.1}$$

where D is the pipe diameter, u_m is the mean velocity, and c_f is a function of the Reynolds number $u_m D/\nu$.

For flow through the passages of a packed bed we could say

$$\frac{\Delta p}{\rho v_e^2} = 2c_f \frac{l_e}{D_h}, \tag{21.5.2}$$

where $D_h = 4m$ is the hydraulic diameter, v_e is the mean effective velocity, and l_e is the effective length of a path through the bed. The coefficient c_f should be a function of the Reynolds number defined by $v_e D_h/\nu$.

It was shown in the previous section that

$$m = \frac{\varepsilon}{(1-\varepsilon)} \frac{D_s}{6\lambda}$$

and

$$v_e = \frac{u_m l_e}{\varepsilon l}.$$

Hence

$$Re = \frac{v_e D_h}{\nu} = \frac{2}{3}\frac{D_s}{\lambda}\frac{l_e}{l}\frac{u_m}{(1-\varepsilon)\nu} \tag{21.5.3}$$

and

$$c_f = \frac{\Delta p}{\rho v_e^2}\frac{D_h}{2l_e} = \frac{1}{3}\left(\frac{l}{l_e}\right)^3 \frac{\varepsilon^3}{\lambda(1-\varepsilon)}\frac{D_s}{l}\frac{\Delta p}{\rho u_m^2}. \tag{21.5.4}$$

The quantity l/l_e should remain at much the same value in all cases and may be ignored in the final definition of the Reynolds number and the friction factor. Thus if we define a modified Reynolds number Re' and a friction coefficient k_f by

$$Re' = \frac{2}{3\lambda}\frac{1}{(1-\varepsilon)}\left(\frac{u_m}{\nu}D_s\right), \tag{21.5.5}$$

$$k_f = \frac{1}{3}\frac{\varepsilon^3}{\lambda(1-\varepsilon)}\frac{D_s}{l}\frac{\Delta p}{\rho u_m^2}, \tag{21.5.6}$$

we should expect k_f to be a function of Re' only. It must, however, be appreciated that the use of these modified coefficients is not universal, some authors preferring to omit the factors $\frac{2}{3}$ and $\frac{1}{3}$.

Substituting into the Carman–Kozeny equation (21.4.6) we find that

$$k_f = \frac{40}{Re'}. \tag{21.5.7}$$

This equation is, however, only applicable in the viscous flow regime. At higher Reynolds numbers inertial effects become important and the pressure gradient becomes proportional to ρu_m^2. The friction coefficient k_f thus tends to a constant value which is found empirically to be about 0.58. The empirical equation

$$k_f = \frac{40}{Re'} + 0.58 \tag{21.5.8}$$

based on the work of Ergun fits the experimental results over a wide range of Reynolds numbers. For spherical particles this can be re-arranged to give

$$\frac{\Delta p}{l} = \frac{180\,\mu u_m(1-\varepsilon)^2}{D_s^2\varepsilon^3} + \frac{1.75\,\rho\mu_m^2(1-\varepsilon)}{D_s\varepsilon^3}. \tag{21.5.9}$$

The general form of the Ergun equation is shown in figure 183 and it is clear that for $Re < c10$ the departure from the viscous relationship given by equation (21.5.7) is negligible.

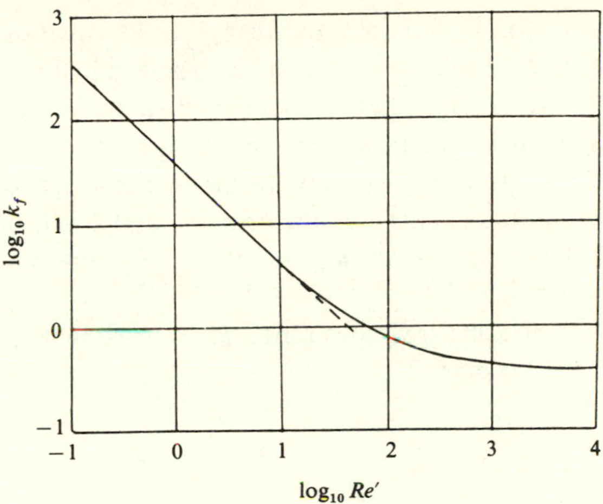

Fig. 183. Dependence of the friction factor k_f on the Reynolds number *Re*.

21.6 Fluidised solids

Starting with the case of flow of a fluid, gas or liquid, upwards through a static packed bed, with mean approach velocity u_m, as illustrated diagrammatically in figure 184, consider what happens if the flow rate is increased. The pressure difference Δp will be related to the superficial

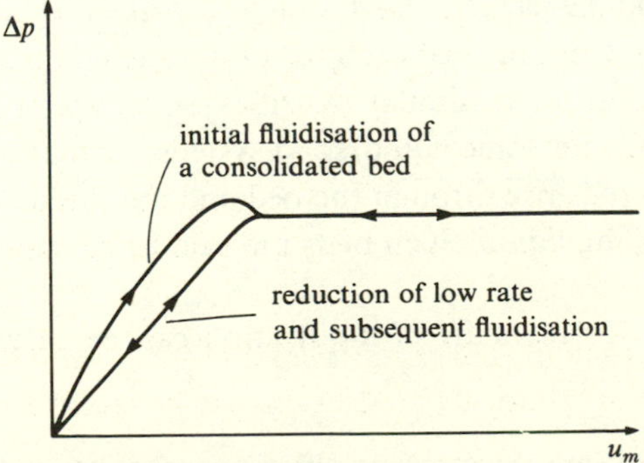

Fig. 184. Dependence of the pressure drop Δp on the velocity u_m through a fluidised bed.

velocity u_m by equation (21.5.6), i.e.

$$\Delta p = 3k_f\lambda \frac{(1-\varepsilon)}{\varepsilon^3}\frac{l}{D_s}\rho u_m^2, \tag{21.6.1}$$

where k_f is a function of the Reynolds number given by equation (21.5.5).

The pressure difference will increase with increasing fluid flow rate until it becomes equal to the weight of the particles per unit area. At this stage the particles are supported by the flow and the bed is said to be fluidised. Further increase in the flow rate causes no further increase in the pressure drop though some variation in the pressure drop is found in velocities close to the minimum fluidising velocity u_{mf} as shown in figure 184, due to wedging of the particles.

Usually the flow is within the viscous range, $Re' < 10$, so that $k_f = 40/Re'$ and equation (21.6.1) becomes

$$\Delta p = 180\frac{\lambda^2 \mu u_m (1-\varepsilon)^2}{D_s^2 \varepsilon^3}. \tag{21.6.2}$$

Under these circumstances Δp increases linearly with u_m up to u_{mf}. At higher Reynolds numbers, as will occur for example with larger particles, the inertial terms become important and the relationship between Δp and u_m becomes concave upwards. More commonly, however, the bed expands slightly as the fluidisation point is approached and with decreasing ε the relationship between Δp and u_m becomes concave downwards. Under these circumstances the bed does not normally re-consolidate on decreasing the flow rate, so that hysteresis in the relationship between Δp and u_m may be found as shown in figure 184.

The behaviour of the bed at velocities above minimum fluidisation depends primarily upon the ratio of the densities of the particles and the fluid. When the fluid and the particles have similar densities, as is often the case with liquid/solid systems, the bed expands uniformly and is said to be particulately fluidised, but for dissimilar densities as occurs in gas/solid systems and exceptionally for some liquid/solid systems such as water and lead shot, particle-free voids rise through the bed and the surface takes on the appearance of a boiling liquid. Such beds are said to be aggregatively fluidised.

The critical velocity for the start of fluidisation can be calculated as follows.

Under fluidising conditions the pressure difference must be related to the weight of solid particles supported by the fluid by

$$\Delta p A = (\rho_s - \rho) g (1 - \varepsilon) A l$$

or

$$\Delta p = (\rho_s - \rho)(1 - \varepsilon) g l, \tag{21.6.3}$$

where ρ_s is the density of the solid and ρ is the density of the fluid.

Up to the point of minimum fluidisation, however, the pressure difference is given by equation (21.6.1) and hence eliminating Δp the minimum fluidising velocity u_{mf} is given by

$$u_{mf}^2 = \frac{(\rho_s - \rho) g D_s}{3 \rho \lambda k_f} \varepsilon_{mf}^3, \tag{21.6.4}$$

where ε_{mf} is the void fraction at minimum fluidisation. If, however, the flow is viscous up to the point of minimum fluidisation, we can use equation (21.6.2) instead of the more general equation (21.6.1), giving

$$u_{mf} = \frac{(\rho_s - \rho) g D_s^2 \varepsilon_{mf}^3}{180\, \mu \lambda^2 (1 - \varepsilon_{mf})}. \tag{21.6.5}$$

The difficulty with the use of equation (21.6.4) and (21.6.5) lies in the uncertainty of the value of ε_{mf}. For spherical particles the loosest stable packing is the cubic array for which $\varepsilon = 0.476$. Using this value in equation (21.6.5) yields

$$u_{mf} = 0.001\,14\, g D_s^2 (\rho_s - \rho)/\mu. \tag{21.6.6}$$

However, this result tends to over-estimate u_{mf} and the empirical results,

$$u_{mf} = 0.000\,81\, g D_s^2 (\rho_s - \rho)/\mu \tag{21.6.7}$$

due to Rowe, and

$$u_{mf} = 0.0007\, Re_{mf}^{-0.063}\, g D_s^2 (\rho_s - \rho)/\mu, \tag{21.6.8}$$

where $Re_{mf} = \rho u_{mf} D_s/\mu$, due to Leva, are to be preferred.

For fluid velocities slightly above the minimum fluidising velocity the bed has a well-defined top surface. As the fluid velocity approaches the terminal settling velocity u_t given by equation (21.3.7), entrainment takes place and the bed loses its well-defined surface. In a bed of mixed particles the entrainment is gradual, the smaller particles being carried away at lower velocities. However, even those particles with terminal velocities below the mean approach velocity are not entrained immediately. Only those particles that are close to the top surface are carried away and the rate of entrainment is controlled by the mixing processes within the bed.

21.7 Expansion of fluidised beds

In a particulately fluidised bed the superficial velocity of the fluid u_m equals the terminal settling velocity of the particles in the bed. Hence from the Richardson & Zaki correlation, equation (21.3.11),

$$u_m = u_t \varepsilon^n,$$

and the height of the bed H can be calculated from the known weight W of the particles

$$W = \rho_s(1-\varepsilon)AH.$$

The bed height is seen to increase steadily with increasing flow rate and tends to infinity as u_m approaches the terminal settling velocity of an individual particle u_t.

The mechanism for bed expansion in an aggregatively fluidised bed is completely different. Instead of a uniform expansion of the bed, regions devoid of particles, known as bubbles, rise through a fluidised phase. Over a considerable range of velocities the behaviour can best be described in terms of the two-phase theory of fluidisation, which postulates that all the fluid in excess of that required for minimum fluidisation passes in the form of bubbles through a particulate phase that has the same void fraction as the bed at minimum fluidisation, ε_{mf}.

Since the void fraction within the particulate phase is constant, its volume must also be constant and the fraction of the bed occupied by bubbles ε_b is given by

$$\varepsilon_b = \frac{(H - H_{mf})}{H}, \tag{21.7.1}$$

where H_{mf} is the bed height at minimum fluidisation.

If the bubble rise velocity is u_A, the quantity of fluid passing through as bubbles is $A\varepsilon_b u_A$ and this is equal to the excess flow rate beyond minimum fluidisation. Hence

$$\varepsilon_b = \frac{H - H_{mf}}{H} = \frac{u_m - u_{mf}}{u_A}. \tag{21.7.2}$$

If the bubbles are not too large, they behave as bubbles in an inviscid fluid and by analogy with equation (19.3.7) the rise velocity u_b is given by

$$u_b = 0.716\sqrt{(gD_E)}, \tag{21.7.3}$$

where D_E is the diameter of the sphere whose volume is that of the bubble.

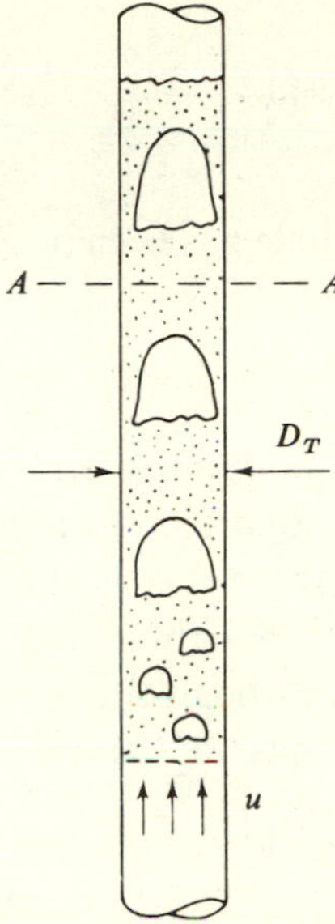

Fig. 185. Slug flow in a fluidised bed.

The slight difference in the constants in equations (21.7.3) and (19.3.7) reflects somewhat different bubble shapes in the two cases, probably due to the absence of surface tension in the fluidised bed. For bubbles in which D_E is comparable with or greater than the tube diameter D_T, wall effects predominate and the bubbles take the form of slugs as shown in figure 185. Equation (19.5.1) applies in this case, i.e.

$$u_b = 0.35\sqrt{(gD_T)}. \tag{21.7.4}$$

Since equation (21.5.3) applies for small bubbles and (21.5.4) for large bubbles, it is appropriate to take the smaller of the two predicted values of u_b.

In the case of a slugging bed the velocity u_b is the velocity of the bubble relative to the particles ahead of it. The particle velocity can be found by considering a control surface such as AA passing between bubbles as shown in figure 185. Particles cross this surface with velocity u_p, and since this part of the bed is in a state of minimum fluidisation the gas velocity exceeds the particle velocity by $u_e = u_{mf}/\varepsilon_{mf}$. Thus the volumetric flow rate across AA is

$$u_p A(1-\varepsilon_{mf}) + \left(u_p + \frac{u_{mf}}{\varepsilon_{mf}}\right) A\varepsilon_{mf},$$

which must equal the total volumetric inflow to the bed $u_m A$. Hence

$$u_p = u_m - u_{mf} \tag{21.7.5}$$

and the absolute rise velocity of a slug is given by

$$u_A = u_b + u_p = u_m - u_{mf} + 0.35\sqrt{(gD_T)}. \tag{21.7.6}$$

A similar correction in the case of a small bubble has been proposed by Farrakhalaee[5], so that in this case,

$$u_A = u_m - u_{mf} + 0.716\sqrt{(gD_E)}. \tag{21.7.7}$$

The arguments leading to this expression are not simple and reference should be made to the original work. However, the correction in the case of small bubbles is relatively unimportant since the value of D_E is rarely known to much precision. In a typical bed not only are the bubbles of a range of sizes but the bubbles tend to grow as they rise through the bed.

In a wide bed where $D_T \gg D_E$ we can use equations (21.7.2) and (21.7.7) to give the bed height H

$$\frac{H - H_{mf}}{H} = \frac{u_m - u_{mf}}{u_m - u_{mf} + 0.716\sqrt{(gD_E)}} \tag{21.7.8}$$

or

$$\frac{H - H_{mf}}{H_{mf}} = \frac{(u_m - u_{mf})}{0.716\sqrt{(gD_E)}}. \tag{21.7.9}$$

The top surface is, however, agitated by the continual bursting of bubbles and H must be interpreted as the average height. For narrower beds with $D_T < D_E$ slugs are formed and the top surface rises and falls in a cyclic manner. Between bubbles the bed surface rises with velocity u_p but when a bubble reaches the surface the particles surrounding it fall freely under gravity as shown in figure 186. If one considers the behaviour of the first bubble following a sudden increase in gas flow rate from the minimum fluidising flow to some higher flow, one can readily show that equation (21.7.2) gives the maximum bed height H_{max}. Thus for a slugging bed

$$\frac{H_{max} - H_{mf}}{H_{max}} = \frac{u_m - u_{mf}}{u_m - u_{mf} + 0.35\sqrt{(gD_T)}} \tag{21.7.10}$$

or

$$\frac{H_{max} - H_{mf}}{H_{mf}} = \frac{u_m - u_{mf}}{0.35\sqrt{(gD_T)}}. \tag{21.7.11}$$

21.8 Heat and mass transfer in fluidised beds

Two different aspects of heat transfer may arise in a fluidised bed. In some cases internal heating or cooling of the bed is carried out by means

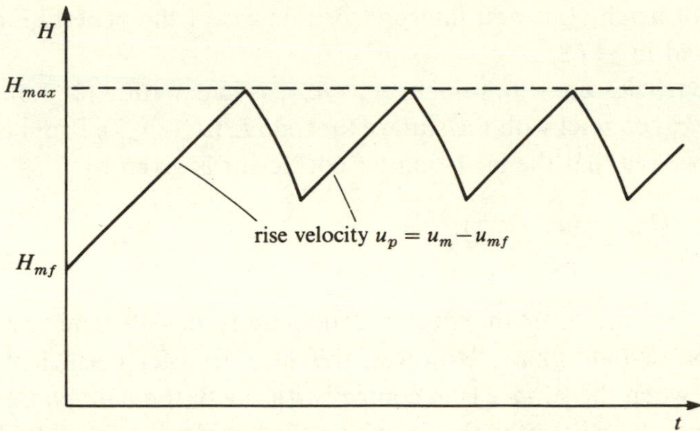

Fig. 186. Fluctuations in the height of a slugging fluidised bed.

of tubes passing through the bed and carrying some fluid. In other cases, if a reaction is taking place, the heat of reaction is removed from or supplied to the bed by the gas stream itself. In the first case the problem is that of heat transfer between the gas solid mixture and the surface of the tubes or the wall of the bed, while in the second case the problem is of heat transfer between the gas and the solid particles in the bed.

For an individual particle in a fluidised bed the heat transfer coefficient between the particle and the gas can be expressed by

$$Nu = \frac{hD_p}{k} = f(Re, Pr), \tag{21.8.1}$$

where Re is some suitably defined Reynolds number such as $D_p u_t/\nu$. This correlation can often be put in the form

$$Nu = 2 + C\,Re^a\,Pr^{\frac{1}{3}}, \tag{21.8.2}$$

where C and a may be treated as constants for a reasonable range. However, except for small particles, for which Re is low and $Nu \simeq 2$, the heat transfer coefficient tends to be large so that thermal equilibrium is rapidly attained and the experimental evaluation of C and a is difficult. Thus for all beds of depth more than a few centimetres the gas in the particulate phase may be assumed to be at the same temperature as the particles. By contrast, the gas in the bubble phase is not in contact with the particles and need not be at the same temperature. Consequently the mean temperature of the gas leaving the bed may differ from that of the particles. This effect is even more important for mass transfer and is considered in greater detail below.

The heat transfer coefficient from a slugging bed to a short vertical surface of height L is best interpreted in terms of the penetration model considered in § 17.8.

The particles rise with velocity $u_m - u_{mf}$, as in equation (21.7.5), and are therefore in contact with the surface for time $L/(u_m - u_{mf})$. From equation (17.8.6) we see that the heat transfer coefficient is given by

$$2\left\{\frac{k_p\rho_p c_p(u_m - u_{mf})}{\pi L}\right\}^{\frac{1}{2}},$$

where k_p, ρ_p, and c_p are the effective conductivity, density, and specific heat of the particulate phase. However, this heat transfer coefficient is only relevant when the surface is in contact with the particulate phase and the heat transfer coefficient to the gas within the bubbles is negligible. The heat transfer coefficient must therefore be multiplied by the fraction of time that the surface is in contact with the particulate phase. Thus the average heat transfer coefficient is given by

$$h = 2\left\{\frac{k_p\rho_p c_p(u_m - u_{mf})}{\pi L}\right\}^{\frac{1}{2}}\left(\frac{u_b}{u_m - u_{mf} + u_b}\right), \tag{21.8.3}$$

where $u_b = 0.35\,\sqrt{(gD_T)}$.

Equation (21.8.3) shows that h is independent of D_p and almost independent of u_m and D_T provided $u_m > 2u_{mf}$. Experimental measurements confirm these predictions and air-fluidised beds of sand have an almost constant value of 350 W m^{-2} K^{-1} for heat transfer to horizontal tubes and 400 W m^{-2} K^{-1} for heat transfer to vertical surfaces. For other materials these values should be multiplied by the ratio of the values of $(k_p\rho_p c_p)^{\frac{1}{2}}$.

Fluidised beds are commonly used to facilitate a chemical reaction between the gas and the particles and for this purpose the particles are often of small diameter. By analogy with equation (21.8.2) the mass transfer coefficient between the particles and the gas can be correlated as follows

$$Sh = \frac{k_g D_p}{\mathscr{D}} = 2 + C\,Re^a\,Sc^{\frac{1}{3}}, \tag{21.8.4}$$

but, for fine particles, $Sh = 2$ is usually accurate enough. However, in the presence of chemical reaction there may be a considerable net flux to or from the particle and the analysis for diffusion with net flux given in § 14.3 is appropriate. Care must also be taken to ensure that the correct chemical model is used. For example, in the combustion of a coal particle the reaction on the particle surface can be

$$C + O_2 \rightarrow CO_2,$$

but under other circumstances the reaction on the surface is

$$C+CO_2 \rightarrow 2CO$$

and the subsequent reaction

$$2CO+O_2 \rightarrow 2CO_2$$

takes place at greater distances from the particle.

When a fluidised bed is used to promote the reaction between a gas and the particles, the conversion within the particulate phase is much greater than that in the bubble phase where there is little or no contact between the gas and the particles. This by-passing via the bubble phase lowers the overall conversion and is of fundamental importance in the design of fluidised bed reactors. The two-phase theory of fluidisation enables this situation to be analysed as follows.

There is a net flow of gas Q' through a bubble and this quantity can be predicted by potential flow analysis. The flow of gas through a spherical bubble turns out to be three times the flow through a comparable area of bed, i.e.

$$Q' = \frac{3\pi}{4} D_E^2 u_{mf}. \tag{21.8.5}$$

For a slugging bed the expression

$$Q' = \frac{\pi}{4} D_T^2 u_{mf} \tag{21.8.6}$$

is believed to be more appropriate. Gas enters the bubble with concentration of the reacting species C_p which is the concentration within the particulate phase at the current height of the bubble. If we denote by C_b the concentration within the bubble at this height, we see that there is a net outflow of reactant equal to

$$Q'(C_b - C_p)$$

due to throughflow. Besides this there will be a loss due to diffusion across the bubble boundary at a rate

$$k_g A(C_b - C_p),$$

where k_g is an appropriate mass transfer coefficient and A is the surface area of the bubble.

Thus the total outflow is

$$(Q' + k_g A)(C_b - C_p) = Q(C_b - C_p), \tag{21.8.7}$$

where Q is simply the sum $Q'+k_gA$. The mass transfer coefficient k_g is not easy to evaluate but fortunately the term Q' often dominates and it is usually sufficiently accurate to assume that

$$Q=\frac{3\pi}{4}D_E^2u_{mf} \quad \text{or} \quad \frac{\pi}{4}D_T^2u_{mf}.$$

A balance on reactant within the bubble gives

$$V\frac{dC_b}{dt}=-Q(C_b-C_p), \tag{21.8.8}$$

where V is the bubble volume. Recalling that the absolute bubble rise velocity is u_A this can be re-expressed as

$$\frac{dC_b}{dz}=-\frac{Q}{Vu_A}(C_b-C_p), \tag{21.8.9}$$

where z is the vertical co-ordinate.

In principle both C_b and C_p are functions of z but in many cases C_p can be treated as a constant. Not only is there efficient mixing within the particulate phase due to the stirring effect of the bubbles but also reactions tend to be rapid within the particulate phase so that C_p is often very much less than C_b.

Taking C_p as constant we can integrate equation (21.8.9) to give

$$\log_e\left(\frac{C_b-C_p}{C_0-C_p}\right)=-\frac{Qz}{Vu_A}, \tag{21.8.10}$$

where C_0 is the concentration in the incoming stream which is also the concentration in the bubble at $z=0$.

Thus the concentration within the bubble phase at the top of the bed C_{bH} is given by

$$C_{bH}=C_p+(C_0-C_p)\,e^{-X}, \tag{21.8.11}$$

where $X=QH/Vu_A$ and is commonly known as the cross-flow factor. It represents the number of times the bubble is swept through by the gas during its passage through the bed. Substitution from equation (21.7.2) shows that X is also given by

$$X=\frac{QH_{mf}}{Vu_b}. \tag{21.8.12}$$

The concentration in the particulate phase can be found from a mass balance. For unit area of bed we have, inflow of reactant through the base $u_{mf}C_0$,

inflow of reactant from the bubbles $(u_m - u_{mf})(C_0 - C_{bH})$,

outflow from the top $u_{mf}C_p$.

The net inflow must equal the rate of reaction which for a first-order reaction will be $H(1-\varepsilon_b)kC_p$ where k is the reaction velocity constant. This can also be written as $H_{mf}kC_p$ since the volume of the particulate phase is the same as the bed volume at minimum fluidisation.

Summing these quantities gives

$$u_{mf}C_0 + (u_m - u_{mf})(C_0 - C_{bH}) = (u_{mf} + H_{mf}k)C_p. \tag{21.8.13}$$

The mean concentration at the outlet from the bed is the weighted average of C_{bH} and C_p, i.e.

$$u_mC_H = (u_m - u_{mf})C_{bH} + u_{mf}C_p. \tag{21.8.14}$$

Elimination of C_{bH} and C_p from equations (21.8.11), (21.8.13), and (21.8.14) gives

$$\frac{C_H}{C_0} = \beta\,\mathrm{e}^{-X} + \frac{(1-\beta\,\mathrm{e}^{-X})^2}{k'+1-\beta\,\mathrm{e}^{-X}}, \tag{21.8.15}$$

where $\beta = (u_m - u_{mf})/u_m$ and $k' = kH_{mf}/u_m$.

It can be seen that even for a very fast reaction for which $k \to \infty$, the least value of the outlet concentration is

$$C_0\beta\,\mathrm{e}^{-X}$$

and this represents an upper limit on the possible conversion caused by the by-passing of the gas in the bubble phase.

References

[1] Clift, Grace, & Weber, *Bubbles, Drops and Particles*, Acad. Press.

[2] Heywood, *Symposium on the Interaction between Fluids and Particles, Inst. Chem. Eng.*, 1962.

[3] Richardson & Zaki, *Trans. Inst. Chem. Eng.*, **32**, 35 (1954).

[4] Wong, *Heat Transfer for Engineers*, Longmans.

[5] Farrakhalaee, *Gas Distribution and Bubble Motion in Fluidized Beds*, Ph.D. Thesis, Cambridge.

22

Mass transfer with chemical reaction

22.1 Introduction

The usual method of removing an unwanted component from a gas stream is by absorption into some liquid. When the solubility of the absorbed species is high, the straightforward physical absorption processes described in §§ 18.6 to 18.12 can be used. However, when there is no suitable solvent it is sometimes possible to increase the absorption capacity by using a liquid with which the component reacts. Perhaps the commonest case is the removal of CO_2 from a gas stream. The solubility in water is too low for physical absorption to be convenient and absorption into hot aqueous K_2CO_3 may be used. The reaction

$$K_2CO_3 + H_2O + CO_2 \rightleftharpoons 2KHCO_3$$

takes place within the liquid, thereby increasing its capacity for CO_2. Other cases of importance are the absorption of SO_2 and H_2S and for these solutions of amines, particularly diethanolamine (DEA) and monoethanolamine (MEA), can be used.

The performance of a porous catalyst pellet can be analysed by methods similar to those used in gas absorption with chemical reaction and it is therefore appropriate to consider both situations together. In the case of a catalyst pellet the reactants diffuse into the pores where they react and the products diffuse out counter-current to the reactants. Such catalysts may be inherently porous such as activated carbon or the zeolites, or in other cases reactive components such as platinum salts are deposited within the pores of a ceramic support.

Both in gas absorption and in catalysis we have diffusion coupled with chemical reaction and this process is governed by equation (14.6.3),

$$\frac{\partial C}{\partial t} = \mathscr{D} \nabla^2 C + R, \tag{22.1.1}$$

where C is the concentration of the diffusing species and R is the rate of creation of that species per unit volume.

The analogous problem of heat conduction with heat generation has been considered in §§13.4 and 13.5. Here, however, the rate of heat generation Q_s is usually determined by factors other than the temperature distribution and the dependence of Q_s on position is generally known *ab initio*. In the mass transfer case the rate of chemical reaction is normally a function of the concentration and the various orders of reaction each require special treatment. The simple case of first-order reaction is considered in §22.2, *n*th-order reactions in §22.3, and zeroth-order reactions in §22.4. The industrially important case when the reaction is first order with respect to both the absorbed species and the absorbant in considered in §22.5 and the analysis of porous catalyst pellets is considered in §§22.6 and 22.7.

22.2 Gas absorption with first-order chemical reaction

As with physical absorption, gas absorption with chemical reaction is often analysed in terms of the film model. The deficiencies of this model have been discussed in §17.7 and all that is necessary at this stage is to repeat the conclusion that the errors in the film model arise mainly from trying to predict the effective laminar layer thickness from considerations of momentum transfer. The effective laminar layer thickness for mass transfer is found to be insensitive to the nature of the diffusing species and we will assume that it is equal to that found for physical absorption. Thus we can use equation (17.2.5),

$$\delta = \frac{\mathscr{D}}{k_L^0} \tag{22.2.1}$$

to predict the effective laminar layer thickness δ from the experimentally measured value of the mass transfer coefficient k_L^0 for physical absorption of the same species.

In the analysis below we will predict the rate of mass transfer N from a gas into a turbulent film of liquid of thickness Δ flowing down a vertical surface. As in the Whitman analysis, discussed in § 18.8, we will assume that there is an effective laminar layer of thickness δ separating the interface from the well-mixed bulk as shown in figure 187. In most of this chapter we will assume that δ is very much less than the total film thickness Δ. We will denote the concentration at the interface by C_i and, in the absence of mass transfer resistance in the gas, this will be in equilibrium with the partial

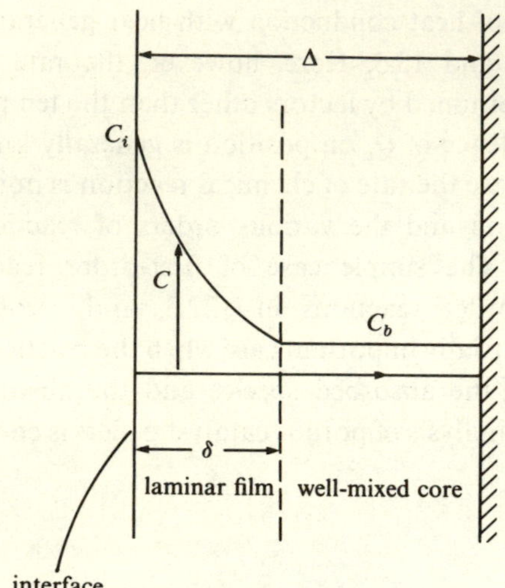

Fig. 187. Concentration profile in a liquid film.

pressure p of the soluble species. Hence,

$$\text{at } x=0, \quad C=C_i=\frac{p}{H}, \tag{22.2.2}$$

where H is the Henry's law constant. If mass transfer within the gas cannot be neglected, the interfacial concentration C_i must be predicted from equations (18.6.19) and (18.6.21).

Within the well-mixed bulk the diffusing species will have concentration C_b but in most cases the reaction within the bulk will reduce this to some small value, particularly when $\Delta \gg \delta$. Thus we can normally say that

$$\text{at } x=\delta, \quad C=0. \tag{22.2.3}$$

We will re-examine this assumption at the end of this section.

For steady uni-directional diffusion through a stagnant film, equation (22.1.1) takes the form,

$$\mathscr{D}\frac{\mathrm{d}^2C}{\mathrm{d}x^2}+R=0, \tag{22.2.4}$$

and for first-order destruction of the diffusing species, $R=-kC$ where k is the reaction velocity constant. Hence

$$\mathscr{D}\frac{\mathrm{d}^2C}{\mathrm{d}x^2}=kC. \tag{22.2.5}$$

The solution to this equation can be expressed either as exponentials or as hyperbolic functions. The latter are more convenient for the present purpose and we obtain

$$C = A \sinh \sqrt{\left(\frac{k}{\mathscr{D}}\right)}x + B \cosh \sqrt{\left(\frac{k}{\mathscr{D}}\right)}x. \tag{22.2.6}$$

Putting in the two boundary conditions, $x=0$, $C=C_i$ and $x=\delta$, $C=0$ (equations (22.2.2) and (22.2.3)), we can evaluate the arbitrary constants A and B giving

$$C = C_i\left\{\cosh \sqrt{\left(\frac{k}{\mathscr{D}}\right)}x - \coth \sqrt{\left(\frac{k}{\mathscr{D}}\right)}\delta \sinh \sqrt{\left(\frac{k}{\mathscr{D}}\right)}x\right\}. \tag{22.2.7}$$

Having evaluated the concentration profile we can now proceed to calculate the absorption rate N, since $N = -\mathscr{D}(\mathrm{d}C/\mathrm{d}x)_{x=0}$. Hence from equation (22.2.7),

$$N = \mathscr{D}C_i \sqrt{\frac{k}{\mathscr{D}}} \coth \sqrt{\left(\frac{k}{\mathscr{D}}\right)}\delta. \tag{22.2.8}$$

It is seen that the gas absorption rate is directly proportional to the interfacial concentration and this result can therefore be expressed in terms of a mass transfer coefficient defined by

$$N = k_L C_i. \tag{22.2.9}$$

Thus

$$k_L = \sqrt{(k\mathscr{D})} \coth \sqrt{\left(\frac{k}{\mathscr{D}}\right)}\delta. \tag{22.2.10}$$

It is worth while at this stage to investigate certain special cases of this basic result. Let us first consider the case of a very slow reaction for which k will be small. Hence the group $\sqrt{(k/\mathscr{D})}\delta$ will also be small and, since $\coth \theta \simeq 1/\theta$ for small θ, equation (22.2.10) becomes

$$k_L \simeq \frac{\mathscr{D}}{\delta} = k_L^0. \tag{22.2.11}$$

Thus the mass transfer coefficient reduces to that measured for physical absorption.

On the other hand, for fast reactions $\sqrt{(k/\mathscr{D})}\delta$ will be large and, since $\coth \theta \simeq 1$ for large θ, equation (22.2.10) becomes

$$k_L \sim \sqrt{(k\mathscr{D})} = k_L^f, \tag{22.2.12}$$

where k_L^f is the mass transfer coefficient for a fast reaction. In this case the mass transfer coefficient is independent of the film thickness δ and this is because the reaction is effectively complete within the film. Since the major deficiency of the film model is some uncertainty about the value of δ, the result for fast reactions is perhaps the most reliable prediction of this analysis and represents one of the few cases where mass transfer coefficients can be predicted with confidence from purely theoretical considerations.

We can use the result for fast reactions as the basis of a non-dimensional presentation of equation (22.2.10). The important group in that equation, $\sqrt{(k/\mathscr{D})}\delta$, may be re-written as k_L^f/k_L^0 and clearly represents the relative importance of chemical and physical absorption. It is commonly called the Hatta number and is usually defined by

$$Ha = \frac{\sqrt{(k\mathscr{D})}}{k_L^0}. \qquad (22.2.13)$$

We can also define an enhancement factor by the ratio of the mass transfer coefficient in the presence of chemical reaction to the mass transfer coefficient for physical absorption,

$$E = \frac{k_L}{k_L^0}. \qquad (22.2.14)$$

With these definitions, equation (22.2.10) becomes

$$E = Ha \coth Ha. \qquad (22.2.15)$$

The relationship between the enhancement factor and the Hatta number is shown in figure 188, from which it can be seen that

$$E \to 1 \quad \text{as } Ha \to 0,$$

$$E \simeq Ha \quad \text{for } Ha > 2.$$

The first of these two results is a dimensionless form of equation (22.2.11) and reflects the fact that for slow reaction the mass transfer coefficient is close to that for physical absorption and the second result is a dimensionless form of the prediction for fast reactions, equation (22.2.12).

We can now reconsider our assumption that the concentration C_b in the well-mixed core is small compared with the interfacial concentration C_i. Considering still the case when the total film thickness Δ is very much larger than the effective laminar layer thickness δ, we can say that the rate of reaction within the bulk, $kC_b(\Delta-\delta) \simeq kC_b\Delta$, must equal the rate of diffusion through the laminar layer which is given approximately by

$$\mathscr{D}\frac{(C_i - C_b)}{\delta}.$$

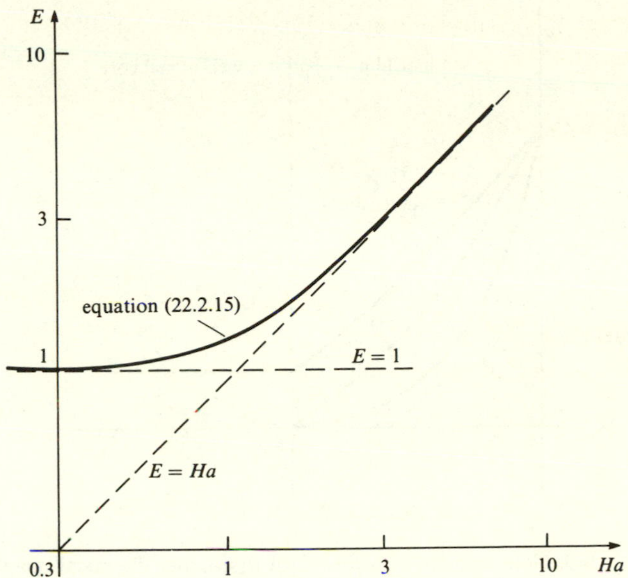

Fig. 188. Enhancement factor E as a function of Hatta number Ha.

Thus

$$\mathscr{D}\frac{(C_i - C_b)}{\delta} = kC_p\Delta$$

or

$$C_b = \frac{C_i}{1 + k\,\delta\Delta/\mathscr{D}} = \frac{C_i}{1 + k\Delta/k_L^0}. \tag{22.2.16}$$

Thus when $k > 9k_L^0/\Delta$, i.e. when $Ha > 3\surd(\delta/\Delta)$, $C_b < 0.1\,C_i$, and the analysis of the preceding parts of this section are valid. For smaller values of Ha, the concentration C_b is appreciable and the rate of absorption is controlled by the rate of chemical reaction within the bulk. Under these circumstances the absorption rate is given approximately by

$$N = kC_b\Delta = \frac{\mathscr{D}kC_i\Delta}{\mathscr{D} + k\,\Delta\delta}. \tag{22.2.17}$$

Thus it is appropriate to define two regimes of mass transfer with chemical reaction. When $Ha < 3\surd(\delta/\Delta)$ we have a *kinetic-controlled regime* in which the concentration C_b is an appreciable fraction of C_i and the absorption rate is given by equation (22.3.17). For $Ha > 3\surd(\delta/\Delta)$ we have a *diffusion-controlled regime*. The bulk concentration C_b is negligible and the enhancement factor is given by equation (22.2.15). This regime is sometimes

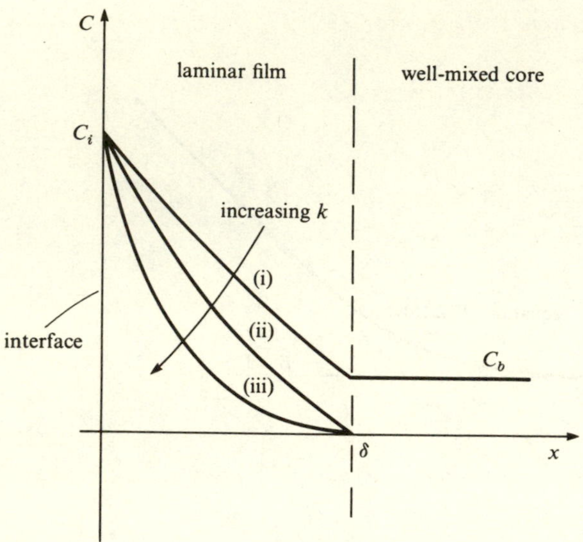

Fig. 189. Concentration profiles with first-order chemical reaction.

sub-divided into one with slow reaction ($Ha<2$) in which the full form of equation (22.2.15) is required and one with fast reaction ($Ha>2$) in which the simpler form

$$E=Ha \tag{22.2.18}$$

is sufficiently accurate.

The concentration profiles for these three cases are given in figure 189. Line (i) shows the profile for the kinetic-controlled regime. Most of the reaction takes place in the bulk and the concentration profile within the laminar film is therefore approximately linear. Line (ii) shows the profile for diffusion control with slow reaction. Reaction takes place both within the laminar film and the well-mixed bulk, though the concentration $C_b \simeq 0$. There is therefore an appreciable concentration gradient at $x=\delta$. The case of fast reaction is given by line (iii). The reaction is effectively complete within the laminar layer and the concentration gradient at $x=\delta$ is approximately zero.

For cases when Δ is not large compared with δ a more accurate result can be obtained by solving equation (22.2.5) subject to the boundary conditions,

$$\text{at} \quad x=\delta, \quad C=C_b, \quad \text{and} \quad -\mathscr{D}\frac{\mathrm{d}C}{\mathrm{d}x}=kC_b(\Delta-b).$$

Such cases are, however, rarely encountered in practice.

The last traces of CO_2 in a gas stream are often removed by absorption into aqueous NaOH. As we will see in §22.5, the process is effectively first order in CO_2 provided the concentration of NaOH is not too small. For 0.5 M NaOH, the reaction velocity constant k is $5\times10^3\ \mathrm{s^{-1}}$ and a typical value of k_L^0 is $1\times10^{-4}\ \mathrm{m\ s^{-1}}$. Since the diffusivity of CO_2 through water is $1.8\times10^{-9}\ \mathrm{m^2\ s^{-1}}$, this latter value implies a laminar layer thickness δ of $(1.8\times10^{-9})/10^{-4}=0.018$ mm which is small compared with a typical film thickness Δ of about 0.5 mm.

The Hatta number under these circumstances is given by $Ha=\surd(k\mathscr{D})/k_L^0=30$ and hence we are in the fast reaction regime so that from equation (22.2.18), $E=Ha=30$ and $k_L=Ek_L^0=3\times10^{-3}\ \mathrm{m\ s^{-1}}$.

If the absorbent had been a $K_2CO_3/KHCO_3$ solution, the reaction velocity constant would have been about $10\ \mathrm{s^{-1}}$ and the Hatta number 1.34. Under these circumstances the full form of equation (22.2.15) is required, giving

$$E=1.34\ \coth 1.34=1.54$$

and

$$k_L=Ek_L^0=1.54\times10^{-4}\ \mathrm{m\ s^{-1}}.$$

In both cases the Hatta number is considerably more than the critical value of $3\surd(\delta/\Delta)=0.57$ for kinetic control and we can therefore deduce that the CO_2 concentration in the bulk is negligible.

22.3 Gas absorption with *n*th-order reaction

Though cases in which the rate of reaction is first order with respect to the diffusing species are by far the most common, other orders of reaction do occur and the case of the general *n*th-order reaction is considered briefly in this section. It must, however, be pointed out that the zeroth-order reaction requires special treatment and this is considered in §22.4 below.

For *n*th-order destruction of the diffusing species,

$$R=-kC^n, \tag{22.3.1}$$

and using the model and assumptions of the previous section we have

$$\mathscr{D}\frac{\mathrm{d}^2C}{\mathrm{d}x^2}=kC^n. \tag{22.3.2}$$

This equation must be solved subject to the boundary conditions

at $x=0, \quad C=C_i,$

at $x=\delta, \quad C=0.$

The solution of equation (22.3.2) is facilitated by multiplying by $\mathrm{d}C/\mathrm{d}x$ and integrating, giving

$$\frac{\mathscr{D}}{2}\left(\frac{\mathrm{d}C}{\mathrm{d}x}\right)^2=\frac{kC^{n+1}}{n+1}+A. \tag{22.3.3}$$

The arbitrary constant A can be found since at $x=0$, $C=C_i$ and $N=-\mathscr{D}(\mathrm{d}C/\mathrm{d}x)$. Thus

$$A=\frac{N^2}{2\mathscr{D}}-\frac{kC_i^{n+1}}{n+1} \tag{22.3.4}$$

and hence on re-arranging equation (22.3.3) we have,

$$\int_0^\delta \mathrm{d}x=\int_{C_i}^0 \frac{\mathrm{d}C}{\left\{\dfrac{2kC^{n+1}}{(n+1)\mathscr{D}}-\dfrac{2kC_i^{n+1}}{(n+1)\mathscr{D}}+\dfrac{N^2}{\mathscr{D}^2}\right\}^{\frac{1}{2}}}. \tag{22.3.5}$$

This last integral must normally be evaluated numerically and this is inconvenient since the value of the absorption rate N is as yet unknown.

However, for fast reactions a simpler result occurs since the reaction is effectively complete within the film and both C and $\mathrm{d}C/\mathrm{d}x$ are zero at $x=\delta$. Thus from equation (22.3.3), $A=0$, and the absorption rate is given by

$$N=-\mathscr{D}\left(\frac{\mathrm{d}C}{\mathrm{d}x}\right)_{x=0}=\sqrt{\left(\frac{2\mathscr{D}kC_i^{n+1}}{n+1}\right)}. \tag{22.3.6}$$

It is seen that this result reduces to equation (22.2.12) when $n=1$.

22.4 Gas absorption with zero-order reaction

Cases exist in which the destruction of the diffusing species is apparently of zero order. This usually occurs when the rate-determining step is first order with respect to some catalyst that is present in only small amounts. This is of particular importance in biochemical processes where the rate of reaction is controlled by the enzyme concentration. For such systems the Michaelis–Menton equation applies and this reduces to a zero-order equation at high substrate concentrations. Other cases of industrial importance include the oxidation of certain liquid hydrocarbons. Here the rate of reaction is determined by the concentration of free radicals and, according to van de Vusse, is effectively of zero order with respect to oxygen over a wide range of concentrations.

Care must, however, always be exercised when dealing with zero-order reactions since all such reactions must become first order with respect to the reacting species as the concentration falls to zero. If this were not so, the reaction would proceed indefinitely and the concentration would become negative. Thus for zero-order destruction we must say that the rate of reaction R has a constant value of $-R_0$ for $C>0$ but $R=0$ when $C=0$.

From equation (22.2.4) we have that

$$\mathscr{D}\frac{\mathrm{d}^2C}{\mathrm{d}x^2}=R_0, \tag{22.4.1}$$

and on integrating twice,

$$\mathscr{D}C=R_0\frac{x^2}{2}+\mathrm{A}\mathrm{x}+\mathrm{B}, \tag{22.4.2}$$

where the arbitrary constants A and B have to be evaluated from appropriate boundary conditions. When doing so we must, however, remember that a zero-order reaction ceases when the concentration falls to zero. Thus the usual boundary conditions of §22.2, i.e.

$$\text{at } x=0, \quad C=C_i,$$

$$\text{at } x=\delta, \quad C=0,$$

are only valid if the concentration is non-zero throughout the laminar film. If, however, the concentration falls to zero within the film at say, $x=z$, the reaction is complete and none of the reacting species diffuses through that plane. Under these circumstances we therefore have

$$\text{at } x=0, \quad C=C_i,$$

$$\text{at } x=z, \quad C=0 \quad \text{and} \quad \frac{\mathrm{d}C}{\mathrm{d}x}=0.$$

Evaluating the arbitrary constants in equation (22.4.2) from the former set of boundary conditions gives

$$A=-\left(\frac{\mathscr{D}C_i}{\delta}+\frac{R_0\delta}{2}\right) \tag{22.4.3}$$

and

$$B=\mathscr{D}C_i, \tag{22.4.4}$$

and from the latter set,

$$A=-\sqrt{(2R_0\mathscr{D}C_i)}, \tag{22.4.5}$$

$$B=\mathscr{D}C_i, \tag{22.4.6}$$

and

$$z=\sqrt{\left(\frac{2\mathscr{D}C_i}{R_0}\right)}. \tag{22.4.7}$$

Clearly the division between the two cases occurs when $z=\delta$ so that equations (22.4.3) and (22.4.4) apply when $\delta<\sqrt{(2\mathscr{D}C_i/R_0)}$ and equations (22.4.5) to (22.4.7) apply when $\delta>\sqrt{(2\mathscr{D}C_i/R_0)}$.

The absorption rate is given by $-\mathscr{D}(\mathrm{d}C/\mathrm{d}x)_{x=0}=-A$. Thus for

$$\delta<\sqrt{\left(\frac{2\mathscr{D}C_i}{R_0}\right)} \quad N=\frac{\mathscr{D}C_i}{\delta}+\frac{R_0\delta}{2}=k_L^0C_i+\frac{R_0\mathscr{D}}{2k_L^0} \tag{22.4.8}$$

and for

$$\delta>\sqrt{\left(\frac{2\mathscr{D}C_i}{R_0}\right)} \quad N=\sqrt{(2R_0\mathscr{D}C_i)}. \tag{22.4.9}$$

The concentration profiles for these two cases are shown in figure 190 and the rate of absorption N is plotted as a function of C_i in figure 191. For $C_i<R_0\delta^2/2D=R_0\mathscr{D}/2k_L^{0^2}$, the rate of absorption is proportional to $\sqrt{C_i}$ and for $C_i>R_0\delta^2/2\mathscr{D}=R_0\mathscr{D}/2k_L^{0^2}$, the absorption rate is linearly related to,

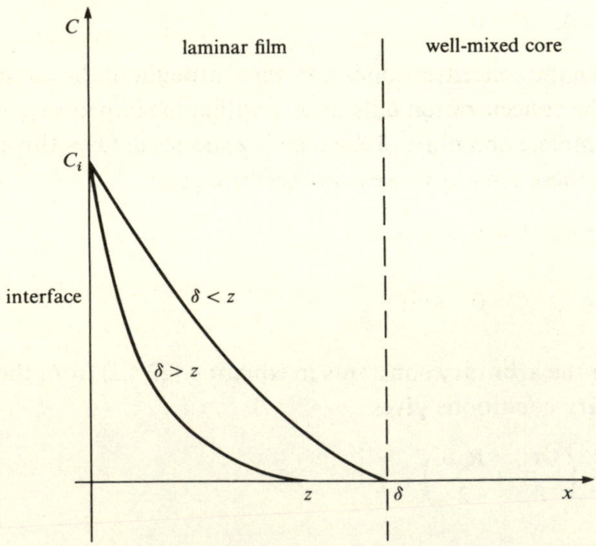

Fig. 190. Concentration profiles with zero-order chemical reaction.

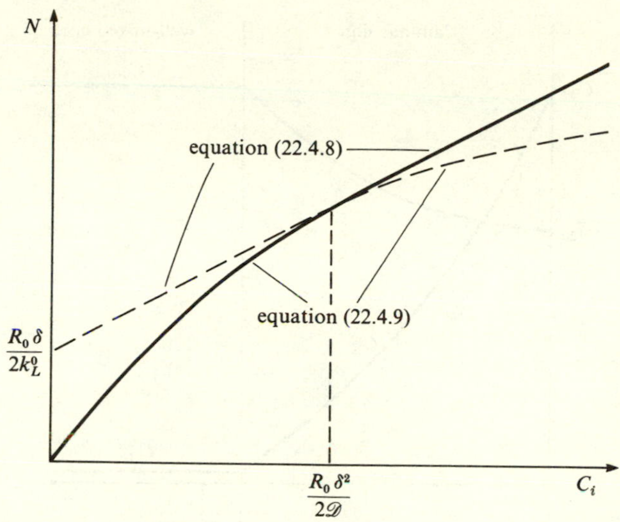

Fig. 191. Absorption rate N as a function of interfacial concentration C_1 for zero-order chemical reaction.

but not directly proportional to, C_i. Thus in neither case can we define an effective mass transfer coefficient and the rate of absorption is of complex order even though the inherent kinetics are of simple, i.e. zero, order.

22.5 Reactions that are first order in two components

Perhaps the commonest situation in industry is when the rate of reaction is first order with respect to both the species being absorbed and to some involatile reactant dissolved in the liquid. The absorption of CO_2 into a $K_2CO_3/KHCO_3$ solution is one such example and in this case the important reaction is

$$CO_2 + H_2O + CO_3^= = 2HCO_3^-.$$

Denoting the absorbed species by A and the involatile reactant by B, we can define a second-order rate constant k_2 by

$$R = -k_2 C_A C_B, \tag{22.5.1}$$

where R is the rate of creation of species A. Here we have given the reaction velocity constant the symbol k_2 to signify that it is a second-order constant with dimensions $m^3\ kmol^{-1}\ s^{-1}$.

Figure 192 shows typical concentration profiles within the film. The concentration C_A falls from C_{Ai} at the interface to the value of C_A within the

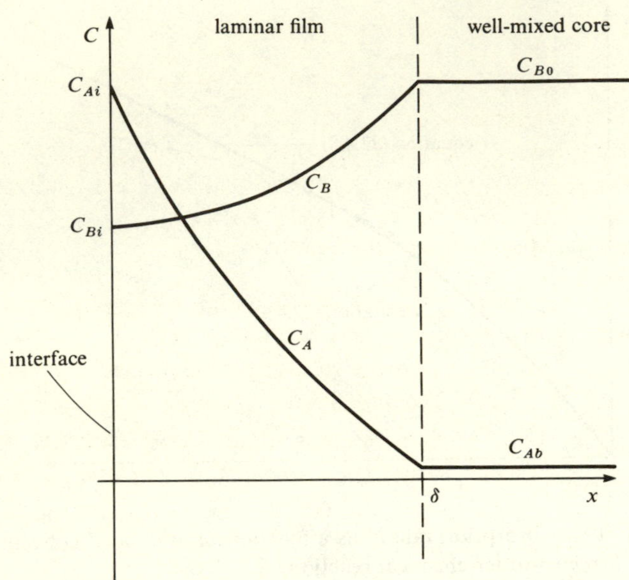

Fig. 192. Concentration profiles for reactions that are first order in two components.

bulk. As discussed in § 22.2, this can usually be taken as zero. The concentration of B in the bulk is denoted by C_{B0} and C_B falls from this value at $x=\delta$ as B diffuses into the film. We are, however, considering the case when B is involatile so that there is no flux of B through the interface and the concentration gradient at the interface is therefore zero.

The reaction between A and B is not necessarily equimolar and we must introduce a stoichiometric coefficient n so that

$$A+nB \rightarrow \text{involatile products}$$

Thus if the rate of creation of species A is R, the rate of creation of species B is nR and hence from equation (22.2.4)

$$\mathscr{D}_A \frac{\mathrm{d}^2 C_A}{\mathrm{d}x^2}=k_2 C_A C_B, \tag{22.5.2}$$

$$\mathscr{D}_B \frac{\mathrm{d}^2 C_B}{\mathrm{d}x^2}=nk_2 C_A C_B, \tag{22.5.3}$$

where $\mathscr{D}_A$ and $\mathscr{D}_B$ are the diffusivities of A and B through the solvent.

This pair of equations is to be solved subject to the boundary conditions

$$x=0, \quad C_A=C_{Ai}, \quad \frac{\mathrm{d}C_B}{\mathrm{d}x}=0, \tag{22.5.4}$$

$$x=\delta, \quad C_A=0, \qquad C_B=C_{B0}. \tag{22.5.5}$$

Unfortunately equations (22.5.2) and (22.5.3) are non-linear and no analytic solution is possible. Van Krevelen & Hoftijzer[1] have, however, solved them numerically and have presented the results graphically. These results will be discussed later, but first we will consider two limiting cases.

Case (i) Species B *in considerable excess*

If the concentration C_{B0} is sufficiently high, the fractional reduction in C_B across the film will be small and we may consider C_B to be effectively constant. Figure 193 illustrates the resulting concentration profiles. This approximation will only be acceptable if $C_{Bi} >$ about 0.9 C_{B0}.

If we consider C_B to be constant, equation (22.5.3) becomes irrelevant and equation (22.5.2) can be written

$$\mathscr{D}_A \frac{\mathrm{d}^2 C_A}{\mathrm{d}x^2} = k_2 C_{B0} C_A. \tag{22.5.6}$$

This is equation (22.2.5) with k replaced by $k_2 C_{B0}$ and the system behaves like one with first-order kinetics. This is the pseudo-first-order regime and by analogy with the analysis of § 22.2 we can say that the enhancement factor E is given by

$$E = Ha \coth Ha, \tag{22.5.7}$$

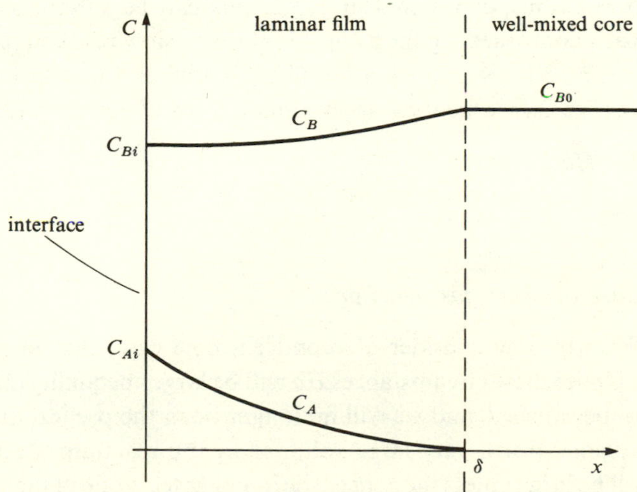

Fig. 193. Concentration profiles with pseudo-first-order chemical reaction.

where the Hatta number is now defined by

$$Ha = \frac{\sqrt{(k_2 C_{B0} \mathscr{D}_A)}}{k_L^0}. \tag{22.5.8}$$

The absorption rate is therefore $k_L^0 C_{Ai} Ha \coth Ha$ and the rate of reaction of B is n times this quantity.

We can now consider the circumstances under which our approximation that $C_B \simeq C_{B0}$ is valid. The rate of reaction of B must equal the rate at which B diffuses into the film, $\mathscr{D}_B(\mathrm{d}C_B/\mathrm{d}x)_{x=\delta}$ and this quantity is given approximately by $\mathscr{D}_B(C_{B0} - C_{Bi})/\delta$. Thus

$$\frac{\mathscr{D}_B(C_{B0} - C_{Bi})}{\delta} \simeq nk_L^0 C_{Ai} Ha \coth Ha. \tag{22.5.9}$$

Our approximation will be valid if $C_{Bi} > 0.9\ C_{B0}$ and therefore we must have that

$$\frac{\mathscr{D}_B C_{B0}}{\delta} > 10\ nk_L^0 C_{Ai} Ha \coth Ha$$

or

$$\frac{\mathscr{D}_B C_{B0}}{\mathscr{D}_A C_{Ai}} > 10\ nHa \coth Ha. \tag{22.5.10}$$

Thus if the ratio C_{B0}/C_{Ai} is large enough the analysis of case (i) will be valid and we will be in the pseudo-first-order reaction regime. As with genuine first-order reactions, considered in §22.2, this can be subdivided into a kinematically controlled regime if $Ha < 3\sqrt{(\delta/\Delta)}$, a slow reaction regime for $3\sqrt{(\delta/\Delta)} < Ha < 2$ for which equation (22.5.7) applies and a fast reaction regime for $Ha > 2$ in which the approximate form of equation (22.5.7),

$$E = Ha,$$

can be used.

Case (ii) Very fast reactions

We will now consider a second extreme case, that of very fast reactions. Under these circumstances Ha will be large, inequality (22.5.10) is unlikely to be satisfied, and we will no longer be in the pseudo-first-order reaction regime. For a very large value of k_2 the reaction rate per unit volume will be large unless the concentration of at least one of the species is small. In the limiting case as $k_2 \rightarrow \infty$ the two species cannot co-exist and the reaction is confined to a narrow reaction zone as shown in figure 194. Since

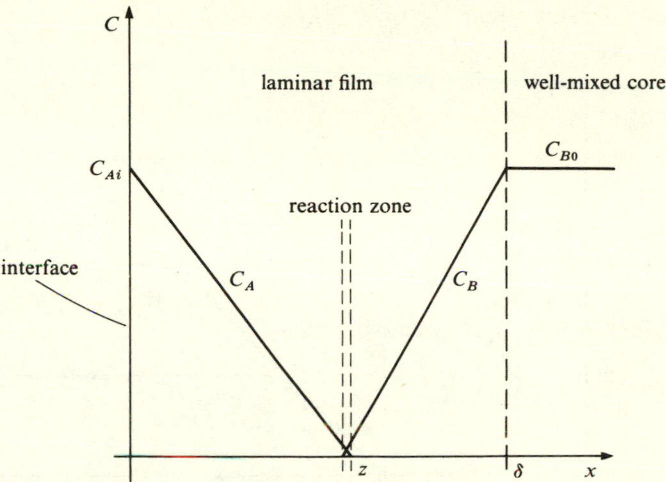

Fig. 194. Concentration profiles with instantaneous chemical reaction.

there is no reaction outside this zone the concentration profiles are linear and the fluxes are given by

$$N_A = \mathscr{D}_A C_{Ai}/z, \tag{22.5.11}$$

$$N_B = \mathscr{D}_B C_{B0}/(\delta - z), \tag{22.5.12}$$

where z is the distance of the reaction zone from the interface. As both species react to completion, $N_B = nN_A$ and hence

$$\frac{\mathscr{D}_B C_{B0}}{(\delta - z)} = \frac{n\mathscr{D}_A C_{Ai}}{z}$$

or

$$\frac{\delta}{z} = 1 + \frac{\mathscr{D}_B C_{B0}}{n\mathscr{D}_A C_{Ai}}. \tag{22.5.13}$$

From this

$$N_A = \frac{\mathscr{D}_A}{\delta} C_{Ai}\left(1 + \frac{\mathscr{D}_B C_{B0}}{n\mathscr{D}_A C_{Ai}}\right) \tag{22.5.14}$$

and the enhancement factor is given by

$$E_i = 1 + \frac{\mathscr{D}_B C_{B0}}{n\mathscr{D}_A C_{Ai}}. \tag{22.5.15}$$

Here the enhancement factor has been given the symbol E_i as it represents the limiting enhancement factor for an instantaneous reaction. It

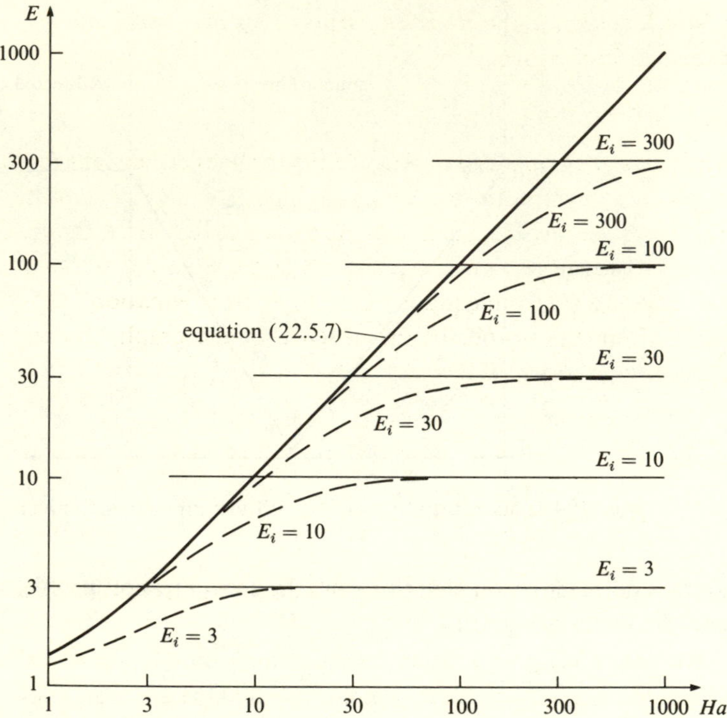

Fig. 195. Enhancement factors for gas absorption with chemical reaction that is first order in both components.

will be noted that E_i does not contain the reaction velocity constant k_2 as this has been assumed to be infinite and consequently E_i is not a function of the Hatta number.

Comparing equation (22.5.15) and inequality (22.5.10) we can say that the criterion for the validity of the pseudo-first-order reaction analysis is that

$$E_i - 1 > 10\, Ha \coth Ha \qquad (22.5.16)$$

and this gives an upper limit to the range of Hatta numbers for that regime.

The results of these two limiting cases may be presented graphically as shown in figure 195. Here the scales are logarithmic with Ha as the abscissa and E as the ordinate. The result of case (i), equation (22.5.7), is represented by a single line with asymptotes $E \to 1$ as $Ha \to 0$ and $E \to Ha$ as $Ha \to \infty$. Case (ii) is represented by a set of horizontal straight lines with E_i as a parameter.

Van Krevelen found that the results of his numerical analysis could be correlated in the form

$$E = f(Ha, E_i). \tag{22.5.17}$$

When presented on the axes of figure 195 the lines of constant E_i were found to become asymptotic to $E = E_i$ at high Ha and asymptotic to $E = Ha \coth Ha$ at low Ha. Such lines are shown dotted in the figure. Copies of the van Krevelen plot are available on a larger scale and all that is required is to evaluate Ha from equation (22.5.8), E_i from equation (22.5.15) and to read E from the graph. In the absence of the graph, the approximate equation due to van Krevelen,

$$E = H \coth H, \tag{22.5.18}$$

where

$$H = Ha \sqrt{\left(\frac{E_i - E}{E_i - 1}\right)}, \tag{22.5.19}$$

may be used. This is reported to agree with the numerical solution to within 10%.

We can now give further consideration to the numerical example presented at the end of § 22.2. There we treated the absorption of CO_2 into 0.5 M NaOH as first order in CO_2 with reaction velocity constant k of $5 \times 10^3\ s^{-1}$. In fact the reaction is also first order with respect to NaOH so that the true reaction velocity constant k_2 is $10^4\ m^3\ kmol^{-1}\ s^{-1}$. Our previous analysis will be correct provided inequality (22.5.10) is satisfied.

The diffusivity of NaOH through water is $1.1 \times 10^{-9}\ m^2\ s^{-1}$ and the stoichiometric coefficient n for the reaction

$$CO_2 + 2NaOH = Na_2CO_3 + H_2O$$

is 2. Thus our analysis is correct provided

$$C_{Ai} < \frac{1.1 \times 10^{-9}}{1.8 \times 10^{-9}} \times \frac{0.5}{10 \times 2 \times 30 \coth 30} = 5.1 \times 10^{-4}\ \text{kmol m}^{-3}.$$

Since the Henry's law constant for CO_2 is $2.56 \times 10^6\ J\ kmol^{-1}$, this value corresponds to an interfacial partial pressure of $5.1 \times 10^{-4} \times 2.56 \times 10^6 = 1300\ N\ m^{-2}$. For higher partial pressures, or lower concentrations of NaOH, the process cannot be treated as pseudo-first order.

We can illustrate the procedure under these circumstances by considering the case when the bulk NaOH concentration C_{B0} is $0.05\ kmol\ m^{-3}$ and the interfacial CO_2 concentration C_{Ai} is $3.9 \times 10^{-3}\ kmol\ m^{-3}$ (corresponding to an interfacial partial pressure of 0.1 bar). We can evaluate the Hatta number from equation (22.5.8),

$$Ha = \frac{\sqrt{(k_2 C_{B0} \mathscr{D}_A)}}{k_L^0} = \frac{(10^4 \times 0.05 \times 1.8 \times 10^{-9})^{\frac{1}{2}}}{10^{-4}} = 9.49,$$

and the enhancement factor for instantaneous reaction from equation (2.5.15),

$$E_i = 1 + \frac{\mathscr{D}_B C_{B0}}{n \mathscr{D}_A C_{Ai}} = 1 + \frac{1.1 \times 10^{-9} \times 0.05}{2 \times 1.8 \times 10^{-9} \times 3.9 \times 10^{-3}} = 4.92.$$

Clearly, inequality (22.5.16) is not satisfied and we must now solve equations (22.5.18) and (22.5.19) simultaneously to find E. By trial and error we obtain $E \simeq H = 4.16$ and hence $k_L = 4.16 \times 10^{-4}$ m s^{-1}.

It will be noted that if $H > 2$, $\coth H \simeq 1$ and equations (22.5.18) and (22.5.19) can be re-arranged to give the quadratic

$$E^2 \left(\frac{E_i - 1}{Ha^2} \right) + E - E_i = 0. \tag{22.5.20}$$

22.6 Porous catalyst slabs

Consider a porous slab of catalyst of thickness $2L$ as shown in figure 196. Within the pores there will be an effective diffusion coefficient $\mathscr{D}$ which may simply be the conventional diffusion coefficient between the reactants and the products. This will be so provided the pore diameter is

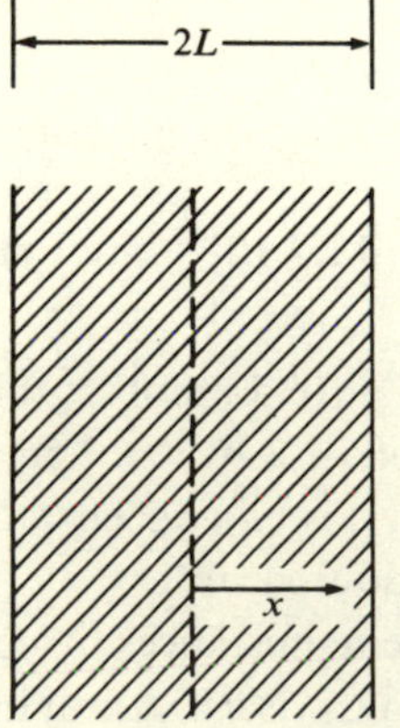

Fig. 196. Slab of porous catalyst.

large compared with the mean free path of the gas molecules. However, for catalysts with fine pores or which operate under reduced pressure the pore diameter may be small compared with the mean free path and molecules collide with the pore wall more frequently than with each other. This process is known as Knudsen diffusion. However, despite the different mechanism the molar flux is still proportional to the concentration gradient and it is possible to define an appropriate diffusion coefficient.

It is usual to define a molar flux in terms of the total cross-sectional area of the catalyst and not just the pore area. Thus in writing down the diffusion equation in the form

$$N = -\mathscr{D} \text{ grad } C, \tag{22.6.1}$$

it must be remembered that the effective diffusion coefficient $\mathscr{D}$ is the void fraction times the true diffusion coefficient which in turn may be either the gas phase or the Knudsen value.

Usually in a catalyst pellet the reaction takes place on the surface of the pore. For first-order kinetics we can say that the reaction rate per unit surface area of the pore R_s is given by $R_s = -k_s C$. Multiplying by s, the surface area per unit volume of catalyst, we have

$$R = -sk_s C \tag{22.6.2}$$

and we can therefore define an effective volumetric reaction velocity constant k by

$$k = sk_s.$$

Thus we can continue to use equation (22.2.5),

$$\mathscr{D}\frac{\mathrm{d}^2 C}{\mathrm{d}x^2} = kC, \tag{22.6.3}$$

provided we remember to use the appropriate values for the constants $\mathscr{D}$ and k.

As in §22.2 we find that this has solution

$$C = A \sinh \sqrt{\left(\frac{k}{\mathscr{D}}\right)}x + B \cosh \sqrt{\left(\frac{k}{\mathscr{D}}\right)}x \tag{22.6.4}$$

and the boundary conditions are now

at $x=0$, $\mathrm{d}C/\mathrm{d}x = 0$ since there can be no diffusion across the plane of symmetry, and

at $x = L$, $C = C^*$, where C^* is the concentration outside the slab.

Alternatively we could use the boundary conditions

at $x = \pm L$, $C = C^*$.

In either case we find that $A = 0$ and

$$B = C^* \operatorname{sech} \sqrt{\left(\frac{k}{\mathscr{D}}\right)} L.$$

Thus

$$C = C^* \frac{\cosh \sqrt{(k/\mathscr{D})}x}{\cosh \sqrt{(k/\mathscr{D})}L}. \tag{22.6.5}$$

We can find the rate of diffusion into the catalyst per unit surface area and this is clearly equal to the reaction rate.

$$N = \mathscr{D}\left(\frac{\mathrm{d}C}{\mathrm{d}x}\right)_{x=L} = C^* \sqrt{(k\mathscr{D})} \tanh \sqrt{\left(\frac{k}{\mathscr{D}}\right)} L. \tag{22.6.6}$$

This is best interpreted in terms of an effectiveness factor η defined as the reaction rate divided by the reaction rate that would have occurred if the concentration had been C^* throughout. This latter quantity is clearly kC^*L and therefore

$$\eta = \frac{1}{L}\sqrt{\frac{\mathscr{D}}{k}} \tanh \sqrt{\left(\frac{k}{\mathscr{D}}\right)} L. \tag{22.6.7}$$

The group $\sqrt{(k/\mathscr{D})}L$ is often called the Thiele modulus and given the symbol h. Comparison with equation (22.2.12) shows that the Thiele modulus is similar in form to a Hatta number.

From equation (22.6.8) we have

$$\eta = \frac{\tanh h}{h} \tag{22.6.8}$$

and this relationship is shown in figure 197. For small h, $\eta \simeq 1$, but for large h, $\eta \simeq 1/h$. These two limiting cases represent the behaviour of thin and thick slabs respectively. In a thin slab the concentration is approximately C^* throughout and the rate of reaction is controlled by the chemical kinetics. For a thick slab the reaction rate is controlled by the speed with which the reactants and products can diffuse through the catalyst.

The constants $\mathscr{D}$ and k often have to be determined experimentally. The most convenient way is to measure the rate of reaction in two slabs, one

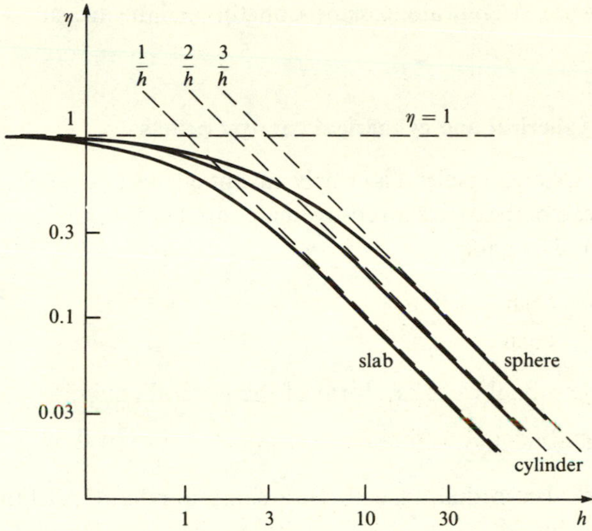

Fig. 197. Effectiveness factor η as a function of Thiele modulus h.

with twice the thickness of the other. If the slabs have thicknesses $2L_1$ and $4L_1$, the Thiele moduli will be h_1 and $2h_1$ where

$$h_1 = \sqrt{\left(\frac{k}{\mathscr{D}}\right)} L_1.$$

The reaction rates per unit area will be

$$Q_1 = C^* k L_1 \eta_1$$

and

$$Q_2 = C^* k 2 L_1 \eta_2.$$

The ratio of these quantities r is therefore given by

$$r = \frac{2\eta_2}{\eta_1},$$

which from equation (22.6.8) can be written

$$r = \frac{\tanh 2h_1}{\tanh h_1}. \tag{22.6.9}$$

Recalling that $\tanh 2\theta = 2 \tanh \theta / (1 + \tanh^2 \theta)$ we find that

$$1 + \tanh^2 h_1 = \frac{2}{r} = \frac{2Q_1}{Q_2}. \tag{22.6.10}$$

Thus h_1 can be readily evaluated and from this and the measured value of Q_1 we can find k and $\mathscr{D}$.

22.7 Spherical and cylindrical catalyst pellets

For a spherical pellet it is clearly advantageous to use spherical co-ordinates instead of the cartesian co-ordinates used in the previous sections of this chapter. The equation

$$\mathscr{D}\frac{\mathrm{d}^2C}{\mathrm{d}x^2}=kC$$

is the one-dimensional cartesian form of the general equation

$$\mathscr{D}\nabla^2 C=kC$$

which in spherical co-ordinates, with the assumption that C is a function of radius only, becomes

$$\frac{\mathscr{D}}{r^2}\frac{\mathrm{d}}{\mathrm{d}r}\left(r^2\frac{\mathrm{d}C}{\mathrm{d}r}\right)=kC. \tag{22.7.1}$$

This equation could be obtained by considering diffusion into and out of a spherical shell or could be obtained from the general transport equations of §§ 13.6 and 14.6.

Equation (22.7.1) is simplified if we make the substitution $C=y/r$. This gives

$$\mathscr{D}\frac{\mathrm{d}^2y}{\mathrm{d}r^2}=ky, \tag{22.7.2}$$

which has the solution

$$y=A\sinh\sqrt{\left(\frac{k}{\mathscr{D}}\right)}r+B\cosh\sqrt{\left(\frac{k}{\mathscr{D}}\right)}r. \tag{22.7.3}$$

Hence

$$C=\frac{A}{r}\sinh\sqrt{\left(\frac{k}{\mathscr{D}}\right)}r+\frac{B}{r}\cosh\sqrt{\left(\frac{k}{\mathscr{D}}\right)}r. \tag{22.7.4}$$

The second term in this expression tends to infinity as r tends to zero and since the concentration C must remain finite, the arbitrary constant B must be zero. The remaining arbitrary constant A can be evaluated from

$$C=C^* \quad \text{at } r=a,$$

where a is the radius of the pellet.

From this we find that

$$C = C^* \frac{a \sinh \sqrt{(k/\mathscr{D})}r}{r \sinh \sqrt{(k/\mathscr{D})}a}. \tag{22.7.5}$$

The total inflow of reactants Q is $4\pi a^2 \mathscr{D}(\mathrm{d}C/\mathrm{d}r)_{r=a}$ and the effectiveness factor η is $Q/\frac{4}{3}\pi a^3 kC^*$.

On differentiating equation (22.7.4) we find that

$$\eta = \frac{3}{h}\left(\frac{1}{\tanh h} - \frac{1}{h}\right), \tag{22.7.6}$$

where the Thiele modulus h is here defined as $\sqrt{(k/\mathscr{D})}a$.

For small h, $\eta \simeq 1$, but for large h, $\eta \simeq 3/h$ as shown in figure 197.

For a long cylindrical rod of catalyst of radius a we must use cylindrical co-ordinates for which

$$\frac{\mathscr{D}}{r}\frac{\mathrm{d}}{\mathrm{d}r}\left(r\frac{\mathrm{d}C}{\mathrm{d}r}\right) = kC. \tag{22.7.7}$$

This equation has solution

$$C = C^* \frac{I_0(\sqrt{(k/\mathscr{D})}r)}{I_0(\sqrt{(k/\mathscr{D})}a)}, \tag{22.7.8}$$

where I_0 is the modified Bessel function of the first kind and zero order. From this it can be shown that

$$\eta = \frac{2I_1(h)}{hI_0(h)}. \tag{22.7.9}$$

In this case $\eta \simeq 1$ for small h and $\eta \simeq 2/h$ for large h.

References

[1] van Krevelen & Hoftijzer, *Trans. Inst. Chem. Eng.*, **32**, S60 (1954).

APPENDIX 1

Vector notation and vector analysis

Vector notation is used in this book because it is concise and easy to understand, and it helps the reader to form a clear picture of what is happening in a three-dimensional flow problem. A *vector quantity* is always represented by a letter in bold-face type, for example:

$$\mathbf{a}=a_x\mathbf{i}+a_y\mathbf{j}+a_z\mathbf{k}, \tag{1}$$

where **i**, **j**, **k** are the unit vectors in the x, y, and z directions, and a_x, a_y, a_z are the scalar components of the vector **a**. Different conventions may be adopted for the scalar components of a vector quantity. As noted in § 1.6, for example, the position vector **r** is denoted by

$$\mathbf{r}=x\mathbf{i}+y\mathbf{j}+z\mathbf{k}$$

and the fluid velocity vector **u** may be expressed either as

$$\mathbf{u}=u_x\mathbf{i}+u_y\mathbf{j}+u_z\mathbf{k} \tag{3}$$

or as

$$\mathbf{u}=u\mathbf{i}+v\mathbf{j}+w\mathbf{k}. \tag{4}$$

The *scalar product* of two vectors **a** and **b** is defined by

$$\mathbf{a}\cdot\mathbf{b}=a_xb_x+a_yb_y+a_zb_z. \tag{5}$$

The *vector product* of two vectors **a** and **b** is defined by

$$\mathbf{a}\times\mathbf{b}=\begin{vmatrix}\mathbf{i} & \mathbf{j} & \mathbf{k}\\ a_x & a_y & a_z\\ b_x & b_y & b_z\end{vmatrix} \tag{6}$$

and it should be noted that the vector product is a vector quantity whose direction is at right angles to both **a** and **b**.

The vector operator *nabla* or *del* is defined by

$$\nabla=\mathbf{i}\frac{\partial}{\partial x}+\mathbf{j}\frac{\partial}{\partial y}+\mathbf{k}\frac{\partial}{\partial z}. \tag{7}$$

The *gradient of a scalar quantity* ϕ is then given by

$$\text{grad}\,\phi=\nabla\phi=\mathbf{i}\frac{\partial\phi}{\partial x}+\mathbf{j}\frac{\partial\phi}{\partial y}+\mathbf{k}\frac{\partial\phi}{\partial z} \tag{8}$$

and the general form of the gradient transport equation, describing the molecular transfer processes of diffusion and heat conduction, may be expressed as in (2.9.1) by

$$\mathbf{f}=-\kappa\,\text{grad}\,\phi. \tag{9}$$

The *divergence* of a vector quantity $\mathbf{a}$ is given by

$$\text{div}\,\mathbf{a}=\nabla\cdot\mathbf{a}=\frac{\partial a_x}{\partial x}+\frac{\partial a_y}{\partial y}+\frac{\partial a_z}{\partial z}. \tag{10}$$

In the case of the fluid velocity vector $\mathbf{u}$ we can write

$$\text{div}\,\mathbf{u}=\nabla\cdot\mathbf{u}=\frac{\partial u}{\partial x}+\frac{\partial v}{\partial y}+\frac{\partial w}{\partial z} \tag{11}$$

and, as noted in (2.3.7) and (5.3.3), div $\mathbf{u}$ represents the volumetric rate of strain, or rate of dilatation, of a material element of fluid. For an incompressible fluid we have div $\mathbf{u}=0$.

The *curl* of a vector quantity $\mathbf{a}$ is given by

$$\text{curl}\,\mathbf{a}=\nabla\times\mathbf{a}=\begin{vmatrix}\mathbf{i} & \mathbf{j} & \mathbf{k}\\ \dfrac{\partial}{\partial x} & \dfrac{\partial}{\partial y} & \dfrac{\partial}{\partial z}\\ a_x & a_y & a_z\end{vmatrix}. \tag{12}$$

The vorticity vector $\boldsymbol{\omega}$ is related to the velocity vector $\mathbf{u}$ by

$$\boldsymbol{\omega}=\text{curl}\,\mathbf{u}=\begin{vmatrix}\mathbf{i} & \mathbf{j} & \mathbf{k}\\ \dfrac{\partial}{\partial x} & \dfrac{\partial}{\partial y} & \dfrac{\partial}{\partial z}\\ u & v & w\end{vmatrix} \tag{13}$$

and hence we obtain the components of $\boldsymbol{\omega}$ as given in (3.8.2).

The operator $(\mathbf{u}\cdot\nabla)$ appears frequently in fluid mechanics and has the meaning $u(\partial/\partial x)+v(\partial/\partial y)+w(\partial/\partial z)$. As shown in § 2.2, $(\mathbf{u}\cdot\nabla)$ represents the rate of change following the motion of the fluid and may be described as the *convective differential*. It may be applied either to a scalar or to a vector property. When applied to the velocity vector $\mathbf{u}$ it gives the convective acceleration of the fluid element $(\mathbf{u}\cdot\nabla)\mathbf{u}$. This is often written simply as $\mathbf{u}\cdot\nabla\mathbf{u}$, but it should be noted that $\nabla\mathbf{u}$ by itself has no meaning.

The second-order differential expression div grad ϕ plays an important role in the analysis of transfer processes. It may be expressed as

$$\text{div grad } \phi = \nabla\cdot\nabla\phi = \nabla^2\phi = \frac{\partial^2\phi}{\partial x^2} + \frac{\partial^2\phi}{\partial y^2} + \frac{\partial^2\phi}{\partial z^2}. \tag{14}$$

The term $\kappa\,\nabla^2\phi$ which appears in the basic diffusion or transport equation (2.9.5) has the physical meaning that it represents the net rate of supply of the transferable property ϕ, per unit volume, as a result of the diffusion law expressed by (9).

The operator ∇^2 may also be applied to a vector quantity, for example to the fluid velocity $\mathbf{u}$, with the meaning that

$$\nabla^2\mathbf{u} = \nabla^2 u\mathbf{i} + \nabla^2 v\mathbf{j} + \nabla^2 w\mathbf{k}. \tag{15}$$

The term $\mu\,\nabla^2\mathbf{u}$, which appears in the Navier–Stokes equation, thus represents the diffusion of the three scalar components of the fluid momentum by the action of viscosity.

The expression $\mathbf{u}\cdot\nabla^2\mathbf{u}$ which arises in the energy equation (5.5.19) has the meaning

$$\mathbf{u}\cdot\nabla^2\mathbf{u} = u\,\nabla^2 u + v\,\nabla^2 v + w\,\nabla^2 w. \tag{16}$$

It is shown in § 5.5 that the term $-\text{div}\,(p\mathbf{u})$ represents the rate at which *flow work* is being done per unit volume. Noting that $(\partial/\partial x)(pu) = p(\partial u/\partial x) + u(\partial p/\partial x)$, it is seen from (11) that

$$\text{div}\,(p\mathbf{u}) = p\,\text{div}\,\mathbf{u} + \mathbf{u}\cdot\nabla p, \tag{17}$$

which is the result quoted in (5.5.6).

The following expansion formulae may also easily be verified:

$$\text{div}\,(\mathbf{a}\times\mathbf{b}) = \mathbf{b}\cdot\text{curl}\,\mathbf{a} - \mathbf{a}\cdot\text{curl}\,\mathbf{b}, \tag{18}$$

$$\text{curl}\,(\mathbf{a}\times\mathbf{b}) = \mathbf{b}\cdot\nabla\mathbf{a} - \mathbf{a}\cdot\nabla\mathbf{b} + \mathbf{a}\,\text{div}\,\mathbf{b} - \mathbf{b}\,\text{div}\,\mathbf{a}, \tag{19}$$

$$\text{grad}\,(\mathbf{a}\cdot\mathbf{b}) = \mathbf{b}\cdot\nabla\mathbf{a} + \mathbf{a}\cdot\nabla\mathbf{b} + \mathbf{b}\times\text{curl}\,\mathbf{a} + \mathbf{a}\times\text{curl}\,\mathbf{b}. \tag{20}$$

From (20), putting $\mathbf{a} = \mathbf{b} = \mathbf{u}$, we have

$$\tfrac{1}{2}\,\text{grad}\,(\mathbf{u}^2) = \mathbf{u}\cdot\nabla\mathbf{u} + \mathbf{u}\times\text{curl}\,\mathbf{u}$$

or

$$\mathbf{u}\cdot\nabla\mathbf{u} = \tfrac{1}{2}\nabla(\mathbf{u}^2) - \mathbf{u}\times\text{curl}\,\mathbf{u}, \tag{21}$$

which is the result quoted in (4.2.4).

We have seen that the second-order differential function div grad ϕ can be expressed by $\nabla^2\phi$ as in (14). Other second-order differential functions are:

$$\text{div curl } \mathbf{u} = \nabla \cdot \nabla \times \mathbf{u} = 0, \tag{22}$$

$$\text{curl grad } \phi = \nabla \times \nabla\phi = 0, \tag{23}$$

$$\begin{aligned} \text{curl curl } \mathbf{u} &= \nabla \times \nabla \times \mathbf{u} = \nabla(\nabla \cdot \mathbf{u}) - \nabla^2\mathbf{u} \\ &= \text{grad div } \mathbf{u} - \nabla^2\mathbf{u}. \end{aligned} \tag{24}$$

It is particularly important to note that div curl **u** is identically equal to zero under all conditions. Thus div $\boldsymbol{\omega}$ is always zero as observed in (3.8.19).

It will be seen from (19) that

$$\text{curl } (\mathbf{u} \times \boldsymbol{\omega}) = \boldsymbol{\omega} \cdot \nabla\mathbf{u} - \mathbf{u} \cdot \nabla\boldsymbol{\omega} + \mathbf{u} \text{ div } \boldsymbol{\omega} - \boldsymbol{\omega} \text{ div } \mathbf{u}.$$

Hence for an incompressible fluid with div $\mathbf{u} = 0$, we have

$$\text{curl } (\mathbf{u} \times \boldsymbol{\omega}) = \boldsymbol{\omega} \cdot \nabla\mathbf{u} - \mathbf{u} \cdot \nabla\boldsymbol{\omega}, \tag{25}$$

which is the result quoted in (5.6.6). It will also be noted from (23) that curl grad ϕ is identically equal to zero. Thus if the velocity **u** can be expressed, under certain conditions, as the gradient of a scalar potential ϕ as in (4.3.1), the vorticity curl **u** must be zero at all points. Thus potential flow is synonymous with irrotational flow.

Frequent use is made of *Gauss's divergence theorem* in the analysis of fluid motion. This states that, if **a** is a continuous vector function of the position vector **r**, and if we consider a closed surface S enclosing a spatial volume V, then

$$\int_S \mathbf{a} \cdot \mathbf{n} \, \mathrm{d}S = \int_V (\text{div } \mathbf{a}) \, \mathrm{d}V, \tag{26}$$

i.e. surface integral of $\mathbf{a}$ = volume integral of div **a**.

A similar principle known as *Stokes's theorem* states that, if we consider a closed curve L forming the boundary of an open surface S, then

$$\int_L \mathbf{a} \cdot \mathrm{d}\mathbf{L} = \int_S (\text{curl } \mathbf{a}) \cdot \mathbf{n} \, \mathrm{d}S, \tag{27}$$

i.e. line integral of $\mathbf{a}$ = surface integral of curl **a**.

APPENDIX 2

Note on the kinetic theory of gases

The transport properties of a gas are considered in chapter 2 in terms of the motion of the gas molecules. The purpose of this appendix is simply to review some of the more basic concepts of the kinetic theory of gases which have a bearing on the analysis outlined in §§ 2.5–2.8. For further information on the subject, however, the reader should refer to one of the specialised texts such as *An Introduction to the Kinetic Theory of Gases* by Sir James Jeans.

One of the basic concepts in the kinetic theory is that of the equipartition of energy. When two different gases are mixed at the same temperature, the average kinetic energy of their molecules will be the same. For monatomic molecules of mass m, the average kinetic energy of a molecule may be expressed as

$$\tfrac{1}{2}m\,\overline{v^2}=\tfrac{1}{2}m(\overline{v_x^2}+\overline{v_y^2}+\overline{v_z^2})=\tfrac{3}{2}kT, \tag{1}$$

where v is the magnitude of the molecular velocity, v_x, x_y, and v_z are the components of the molecular velocity, k is Boltzmann's constant, and T is the thermodynamic temperature. We are assuming here that the molecules are effectively 'point masses' with kinetic energy of translation but with no rotational or vibrational energy. Since the mean values of v_x^2, v_y^2, and v_z^2 must be equal, it will be seen from (1) that the energy for each degree of freedom is given by $\tfrac{1}{2}kT$.

The distribution of the molecular velocities is governed by Maxwell's law. We will simply state the result here that the number of molecules per unit volume whose thermal velocities lie within the range v to $v+\mathrm{d}v$ is given by

$$n_0 f(v)\,\mathrm{d}v=4\pi n_0\left(\frac{m}{2\pi kT}\right)^{\frac{3}{2}}\mathrm{e}^{-mv^2/2kT}v^2\,\mathrm{d}v, \tag{2}$$

where n_0 = total number of molecules per unit volume; m = mass of one molecule; k = Boltzmann's constant = 1.3805×10^{-23} J K^{-1}. The mathematical expression $f(v)$ is known as the velocity distribution function. It

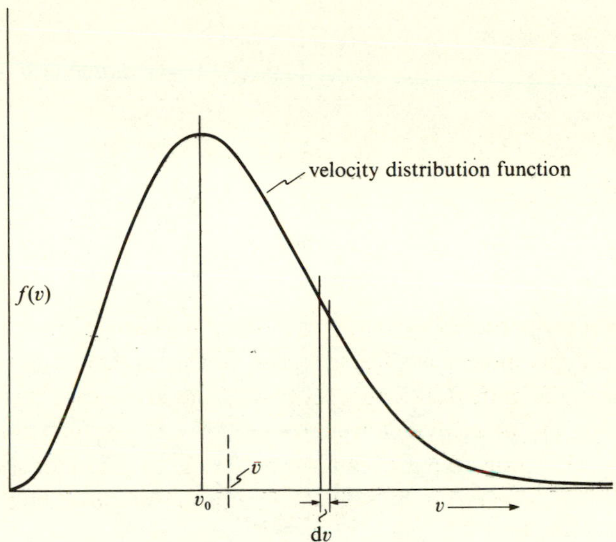

Fig. 198. Maxwellian velocity distribution.

will be noted that $\int_0^\infty f(v)\,dv = 1$. The Maxwellian velocity distribution is indicated in figure 198. The *most probable velocity* v_0 is determined by differentiating the expression for $f(v)$ to find a maximum, with the result that

$$v_0 = \left(\frac{2kT}{m}\right)^{\frac{1}{2}}. \tag{3}$$

We may thus re-write equation (2) in the form

$$f(v)\,dv = \frac{4}{\sqrt{\pi}}\left(\frac{v}{v_0}\right)^2 e^{-(v/v_0)^2}\, d\left(\frac{v}{v_0}\right). \tag{4}$$

The mean value of v^2 may be calculated from (2) or (4) and is found to be

$$\overline{v^2} = \frac{3kT}{m} = \frac{3}{2}v_0^2 \tag{5}$$

and it will be noted that $\frac{1}{2}m\,\overline{v^2} = \frac{3}{2}kT$, which is consistent with (1). The mean value of the magnitude of the thermal velocity, or *average speed* $\bar{v}$, may also be calculated from (2) and is found to be

$$\bar{v} = \left(\frac{8kT}{\pi m}\right)^{\frac{1}{2}} = 1.128\, v_0. \tag{6}$$

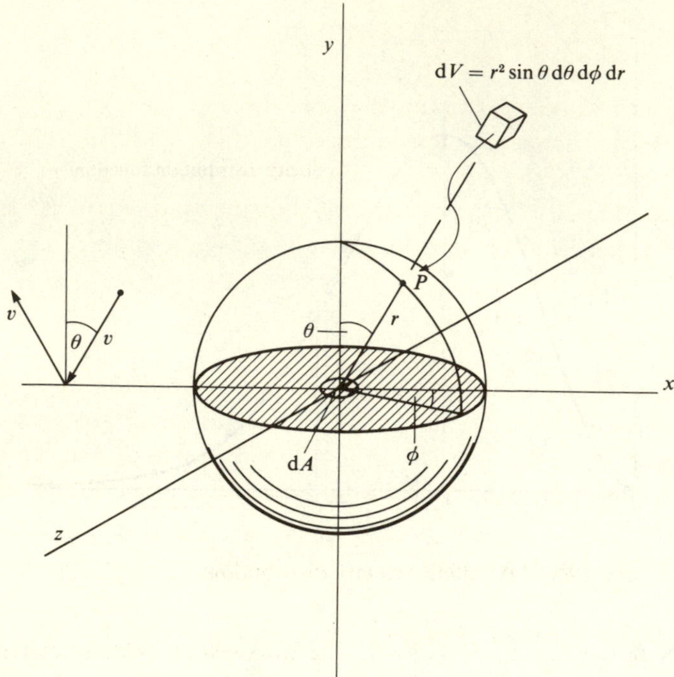

Fig. 199. Impact of molecules with a wall.

We can calculate the pressure in a gas on the basis of the kinetic theory. Referring to figure 199, we will suppose that the xz plane at $y=0$ forms a solid wall or boundary. Molecules coming from above this plane will hit the wall and bounce off again. We will consider first of all only those molecules having velocity magnitudes in the range v to $v+\mathrm{d}v$. All such molecules which are in a position to strike the small target area $\mathrm{d}A$ situated at the origin, as indicated in figure 199, within an interval of time between $t=0$ and $t=\mathrm{d}t$, must be contained within the hemisphere of radius $r=v\,\mathrm{d}t$ drawn about the origin at time $t=0$. We now focus attention on a small element of volume $\mathrm{d}V$ located at radius r within this hemisphere, as indicated in the figure at point P. Expressed in spherical polar co-ordinates r, θ, ϕ, the element of volume $\mathrm{d}V=r^2 \sin\theta\,\mathrm{d}\theta\,\mathrm{d}\phi\,\mathrm{d}r$. Since all directions are equally probable for the molecules contained within the volume element at point P, the proportion of such molecules which are targeted on the element of area $\mathrm{d}A$ in the xz plane will be given by the fractional solid angle $(\mathrm{d}A\cos\theta)/4\pi r^2$ subtended by $\mathrm{d}A$ at the point P.

Hence the number of molecules in the volume element $\mathrm{d}V$ which will hit the area $\mathrm{d}A$ in time $\mathrm{d}t$ is given by

$$\frac{\mathrm{d}A \cos\theta}{4\pi r^2} n_0 f(v)\,\mathrm{d}v\, r^2 \sin\theta\, \mathrm{d}\theta\, \mathrm{d}\phi\, \mathrm{d}r. \tag{7}$$

Each such molecule striking the area $\mathrm{d}A$ and bouncing off again will experience a change of y momentum equal to $2mv \cos\theta$. The total change of y momentum of the molecules contained at time $t=0$ within the volume element $\mathrm{d}V$, and whose velocities are within the band v to $v+\mathrm{d}v$, striking the area $\mathrm{d}A$ in time $\mathrm{d}t$ is therefore given by the expression:

$$\frac{\mathrm{d}A}{2\pi} mn_0 vf(v)\,\mathrm{d}v \sin\theta \cos^2\theta\, \mathrm{d}\theta\, \mathrm{d}\phi\, \mathrm{d}r.$$

In order to find the total change of y momentum of *all* the molecules hitting the target area $\mathrm{d}A$ in time $\mathrm{d}t$, we have to integrate the above expression over a range of r from 0 to $v\,\mathrm{d}t$, ϕ from 0 to 2π, θ from 0 to $\frac{1}{2}\pi$, and v from 0 to ∞. Hence integrating successively with respect to r, ϕ, and θ, and noting that

$$\int_0^{\pi/2} \sin\theta \cos^2\theta\, \mathrm{d}\theta = \left[-\frac{\cos^3\theta}{3}\right]_0^{\pi/2} = \frac{1}{3}$$

we have for the total change of y momentum in time $\mathrm{d}t$:

$$\frac{mn_0}{3}\mathrm{d}A\, \mathrm{d}t \int_0^\infty v^2 f(v)\,\mathrm{d}v = \frac{mn_0}{3}\overline{v^2}\,\mathrm{d}A\, \mathrm{d}t. \tag{8}$$

It follows that the *rate of change* of y momentum of the molecules hitting unit area, which by definition is the pressure exerted by the gas molecules on the plane, is given by

$$p = \frac{mn_0}{3}\overline{v^2}. \tag{9}$$

If we now substitute for $\overline{v^2}$ from (5) we get

$$p = n_0 kT, \tag{10}$$

which is the result quoted in (1.2.3).

We will now consider the situation in which there is *no* solid boundary in the xz plane at $y=0$ and where the molecules are free to cross the plane in either direction. Referring to figure 199 again, but noting that the xz plane is now transparent to the molecules, we can say that the number of molecules in the volume element $\mathrm{d}V$, and in the velocity band v to $v+\mathrm{d}v$, which are targeted on the area $\mathrm{d}A$ in the xz plane, is given by the same expression (7) as before. These molecules will now cross the area $\mathrm{d}A$ within time $\mathrm{d}t$ provided the value of r, which specifies the location of the element $\mathrm{d}V$, is less than $v\,\mathrm{d}t$.

The movement of molecules in one direction across the xz plane from above to below, through the area $\mathrm{d}A$ in the interval of time $\mathrm{d}t$, is obtained by integrating the expression (7) over a range of r from 0 to $v\,\mathrm{d}t$, ϕ from 0 to 2π, θ from 0 to $\frac{1}{2}\pi$, and v from 0 to ∞. Hence, carrying out the first three steps of this integration with respect to r, ϕ, and θ, and noting that

$$\int_0^{\pi/2} \sin\theta\cos\theta\,\mathrm{d}\theta = \left[\frac{\sin^2\theta}{2}\right]_0^{\pi/2} = \frac{1}{2}$$

we have for the transfer of molecules in the velocity band v to $v+\mathrm{d}v$, in one direction through area $\mathrm{d}A$ in time $\mathrm{d}t$:

$$\frac{n_0}{4}\,\mathrm{d}A\,\mathrm{d}t v f(v)\,\mathrm{d}v,$$

i.e.

$$\text{rate of flow per unit area} = \tfrac{1}{4} n_0 v f(v)\,\mathrm{d}v, \tag{11}$$

which is the result quoted in (2.5.1). The total flux in one direction is then obtained by integrating with respect to v from 0 to ∞. Hence

$$\text{molecular flow rate per m}^2 = \frac{n_0}{4}\int_0^\infty v f(v)\,\mathrm{d}v = \frac{n_0}{4}\,\bar{v}, \tag{12}$$

where $\bar{v}$ is the mean value of the molecular velocity magnitude as given by (6).

It is obvious that, under steady state conditions, the flow of molecules from above to below the xz plane must be equal and opposite to the flow from below to above the plane, i.e. the *net flux* is zero.

APPENDIX 3

The rate of strain in a fluid

In order to investigate the rate of strain in a fluid we need to consider the relative motion in the neighbourhood of a point. Taking a point in space specified by the position vector **r**, where the fluid velocity is **u**, we can say that the velocity will be $\mathbf{u}+\mathrm{d}\mathbf{u}$ at a neighbouring point specified by $\mathbf{r}+\mathrm{d}\mathbf{r}$, and that the differential velocity will be $\mathrm{d}\mathbf{u}=(\mathrm{d}\mathbf{r}\cdot\nabla)\mathbf{u}$. We can express the components of the differential velocity as:

$$\mathrm{d}u=\frac{\partial u}{\partial x}\,\mathrm{d}x+\frac{\partial u}{\partial y}\,\mathrm{d}y+\frac{\partial u}{\partial z}\,\mathrm{d}z,$$

with similar expressions for $\mathrm{d}v$ and $\mathrm{d}w$. These may be written alternatively in the following form:

$$\mathrm{d}u=e_{xx}\,\mathrm{d}x+e_{yx}\,\mathrm{d}y+e_{zx}\,\mathrm{d}z+\tfrac{1}{2}(0-\omega_z\,\mathrm{d}y+\omega_y\,\mathrm{d}z)$$

$$\mathrm{d}v=e_{xy}\,\mathrm{d}x+e_{yy}\,\mathrm{d}y+e_{zy}\,\mathrm{d}z+\tfrac{1}{2}(\omega_z\,\mathrm{d}x+0-\omega_x\,\mathrm{d}z)$$

$$\mathrm{d}w=e_{xz}\,\mathrm{d}x+e_{yz}\,\mathrm{d}y+e_{zz}\,\mathrm{d}z+\tfrac{1}{2}(-\omega_y\,\mathrm{d}x+\omega_x\,\mathrm{d}y+0),$$

where e_{xx}, e_{xy}, etc., are the rate of strain components as given in (5.3.2), and ω_x, ω_y, ω_z are the components of the vorticity vector $\boldsymbol{\omega}$ as given in (3.8.2). We can describe the situation more neatly, however, by saying that the general velocity gradient tensor $(\partial u_i/\partial x_j)$ may be written as the sum of a *symmetric tensor*

$$e_{ij}=\frac{1}{2}\left(\frac{\partial u_i}{\partial x_j}+\frac{\partial u_j}{\partial x_i}\right) \qquad (1)$$

and an *antisymmetric tensor*

$$\xi_{ij}=\frac{1}{2}\left(\frac{\partial u_i}{\partial x_j}-\frac{\partial u_j}{\partial x_i}\right). \qquad (2)$$

The symmetric tensor (1) represents the *rate of strain* of a fluid element, including both extension and shearing, while the antisymmetric tensor (2) represents *rotation* of the element. The antisymmetric tensor is related to the vorticity of the fluid by the expression

$$\xi_{ij} = -\tfrac{1}{2}\varepsilon_{ijk}\omega_k, \tag{3}$$

where ε_{ijk} is the third-order Cartesian tensor having the properties that

$$\begin{aligned}\varepsilon_{ijk} &= 1 \quad \text{if } i, j, k \text{ is an even permutation of } 1, 2, 3,\\ &= -1 \quad \text{if } i, j, k \text{ is an odd permutation of } 1, 2, 3,\\ &= 0 \quad \text{if } i, j, k \text{ are not all different.}\end{aligned}$$

It will be seen from (3) that the rate of rotation, or angular velocity, of a fluid element is equal to half the magnitude of the vorticity vector, which confirms the statement to this effect made in § 3.8.

In the case of a Newtonian fluid it is assumed that there is a simple linear relationship between the deviatoric stress tensor τ_{ij} and the velocity gradient tensor $(\partial u_i/\partial x_j)$. Considerations of symmetry, however, will show that the antisymmetric tensor (2) cannot be involved in the relationship. Noting also that $e_{ij} = e_{ji}$, the relationship consequently reduces to

$$\tau_{ij} = 2\mu e_{ij} + \mu'\delta_{ij}e_{kk}, \tag{4}$$

where μ and μ' are coefficients; and δ_{ij} is the Kronecker delta. Noting from (5.3.3) that $e_{kk} = e_{xx} + e_{yy} + e_{zz} = \operatorname{div} \mathbf{u}$, we can express (4) in the form

$$\tau_{ij} = 2\mu e_{ij} + \mu'\delta_{ij} \operatorname{div} \mathbf{u}. \tag{5}$$

We can now invoke the condition, which will be evident from (5.2.1) and (5.2.2), that the sum of the three normal stress terms in the deviatoric stress tensor must be zero, i.e. we must have

$$\tau_{xx} + \tau_{yy} + \tau_{zz} = 0.$$

It follows from (5) that

$$2\mu(e_{xx} + e_{yy} + e_{zz}) + 3\mu' \operatorname{div} \mathbf{u} = 0$$

or

$$2\mu \operatorname{div} \mathbf{u} + 3\mu' \operatorname{div} \mathbf{u} = 0.$$

Hence we must have $\mu' = -\tfrac{2}{3}\mu$ and the final conclusion is that

$$\tau_{ij} = 2\mu(e_{ij} - \tfrac{1}{3}\delta_{ij} \operatorname{div} \mathbf{u}) \tag{6}$$

which is the result quoted in (5.3.4).

BIBLIOGRAPHY

Although a comprehensive list of original papers and specialised texts relating to fluid mechanics and transfer processes would be of quite prohibitive length, the reader may find it helpful to have a short list of books to which reference may be made for a fuller treatment of individual aspects of the subject. The authors accordingly suggest the following publications for further reading:

Batchelor, G. K. *Theory of Homogeneous Turbulence* (Cambridge University Press, 1953)

Batchelor, G. K. *An Introduction to Fluid Dynamics* (Cambridge University Press, 1967)

Bird, R. B., Stewart, W. E. & Lightfoot, E. W. *Transport Phenomena* (Wiley, 1960)

Bradshaw, P. *Introduction to Turbulence and its Measurement* (Pergamon, 1971)

Bradshaw, P. (ed.) *Turbulence* (Springer Verlag, 1976)

Butterworth, D. & Hewitt, G. I. *Two Phase Flow and Heat Transfer* (Oxford University Press, 1977)

Carslaw, H. S. & Jaeger, J. C. *Conduction of Heat in Solids* (Oxford University Press, 1973)

Chisholm, D. *Two Phase Flow* (George Godwin, 1983)

Coulson, J. M. & Richardson, J. I. *Chemical Engineering*, Vols I and II (Pergamon, 1984)

Crank, J. *Mathematics of Diffusion* (Clarendon Press, 1984)

Cussler, E. L. *Diffusion: Mass Transfer in Fluid Systems* (Cambridge University Press, 1984)

Danckwerts, P. V. *Gas Liquid Reactions* (McGraw-Hill, 1970)

Davidson, J. F. & Harrison, D. *Fluidised Particles* (Cambridge University Press, 1979)

Dixon, S. L. *Fluid Mechanics and Thermodynamics of Turbomachinery* (Pergamon, 1978)

Eckert, E. R. & Drake, R. M. *Heat and Mass Transfer* (McGraw-Hill, 1959)

Goldstein, S. (ed.) *Modern Developments in Fluid Dynamics* (Oxford, 1950)

Horlock, J. H. S. *Axial Flow Compressors* (Butterworths, 1958)

Hunsaker, J. C. & Rightmire, B. G. *Engineering Applications of Fluid Mechanics* (McGraw-Hill, 1947)

Jeans, J. *Introduction to the Kinetic Theory of Gases* (Cambridge University Press, 1946)
Lighthill, M. J. *Waves in Fluids* (Cambridge University Press, 1978)
McAdams, W. H. *Heat Transmission* (McGraw-Hill, 1954)
Prandtl, L. *The Essentials of Fluid Dynamics* (Blackie, 1967)
Schlichting, H. *Boundary Layer Theory* (McGraw-Hill, 1979)
Shapiro, R. H. *The Dynamics and Thermodynamics of Compressible Fluid Flow* (Ronald Press, 1953)
Sherwood, T. K., Pigford, R. L. & Wilke, C. R. *Mass Transfer* (McGraw-Hill, 1975)
Tabor, D. *Gases, Liquids and Solids*, 2nd edn (Cambridge University Press, 1979)
Taylor, G. I. *Scientific Papers*, Vol. 2 (Cambridge University Press, 1960)
Townsend, A. A. *The Structure of Turbulent Shear Flow* (Cambridge University Press, 1977)
Tritton, D. J. *Physical Fluid Dynamics* (Van Nostrand, 1977)
Wallis, G. B. *One Dimensional Two-Phase Flow* (McGraw-Hill, 1969)
Wong, H. J. *Heat Transfer for Engineers* (Longman, 1977)

INDEX